Developments in Atmospheric Science, 1

# Structure and Dynamics of the Upper Atmosphere

Developments in Atmospheric Science, 1

# STRUCTURE AND DYNAMICS OF THE UPPER ATMOSPHERE

*Proceedings of the 2nd Course of the International School of Atmospheric Physics, "Ettore Majorana" Centre for Scientific Culture, held in Erice (Italy), 13–27 June, 1971*

edited by

## FRANCO VERNIANI

*Institute for Atmospheric Physics of the Italian National Research Council, Bologna; and University of Bologna, Italy*

**ELSEVIER SCIENTIFIC PUBLISHING COMPANY**
**Amsterdam — Oxford — New York 1974**

ELSEVIER SCIENTIFIC PUBLISHING COMPANY
335 JAN VAN GALENSTRAAT
P.O. BOX 211, AMSTERDAM, THE NETHERLANDS

AMERICAN ELSEVIER PUBLISHING COMPANY, INC.
52 VANDERBILT AVENUE
NEW YORK, NEW YORK 10017

Library of Congress Card Number: 72-97437

ISBN 0-444-41105-4

With 339 illustrations and 40 tables

Copyright © 1974 by Elsevier Scientific Publishing Company, Amsterdam

Printed in The Netherlands

# Foreword

Thanks to the initiative of Professor A. Zichichi, Director of the "Ettore Majorana" Centre for Scientific Culture, it was possible to set up the International School of Atmospheric Physics and hold its first Course in June 1970. This was followed in June 1971 by a second Course on "Structure and Dynamics of the Upper Atmosphere", that was sponsored by the Italian National Research Council, the Ministry for Public Education, the Ministry for Scientific and Technological Research and the Sicilian Regional Government.

The Course aimed at providing both a comprehensive survey of basic topics like photochemical atmosphere models, tides and gravity waves in the atmosphere, the structure of the various ionospheric regions, the source of atmospheric electrification and a review of some more specialized topics of great current interest and intrinsic importance, such as global observations of the upper atmosphere with meteorological sounding rockets and with remote sensing techniques from satellites, composition studies in the thermosphere by means of mass spectrometers, optical techniques for temperature determination, and radio meteor studies of winds and turbulence in the upper atmosphere.

The Course included ten cycles of lectures and ten seminars for a total of 50 hours. It was attended by 52 participants coming from 17 countries, namely: Argentina, Australia, Belgium, France, Germany, Hong-Kong, India, Israel, Italy, The Netherlands, Norway, Poland, Spain, Sweden, Turkey, The United Kingdom and The United States.

This volume includes most of the material that was presented and discussed at Erice, so I hope the readers will enjoy the book as much as the participants enjoyed attending lectures, seminars and discussions.

Putting together this book was quite a time-consuming task. Mrs. Angelica Ciampi helped me in the revision of the manuscripts, as far as the English language was concerned, Mrs. Maria Teresa Tibaldi typed most of the final typescripts and checked figures and references together with Mrs. Renata Malossi. I wish to thank them very warmly for their invaluable help. Mrs. Tibaldi also acted as secretary of the School and I am grateful to her for the intelligent cooperation in the various stages of the course organization.

Finally, I should like to express my gratitude to Professor Zichichi, Director of the "E. Majorana" Centre, for giving me the possibility to organize the International School of Atmospheric Physics.

Bologna, July 1972

F. VERNIANI

# Contributors

Antonio ELENA — Geophysical and Geodetic Institute, University of Genoa, Italy

Giorgio FIOCCO — ESRIN, European Space Research Institute, Frascati (Rome), and University of Florence, Italy

Ewald HARNISCHMACHER — Ionospheric Institute, Breisach, Germany

Eigil HESSTVEDT — Institute of Geophysics, University of Oslo, Blindern, Norway

Richard S. LINDZEN — Department of Geophysical Sciences, Harvard University, Cambridge, Massachusetts, U.S.A.

Franco MARIANI — Institute of Physics, University of L'Aquila, CNR Laboratory for Plasma in Space, Rome, Italy

Heinz G. MULLER — Department of Physics, University of Sheffield, Great Britain

William NORDBERG — NASA Goddard Space Flight Center, Greenbelt, Maryland, U.S.A.

Giovanni PERONA — Institute of Electronics and Telecommunications, Polytechnic of Turin, Italy

Henry RISHBETH — Radio and Space Research Station, Slough, Great Britain

Gian Carlo RUMI — National Electrotechnical Institute "G. Ferraris", Turin, Italy

Jacek WALCZEWSKI — Hydro-Meteorological Institute, Krakow, Poland

Willis L. WEBB — Atmospheric Science Laboratory, U.S. Army Electronics Command, White Sands Missile Range, New Mexico, U.S.A.

Ulf von ZAHN — Institute of Physics, University of Bonn, Germany

## Co-authors

Glauco BENEDETTI-MICHELANGELI — ESRIN, European Space Research Institute, Frascati (Rome), Italy

Mario BOSSOLASCO — Geophysical and Geodetic Institute, University of Genoa, Italy

Giorgio CAPPUCCIO — ESRIN, European Space Research Institute, Frascati (Rome); now at the Institute for Atmospheric Physics, Rome, Italy

Fernando CONGEDUTI — ESRIN, European Space Research Institute, Frascati (Rome), Italy

Donald F. HEATH                    NASA Goddard Space Flight Center, Greenbelt, Mary-
                                   land, U.S.A.

Ernest HILSENRATH                  NASA Goddard Space Flight Center, Greenbelt, Mary-
                                   land, U.S.A.

Arlin J. KRUEGER                   NASA Goddard Space Flight Center, Greenbelt, Mary-
                                   land, U.S.A.

Cuddapah PRABHAKARA                NASA Goddard Space Flight Center, Greenbelt, Mary-
                                   land, U.S.A.

John S. THEON                      NASA Goddard Space Flight Center, Greenbelt, Mary-
                                   land, U.S.A.

# Contents

# CONTENTS

*Opening Speech*:
# IMPACT OF MODERN TECHNOLOGY ON THE ATMOSPHERIC SCIENCES

GIORGIO FIOCCO

*Università di Firenze and European Space Research Institute (ESRIN), Frascati (Italy)*

I am going to restrict myself to some opening remarks that will be neither too specific nor too technical. I have chosen a title which is vague enough, perhaps a little banal, but which I hope describes some general aspects of the work in the atmospheric sciences.

The impact of technology means that a very substantial level of resources is now available to disciplines, some of which, only one or two decades ago, utilized rather simple instrumentation and, often, empirical methods of analysis. On the other hand, it may also mean that a very large part of those resources is used in solving technical problems, when a certain amount of ingenuity in the definition of the scientific aspects of the problem might have saved time and effort.

In our culture it is often not clear whether the offer of a product or a service is in response to a need, or whether demand is being artificially created. Judging the validity of scientific research is not simple. A criterion should be based on a reasonable compromise among originality, usefulness and cost. This does not mean that a piece of work should be trivial or useless as long as it is expensive. Rather the opposite! As far as usefulness goes, the atmospheric sciences come out rather honourably.

I assume that many of those attending this school are, like myself, from Western Europe. Since for the last several years most of the scientific effort has been carried out in the U.S. and the U.S.S.R., the choice among originality, usefulness and cost for the European scientist has been, and still is, more limited. With low cost being a sort of boundary condition, let us try at least to be original and useful.

In the atmospheric sciences, as in other sciences that have emerged from the stage of taxonomy, results are obtained by a competition between (analytic, numerical) models and experiments. One practical example is weather prediction. Numerical models are developed by digital computers; data are collected by a worldwide network, and now also by satellites. At least twice a day several weather bureaus turn out forecasts, the validity of which is tested by a multitude of users. Originality may not be too high, and individual contributions may disappear in much routine work, but the usefulness is great and the cost is high.

Although the mathematical and physical sophistication of the model utilized is not great, the organization necessary to carry out the scheme is very complex.

It appears, by the way, that the prediction of an experienced forecaster is not much worse than that carried out by the computer. On the basis of the preliminary computer analysis, the forecaster can make an even better prediction. Thus, one wonders whether some aspects of this extremely laborous computation could be replaced either by more powerful mathematical techniques or by better understanding of how the human mind works.

In practice, now, progress in weather forecasting is closely related to progress in computers. But notice that the cost of a commercially available, advanced computer of medium size can be $10^7$\$ or a monthly rental of approximately $2 \cdot 10^5$\$. After considering, for example, that a reduction by a factor of 2 in the size of the computational grid involves an increase by a factor close to $2^4$ in the computer size, it becomes obvious that many of the problems in the atmospheric sciences cannot be solved by sheer strength of computation in the foreseeable future. In this spiral of high costs, the question of international cooperation also becomes essential. At the moment there are more than ten agencies in the world that carry out detailed weather predictions. Several of these are in Europe; in this case there is an evident overlap even for short-term predictions. Therefore, the establishment of common centers is becoming mandatory. Of course, the establishment of large facilities should make global weather predictions possible and allow some of the available computer time to be dedicated to research. Thus, it should be possible to study in detail, for instance, the kind of problems that Professors Hesstvedt and Lindzen discuss in their contributions.

When global predictions are considered, one of the difficulties is lack of adequate data, especially for the Southern hemisphere, the oceans, and the less populated areas. Satellites perform very well as far as global coverage is concerned. The instruments carried by the present generation of meteorological satellites permit: (1) Continuous observation of the earth's cloud cover; (2) Infrared sounding of the atmosphere to allow quantitative deviation of vertical temperature profiles; (3) Infrared scanning to study the earth's radiative budget.

With this basic instrumentation, aided also by rocket campaigns, a fairly complete description of the mesospheric and stratospheric meteorology has emerged. The measurements at tropospheric levels are more difficult to interpret. Synoptic observations of cloud formations are of interest to the weather forecaster. They permit, for instance, the early detection of severe storms, especially since the area observed may not be adequately covered by ground stations. This information is of a qualitative nature and some scientists have tended to de-emphasize its importance. Cloud movements are, of course, an aid to the measurement of wind velocity, even though doubts may remain in some cases as to whether clouds move at the same velocity as the wind. Cloud heights, also, are not measured directly.

As we shall see in Dr. Nordberg's lectures, the measurement of atmospheric temperature profiles from the $CO_2$ 15-$\mu$ emission band was one of the important results. The contributions to the total spectral density measured on the satellite come from different height levels, so that for each portion of the spectrum a different weighting function exists. The problem of data reduction involves the inversion of raw data—an operation which can be ambiguous. The contribution of clouds and aerosols sets a limit to the present accuracy of the technique. The problem arises of making observations through holes in clouds, or making use of different spectral regions.

The possibility of detecting minor atmospheric constituents is still at the experimental stage. This is also of importance to the circulation of pollutants and other ecological problems. I suspect that a good deal of work is still necessary in order to develop adequate sensors and to prepare the atmospheric models necessary for the interpretation of data. The study of more complex compounds, such as the polymers of water, clusters and aerosols, is also attracting great interest. Satellites can be used to communicate with buoys, to track floating balloons, and to relay data from ground stations and data centers. But just to give some idea about cost: an estimate for a satellite including the development of the sensors, ground stations, etc., may run over $5 \cdot 10^7$\$.

A rapid survey like this should not neglect the use of the manned orbiting laboratory, especially with regard to the testing of particularly complex instrumentation: but here, the question of costs may really get out of hand.

Last, but not least, I should call your attention to the important work that can be done with relatively inexpensive ground-based instruments, such as the meteor radars. In this connection, I will give a seminar on the laser as a tool for atmospheric studies. This is, of course, only one of the main tools that the atmospheric scientist has at his disposal today.

Let us remember that knowledge about the earth's atmosphere is important in many facets of human activity, from strictly economic ones to more speculative ones. For instance, exact geodetic surveys carried out from satellites equipped with corner reflectors, or from the moon itself, are limited by our knowledge of the refractive index of air. Fluctuations in air properties should be better known if accuracies of the order of a few cm, which are technically feasible, are to be achieved. I do not have to expound further on the vast importance of accurate climatogical studies and weather predictions. Therefore, having established proper priorities, it is clear that the field of atmospheric study is not only scientifically valid, but also of great practical relevance.

# PHOTOCHEMICAL ATMOSPHERE MODELS

EIGIL HESSTVEDT

*Institute of Geophysics, University of Oslo, Blindern (Norway)*

## SUMMARY

The first section describes how a photochemical atmosphere model is built up. Definitions are given of basic concepts such as reaction and dissociation rate coefficients and lifetimes. The pure oxygen atmosphere model is used as an example. The necessity of considering atmospheric transport together with chemistry is emphasized.

In section 2 the principles given in section 1 are applied to an oxygen–hydrogen model in order to study the composition of the upper stratosphere and lower mesosphere. The computation of ozone is in fair agreement with observational dat and turns out to be of particular interest.

## 1. INTRODUCTION TO PHOTOCHEMICAL ATMOSPHERE MODELS

An efficient study of the upper atmosphere depends, to a large extent, upon our knowledge of its chemical composition. During the last few years a series of measurements has provided us with useful information, but the situation is still far from satisfactory, even for such important trace components as atomic oxygen, ozone, nitric oxide and water vapor.

In addition to experimental results, useful information can be obtained from photochemical models. In such models, one tries to simulate the chemistry of the real atmosphere by specifying a set of chemical reactions which are believed to be important for the problem to be studied. In the more recent models, the effect of vertical air transport has been included as well. There are two kinds of reactions:

(a) chemical reactions, e.g.:

$$O + O + M \rightarrow O_2 + M \qquad k_1 = 1.5 \cdot 10^{-34} \exp(1.4/RT)\,cm^6\,sec^{-1} \qquad (1)$$

$$O + O_2 + M \rightarrow O_3 + M \qquad k_2 = 5.8 \cdot 10^{-35} \exp(1.5/RT)\,cm^6\,sec^{-1} \qquad (2)$$

$$O + O_3 \rightarrow 2O_2 \qquad k_3 = 3.3 \cdot 10^{-11} \exp(-4.2/RT)\,cm^3\,sec^{-1} \qquad (3)$$

where M denotes an arbitrary air molecule;
   (b) photodissociation reactions, e.g.:

$$O_2 + h\nu \rightarrow 2O \qquad\qquad J_2 \qquad\qquad (4)$$

$$O_3 + h\nu \rightarrow O + O_2 \qquad J_3 \qquad\qquad (5)$$

On the basis of this system of reactions it is feasible to define a model which is often referred to as the pure oxygen atmosphere model. This model gives us useful information about how the available oxygen is distributed over the three components: O, $O_2$ and $O_3$. But before setting up the model in a mathematical form, the concepts "reaction rate constant" and "dissociation rate constant" have to be defined.

   The rate of a reaction is proportional to the number of collisions between the reacting partners per unit volume and time. For example, the rate of reaction 3 is proportional to the number of collisions between oxygen atoms and ozone molecules, i.e., to the product of their number densities $[O]\cdot[O_3]$. (Brackets are used to denote the number density of a component.) The loss of oxygen atoms from reactions 1, 2 and 3 is, therefore:

$$k_1[M] \cdot [O]^2 + k_2[M] \cdot [O_2] \cdot [O] + k_3[O_3] \cdot [O] \qquad\qquad (6)$$

The factors $k_1$, $k_2$ and $k_3$ are called reaction rate coefficients and are measured in the laboratory. Such measurements are usually difficult to perform and the uncertainty may be quite great for many of the reactions used in the models.

   The dissociation rate of a component is proportional to the number density of the component. For example, the dissociation rate of reaction 4 is $J_2[O_2]$. The proportionality factor $J$ is called the dissociation rate coefficient and is computed on the basis of the intensity of solar radiation, mainly in the ultraviolet, but also in the visible range.

   The depletion of solar energy is given by:

$$\frac{dI_\lambda}{I_\lambda} = -k_\lambda \cdot \rho \cdot ds = -\sigma_\lambda \cdot n \cdot ds \qquad\qquad (7)$$

where $I_\lambda$ is the intensity of solar radiation, $k_\lambda$ the absorption coefficient, $\rho$ the density and $n$ the number density of the absorbent, $ds$ an infinitesimal path length. The absorption cross-section $\sigma_\lambda$ depends strongly upon the wavelength $\lambda$. For wavelengths longer than about 1200 Å, $O_2$ and $O_3$ are the only important absorbents in the atmosphere. The intensity of solar radiation for a given optical depth may then be expressed as:

$$I_\lambda = I_{\lambda,0} \cdot \exp[-\sigma_{2,\lambda} \cdot \Sigma_2 - \sigma_{3,\lambda} \cdot \Sigma_3] \qquad\qquad (8)$$

where $\sigma_{2,\lambda}$ and $\sigma_{3,\lambda}$ are the absorption cross-sections of $O_2$ and $O_3$, $\Sigma_2$ and $\Sigma_3$ are the total numbers of $O_2$ and $O_3$ molecules per cm² along the ray's path, from the top of the atmosphere, where the intensity is $I_{\lambda,0}$, to the point considered.

The dissociation rate coefficient may now be expressed as:

$$J = \int_{\lambda=0}^{\infty} \sigma_\lambda \cdot I_\lambda \cdot d\lambda \tag{9}$$

The time variations of the three components, $O$, $O_2$ and $O_3$ are the algebraic sums of their production and loss terms:

$$\frac{\partial[O]}{\partial t} = 2J_2 \cdot [O_2] + J_3 \cdot [O_3] - 2k_1[M] \cdot [O]^2 - k_2[M] \cdot [O_2] \cdot [O]$$
$$- k_3[O_3] \cdot [O] \tag{10}$$

$$\frac{\partial[O_2]}{\partial t} = J_3[O_3] + k_1[M] \cdot [O]^2 + 2k_3[O] \cdot [O_3] - J_2[O_2]$$
$$- k_2[M] \cdot [O] \cdot [O_2] \tag{11}$$

$$\frac{\partial[O_3]}{\partial t} = k_2[M] \cdot [O_2] \cdot [O] - J_3[O_3] - k_3[O] \cdot [O_3] \tag{12}$$

If it is assumed that the total amount of oxygen at a given level is a constant fraction (21%) of the total number density, one of these expressions may be replaced by:

$$\frac{1}{2}[O] + [O_2] + \frac{3}{2}[O_3] = 0.21[M] \tag{13}$$

Number densities corresponding to photochemical equilibrium are computed from (10), (12) and (13) by introducing the steady state concept $\partial[O]/\partial t = \partial[O_3]/\partial t = 0$. On the basis of the equilibrium values, one concludes that reactions 1 and 5 contribute the largest terms in (10) and (12). The physical significance of this is that an unbalance in the system will result in a rapid exchange between $O$ and $O_3$, followed by an exchange between $O$ and $O_3$ on one hand and $O_2$ on the other. Chapman introduced the useful concept "odd oxygen", $[O] + [O_3]$. The time variation of odd oxygen is relatively slow compared to the variation of $O$ and $O_3$ separately:

$$\frac{\partial([O] + [O_3])}{\partial t} = 2J_2[O_2] - 2k_3[O] \cdot [O_3] - 2k_1[M] \cdot [O]^2 \tag{14}$$

It is convenient to use (14) instead of (10) above approximately 60 km, and instead of (12) below that level.

The expressions (10), (11) and (12) are of the form:

$$\frac{\partial n}{\partial t} = P - Qn - Rn^2 \tag{15}$$

where $P$ is the production term and $(Qn + Rn^2)$ is the loss term. If $P$, $Q$ and $R$ are constant or vary much more slowly than $n$, eq. 15 may be integrated:

$$\frac{n - n_e}{n - n_e'} = \frac{n_0 - n_e}{n_0 - n_e'} \exp[-(Q^2 + 4RP)^{1/2} \cdot t] \qquad (16)$$

for $R \neq 0$, and:

$$n - n_e = (n_0 - n_e)\exp(-Qt) \qquad (17)$$

for $R = 0$. The lifetime of the component is then $(Q^2 + 4RP)^{-1/2}$. In the case of the pure oxygen atmosphere model, $P$, $Q$ and $R$ are not constant, but a comparison of the lifetimes, as defined above, shows quite clearly that ozone has a very short lifetime ($J_3^{-1} \simeq 100$ sec) in the upper mesosphere and lower thermosphere, whereas atomic oxygen has a very short lifetime in the stratosphere (1 sec at 40 km, decreasing downwards). Therefore, calculations will follow different patterns for these two parts of the atmosphere.

In the stratosphere atomic oxygen is in photochemical equilibrium. Since reaction 1 is unimportant and may be neglected, we have:

$$[O] \simeq \frac{J_3[O_3] + 2J_2[O_2]}{k_2[M] \cdot [O_2] + k_3[O_3]} \qquad (18)$$

Since $[O_2] \gg ([O] + 3[O_3])/2$ and $[O_3] \gg [O]$, one obtains from (14) and (18) the following expression:

$$\frac{\partial[O_3]}{\partial t} = \frac{2J_2[O_2] \cdot k_2[M] \cdot [O_2] - 2J_2[O_2] \cdot k_3[O_3] - 2k_3[O_3] \cdot J_3[O_3]}{k_2[M] \cdot [O_2] + k_3[O_3]}$$

$$\simeq 2J_2[O_2] - 2\frac{J_3 k_3}{k_2[M] \cdot [O_2]} \cdot [O_3]^2 \qquad (19)$$

The equilibrium concentration of ozone in the stratosphere is then:

$$[O_3]_e = [O_2] \cdot \sqrt{\frac{J_2 k_2[M]}{J_3 k_3}} \qquad (20)$$

and the lifetime is:

$$\tau_3 = \sqrt{\frac{k_2[M]}{16 J_2 J_3 k_3}} \qquad (21)$$

Since $[M]$ increases and $J_2$ and $J_3$ decrease as one goes down in the ozone layer, the lifetime of ozone becomes very long in the lower part of the ozone layer. It is less than one day at 40 km, one week at 30 km and several months at 20 km. Evidently the photochemistry of ozone is so slow in the lower part of the ozone layer that transport processes become more important and induce significant departures from photochemical values.

Contrary to what is the case in the ozone layer, atomic oxygen has a long lifetime in the thermosphere. On the other hand, ozone has a short lifetime there and is consequently very close to photochemical equilibrium:

$$[O_3] \simeq \frac{k_2[M] \cdot [O_2] \cdot [O]}{J_3 + k_3[O]} \tag{22}$$

Since $[O] \gg [O_3]$ at these levels, (14) gives:

$$\frac{\partial [O]}{\partial t} \simeq 2J_2 \cdot [O_2] - 2\left(k_1[M] + \frac{k_3 k_2[M] \cdot [O_2]}{J_3 + k_3[O]}\right) \cdot [O]^2 \tag{23}$$

Number densities of O and $O_2$ may then be obtained from (13) and (23). Such a simple computation predicts that $O_2$ is dissociated to such a degree above 95 km, that O becomes the most abundant oxygen allotrope. According to the model, $O_2$ should decrease very rapidly with height. This is inconsistent with the measurements which indicate that $[O] \simeq [O_2]$ at 120 km. The reason for this disagreement is that atmospheric transport processes were neglected in this model. The lifetime of atomic and molecular oxygen is about one week at 90 km and one month at 100 km.

Since the horizontal gradients of atomic oxygen are relatively small, its height distribution is primarily determined by vertical transport. Since so little is known about the vertical velocities in the thermosphere, we shall here only consider vertical diffusion, both molecular and turbulent. The vertical flux $F$ of a component may be expressed as:

$$F = -D\left(\frac{\partial n}{\partial z} + \frac{n}{H} + \frac{n}{T}\frac{\partial T}{\partial z}\right) - K_z\left(\frac{\partial n}{\partial z} + \frac{n}{H_a} + \frac{n}{T}\frac{\partial T}{\partial z}\right) \tag{24}$$

where $D$ is the molecular diffusion coefficient, $K_z$ the vertical eddy diffusion coefficient, $H$ the scale height of the component and $H_a$ the scale height of air. The continuity equation for the component may then be expressed as:

$$\frac{\partial n}{\partial t} = P - Qn - \frac{\partial F}{\partial z} \tag{25}$$

The eddy diffusion coefficient varies with height, latitude and season. Although our information on $K_z$ is insufficient, it is important to note that variations of $K_z$ within "reasonable" limits give theoretical values of $[O]/[O_2]$ which are, generally speaking, consistent with observations. But if one wants to study the oxygen profiles in more detail, better information is needed on $K_z$ as well as on the vertical velocity.

We have concentrated on the pure oxygen model, because such a model is relatively easy to survey, being based on only five reactions. Similar models have been used to study the distribution of other components. A commonly used model is the oxygen–hydrogen model which has principally been used to study the vertical distribution of water vapor. Such a model is described in the next section, where a computation is made for the 35–70-km region. Models including nitrogen have been defined to study the distribution of nitric oxide. In the future, similar models will be used to study other minor constituents of interest. However, the basis for most of these models will remain the pure oxygen model.

## 2. A TIME–DEPENDENT PHOTOCHEMICAL MODEL OF THE UPPER STRATOSPHERE AND LOWER MESOSPHERE

### 2.1. Introduction

During the last few years a series of ozone measurements has been made for the 35–70-km region. Daytime and night-time measurements confirm that there is, in the upper part of this region, a pronounced diurnal variation, as predicted by theoretical models (Hunt, 1965; London, 1967). Such models also show that there are, near sunrise and sunset, rapid variations which cannot easily be observed directly. It is the aim of this section to describe, on the basis of an oxygen–hydrogen atmosphere model, the diurnal variation of ozone.

### 2.2. The photochemical model

The photochemical model used in this study is a conventional oxygen–hydrogen model. Ten components are considered: $O(^3P)$, $O(^1D)$, $O_2$, $O_3$, OH, $HO_2$, $H_2O_2$, $H_2$, $H_2O$ and H. Among these components, 38 reactions, listed in Table I, are considered. Not all these reactions are of importance for our present problem. It turns out that reactions 1, 4b, 17, 19, 26, 28, 30, 31, 32, 33, 35 and 36 contribute only slightly to the photochemistry of the system and may therefore be disregarded.

Temperatures and air pressure data are taken from Cole et al. (1965). Data for the flux of solar radiation and absorption cross-sections for $O_2$ and $O_3$ have been summarized by Ackerman (1971). These data are used in our model. Absorption cross-sections for $O_2$ in the Schumann–Runge bands are taken from Huffman (1968). Absorption cross-sections of $H_2O$ are taken from Thompson et al. (1963), and of $H_2O_2$ from Schumb et al. (1955).

### 2.3. Results

A preliminary study of a photochemical equilibrium model gives a good indication of the lifetimes of the ten components. $O(^1D)$ has always a very short lifetime ($\sim 10^{-6}$ seconds) and will therefore always be in photochemical equilibrium with ozone. $H_2$ and $H_2O$, the most important hydrogen components, have lifetimes of the order of years at these levels, and turbulent mixing will establish constant mixing ratios in the vertical. In agreement with generally accepted ideas, we shall

TABLE I

Chemical reactions used in this model

| | Reaction | Rate constant |
|---|---|---|
| (1) | $O(^3P) + O(^3P) + M \rightarrow O_2 + M$ | $k_1 = 1.5 \cdot 10^{-34} \exp(1.4/RT)$ |
| (2) | $O(^3P) + O_2 + M \rightarrow O_3 + M$ | $k_2 = 5.8 \cdot 10^{-35} \exp(1.5/RT)$ |
| (3) | $O(^3P) + O_3 \rightarrow 2O_2$ | $k_3 = 3.3 \cdot 10^{-11} \exp(-4.2/RT)$ |
| (4a) | $O_2 + h\nu \rightarrow O(^3P) + O(^3P)$ | $J_{2a}(1750\text{Å} < \lambda < 2424\text{Å})$ |
| (4b) | $O_2 + h\nu \rightarrow O(^3P) + O(^1D)$ | $J_{2b}(\lambda < 1750\text{Å})$ |
| (5a) | $O_3 + h\nu \rightarrow O(^3P) + O_2$ | $J_{3a}(\lambda > 3100\text{Å})$ |
| (5b) | $O_3 + h\nu \rightarrow O(^1D) + O_2$ | $J_{3b}(\lambda < 3100\text{Å})$ |
| (6) | $OH + O(^3P) \rightarrow H + O_2$ | $k_6 = 5 \cdot 10^{-11}$ |
| (7) | $HO_2 + O(^3P) \rightarrow OH + O_2$ | $k_7 = 10^{-11}$ |
| (8) | $H + O_2 + M \rightarrow HO_2 + M$ | $k_8 = 3 \cdot 10^{-32}(273/T)^{1.3}$ |
| (9) | $H + O_3 \rightarrow OH + O_2$ | $k_9 = 2.6 \cdot 10^{-11}$ |
| (10) | $OH + HO_2 \rightarrow H_2O + O_2$ | $k_{10} = 10^{-11}$ |
| (11) | $H_2O_2 + h\nu \rightarrow 2OH$ | $J_{H_2O_2}(1875\text{Å} < \lambda < 3825\text{Å})$ |
| (12) | $O(^3P) + H_2O_2 \rightarrow OH + HO_2$ | $k_{12} = 10^{-15}$ |
| (13) | $HO_2 + HO_2 \rightarrow H_2O_2 + O_2$ | $k_{13} = 1.5 \cdot 10^{-12}$ |
| (14) | $OH + H_2O_2 \rightarrow H_2O + HO_2$ | $k_{14} = 4 \cdot 10^{-13}$ |
| (15) | $OH + OH \rightarrow H_2O + O(^3P)$ | $k_{15} = 2 \cdot 10^{-12}$ |
| (16) | $H_2O + h\nu \rightarrow OH + H$ | $J_{H_2O}(1350\text{Å} < \lambda < 2375\text{Å} + Ly\alpha)$ |
| (17) | $H + HO_2 \rightarrow H_2O + O(^3P)$ | $k_{17} = 2 \cdot 10^{-10} \exp(-4/RT)$ |
| (18) | $H + HO_2 \rightarrow H_2 + O_2$ | $k_{18} = 3 \cdot 10^{-12}$ |
| (19) | $H + H + M \rightarrow H_2 + M$ | $k_{19} = 1.2 \cdot 10^{-32}(273/T)^{0.7}$ |
| (20) | $O(^1D) + M \rightarrow O(^3P) + M$ | $k_{20} = 3.8 \cdot 10^{-11}$ |
| (21) | $O(^1D) + H_2 \rightarrow OH + H$ | $k_{21} = 10^{-11}$ |
| (22) | $O(^3P) + H_2 \rightarrow OH + H$ | $k_{22} = 7 \cdot 10^{-11} \exp(-10.2/RT)$ |
| (23) | $HO_2 + O_3 \rightarrow OH + 2O_2$ | $k_{23} = 10^{-17}$ |
| (24) | $OH + O_3 \rightarrow HO_2 + O_2$ | $k_{24} = 2.5 \cdot 10^{-11} \exp(-3/RT)$ |
| (25) | $H + H_2O_2 \rightarrow H_2 + HO_2$ | $k_{25} = 10^{-13}$ |
| (26) | $H + O(^3P) + M \rightarrow OH + M$ | $k_{26} = 8 \cdot 10^{-33}$ |
| (27) | $H + O_3 \rightarrow HO_2 + O(^3P)$ | $k_{27} = 2 \cdot 10^{-10} \exp(-4/RT)$ |
| (28) | $H + O_2 \rightarrow OH + O(^3P)$ | $k_{28} = 1 \cdot 10^{-9} \exp(-16.8/RT)$ |
| (29) | $O(^1D) + H_2O \rightarrow 2OH$ | $k_{29} = 3 \cdot 10^{-10}$ |
| (30) | $H + HO_2 \rightarrow 2OH$ | $k_{30} = 3 \cdot 10^{-12}$ |
| (31) | $H + OH + M \rightarrow H_2O + M$ | $k_{31} = 2.5 \cdot 10^{-31}$ |
| (32) | $O(^3P) + OH + M \rightarrow HO_2 + M$ | $k_{32} = 1.4 \cdot 10^{-31}$ |
| (33) | $H + OH \rightarrow H_2 + O(^3P)$ | $k_{33} = 3 \cdot 10^{-11} \exp(-8.3/RT)$ |
| (34) | $H_2 + OH \rightarrow H_2O + H$ | $k_{34} = 10^{-10} \exp(-5.9/RT)$ |
| (35) | $O(^1D) + O_3 \rightarrow 2O_2$ | $k_{35} = 3 \cdot 10^{-10}$ |
| (36) | $HO_2 + h\nu \rightarrow OH + O(^3P)$ | $J_{HO_2} = 1.4 \cdot 10^{-4}$ |

assume:

$$[O_2] = 0.2095 \cdot [M]$$

$$[H_2] = 5 \cdot 10^{-7} \cdot [M]$$

$$[H_2O] = 6.5 \cdot 10^{-6} \cdot [M]$$

As is often done in oxygen-hydrogen models, we shall consider the two functions:

$$[\tilde{O}] = [O(^1D)] + [O(^3P)] + [O_3] = [O] + [O_3]$$

$$[\tilde{H}] = [H] + [HO_2] + [OH]$$

On the basis of the reactions given above, we may, for each component, express the time variation in number density $[X]$ due to photochemistry by an equation of the form:

$$\frac{\partial[X]}{\partial t} = P - Q \cdot [X] - R \cdot [X]^2$$

where $P$ is the production and $(Q \cdot [X] + R \cdot [X]^2)$ is the destruction term. If $P$, $Q$ and $R$ were constants, the lifetime of $X$ would be $\tau = (Q^2 + 4R \cdot P)^{-1/2}$. In our case $P$, $Q$ and $R$ are not constants; however, quite often the dominating terms in $P$, $Q$ and $R$ vary more slowly than $[X]$ itself. The quantity $\tau$ will therefore in most cases be equal to the lifetime of the component. A survey of the lifetimes, defined in this way, is given in Fig. 1a for daytime (noon) and in Fig. 1b for night-time (midnight) conditions.

We can see that, particularly during the day, $H_2O_2$ and $\tilde{H}$ have much shorter lifetimes than the characteristic time for eddy transport. $H_2O_2$ and $\tilde{H}$ may therefore be computed without considering the transport terms. The internal equilibrium between OH, $HO_2$, and H is always established in less than 10 sec during the day and in less than 400 sec during the night. $\tilde{O}$ has a long lifetime ($\sim$ 50 days) at night, but during the day the lifetime is from 1 to 40 h, i.e., much shorter than the diffusion time. Internal equilibrium between O and $O_3$ is established in less than 100 sec.

We may then conclude that for all components, except $H_2$ and $H_2O$, the photochemical lifetimes are considerably shorter than the characteristic time for eddy transport. This simplifies the computations since the only interaction between different levels lies in the amount of ozone in the overlying atmosphere. Therefore we are faced with the mathematical problem of integrating a system of six non-linear partial differential equations. In general, this is a difficult problem; however, in our case a straightforward integration leads to a stable solution. Time steps of one minute are used (20 sec near sunrise and sunset), and the use of 30 sec gives no changes in the results. Dissociation rates are computed at intervals of 30 min; before

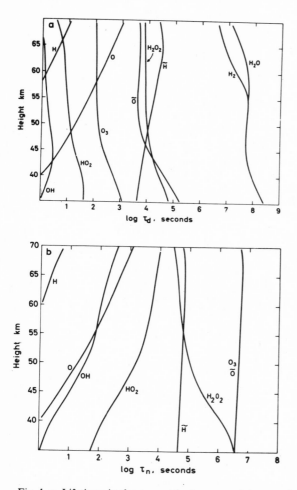

Fig. 1. a. Lifetimes in the oxygen–hydrogen model atmosphere, daytime (noon) conditions. b. Lifetimes in the oxygen–hydrogen model atmosphere, night-time (midnight) conditions.

sunset and after sunrise new values are computed at intervals of 10 min. As initial values, daytime equilibrium concentrations are used, and it is convenient to start the integration at sunset. Convergence (i.e., a repetition of a 24-h cycle) is reached after two days.

When small terms are neglected we may write with a sufficient degree of accuracy:

$$[O(^1D)] = \frac{J_{3b} \cdot [O_3]}{k_{20} \cdot [M]}$$

and:

$$\frac{\partial[\tilde{O}]}{\partial t} = 2J_2 \cdot [O_2] - (2k_3 \cdot [O] + k_9 \cdot [H] + k_{24} \cdot [OH] + k_{23}[HO_2]) \cdot [O_3]$$
$$- (k_6 \cdot [OH] + k_7 \cdot [HO_2]) \cdot [O]$$

During the night, only reactions 23 and 24 are important. In addition to this equation we must use:

$$\frac{\partial[O]}{\partial t} = 2J_2 \cdot [O_2] + J_3 \cdot [O_3] + k_{15}[OH]^2 + k_{27}[H] \cdot [O_3]$$
$$- (k_2 \cdot [M] \cdot [O_2] + k_6 \cdot [OH] + k_7 \cdot [HO_2] + k_3 \cdot [O_3]) \cdot [O]$$

except above 55 km during the day (corresponding to $[O_3] < [O]$), when it is more convenient to use:

$$\frac{\partial[O_3]}{\partial t} = k_2[M] \cdot [O_2] \cdot [O] - (J_3 + k_9 \cdot [H]) \cdot [O_3]$$

The last two equations are integrated analytically over the timestep. (Before integrating, substitution is made from $[O] + [O_3] = [\tilde{O}]$.)

Similarly we may write for the hydrogen cycle:

$$\frac{\partial[\tilde{H}]}{\partial t} = 2(J_{H_2O_2} + k_{12} \cdot [O]) \cdot [H_2O_2] + (J_{H_2O} + k_{29} \cdot [O(^1D)]) \cdot [H_2O]$$
$$+ k_{21} \cdot [O(^1D)] + k_{22} \cdot [O]) \cdot [H_2] - 2(k_{10} \cdot [OH] \cdot [HO_2]$$
$$+ k_{13} \cdot [HO_2] \cdot [HO_2] + k_{15}[OH] \cdot [H] + k_{18} \cdot [H] \cdot [HO_2])$$

At night only reactions 10 and 13 (see Table I) are of importance. Since OH always has a shorter lifetime than $HO_2$, we may integrate:

$$\frac{\partial[OH]}{\partial t} = k_7 \cdot [HO_2] \cdot [O] + k_9 \cdot [H] \cdot [O_3] + k_{23}[HO_2] \cdot [O_3]$$
$$- (k_6 \cdot [O] + k_{24}[O_3]) + [OH]$$

analytically over the timestep. During the day OH will always be in photochemical equilibrium, and even at night the departures from equilibrium are very small. In addition to the equation for $\tilde{H}$ and OH, we may use for $[H] > [HO_2]$ (i.e., above 60 km during the day):

$$\frac{\partial[HO_2]}{\partial t} = k_8 \cdot [M] \cdot [O_2] \cdot [H] - k_7[O] \cdot [HO_2]$$

which may be integrated analytically over the timestep. (Substitution is made from $[H] = [\tilde{H}] - [HO_2] - [OH]$.) Alternatively, when $[HO_2] > [H]$ (i.e., at night and below 60 km during the day), we use:

$$\frac{\partial[\mathrm{H}]}{\partial t} = (k_6 \cdot [\mathrm{O}] + k_{34}[\mathrm{H_2}])[\mathrm{OH}] - (k_8[\mathrm{M}][\mathrm{O_2}] + k_9[\mathrm{O_3}]) \cdot [\mathrm{H}]$$

which is integrated over the timestep. Departures from photochemical equilibrium are negligibly small. For $H_2O_2$ we have:

$$\frac{\partial[\mathrm{H_2O_2}]}{\partial t} = K_{13} \cdot [\mathrm{HO_2}]^2 - (J_{\mathrm{H_2O_2}} + k_{12} \cdot [\mathrm{O}] + k_{14}[\mathrm{OH}]) \cdot [\mathrm{H_2O_2}]$$

where only reaction 13 (Table I) is important at night.

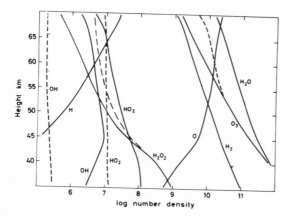

Fig. 2. Chemical composition of the oxygen–hydrogen atmosphere model. Noon values are given by solid curves; when there is a diurnal variation, midnight values are given by broken curves.

The resulting composition is shown in Fig. 2 for 45° latitude, summer. Generally speaking, the results do not differ much from those obtained in previous models.

Of particular interest is the concentration of ozone and its variation with height, latitude, season and time of day. The variation with latitude and season is given in Table IIA for noon conditions, and in Table IIB for midnight conditions. Generally speaking, the ozone concentrations increase from the winter pole to the summer pole. The variations with latitude and season are not great; in most cases they are comprised within a factor of 2. The theoretical values may be compared with the observations of Hilsenrath (1971). Such a comparison is demonstrated in Fig. 3a for a mid-latitude summer case and in Fig. 3b for a high latitude winter case. Fig. 3a shows very good agreement between theory (for 45°N, summer) and observations from Wallops Island (38°N, spring) for all heights down to about 43 km. Lower down, the theory predicts higher ozone values than are actually observed. The reason for this disagreement is not quite clear. There is a possibility that better agreement would be obtained if nitrogen reactions were considered.

PHOTOCHEMICAL ATMOSPHERE MODELS

TABLE IIA

Theoretical values of ozone number densities at noon as a function of latitude

| Height | Winter | | | | Summer | | | |
|--------|--------|--------|--------|--------|--------|--------|--------|--------|
| (km) | 60° | 45° | 30° | 0° | 30° | 45° | 60° | 75° |
| 70 | 2.7+8 | 5.5+8 | 9.2+8 | 1.1+9 | 1.2+9 | 1.3+9 | 1.4+9 | 1.0+9 |
| 65 | 1.3+9 | 1.7+9 | 2.3+9 | 2.5+9 | 2.7+9 | 3.1+9 | 3.3+9 | 3.1+9 |
| 60 | 3.7+9 | 4.6+9 | 5.9+9 | 6.7+9 | 7.1+9 | 8.2+9 | 8.1+9 | 7.8+9 |
| 55 | 9.6+9 | 1.2+10 | 1.6+10 | 1.9+10 | 2.0+10 | 2.1+10 | 2.1+10 | 2.0+10 |
| 50 | 3.7+10 | 3.7+10 | 4.5+10 | 5.1+10 | 5.2+10 | 5.5+10 | 5.6+10 | 5.7+10 |
| 45 | 1.7+11 | 1.7+11 | 1.8+11 | 1.9+11 | 1.9+11 | 2.0+11 | 1.9+11 | 1.9+11 |
| 40 | 4.9+11 | 6.3+11 | 6.8+11 | 7.7+11 | 7.8+11 | 7.7+11 | 7.0+11 | 6.3+11 |
| 35 | 1.2+12 | 1.4+12 | 1.5+12 | 1.9+12 | 1.9+12 | 1.8+12 | 1.7+12 | 1.4+12 |

TABLE IIB

Theoretical values of ozone number densities at midnight as a function of latitude

| Height | Winter | | | | Summer | | |
|--------|--------|--------|--------|--------|--------|--------|--------|
| (km) | 60° | 45° | 30° | 0° | 30° | 45° | 60° |
| 70 | 7.7+9 | 6.1+9 | 5.5+9 | 4.1+9 | 4.1+9 | 5.1+9 | 5.9+9 |
| 65 | 1.5+10 | 1.0+10 | 9.7+9 | 9.1+9 | 9.3+9 | 1.0+10 | 1.1+10 |
| 60 | 1.4+10 | 1.4+10 | 1.5+10 | 1.5+10 | 1.5+10 | 1.6+10 | 1.6+10 |
| 55 | 1.7+10 | 2.1+10 | 2.4+10 | 2.6+10 | 2.6+10 | 2.7+10 | 2.6+10 |
| 50 | 4.0+10 | 4.4+10 | 5.2+10 | 5.7+10 | 5.9+10 | 6.1+10 | 6.4+10 |
| 45 | 1.7+11 | 1.7+11 | 1.8+11 | 1.9+11 | 2.0+11 | 2.0+11 | 1.9+11 |
| 40 | 4.8+11 | 6.3+11 | 6.8+11 | 7.5+11 | 7.7+11 | 7.6+11 | 6.8+11 |
| 35 | 1.2+12 | 1.4+12 | 1.5+12 | 1.9+12 | 1.9+12 | 1.8+12 | 1.7+12 |

Fig. 3b shows a comparison of two observations in December from Point Barrow (71°N), and theoretical values for 60°N, winter, night. Fair agreement exists up to about 55 km; higher up the theory predicts higher values than were actually observed. However, it should be remembered that night-time concentrations at 60°N do not apply to the polar winter night. When ozone is transported from the sunlit region into the polar winter night, it immediately attains the night-time value for that latitude and then remains constant for some time, since the lifetime for night-time conditions is about $4 \cdot 10^6$ sec. However, after about one day $HO_2$ and OH will decrease. The lifetime of ozone will then increase and soon become longer than the characteristic time for eddy diffusion. This means that, for this case, eddy diffusion must be considered together with photochemistry.

The diurnal variation is negligible at (and below) 50 km and beccmes more and more significant the higher up we go, as it is shown in Fig. 4a and 4b. Immediately

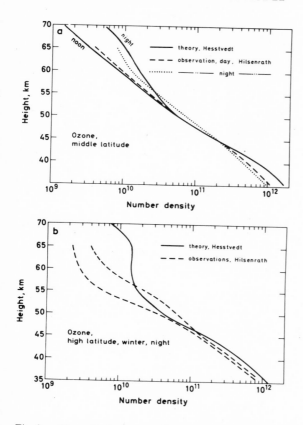

Fig. 3. a. Theoretical ozone profiles for 45° latitude summer, compared to observed profiles for Wallops Island (38°N), spring. b. Theoretical ozone night-time profile for 60° latitude compared to profiles observed at Point Barrow (71°N), winter, night.

after sunrise a sudden drop takes place, followed by a slower increase during the day. At sunset we have a sudden increase to the almost constant night-time value.

*2.4. Conclusion*

The fair agreement between observed and computed values of the ozone concentration supports the view that theoretical models are a most efficient tool for studying details in the ozone distribution, such as its variation with time and latitude. Future measurements should be directed towards the upper part of the region studied here, i.e., from about 65 km and up. Furthermore, observations from high latitudes in the winter are of interest.

It was mentioned that the theoretical model gives too much ozone below about 43 km. There are two causes of this discrepancy. Many of the reaction rates are not

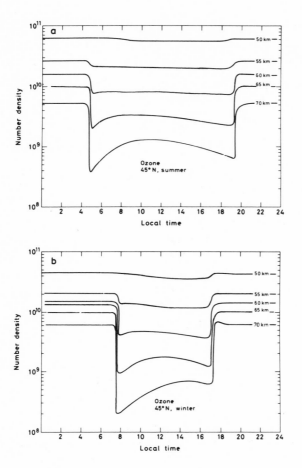

Fig. 4. a. Diurnal variation of ozone, 45°N, summer. b. Diurnal variation of ozone, 45°N, winter.

very well known: of course, errors in $k_2$, $k_3$, $J_2$ and $J_3$ will influence the results; moreover, other reaction rates play an important role in this connection. However, the failure of the model arises mainly from the fact that we have disregarded reactions involving nitrogen. A thorough analysis of the influence of nitrogen oxides appears to be a fruitful path in future theoretical studies of atmospheric ozone.

## REFERENCES

Ackerman, M., Ultraviolet solar radiation related to mesospheric processes. In: *Mesospheric Models and Related Experiments* (G. Fiocco, Editor). Reidel, Dordrecht, pp. 149–159, 1971.
Cole, A. E., Court, A. and Kantor, A. J., Model atmospheres. In: *Handbook of Geophysics and Space Environments* (S. Valley, Editor). McGraw-Hill, New York, N.Y., pp. 2/1–22, 1965.

Hilsenrath, E., Ozone measurements in the mesosphere and stratosphere during two significant geophysical events. *J.Atmos.Sci.*, **28**:295–297, 1971.

Huffman, R. E., Absorption cross sections of atmospheric gases for use in aeronomy. *Symposium on laboratory measurements of aeronomic interest*. York University, Toronto, Canada, pp. 95–120, 1968.

Hunt, B. G., A non-equilibrium investigation into the diurnal photochemical atomic oxygen and ozone variations in the mesosphere. *J.Atmos.Terr.Phys.*, **27**:133–144, 1965.

London, J., The average distribution and time variations of ozone in the stratosphere and mesosphere. In: *Space Research VII* (*Vol. I*) (R.L. Smith-Rose, Editor). North-Holland, Amsterdam, pp. 172–185, 1967.

Schumb, W. C., Gatterfield, C. N. and Wentworth, R. L., *Hydrogen Peroxide*. Reinhold, New York, N.Y., 287 pp., 1955.

Thompson, B. A., Harteck, P. and Reeves, R. R., Ultraviolet absorption coefficient of $CO_2$, CO, $O_2$, $H_2O$, $N_2O$, $NH_3$, NO, $SO_2$ and $CH_4$ between 1850 and 4000 Å. *J.Geophys.Res.*, **68**:6431–6436, 1963.

# TIDES AND INTERNAL GRAVITY WAVES IN THE ATMOSPHERE

RICHARD S. LINDZEN[*]

*Department of Geophysical Sciences, Harvard University, Cambridge, Mass. (U.S.A.)*

## SUMMARY

The history, data and theory for tides and internal gravity waves in the atmosphere are comprehensively reviewed. The relationship between tides and internal waves is explored in detail. Special attention is given to the effects of dissipation, and to tides in the thermosphere.

## 1. HISTORY OF THE STUDY OF TIDES AND ELEMENTARY PHYSICS OF INTERNAL GRAVITY WAVES

### 1.1. Introduction

This paper in four parts will, I hope, serve to explain our present understanding of atmospheric tides and of internal gravity waves, tides being a particular example of the latter. With only small modifications, this paper is taken from reports and lectures already published: namely a monograph by the late Sydney Chapman and myself (Chapman and Lindzen, 1970), a set of lectures for the American Mathematical Society (Lindzen, 1971a), and two shorter lectures delivered at the Fourth Esrin-Esrolab Meeting in Frascati (Lindzen, 1971b) and at the Marshall Space Flight Center (Lindzen, 1971c).

Atmospheric tides refer to those oscillations in any atmospheric field whose periods are integral fractions of either a lunar or a solar day. The bulk of our concern, moreover, will be restricted to global scale, migrating tides. By migrating tides we mean tides which depend on local time; i.e., tides which follow the apparent motion of the sun or moon. Atmospheric tides are excited not only by the tidal gravitational potential of the sun and moon, but also (and as it turns out, to a larger extent) by the daily variations in solar heating. The atmosphere responds to tidal or thermotidal excitation by means of internal gravity waves.

There are a number of reasons for studying atmospheric tides. To begin with, we know their frequencies and zonal wavenumbers precisely, and somewhat less

---

[*] Alfred P. Sloan Foundation Fellow.

precisely the magnitude and distribution of their excitation. We may, therefore, calculate the atmospheric response, and isolate that response in the data. In other words, they are a good vehicle for checking theories of internal waves in the atmosphere. Another reason for studying atmospheric tides is that they form a substantial part of the total meteorology of the upper atmosphere—at least above 50 km. However, the historical reasons for studying atmospheric tides were primarily intellectual. Indeed, both observations in the upper atmosphere, and the recognition that atmospheric tides are internal gravity waves are primarily products of the last decade.

We shall begin with a review of the history of the study of tides. This will be followed by an introduction to the physics of internal gravity waves in general.

## 1.2. History

Tidal oscillations in the atmosphere (apart from sea breezes, which marginally fit our definition) were not recognized until the development of the barometer by Torricelli (ca. 1643). By contrast, sea tides were first reported before 320 B.C. by Pytheas of Marseilles who, in a voyage of exploration, circumnavigated Britain. In the course of this voyage he noticed the approximately twice-daily progression of high and low water, whose phase appeared to parallel the time of the moon's transit. It was early suggested that the tides were due to some force of attraction emanating from the moon, but it awaited Newton to explain why the lunar sea tides were semidiurnal. Newton realized that the lunar tidal potential would excite oscillations in the atmosphere, as well as in the sea. However, he thought the atmospheric oscillations would be too small to be detected. Given the amount of data available in northern Europe in the 17th century, he was certainly correct. However, the observation of atmospheric tides in the tropics proved to be an easy matter. In Fig. 1 we can see time records of surface pressure at two stations, Potsdam (52°N) and Batavia (now Jakarta, 6°S). The record at Potsdam is dominated by the moving highs and lows of midlatitude weather systems. At Batavia, however, an extraordinarily regular semidiurnal oscillation is evident. However, in contrast to sea tides, the period is one half of a solar day. Laplace was already aware of this observational fact. Since the lunar tidal potential is about twice as large as that due to the sun, Laplace quite properly concluded that the atmospheric tide was due to the thermal action of the sun, but he saw little hope of constructing a theory of thermally excited oscillations (Laplace, 1799).

Lord Kelvin in the 1870's collected surface pressure data from thirty locations and harmonically analyzed these data for the diurnal (24 h), semidiurnal (12 h) and terdiurnal (8 h) components. In agreement with earlier findings, he found that the oscillation was solar and that the semidiurnal component was substantially stronger than the diurnal component. In reporting these results, Kelvin (1882) concurred

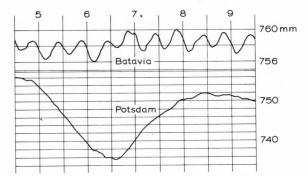

Fig. 1. Barometric variations (on twofold different scales) at Batavia (6°S) and Potsdam (52°N) during November 1919. (After Bartels, 1928.)

with Laplace that the atmospheric tides are thermally excited. He pointed out, however, that unlike gravitational excitation, thermal excitation will be primarily diurnal rather than semidiurnal. To explain the dominance of the semidiurnal surface pressure oscillation, Kelvin suggested that the atmosphere had a suitable free oscillation with a period very near to 12 h. Kelvin's resonance hypothesis dominated thinking in this field for nearly sixty years. Rayleigh (1890) and Margules (1890, 1892, 1893) began the earliest searches for an atmospheric resonance. Margules showed that the atmosphere would be resonant to semidiurnal forcing if the atmosphere behaved like a homogeneous fluid with a depth of 7.85 km. This is known as the *atmosphere's equivalent depth*. Its physical and mathematical meaning will be explained when we turn to the theory of atmospheric tides. For the moment, it suffices to say that the equivalent depth of an atmosphere depends on the thermodynamic state of the atmosphere, and most significantly upon the mean thermal structure of the atmosphere. Both Rayleigh and Margules made rather crude and unrealistic assumptions concerning these matters, which made their results uncertain. Nevertheless, they each concluded that resonance was a real possibility. Lamb (1910, 1916) investigated the matter in greater detail. He found that for either isothermal atmospheres wherein density variations occur isothermally, or for an atmosphere in adiabatic equilibrium, the equivalent depth of the atmosphere will be very nearly resonant. In addition, Lamb showed that in an atmosphere where the mean temperature varied linearly with height, but not adiabatically, there would be an infinite number of equivalent depths—thus greatly enhancing the possibilities for resonance. This particular finding was essentially forgotten for twenty years. Finally, Lamb suggested the possibility that the solar semidiurnal atmospheric tide might indeed be gravitationally excited, since that would require a resonance magnification of about 70 which, in turn, would require that the resonant periods be within 2 min of 12 h. The lunar period would be 12 h

26 min. Arguing against this was, as Lamb noted, the intrinsic unlikelihood of such close agreement in period, and the fact that the observed surface pressure oscillation's phase led, rather than lagged, the transits of the sun. Chapman (1924) argued that the resonance theory might still be correct. The existence of a thermal excitation comparable in magnitude to the gravitational excitation could explain the phase. In 1932, G.I. Taylor put forth a rather significant criticism of the resonance theory. If the atmosphere had an equivalent depth, $h$, then atmospheric disturbances excited by explosions, earthquakes, etc. should travel at a speed, $\sqrt{gh}$. Data from the Krakatoa eruption of 1883 showed that the atmospheric pulse traveled at a speed of 319 m sec$^{-1}$, corresponding to $h = 10.4$ km—a value too far from 7.85 km to produce resonance. Later, Taylor (1936) re-established Lamb's result (though more rigorously) that the atmosphere might have several equivalent depths—thus allowing some remaining hope for the, by now much modified, Kelvin resonance hypothesis. This hope received an immense boost from the work of Pekeris (1937). Pekeris investigated the equivalent depths of a variety of atmospheres with relatively complicated thermal structures, and found that for an atmosphere whose temperature increased above the tropopause (ca. 12 km) to a high value (350°K) near 50 km, and then decreased upwards to a low value, there existed a second equivalent depth close to 7.9 km. More convincing was the fact that the assumed temperature profile agreed excellently with a profile proposed by Martyn and Pulley (1936), on the basis of then recent meteor and anomalous sound observations. It was almost as though Pekeris had adduced the atmosphere's thermal structure from tidal data at the earth's surface. His results, moreover, appeared to explain other observations of ionospheric and geomagnetic tidal variations. The vindication of the resonance theory seemed well-nigh complete. Pekeris countered Taylor's early criticism by showing that a low-level disturbance would primarily excite the faster mode associated with $h = 10.4$ km. A re-examination of the Krakatoa evidence by Pekeris even showed some evidence for the existence of the slower mode, for which $h = 7.9$ km.

For the next fifteen years most research on this subject was devoted to the refinement and interpretation of Pekeris' work. This work is comprehensively reviewed by Wilkes (1949). Unfortunately, in the aftermath of World War II numerous rocket probings of atmospheric temperature were made and these showed a different temperature structure from that proposed by Martyn and Pulley. In particular, the temperature maximum at 50 km was much cooler (about 280°K instead of 350°K). In addition, the temperature decline above 50 km ended at 80 km above which the temperature again increases, reaching very high values (600°K–1400°K) above 150 km. For the new temperature profiles, the atmosphere no longer had a second equivalent depth, and the magnification of the solar semidiurnal tide was no longer sufficient to account for the observed semidiurnal tide on the basis of any realistic combination of gravitational excitation, and excitation due to the upward diffusion of the daily variation of surface temperature (Jacchia and Kopal, 1952).

With the demise of the resonance theory, the search began for additional thermal sources. Although most of the sun's radiation is absorbed by the earth's surface, about 10% is absorbed directly by the atmosphere, and this appeared a likely source of excitation. Siebert (1961) investigated that effectiveness of insolation absorption by water vapor in the troposphere, and found that it could account for one-third of the observed semidiurnal surface pressure oscillation. This was far more than could be accounted for by gravitational excitation or surface heating. Siebert also investigated the effectiveness of insolation absorption by ozone in the mesosphere. He found its effect to be relatively small. As it turns out, this last conclusion followed primarily from Siebert's use of an exceedingly unrealistic basic temperature distribution for the atmosphere above the lower troposphere. Butler and Small (1963) corrected this error, and showed that ozone absorption indeed accounted for the remaining two-thirds of the surface semidiurnal oscillation.[*] At this point one was forced to return to Kelvin's original question: why wasn't the diurnal oscillation stronger than the semidiurnal. The situation had, moreover, become more complicated. Data above the ground up to about 100 km showed that above the ground the diurnal oscillations were as strong and often stronger than semidiurnal oscillations. Lindzen (1967) carried out theoretical calculations for the diurnal tide which provided satisfactory answers for the observational facts. Over half the globe (polewards of ±30° latitude), 24 h is longer than the local pendulum day, and under these circumstances a 24-h oscillation is incapable of propagating vertically. Because of this, it turns out that about 80% of the diurnal excitation goes into physically trapped modes which cannot propagate disturbances aloft to the ground. The atmospheric response to these modes in the neighborhood of the excitation is, however, substantial. (These trapped diurnal modes were discovered independently by Lindzen, 1966 and Kato, 1966.) In addition, there exist (primarily within ±30° latitude) diurnal modes which propagate vertically. However, as one could deduce from the dispersive properties of internal gravity waves, the long period and the restricted latitude scale of these waves causes them to have relatively short vertical wavelengths (25 km and less). They are, therefore, subject to some destructive interference effects. While Butler and Small suggested that this could explain the relatively small amplitude of the diurnal tide, subsequent calculations show that this effect would be inadequate. What really proves to be important is that the propagating modes receive only 20% of the excitation.

After almost a century, Kelvin's question seems satisfactorily answered. However, as we shall see, numerous other problems remain in atmospheric tides.

---

[*] The problem of the semidiurnal oscillation at the ground is not yet completely settled. There remains a discrepancy of one half hour in phase between theory and observation.

*1.3. Elementary physics of internal gravity waves*

Reviews concerning the subject of internal gravity waves may be found in Hines (1960, 1963) and Jones (1971). Some more recent developments are presented in Pitteway and Hines (1963), Booker and Bretherton (1967) and in Lindzen (1970). Generally, the subjects of internal gravity waves and atmospheric tides are presented separately. However, the two subjects are hardly unrelated. Atmospheric tides are, as I have already mentioned, essentially internal gravity waves (modified by the earth's sphericity and rotation) for which the periods (integral fractions of a day) are known precisely and for which the excitations, while not perfectly known, are better known than they are for most other gravity waves.

In this section I shall present an essentially non-mathematical discussion of internal gravity waves which will, I hope, leave the reader with an intuitive understanding of the subject that will help him follow more advanced studies of both internal gravity waves and tides.

Any true wave must be associated with some restoring mechanism. For acoustic waves, the restoring force arises from the compressibility of the air; for internal gravity waves, it is the buoyancy force exerted on a displaced fluid element in a stably stratified fluid.

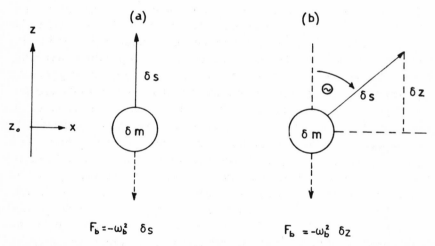

Fig. 2. Schematic description of fluid elements, their displacements, and buoyancy forces per unit mass.

Consider an element of fluid at some level $z_o$, in a fluid where density $\rho$ is decreasing with height at a rate $-d\rho/dz$. The situation is depicted in Fig. 2a. The mass of the fluid element is:

$$\delta m = \rho(z_o)\delta v$$

where $\delta v$ = volume of fluid element. If we displace $\delta m$ a small distance $\delta s$, it will be subject to a buoyancy force:

$$-g\left[\rho(z_o) - \left(\rho(z_o) + \frac{d\rho(z_o)}{dz}\delta s\right)\right]\delta v$$

acting to return $\delta m$ to $z_o$; $g$ is the acceleration of gravity. Variations of $\delta v$ due to compressibility have, for the moment, been neglected. The equation of motion for $\delta m$ is:

$$\rho(z_o)\delta v\frac{d^2\delta s}{dt^2} = g\frac{d\rho}{dz}\delta s\delta v$$

or:

$$\frac{d^2\delta s}{dt^2} = \frac{g}{\rho(z_o)}\frac{d\rho}{dz}\delta s \tag{1}$$

In a stably stratified, incompressible fluid $d\rho/dz < 0$. Hence, eq. 1 describes a harmonic oscillation with a frequency $\omega_b$, given by:

$$\omega_b^2 = -\frac{g}{\rho}\frac{d\rho}{dz} \tag{2}$$

$\omega_b$ is known as the Brunt–Vaisala frequency. The effect of compressibility is to change our expression for $\omega_b$ to the following:

$$\omega_b^2 = \frac{g}{T}\left(\frac{dT}{dz} + \frac{g}{c_p}\right) \tag{3}$$

where $T$ is the temperature of the ambient fluid.

Let us designate the buoyancy force per unit volume on a displaced fluid element as:

$$F_b = -\omega_b^2\delta s \tag{4a}$$

$F_b$ is directed vertically. Now consider a fluid element that is somehow constrained to move at some angle $\theta$, with respect to the vertical (viz., Fig. 2b). The force exerted on this fluid element will be the projection of the buoyancy force:

$$F = -\omega_b^2\delta z\cos\theta = -\omega_b^2\cos^2\theta\delta s \tag{4b}$$

and the element will oscillate with a frequency, $\omega$, given by:

$$\omega^2 = \omega_b^2\cos^2\theta \tag{5}$$

A real fluid is, of course, continuous and cannot be thought of in terms of isolatable elements. However, when we excite a real fluid at a frequency $\omega$ (where $\omega^2 < \omega_b^2$), the pressure forces do indeed constrain the fluid to move at an angle $\theta$ with respect to the vertical, such that the projection of the buoyancy force corresponds to the restoring force appropriate to the frequency $\omega$. The situation becomes more concrete when we consider waves excited by a corrugated bottom, moving at a constant speed $c$ (viz., Fig. 3). The horizontal wavelength imposed by the corrugated bottom is $L_H$, which can also be expressed in terms of horizontal wavenumber $k = 2\pi/L_H$.

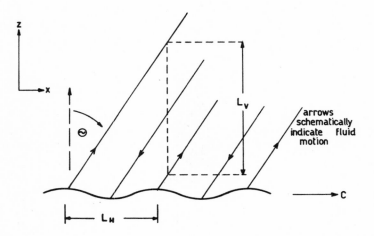

Fig. 3. Schematic description of motions excited in a stably stratified fluid by the motion (at speed, $c$) of a corrugated bottom.

Moving the bottom at a speed $c$, gives rise to a local oscillation with frequency:

$$\omega = kc \tag{6}$$

The angle $\theta$ is given by eq. 5. If the fluid flow shown in Fig. 3 were to be plotted as a function of altitude at a fixed time and horizontal position, then we would see a sinusoidal structure with a vertical wavelength $L_v$ (or vertical wavenumber $n = 2\pi/L_v$). Now:

$$\left(\frac{L_H}{L_v}\right)^2 = \left(\frac{n}{k}\right)^2 = \tan^2\theta \tag{7}$$

Combining (7) and (5) gives, approximately, the correct dispersion relation for internal gravity waves:

$$\left(\frac{n}{k}\right)^2 = \frac{\sin^2\theta}{\cos^2\theta} = \frac{1-\cos^2\theta}{\sin^2\theta} = \frac{1-\left(\dfrac{\omega}{\omega_b}\right)^2}{\left(\dfrac{\omega}{\omega_b}\right)^2} \cdot \tag{8a}$$

or:

$$n^2 = \frac{1-\left(\dfrac{\omega}{\omega_b}\right)^2}{\left(\dfrac{\omega}{\omega_b}\right)^2} k^2 \tag{8b}$$

Thus far, we are considering oscillations of the form:

$$e^{i(\omega t + kx + nz)} \tag{9}$$

where: $z$ = altitude; $x$ = horizontal distance; and $t$ = time. Such oscillations are waves which propagate upward with (in the absence of dissipation or mean flow) constant energy density. Since density is decreasing with height, the constancy of energy density (consisting in terms like $\frac{1}{2}\rho u^2$, where $u$ = oscillatory horizontal velocity) requires that oscillatory fields increase in amplitude as $1/\rho^{1/2}$. The relations obtained thus far are appropriate only for isothermal atmospheres, where $\omega_b^2$ is a constant. Because of the requirement of energy constancy, (9) must be replaced by:

$$e^{z/2H} e^{i(\omega t + kx + nz)} \tag{10}$$

for an isothermal atmosphere where:

$$H = \frac{RT}{g} \tag{11}$$

where $R$ is the gas constant for air and $H$ is the atmospheric scale height. Because the $z$-dependence is no longer purely sinusoidal, (8b) must be slightly modified:

$$n^2 = \frac{1-\left(\dfrac{\omega}{\omega_b}\right)^2}{\left(\dfrac{\omega}{\omega_b}\right)^2} k^2 - \frac{1}{4H^2} \tag{12}$$

Incidentally, we may see from Fig. 3 why *upward* energy flux for internal gravity waves implies *downward* phase progression. If the bottom has to do work on the fluid, it must push the fluid; therefore, the fluid motions must tilt to the right of the vertical, if $c$ is positive. In this case, a given phase appears first aloft, and then progresses downward.

Essential to the existence of internal gravity waves is the ability of fluid elements to retain their buoyancy (adiabaticity), and to move freely *across* isobars. The latter is inhibited by various frictional processes, and also by rotation which tends to cause a fluid to move *parallel* to isobars. In a plane rotating system, the dispersion relation for internal gravity waves becomes:

$$n^2 = \frac{\omega_b^2 \left[ 1 - \left( \frac{\omega}{\omega_b} \right)^2 \right]}{\omega^2 - f^2} k^2 - \frac{1}{4H^2} \tag{13}$$

where $f = 2\,\Omega$, and $\Omega$ is the rotation rate of the fluid. From (13) we see that the vertical propagation of an internal gravity wave requires:

(I)     $f^2 < \omega^2 < \omega_b^2$

(II)    $k^2 > \dfrac{1}{4H^2} \dfrac{\omega^2 - f^2}{\omega_b^2 \left[ 1 - \left( \dfrac{\omega}{\omega_b} \right)^2 \right]}$

The requirement $\omega^2 > f^2$ is particularly important for atmospheric tides. On a rotating sphere we must consider $\Omega \sin \varphi$ (where $\varphi$ = latitude), the vertical component of the rotation vector, in place of $\Omega$. Thus $f$ varies from zero at the equator to $2\pi/12$ h at the poles. Hence, internal gravity waves with periods *shorter* than 12 h can propagate vertically anywhere, but for longer periods, internal gravity waves propagate vertically in regions increasingly confined to the vicinity of the equator. Thus, the diurnal tidal modes consist in vertically propagating modes confined primarily to the region between $\pm 30°$ latitude, and exponentially trapped modes polewards of these latitudes.

Finally, some comment must be made about the excitation of gravity waves (including tides). In our example, we considered excitation by a moving, corrugated boundary. While pedagogically convenient, this was an obviously unrealistic example. Nevertheless, it contains the element which is essential to any excitation— it causes time varying vertical displacements of material surfaces. More practical ways of doing this are by fluid motions over stationary orography* by explosions and volcanic eruptions, or by the daily variations in the absorption of sunlight by the atmosphere due to the rotation of the earth. The list is hardly exhaustive. In a stably stratified atmosphere, where dissipation is not dominant, any excitation with frequency components between $\omega_b$ and $f$ will excite internal gravity waves.

---

*  By means of a shift of coordinates, this is equivalent to our idealized excitation; examples in nature are waves in the lee of wind flowing over mountains. Although mountains do not consist in infinite sinusoidal corrugations, their effects can be simulated by Fourier synthesis.

## 2. ATMOSPHERIC TIDES: DATA

Before proceeding to the mathematical theory of atmospheric tides, it is advisable to present a description of the phenomena about which we propose to theorize. In the short time allowed us we can hardly hope to do justice to the presentation of the data. Let it suffice to say that at many stages our observational picture is based on inadequate data; in almost all cases, the analyses of data required the extrication of fourier components from noisy data, and in some instances even the observational instruments have introduced uncertainties. Details of these matters may be found in Chapman and Lindzen (1970).

Until recently, almost all data analyses for atmospheric tides were based on surface pressure data. Although tidal oscillations in surface pressure are generally small, at quite a few stations we have as much as 50–100 years of hourly or bi-hourly data. As a result, even today, our best tidal data are for surface pressure. Fig. 4 shows the amplitude and phase of the solar semidiurnal oscillation over the globe; it was prepared by Haurwitz (1956), on the basis of data from 296 stations. The phase over most of the globe is relatively constant, implying the dominance of the migrating semidiurnal tide, but other components are found as well (the most significant of which is the semidiurnal standing oscillation, viz., Fig. 5). If we let $t =$ local time nondimensionalized by the solar day, then, according to Haurwitz, the solar semidiurnal tide is well represented by:

$$S_2(p) = 1.16 \text{ mbar } \sin^3\theta \, \sin(2t + 158°) + 0.085 \text{ mbar } P_2(\theta) \, \sin(2t_u + 118°) \quad (14)$$

where $\theta =$ colatitude, $t_u =$ Greenwich time, and $P_2(\theta) = 1/2(3\cos^2\theta - 1)$. One of the remarkable features of $S_2(p)$ is the fact that it hardly varies with season. This can be seen from Fig. 6. The situation is more difficult for $S_1(p)$. It varies with season, it is weak, and it is strongly polluted by non-migrating diurnal oscillations (viz., Fig. 7). There are values of $s$ with amplitudes as large as $1/4$ of that pertaining to $s = 1$. For $S_2(p)$, $s = 2$ was twenty times as large as its competitor. Moreover, large values of $s$, being associated with a small scale (large gradients), produce larger winds for a given amplitude of pressure oscillation than $s = 1$. We will return to this later. According to Haurwitz (1965), $S_1(p)$ is roughly representable as follows:

$$S_1^1(p) = 593 \ \mu\text{bar } \sin^3\theta \, \sin(t + 12°) \quad (15)$$

Data have also been analyzed for small terdiurnal and higher harmonics. Even $L_2(p)$ has been isolated. As we can see in Fig. 8, the amplitude of $L_2(p)$ is about $1/20$ of $S_2(p)$. $L_2(p)$ also has a peculiar seasonal variation, as can be seen in Fig. 9. The seasonal variation occurs with both the northern and southern hemispheres in phase.

$S_2(p)$ equilines: amplitude ($s_2$; unit $10^{-2}$ mb) and phase ($\sigma_2$)

Fig. 4.

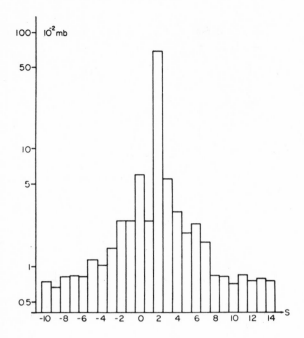

Fig. 5. The amplitudes (on a logarithmic scale, and averaged over the latitudes 80°N to 70°S) of the semidiurnal pressure waves, parts of $S_2(p)$, of the type $y_s \sin(2t_u + s\varphi + \sigma_s)$, where $t_u$ signifies universal mean solar time. (After Kertz, 1956.)

Data above the surface are rarer and less accurate, but some are available. At a few stations there are sufficiently frequent balloon ascents to permit tidal analyses. The results of one such analysis are shown in Table I. The results are sufficient to at least establish orders of magnitude. In the troposphere, horizontal wind oscillations have amplitudes $\sim 10$ cm/sec. In the stratosphere, amplitudes are about 50 cm/sec. However, data at a single station do not permit separating migrating tides from other components. At many stations, balloon soundings are made twice a day at 12-h intervals. The soundings are made, moreover, simultaneously at all stations (i.e., at the same universal time). Thus if one subtracts the average of measurements taken at 1200 GMT from the average of measurements taken at 0000 GMT, one should obtain a fair approximation to the diurnal component of the flow field at 0000 GMT. If the migrating tide dominates the wind oscillation, then we should see a clear zonal wave number 1 pattern. The results are shown in Fig. 10, 11 and 12 (Wallace and Hartranft, 1966). We can see that the diurnal flow field at 700 mbar

Fig. 4. World maps showing equilines of (below) the amplitude (units of $10^{-2}$ mbar) and (above) the phase relative to local mean time of the solar semidiurnal surface pressure oscillation. (After Haurwitz, 1956.)

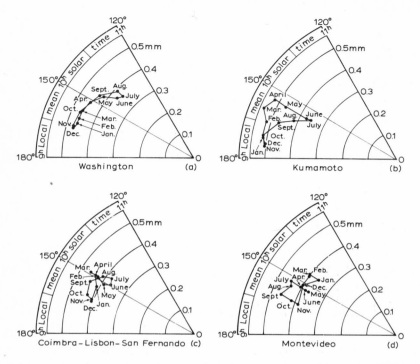

Fig. 6. Harmonic dials showing the amplitude and phase of $S_2(p)$ for each calendar month, for four widely spaced stations in middle latitudes: (a) Washington, D.C.; (b) Kumamoto; (c) mean of Coimbra, Lisbon and San Fernando; (d) Montevideo (Uruguay). (After Chapman, 1951).

is dominated by gyres, associated with relatively small orographic features. At 60 mbar, orographic gyres remain, but they are associated with large features like the Pacific Ocean. By 15 mbar, we have a simple wavenumber 1 pattern. This suggests that orographic effects have died out by 30 km, and that diurnal oscillations above this level will be representative of the migrating tide.

In the region between 30 and 60 km, most of our data come from meteorological rocket soundings. These are comparatively infrequent and the method of analysis becomes, a priori, a serious problem. However, it turns out that results of different analyses appear to be compatible (at least for the diurnal component) because tidal winds at these heights are already a very significant part of the total wind (at least in the north–south direction). This is seen in Fig. 13, where we show the southerly wind as measured over a period of 51 h at White Sands, N.M. Analyses of tidal winds at various latitudes are now available. Fig. 14 shows the phase and amplitude of the semidiurnal oscillation at about 30°N. Below 50 km, the results appear quite uncertain (Reed, 1967). In Fig. 15 and 16 we see the diurnal component at 61°N and at 20°N respectively. Amplitudes are of the order of 10 m sec$^{-1}$ at 60 km, but phase at 20°N is more variable than at 61°N (Reed et al., 1969).

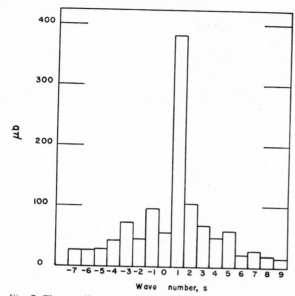

Fig. 7. The amplitudes (on a logarithmic scale, and averaged over the latitudes from the north pole to 60°S) of the diurnal pressure waves, parts of $S_1(p)$, of the type $y_s\sin(t_u + s\varphi + \sigma_s)$. (After Haurwitz, 1965.)

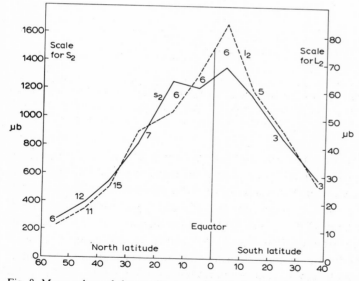

Fig. 8. Mean values of the amplitudes $s_2$ (full line) and $l_2$ (broken line) of the annual mean solar and lunar semidiurnal air-tides in barometric pressure, $S_2(p)$ and $L_2(p)$, for 10° belts of latitude. The numbers beside each point show from how many stations that point was determined. (After Chapman and Westfold, 1956.)

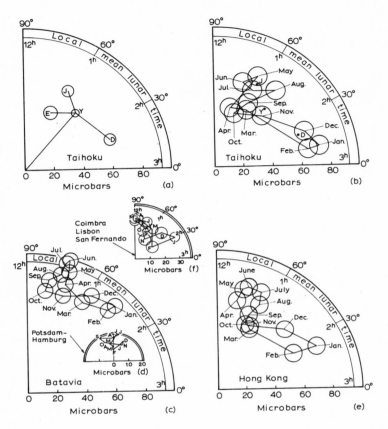

Fig. 9. Harmonic dials, with probable error circles, indicating the changes of the lunar semidiurnal air-tide in the barometric pressure in the course of a year: (a) Annual ($Y$) and four-monthly seasonal ($J$, $E,D$) determinations for Taihoku, Formosa (now Taipei, Taiwan) (1897–1932). Also five sets of twelve monthly-mean dial points. See Table I for particulars of the seven stations. (After Chapman, 1951.)

Between 60 km and 80 km, there are too little data for tidal analyses. Between 80 and 105 km, there is a growing body of data from the observation of ionized meteor trails by doppler radar. Most of these data are for vertically averaged wind over the whole range 80–105 km. Some such data for Jodrell Bank (Greenhow and Neufeld, 1961) (58°N) and Adelaide (Elford, 1959) (35°S) are shown in Fig. 17 and 18. Typical magnitudes are around 20 m sec$^{-1}$. All quantities are subject to large seasonal fluctuations and error circles. At Adelaide, diurnal oscillations predominate, while at Jodrell Bank semidiurnal oscillations predominate; at both stations tidal winds appear to exceed other winds. In comparing these observations with theory, due caution must be exercised in interpreting averages over such great depths.

TABLE I

Diurnal and semidiurnal variations of the eastward and northward components of the wind at Terceira, Azores

| Mean pressure (mbar) | Variation of eastward wind component | | | | | | Variation of northward wind component | | | | | |
|---|---|---|---|---|---|---|---|---|---|---|---|---|
| | diurnal | | | semidiurnal | | | diurnal | | | semidiurnal | | |
| | $s_1$ | $\sigma_1$ | P.E. | $s_2$ | $\sigma_2$ | P.E. | $s_1$ | $\sigma_1$ | P.E. | $s_2$ | $\sigma_2$ | P.E. |
| Ground | 2 | 75 | 6 | 8 | 324 | 7 | 7 | 341 | 8 | 21 | 52 | 6 |
| 1000 | 4 | 115 | 9 | 12 | 298 | 6 | 6 | 337 | 6 | 17 | 52 | 6 |
| 950 | 2 | 154 | 8 | 14 | 317 | 8 | 21 | 271 | 11 | 22 | 35 | 6 |
| 900 | 2 | 248 | 3 | 19 | 292 | 6 | 32 | 272 | 9 | 23 | 31 | 8 |
| 850 | 8 | 257 | 11 | 14 | 266 | 7 | 25 | 256 | 13 | 23 | 14 | 8 |
| 800 | 4 | 322 | 10 | 22 | 313 | 8 | 18 | 265 | 11 | 31 | 359 | 8 |
| 750 | 22 | 145 | 14 | 22 | 278 | 10 | 16 | 251 | 15 | 29 | 12 | 14 |
| 700 | 5 | 304 | 11 | 18 | 304 | 8 | 9 | 4 | 11 | 22 | 33 | 9 |
| 650 | 4 | 255 | 11 | 20 | 292 | 5 | 13 | 318 | 9 | 23 | 50 | 8 |
| 600 | 8 | 63 | 12 | 20 | 272 | 6 | 12 | 281 | 11 | 16 | 1 | 7 |
| 550 | 20 | 159 | 10 | 31 | 327 | 9 | 18 | 249 | 13 | 16 | 7 | 10 |
| 500 | 20 | 124 | 9 | 25 | 276 | 10 | 15 | 317 | 15 | 15 | 63 | 9 |
| 450 | 17 | 76 | 11 | 26 | 295 | 10 | 22 | 295 | 16 | 20 | 346 | 10 |
| 400 | 18 | 1 | 8 | 28 | 291 | 10 | 14 | 342 | 17 | 10 | 317 | 12 |
| 350 | 19 | 258 | 12 | 42 | 291 | 9 | 13 | 257 | 19 | 16 | 319 | 9 |
| 300 | 24 | 193 | 18 | 51 | 292 | 14 | 8 | 247 | 21 | 14 | 4 | 13 |
| 250 | 52 | 177 | 16 | 26 | 245 | 12 | 52 | 267 | 17 | 8 | 285 | 12 |
| 200 | 56 | 153 | 15 | 46 | 267 | 11 | 18 | 238 | 14 | 15 | 338 | 12 |
| 175 | 13 | 164 | 13 | 37 | 278 | 14 | 34 | 241 | 13 | 25 | 339 | 10 |
| 150 | 16 | 186 | 11 | 39 | 300 | 11 | 18 | 185 | 14 | 52 | 18 | 9 |
| 125 | 14 | 127 | 8 | 29 | 242 | 9 | 5 | 241 | 7 | 23 | 4 | 8 |
| 100 | 27 | 112 | 10 | 55 | 280 | 10 | 34 | 153 | 11 | 28 | 19 | 6 |
| 80 | 31 | 111 | 11 | 37 | 280 | 9 | 21 | 194 | 9 | 28 | 21 | 7 |
| 60 | 34 | 109 | 11 | 36 | 262 | 8 | 40 | 196 | 9 | 27 | 5 | 10 |
| 50 | 19 | 132 | 9 | 41 | 256 | 8 | 34 | 236 | 10 | 42 | 356 | 7 |
| 40 | 6 | 96 | 11 | 44 | 263 | 12 | 27 | 235 | 11 | 49 | 4 | 8 |
| 30 | 23 | 181 | 11 | 67 | 280 | 13 | 21 | 221 | 10 | 65 | 17 | 11 |
| 20 | 30 | 147 | 12 | 62 | 295 | 12 | 64 | 235 | 13 | 60 | 36 | 14 |
| 15 | 25 | 114 | 23 | 91 | 303 | 20 | 66 | 238 | 14 | 61 | 30 | 10 |
| 10 | – | – | – | – | – | – | – | – | – | – | – | – |

Annual mean values of the amplitude, $s$ (in cm/sec), and phase, $\sigma$ (in degrees). P.E. is the radius of the probable error circle of the annual means. Time of maximum wind is related to $\sigma$ by the expression $t_{max.} = (450° - \alpha)/15n$, where $n = 1$ for diurnal variations, and $n = 2$ for semidiurnal variations. (After Harris et al., 1962.)

Between 90 and 130 km (and higher) wind data can be obtained by visually tracking luminous vapor trails emitted from rockets. In most cases this is possible only in twilight at sunrise and sundown. Hines (1966) used such data to form 12-h wind differences, which seemed likely to indicate the diurnal contribution to the total wind at dawn at Wallops Island (38°N). His result is shown in Fig. 19. There

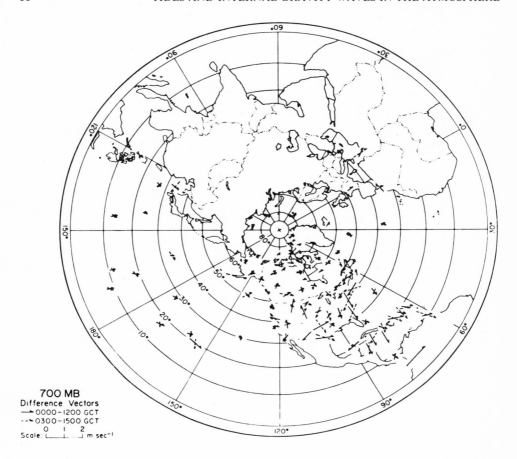

Fig. 10. Annual average wind differences 0000–1200 GMT (solid) and 0300–1500 GMT (dashed) at 700 mbar plotted in vector form. The length scale is given in the figure. (After Wallace and Hartranft, 1969.)

is an evident rotation of the wind vector with height, characteristic of an internal wave with a vertical wavelength of about 20 km. Amplitude appears to grow with height up to 105 km, and then to decay. However, data from above 200 km (mostly from the analysis of satellite drag observations), show an immense daily variation in the thermosphere (Harris and Priester, 1965). This is seen in Fig. 20, where the variation with time of the height of constant density surfaces is shown. Day-night variations in density of almost an order of magnitude are found at 600 km. There are problems with this observational technique and real variations could be much larger.

Having familiarized ourselves with atmospheric tides as they are actually observed, we will now proceed to their mathematical theory.

Fig. 11. Annual average wind differences 0000–1200 GMT at 60 mbar plotted in vector form. The length scale is given in the figure. (After Wallace and Hartranft, 1969.)

## 3. ATMOSPHERIC TIDES: THEORY

### 3.1. Assumptions, approximations and equations

The theory of atmospheric tides consists both in the calculation of the atmospheric response to arbitrary gravitational and thermal forcing, and in the specification of the sources of excitation. Historically, the framework for doing the former was developed first, and we shall follow the same order in these lectures.

Our starting point in calculating the atmosphere's response to tidal (or thermotidal) excitation are the Navier–Stokes equations for a perfect gas in local thermody-

**15 MB**
Difference Vectors
→ 0000-1200 GCT
Scale: └─┴─┴─┴─┘ m sec⁻¹
     0 1 2 3 4

Fig. 12. Annual average wind differences 0000–1200 GMT at 15 mbar plotted in vector form. The length scale is given in the figure. (After Wallace and Hartranft, 1969.)

namic equilibrium. This starting point already involves a substantial number of assumptions and approximations; but they are of the sort that experience has shown are almost completely untroublesome.

There are a number of assumptions which are also untroublesome, but which call for more justification.

(1) The gas constant for air is a constant. This is very nearly so up to 90 km, but by 200 km a 10% error is introduced.

(2) The atmosphere is thin compared with the earth's radius. If we write the distance from the earth's center as $r = a + z$ where $a$ = radius of solid earth, and $z$ = distance from earth's surface, then we shall neglect terms in our equations of $O(z/a)$. It appears that for oscillations of planetary scales and tidal periods, this assumption also permits us to neglect the component of the earth's rotation vector parallel to the earth's surface (Phillips, 1966, 1968).

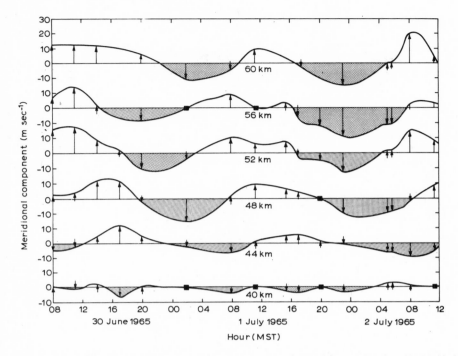

Fig. 13. Meridional wind components $u$ in m/sec, averaged over 4 km centered at 40, 44, 48, 52, 56 and 60 km. Positive values indicate a south to north flow. (After Beyers et al., 1966.)

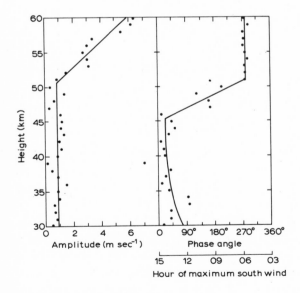

Fig. 14. Phase and amplitude of the semidiurnal variation of the meridional wind component $u$ at 30°N, based on data for White Sands (32.4°N) and Cape Kennedy (28.5°N). (After Reed, 1967.)

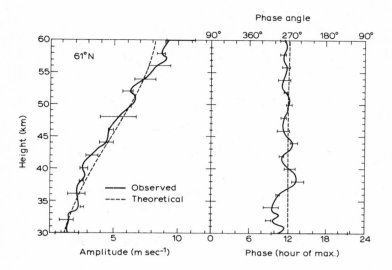

Fig. 15. Amplitude and phase of the diurnal variation of the meridional wind component $u$ at 61°N. Phase angle, in accordance with the usual convention, gives the degrees in advance of the origin (chosen as midnight) at which the sine curve crosses from $-$ to $+$. (After Reed et al., 1969.)

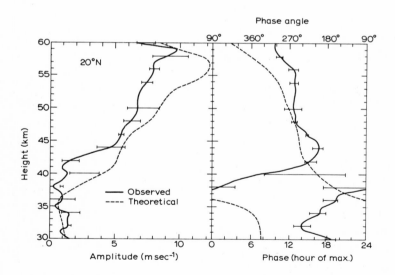

Fig. 16. Amplitude and phase of the diurnal variation of the meridional wind component $u$ at 20°N. (After Reed et al., 1969.)

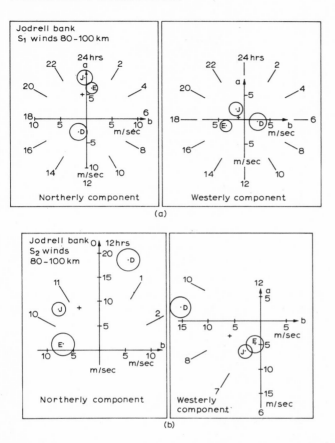

Fig. 17. Harmonic dials for the mean northerly and westerly components of (a) the diurnal and (b) the semidiurnal wind variations at Jodrell Bank at 80–100 km. Crosses indicate annual mean values, dots seasonal values. Circles show probable errors of the seasonal means. (From Haurwitz, 1964.)

(3) Tidal oscillations are in hydrostatic equilibrium. This amounts to assuming that vertical fluid accelerations will be much less than 10 m/sec$^2$ (Yanowitch, 1966).

(4) The ellipticity of the earth is ignored; the earth is assumed to be a sphere. Although the above can be justified at some length, we will omit this in order to move on to those approximations which are most important in limiting the utility of conventional tidal theory. One of these is:

(5) The earth's surface topography (land-sea distribution, mountain ranges, etc.) is ignored. The earth's surface is assumed to be a smooth rigid sphere where $w$ (the vertical velocity) is zero. Related to this assumption is the assumption that the distribution of radiation absorbing gases is independent of longitude.

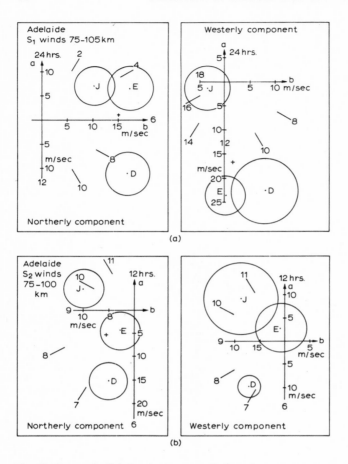

Fig. 18. Harmonic dials for the mean northerly and westerly components of (a) the diurnal and (b) the semidiurnal wind variations at Adelaide, Australia, at 80–100 km. Crosses indicate annual mean values, dots seasonal values. Circles show probable errors of the seasonal means. (From Haurwitz, 1964.)

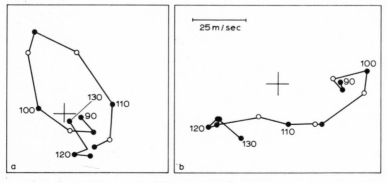

Fig. 19. Vectograms showing (a), the diurnal tide at dawn and (b), the prevailing wind plus the semidiurnal tide at its dawndusk phase, as functions of height. (After Hines, 1966.)

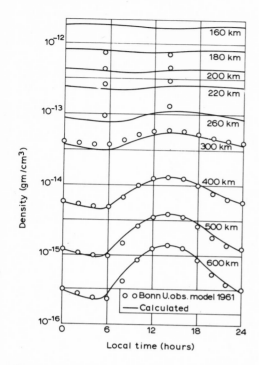

Fig. 20. Diurnal variation of density for selected altitudes from 160 to 600 km. The solid curves give the values calculated with Table I. The circles are densities taken from the Bonn University Observatory observational model of 1961.

At this point our equations are the following:

$$\rho\left(\frac{\partial u}{\partial t} + \vec{u} \cdot \nabla u + \frac{wu}{a} - v^2\frac{\cot\theta}{a} - 2\omega v \cos\theta\right) = -\frac{1}{a}\frac{\partial p}{\partial\theta} - \frac{\rho}{a}\frac{\partial\Omega}{\partial\theta} - F_\theta \qquad (16a)$$

$$\rho\left(\frac{\partial v}{\partial t} + \vec{u} \cdot \nabla v + \frac{wv}{a} + uv\frac{\cot\theta}{a} + 2\omega u \cos\theta\right) = -\frac{1}{a\sin\theta}\frac{\partial p}{\partial\varphi} - \frac{\rho}{a\sin\theta}\frac{\partial\Omega}{\partial\varphi} - F_\varphi$$

$$(17)$$

$$\frac{\partial p}{\partial z} = -\rho\frac{\partial\Omega}{\partial z} \qquad (18)$$

$$\frac{\partial\rho}{\partial t} + \vec{u} \cdot \nabla\rho + \rho\nabla \cdot \vec{u} = 0 \qquad (19)$$

$$\rho c_v\frac{\mathrm{d}T}{\mathrm{d}t} = RT\frac{\mathrm{d}\rho}{\mathrm{d}t} + J - C \qquad (20)$$

$$p = \rho RT \qquad (21)$$

where:

$$\vec{u} \cdot \nabla = \frac{u}{a} \frac{\partial}{\partial \theta} + \frac{v}{a \sin \theta} \frac{\partial}{\partial \varphi} + w \frac{\partial}{\partial z}$$

$$\nabla \cdot \vec{u} = \frac{1}{a \sin \theta} \frac{\partial}{\partial \theta} (u \sin \theta) + \frac{1}{a \sin \theta} \frac{\partial v}{\partial \varphi} + \frac{\partial w}{\partial z}$$

$$\frac{d}{dt} = \frac{\partial}{\partial t} + \vec{u} \cdot \nabla$$

and: $\theta$ = colatitude; $\varphi$ = longitude; $t$ = time; $u$ = northerly velocity; $v$ = westerly velocity; $w$ = vertical velocity; $T$ = temperature; $p$ = pressure; $\rho$ = density; $\Omega$ = gravitational potential; $\omega$ = earth's rotation rate; $R$ = gas constant for air = $2.871 \cdot 10^6$ erg g$^{-1}$ deg$^{-1}$; $c_v$ = heat capacity at constant volume; $F_\theta$ = frictional force in $\theta$ direction; $F_\varphi$ = frictional force in $\varphi$ direction; $J$ = thermotidal heating; $C$ = heat diffusion and radiative cooling.

The form of $F_\theta$, $F_\varphi$ and $C$ for molecular dissipation may be found in Goldstein (1938). Other dissipative processes (eddy diffusion, ion drag, infrared cooling) are described in Lindzen (1970). In general the expressions are extremely complicated, and in the case of eddy diffusion and infrared cooling, quite uncertain. We next adopt the following:

(6) $F_\theta$, $F_\varphi$ and $C$ are neglected. It is a very complicated matter to systematically justify this assumption a priori, but we will check it a posteriori.

Two important simplifications remain.

(7) Tidal oscillations are taken to be linearizable perturbations on a mean flow; i.e., if we write:

$$\begin{bmatrix} u \\ v \\ w \\ T \\ p \\ \rho \end{bmatrix} = \begin{bmatrix} u_o \\ v_o \\ w_o \\ T_o \\ p_o \\ \rho_o \end{bmatrix} + \begin{bmatrix} u' \\ v' \\ w' \\ \delta T \\ \delta p \\ \delta \rho \end{bmatrix} \quad (22)$$

$$\text{mean} \qquad \text{tidal}$$
$$\text{fields} \qquad \text{oscillations}$$

then we shall neglect quadratic and higher order terms in the tidal perturbations. Let us explicitly consider equation 16a:

$$(\rho_o + \delta\rho)\left(\frac{\partial u'}{\partial t} + \vec{u}_o \cdot \nabla u' + \vec{u}' \cdot \nabla u_o + \vec{u}' \cdot \nabla \vec{u}' + \vec{u}_o \cdot \nabla \vec{u}_o + \frac{w'u_o}{a}\right.$$

$$\left. + \frac{w_o u'}{a} + \frac{w_o u_o}{a} + \frac{w'u'}{a} - (v_o^2 + 2v_o v' + v'^2)\frac{\cot\theta}{a} - 2\omega\cos\theta(v_o + v')\right)$$

$$= -\frac{1}{a}\frac{\partial p_o}{\partial\theta} - \frac{1}{a}\frac{\partial\delta p}{\partial\theta} - \frac{(\rho_o + \delta\rho)}{a}\frac{\partial\Omega}{\partial\theta} \qquad (16b)$$

Now $\Omega$ is considered to be of the form:

$$\Omega = \Omega_o(z) + \Omega'(\theta,\varphi,z,t)$$

where $\Omega'$ is a tidal forcing. If primed quantities are set to zero, mean terms must satisfy our equations exactly. Thus, (16b) becomes:

$$(\rho_o + \delta\rho)\left(\frac{\partial u'}{\partial t} + \vec{u}_o \cdot \nabla u' + \vec{u}' \cdot \nabla u_o + \vec{u}' \cdot \nabla u'\right.$$

$$\left. + \frac{w'u_o}{a} + \frac{w_o u'}{a} + \frac{w'u'}{a} - (2v_o v' + v'^2)\frac{\cot\theta}{a} - 2\omega v'\cos\theta\right)$$

$$+ \delta\rho\left(\vec{u}_o \cdot \nabla u_o + \frac{w_o u_o}{a} - v_o^2\frac{\cot\theta}{a} - 2\omega v_o\cos\theta\right) = -\frac{1}{a}\frac{\partial p}{\partial\theta} - \frac{(\rho_o + \delta\rho)}{a}\frac{\partial\Omega'}{\partial\theta}$$

$$(16c)$$

and linearization leads to:

$$\rho_o\left(\frac{\partial u'}{\partial t} + \vec{u}_o \cdot \nabla u' + \vec{u}' \cdot \nabla u_o + \frac{w'u_o}{a} + \frac{w_o u'}{a} - 2v_o v'\frac{\cot\theta}{a} - 2\omega v'\cos\theta\right)$$

$$+ \delta\rho\left(\vec{u}_o \cdot \nabla u_o + \frac{w_o u_o}{a} - v_o^2\frac{\cot\theta}{a} - 2\omega v_o\cos\theta\right) = -\frac{1}{a}\frac{\partial\delta p}{\partial\theta} - \frac{\rho_o}{a}\frac{\partial\Omega'}{\partial\theta} \quad (16d)$$

Similar results are obtained for eq. 17–21. Linearization clearly requires "small" tidal amplitudes, though the definition of small is not so clear, especially since it is not evident which of the neglected nonlinear terms will be most important. As with our other approximations, we will check linearity a posteriori. However, at the least, we expect that linearity requires that forcing be sufficiently small. Observations suggest, moreover, that linearity occurs in the troposphere where $|\delta p/p_o| \sim 10^{-3}$. We shall see, however, that neither of these considerations is sufficient to ensure linearity at all levels—at least when dissipative processes are neglected. Finally, it is clear that for arbitrary choices for the basic state, the solution of six coupled equations like eq. 25 is likely to prove intractable. This leads us to our last major approximation.

(8) $u_o = 0$ and $p_o$, $\rho_o$, and $T_o$ are independent of latitude and longitude. In a rotating fluid, the assumptions for $p_o$, $\rho_o$, and $T_o$ imply $\vec{u}_o = 0$. In general

this approximation may seem initially poor, since it is commonly supposed that linearization requires $|\vec{u}'| \ll |\vec{u}_o|$. This is simply untrue. For example if in eq. 16c $|\vec{u}' \cdot \nabla u'| \ll |\partial u'/\partial t|$, then we drop $\vec{u}' \cdot \nabla \vec{u}'$, even if $|\vec{u}'| > |\vec{u}_o|$. What does prove important is the ratio of $|\vec{u}_o|$ to the phase speed of the tide. For migrating tides, this is the linear rotation speed of the earth's surface, which at the equator is about 400 m/sec. 400 m/sec is certainly much greater than $|\vec{u}_o|$, and hence the neglect of $|\vec{u}_o|$ is at least plausible. The matter will be discussed in greater detail later.

Having made all the above assumption and/or approximations, we are left with a basic state where:

$$\frac{\partial p_o}{\partial z} = -\rho_o g \tag{23}$$

and:

$$p_o = \rho_o R T_o \tag{24}$$

$T_o(z)$ will be specified and eq. 23 and 24 yield:

$$p_o = p_o(o) e^{-x} \tag{25}$$

$$\rho_o = p_o / R T_o \tag{26}$$

where:

$$x = \int_0^z \frac{\mathrm{d}z}{H} \tag{27}$$

and:

$$H = \frac{R T_o}{g} \tag{28}$$

$H$ is known as the *local scale height* while $x$ is the height in the scale heights. For our perturbation equations, we have:

$$\frac{\partial u'}{\partial t} - 2\omega v' \cos\theta = -\frac{1}{a}\frac{\partial}{\partial\theta}\left(\frac{\delta p}{\rho_o} + \Omega'\right) \tag{29}$$

$$\frac{\partial v'}{\partial t} + 2\omega u' \cos\theta = -\frac{1}{a\sin\theta}\frac{\partial}{\partial\varphi}\left(\frac{\delta p}{\rho_o} + \Omega'\right) \tag{30}$$

$$\frac{\partial \delta p}{\partial z} = -g\delta\rho - \rho_o\frac{\partial\Omega'}{\partial z} \tag{31}$$

$$\frac{\mathrm{d}\rho}{\mathrm{d}t} + \rho_o X = 0 \tag{32}$$

$$c_v \frac{dT}{dt} = \frac{RT_0}{\rho_0} \frac{d\rho}{dt} + J' \tag{33}$$

$$\frac{\delta p}{p_0} = \frac{\delta\rho}{\rho_0} + \frac{\delta T}{T_0} \tag{34}$$

where: $\chi \equiv \nabla \cdot \vec{u}'$ and: $df/dt = \partial \delta f/\partial t + w'(df_0/dt)$. Our aim is to reduce eq. 29–34 to a single equation in a single unknown. Several steps of this reduction are worth noting.

First we eliminate $\delta T$ using (34). (33) becomes:

$$\frac{dp}{dt} = \gamma g H \frac{d\rho}{dt} + \rho_0(\gamma - 1)J' \tag{35}$$

where: $\gamma = c_p/c_v$ and $c_p = R + c_v$. Next we assume solutions of the form:

$$e^{i(\sigma t + s\varphi)} \tag{36}$$

where: $\sigma = 2\pi/\tau$; $\tau =$ solar or lunar day$/m$; $m =$ an integer; $s = 0, \pm 1, \pm 2, \ldots$ etc. With solutions of this form: $\partial/\partial t \to i\sigma$ and $\partial/\partial\varphi \to is$, eq. 29 and 30 become:

$$i\sigma u' - 2\omega v' \cos\theta = -\frac{1}{a} \frac{\partial}{\partial\theta}\left(\frac{\delta p}{\rho_0} + \Omega'\right) \tag{37}$$

$$i\sigma v' + 2\omega u' \cos\theta = -\frac{is}{a \sin\theta}\left(\frac{\delta p}{\rho_0} + \Omega'\right) \tag{38}$$

Equations 37 and 38 are merely algebraic equations in $u'$ and $v'$, which may be solved in terms of $\delta p$ and its $\theta$-derivative, yielding:

$$u' = \frac{i\sigma}{4a^2\omega^2(f^2 - \cos^2\theta)}\left(\frac{\partial}{\partial\theta} + \frac{s \cot\theta}{f}\right)\left(\frac{\delta p}{\rho_0} + \Omega'\right) \tag{39}$$

$$v' = \frac{-\sigma}{4a^2\omega^2(f^2 - \cos^2\theta)}\left(\frac{\cos\theta}{f} \frac{\partial}{\partial\theta} + \frac{s}{\sin\theta}\right)\left(\frac{\delta p}{\rho_0} + \Omega'\right) \tag{40}$$

where: $f \equiv \sigma/2\omega$. Now $u'$ and $v'$ enter the remaining equations only through the velocity divergence, which can now be written, using (39) and (40):

$$\chi = \frac{i\sigma}{4a^2\omega^2}F\left(\frac{\delta p}{\rho_0} + \Omega'\right) + \frac{\partial w'}{\partial z} \tag{41}$$

where:

$$F = \frac{1}{\sin\theta} \frac{\partial}{\partial\theta}\left(\frac{\sin\theta}{f^2 - \cos^2\theta} \frac{\partial}{\partial\theta}\right) - \frac{1}{f^2 - \cos^2\theta}\left(\frac{s}{f} \frac{f^2 + \cos^2\theta}{f^2 - \cos^2\theta} + \frac{s^2}{\sin^2\theta}\right) \tag{42}$$

Now eq. 41, 35, 32 and 31 form four equations in four unknowns, $\delta p$, $\chi$, $\delta\rho$ and $w$. As it turns out, the most convenient variable to solve for is:

$$G = -\frac{1}{\gamma p_0} \frac{dp}{dt} \tag{43}$$

After some manipulation, one obtains:

$$H\frac{\partial^2 G}{\partial z^2} + \left(\frac{dH}{dz} - 1\right)\frac{\partial G}{\partial z} = \frac{g}{4a^2\omega^2}F\left[\left(\frac{dH}{dz} + \kappa\right)G - \frac{\kappa J}{\gamma g H}\right] \qquad (44)$$

where: $\kappa = (\gamma - 1)/\gamma$. In terms of its mathematical symptoms, (44) is a nearly fatal case. It is a mixed elliptic–hyperbolic equation with singularities at $\cos\theta = \pm f, \pm 1$. However, in practice its solution is straightforward, if not simple. Let us recall that $F$ is an operator involving only $\theta$, while the remainder of (44) depends only on $z$. We may therefore attempt to solve (44) by the method of separation of variables.

Let us assume $G$ may be written as:

$$G = \sum_{\text{all } n} L_n(z)\Theta_n(\theta) \qquad (45)$$

and let us assume further that $\{\Theta_n\}$ all $n$ is complete. Then we may write:

$$J = \sum_{\text{all } n} J_n(z)\Theta_n(\theta) \qquad (46)$$

Substituting (45) and (46) into (44), we get:

$$F[\Theta_n] = -\frac{4a^2\omega^2}{ghn}\Theta_n \qquad (47)$$

and:

$$H\frac{d^2 L_n}{dz^2} + \left(\frac{dH}{dz} - 1\right)\frac{dL_n}{dz} + \frac{1}{h_n}\left(\frac{dH}{dz} + \kappa\right)L_n = \frac{\kappa}{\gamma g H h_n}J_n \qquad (48)$$

The separation constant for the $n^{\text{th}}$ mode is $4a^2\omega^2/gh_n$, where $h_n$ is called the *equivalent depth of the $n^{\text{th}}$ mode*. It is written in this fashion in order to make evident that (47) is the same as Laplace's Tidal Equation for a spherical fluid shell of depth $h_n$. We shall give a more meaningful interpretation of $h_n$ shortly. If we require that $\theta_n$ be bounded at the poles, then (47) defines an eigenfunction-eigenvalue problem, where the eigenfunctions are called *Hough Functions*, and the eigenvalues are expressed in terms of $h_n$. The full solution associated with a given Hough Function is called a *Hough Mode*. The vertical structure of a given Hough Mode is given by the solution of eq. 48. Eq. 48 assumes a more easily interpreted form, if we replace $z$ by $x$ as the independent variable and replace $L_n$ by:

$$y_n = e^{-x/2}L_n$$

Then eq. 48 becomes:

$$\frac{d^2 y_n}{dx^2} + \left[\frac{1}{h_n}\left(\kappa H + \frac{dH}{dx}\right) - \frac{1}{4}\right]y_n = \frac{\kappa J_n}{\gamma g h_n}e^{-x/2} \qquad (49)$$

Once the $\theta_n$'s and the associated $y_n$'s are obtained, we may easily solve for the other fields.

$$
\begin{bmatrix} \delta p \\ \delta \rho \\ \delta T \\ w' \end{bmatrix} = \sum_n \begin{bmatrix} \delta p_n(x) \\ \delta \rho_n(x) \\ \delta T_n(x) \\ w_n(x) \end{bmatrix} \Theta_n(\theta)
\tag{50}
$$

$$
u' = \sum_n u_n(x) U_n(\theta)
\tag{51}
$$

$$
v' = \sum_n v_n(x) V_n(\theta)
\tag{52}
$$

where:

$$
\delta p_n = \frac{p_0(0)}{H(x)} \left[ -\frac{\Omega_n}{g} e^{-x} + \frac{\gamma h_n e^{-x/2}}{i\sigma} \left( \frac{dy_n}{dx} - \frac{y_n}{2} \right) \right]
\tag{53}
$$

$$
\delta \rho_n = \frac{p_0(0)}{(gH)^2} \left\{ -\Omega_n e^{-x} \left( 1 + \frac{1}{H} \frac{dH}{dx} \right) \right.
$$
$$
\left. + \frac{\gamma g h_n}{i\sigma} e^{-x/2} \left[ \left( 1 + \frac{1}{H} \frac{dH}{dx} \right) \left( \frac{dy_n}{dx} - \frac{1}{2} \right) + \frac{H}{h_n} \left( \kappa + \frac{1}{H} \frac{dH}{dx} \right) y_n \right] - \frac{\kappa J_n}{i\sigma} \right\}
\tag{54}
$$

$$
\delta T_n = \frac{1}{R} \left\{ \frac{\Omega_n}{H} \frac{dH}{dx} - \frac{\gamma g h_n}{i\sigma} e^{x/2} \left[ \frac{\kappa H}{h_n} + \frac{1}{H} \frac{dH}{dx} \left( \frac{d}{dx} + \frac{H}{h_n} - \frac{1}{2} \right) \right] y_n + \frac{\kappa J_n}{i\sigma} \right\}
\tag{55}
$$

$$
w_n = -\frac{i\sigma}{g} \Omega_n + \gamma h_n e^{x/2} \left[ \frac{dy_n}{dx} + \left( \frac{H}{h_n} - \frac{1}{2} \right) y_n \right]
\tag{56}
$$

$$
u_n = \frac{\gamma g h_n e^{x/2}}{4a\omega^2} \left( \frac{dy_n}{dx} - \frac{1}{2} y_n \right)
\tag{57}
$$

$$
v_n = \frac{i \gamma g h_n e^{x/2}}{4a\omega^2} \left( \frac{dy_n}{dx} - \frac{1}{2} y_n \right)
\tag{58}
$$

$$
U_n = \frac{1}{f^2 - \cos^2\theta} \left( \frac{d}{d\theta} + \frac{s \cot\theta}{f} \right) \Theta_n
\tag{59}
$$

and:

$$
V_n = \frac{1}{f^2 - \cos^2\theta} \left( \frac{\cos\theta}{f} \frac{d}{d\theta} + \frac{s}{\sin\theta} \right) \Theta_n
\tag{60}
$$

An expansion for $\Omega'$ of the form:

$$
\Omega' = \sum_n \Omega_n(x) \Theta_n
\tag{61}
$$

was assumed.

The solution of (49) requires two boundary conditions. One is obtained from (56), and the requirement that: $w = 0$ at $z = x = 0$, namely:

$$\frac{dy_n}{dx} + \left(\frac{H}{h_n} - \frac{1}{2}\right) y_n = \frac{i\sigma}{\gamma g h_n} \Omega_n \quad \text{at } x = 0 \tag{62}$$

As a second boundary condition it often suffices to require boundedness as $x \to \infty$. However, if the top of our atmosphere is isothermal, then (49) has solutions of the form:

$$y_n \sim A e^{i\lambda x} + B e^{-i\lambda x}$$

where:

$$\lambda = \left(\frac{\kappa H}{h_n} - \frac{1}{4}\right)^{1/2}$$

If $\lambda$ is real, then $y_n$ is bounded for any choice of $A$ and $B$. In this instance a radiation condition is imposed: i.e., it is required that there be no incoming energy from infinity. It can be shown that upward energy flux is associated with downward phase speed. The radiation condition therefore implies $B = 0$.

In passing it should be noted that if $J_n = \Omega_n = 0$, then the only solution to (49) is generally $y_n = 0$. There can exist values of $h$ for which (49) can have nontrivial solutions (usually only one value). These are known as the *equivalent depths of the atmosphere*. For realistic atmospheres and hydrostatic waves, the atmosphere has only one equivalent depth, 10.4 km (see history). If an $h_n$ is equal to the equivalent depth of the atmosphere, then we have a resonance.

Before going on to the methods of solving eq. 49 and 47, I would like to show the meaning of a mode's equivalent depth for some very simple cases. We must first note that all information about our geometric configuration is contained in eq. 47. Eq. 49, apart from its dependence on $h_n$, depends only on the atmosphere's thermal and thermodynamic structure. Although the $h_n$'s might be different, eq. 49 would be the same for waves on a non-rotating sphere, a rotating plane or even a non-rotating plane. It proves useful to study the counterparts of (47) for non-rotating and rotating planar atmospheres. For a non-rotating planar atmosphere, eq. 29 and 30 are replaced by:

$$\frac{\partial u'}{\partial t} = -\frac{\partial}{\partial x}\frac{\delta p}{\rho_0} \tag{63}$$

$$\frac{\partial v'}{\partial t} = -\frac{\partial}{\partial y}\frac{\delta p}{\rho_0} \tag{64}$$

We shall consider solutions of the form:

$$e^{i(\sigma t + kx + my)} \tag{65}$$

(63) and (64) become:

$$u' = -\frac{k}{\sigma}\frac{\delta p}{\rho_0} \tag{66}$$

$$v' = -\frac{m}{\sigma}\frac{\delta p}{\rho_0} \tag{67}$$

and:

$$\chi = \frac{\partial u'}{\partial x} + \frac{\partial v'}{\partial y} + \frac{\partial w'}{\partial z} = \frac{i\sigma}{4a^2\omega^2}F\left(\frac{\delta p}{\rho_0}\right) + \frac{\partial w'}{\partial z}$$

$$= -\frac{i\sigma}{gh}\frac{\delta p}{\rho_0} + \frac{\partial w'}{\partial z} = -\frac{i}{\sigma}(k^2 + m^2)\frac{\delta p}{\rho_0} + \frac{\partial w'}{\partial z} \tag{68}$$

Thus, by analogy with our earlier procedure we find:

$$gh = \frac{\sigma^2}{k^2 + m^2} \tag{69}$$

$h$ is thus a measure of the square of the wave's horizontal wavelength. Another interpretation (closely related to the name equivalent depth) is that $h$ is the depth of a homogeneous fluid layer in which the phase speed of gravity waves is $\sigma/\sqrt{k^2 + m^2}$. The identification of $h$ with the thickness of a fluid can, however, be misleading. On a rotating plane, (29) and (30) are replaced by:

$$\frac{\partial u'}{\partial t} - 2\omega v = -\frac{\partial}{\partial x}\left(\frac{\delta p}{\rho_0}\right) \tag{70}$$

$$\frac{\partial v'}{\partial t} + 2\omega u = -\frac{\partial}{\partial y}\left(\frac{\delta p}{\rho_0}\right) \tag{71}$$

Assuming solutions of the form of (65), again we find:

$$(4\omega^2 - \sigma^2)u' = \sigma k\frac{\delta p}{\rho_0} - im2\omega\frac{\delta p}{\rho_0} \tag{72}$$

$$(4\omega^2 - \sigma^2)v' = \sigma m\frac{\delta p}{\rho_0} + ik2\omega\frac{\delta p}{\rho_0} \tag{73}$$

and:

$$\chi - \frac{\partial w'}{\partial z} = (4\omega^2 - \sigma^2)^{-1}[i\sigma(k^2 + m^2)]\frac{\delta p}{\rho_0} = -\frac{i\sigma}{gh}\frac{\delta p}{\rho_0} \tag{74}$$

or:

$$gh = \frac{\sigma^2 - 4\omega^2}{k^2 + m^2} \tag{75}$$

Once again $h$ is a measure of the square of the horizontal wavelength. However, the interpretation of $h$ as a depth runs into trouble, since $h$ can now be negative. (It is interesting to note that the name equivalent depth effectively prevented the discovery of negative equivalent depths until about 6 years ago.) This occurs whenever $\sigma^2 < 4\omega^2$. The depth interpretation breaks down because in a rotating fluid there is a minimum phase speed regardless of fluid depth. For atmospheric waves, negative equivalent depths are perfectly meaningful. Consider eq. 49 (for simplicity, consider an isothermal atmosphere away from excitation sources). When $h_n$ is negative, (49) will have exponential solutions of the form:

$$y_n \sim e^{-\mu x} \tag{76}$$

where:

$$\mu = \left( \frac{1}{4} - \frac{1}{h_n}(\kappa H) \right)^{1/2} < \frac{1}{2} \tag{77}$$

Thus the rate of decay of $y_n$ will exceed the $e^{x/2}$ growth indicated in eq. 55–58, leading to actual decay of amplitude. For $0 < h_n < 4\kappa H$, solutions will be oscillatory in the vertical with amplitudes growing as $e^{x/2}$. For $h_n > 4\kappa H$, there is no phase change with height, but amplitudes will grow exponentially with height, though more slowly than $e^{x/2}$.

Incidentally, it is easily shown that the equivalent depth of an isothermal atmosphere is $h = \gamma H$, where $\gamma = 1.4$ and $H \sim 8$ km.

As in the above examples, $h_n$, on a rotating sphere, is a measure of the square of some characteristic horizontal scale. Also, there will be some negative equivalent depths for any frequency, for which $\sigma^2 < 4\omega^2 \cos^2\theta$ for some $\theta$.

## 3.2. Mathematical methods of solution

### 3.2.1. Laplace's tidal equation
Laplace's tidal equation is given by eq. 42 and 47. It may be rewritten:

$$\frac{d}{d\mu}\left( \frac{1-\mu^2}{f^2-\mu^2} \frac{d\Theta_n}{d\mu} \right) - \frac{1}{(f^2-\mu^2)}\left[ \frac{s}{f} \frac{f^2+\mu^2}{f^2-\mu^2} + \frac{s^2}{1-\mu^2} \right]\Theta_n + \frac{4a^2\omega^2}{gh_n}\Theta_n = 0 \tag{78}$$

where $\mu = \cos\theta$.

There are many theorems which can and have been proven about the solutions to (78). These can be found in Hough (1898), Longuet-Higgins (1967), and Flattery (1967). We shall state without proof the most important of these:

(1) All solutions are bounded at $f = \pm\mu$; the singularities at these points are in Ince's terminology "apparent."
(2) The boundary condition that $\theta_n$ be bounded at $\mu = \pm1$ leads to an

eigenfunction–eigenvalue problem (analogous to the equation for Associated Legendre Polynomials).

(3) For a given choice of $s$, the eigenfunctions $\theta_n$ are orthogonal.

(4) The eigenvalues $h_n$ are real.

(5) If $f \neq \pm 1$, then $\theta_n$ behaves as $(1 - \mu^2)^{s/2}$ at the poles; if $f = \pm 1$, then $\theta_n$ behaves as $(1 - \mu^2)^{s/2+1}$ at the poles.

Apart from a few special cases, eq. 78 has no simple closed form solutions*, and must be solved approximately. There are many approaches to this problem, including the use of WKB methods (Golitsyn and Dikii, 1966).

The most common approach is to assume $\theta_n$ may be expanded in a series of Associated Legendre Polynomials:

$$\Theta_n^{\sigma,s} = \sum_{m=s}^{\infty} C_{n,m}^{\sigma,s} P_m^s(\mu) \tag{79}$$

where $P_m^s(\mu)$ is defined as in Whittaker and Watson (1927). The substitution of (79) into (78) leads to a set of relations of the following form:

$$A_m^{\sigma,s} C_{n,m-2} + B_m^{\sigma,s}(h)C_{n,m} + D_m^{\sigma,s} C_{n,m+2} = 0, \quad m = s, s+1, \ldots \tag{80}$$

In order for (80) to have a solution, the infinite determinant of its coefficients must equal zero. This condition leads to an infinite number of eigensolutions $\{h_n\}$, and for each of these one obtains a set of coefficients $\{C_{n,m}\}_{m=s,\infty}$. Since (80) is homogeneous, the coefficients are determined only to within an arbitrary factor. One way to assign this factor is to assume that $\{\Theta_n\}$ are orthonormal, in which case:

$$\int_{-1}^{1} \Theta_n \Theta_m \, d\mu = \delta_{nm} \tag{81}$$

where: $\delta_{nm} = 1, n = m$ and $\delta_{nm} = 0, n \neq m$.

The reader is referred to Lindzen (1971a) for specific details.

### 3.2.2. Vertical structure equation

The vertical structure depends on the distribution of the mean temperature, and both the thermal and gravitational excitation. Apart from a resonant value of $h_n$, the vertical solution will be zero without excitation. For realistic mean distributions of temperature and excitation, there are no simple closed form solutions.** However, extremely effective numerical approaches are available. In particular, a finite difference scheme described by Bruce et al. (Carrier and Pearson, 1968) has proven useful.

---

* There are three exceptions to this statement: (1) when $s = 0$, $f = 1$, there are trigonometric solutions; (2) when rotation is neglected, solutions are spherical harmonics; (3) when $h = \infty$, solutions are the sum of two spherical harmonics. Details and further references may be found in Flattery (1967).

** There are, of course, simplified atmospheres for which analytic solutions have been found. Some of these are cited in section 1. However, even for isothermal atmospheres, one must use Green Function techniques to handle distributed heat sources.

Let us repeat (49):

$$\frac{d^2 y_n}{dx^2} + \left[ \frac{1}{h_n} \left( \kappa H + \frac{dH}{dx} \right) - \frac{1}{4} \right] y_n = \frac{\kappa J_n}{\gamma g h_n} e^{-x/2} \tag{82}$$

and (62):

$$\frac{dy_n}{dx} + \left( \frac{H}{h_n} - \frac{1}{2} \right) y_n = \frac{i\sigma}{\gamma g h_n} \Omega_n \quad \text{at } x = 0 \tag{83}$$

Let us now divide our $x$-domain into a number of discrete levels, $x_0, x_1, x_2, \ldots$ where $x_o = 0$, and the remaining levels are equally spaced into a separation $\delta x$. At $x = x_m$

$$\frac{d^2 y_n(x_m)}{dx^2} \approx \frac{y_n(x_{m+1}) - 2y_n(x_m) + y_n(x_{m-1})}{(\delta x)^2} \tag{84}$$

Substituting (84) into (82) we get:

$$A_m y_n(x_{m+1}) + B_m y_n(x_m) + C_m y_n(x_{m-1}) = D_m \tag{85}$$

where:

$$A_m = 1$$

$$B_m = -\left\{ 2 + \frac{(\delta x)^2}{4} \left[ 1 - \frac{4}{h_n} \left( \kappa H(x_m) + \frac{dH}{dx} \bigg|_{x=x_m} \right) \right] \right\}$$

$$C_m = 1$$

and:

$$D_m = (\delta x)^2 \frac{\kappa J_n(x_m)}{\gamma g h_n}$$

Eq. 85 is solved as follows. Let:

$$y_n(x_m) = \alpha_m y_n(x_{m+1}) + \beta_m \tag{86}$$

where $\alpha_m$ and $\beta_m$ are new variables. Similarly:

$$y_n(x_{m-1}) = \alpha_{m-1} y_n(x_m) + \beta_{m-1}. \tag{87}$$

Substituting (87) into (85) we get:

$$y_n(x_m) = -\frac{A_m y_n(x_{m+1})}{B_m + \alpha_{m-1} C_m} + \frac{D_m - \beta_{m-1} C_m}{B_m + \alpha_{m-1} C_m} \tag{88}$$

and comparing (88) with (86) we get:

$$\alpha_m = -\frac{A_m}{B_m + \alpha_{m-1} C_m} \tag{89}$$

and:

$$\beta_m = \frac{D_m - \beta_{m-1} C_m}{B_m + \alpha_{m-1} C_m} \tag{90}$$

Now, once we know $\alpha_0$ and $\beta_0$, we can solve for all $\alpha_m$'s and $\beta_m$'s. $\alpha_0$ and $\beta_0$ can be obtained from (83), the lower boundary condition. For convenience, let us approximate (83) by a one-sided difference:

$$\frac{y_n(x_1) - y_n(x_0)}{\delta x} + \left(\frac{H(x_0)}{h_n} - \frac{1}{2}\right) y_n(x_0) = \frac{i\sigma\Omega_n(x_0)}{\gamma g h_n} \tag{91}$$

(91) may be rewritten:

$$y_n(x_0) = \frac{1}{1 - \left(\frac{H(x_0)}{h_n} - \frac{1}{2}\right)} y_n(x_1) - \frac{i\sigma\Omega_n(x_0)\delta x}{\gamma g h_n\left[1 - \left(\frac{H(x_0)}{h_n} - \frac{1}{2}\right)\delta x\right]} \tag{92}$$

and comparing (92) with (86):

$$\alpha_0 = \frac{1}{1 - \left(\frac{H(x_0)}{h_n} - \frac{1}{2}\right)\delta x} \tag{93}$$

$$\beta_0 = \frac{-i\sigma\Omega_n(x_0)\delta x}{\gamma g h_n\left[1 - \left(\frac{H(x_0)}{h_n} - \frac{1}{2}\right)\delta x\right]} \tag{94}$$

Finally, in order to solve for $y_n$, we must know $y_n$ at some high level, at which point (86) may be used to determine $y_n$ at all lower levels. Sufficiently high in the atmosphere $H$ and $T_0$ are constants. As an example of the use of the upper boundary condition, we will consider $h_n < 4\kappa H_\infty$, where $H_\infty$ is the constant $H$ in the high thermosphere. Then, from the radiation condition:

$$y_n = Ae^{i\lambda x} \tag{95}$$

in the high thermosphere where:

$$\lambda = \left(\frac{\kappa H_\infty}{h_n} - \frac{1}{4}\right)^{1/2} \tag{96}$$

Differentiating (95) we get:

$$\frac{dy_n}{dx} = i\lambda y_n \tag{97}$$

and approximating (97) by finite differences we get:

$$\frac{y_n(x_m) - y_n(x_{m-2})}{2\delta x} = i\lambda y_n(x_{m-1}) \tag{98}$$

where $x_m$ is taken to be our top level. Now from (86) we have:

$$y_n(x_{m-2}) = \alpha_{m-2} y_n(x_{m-1}) + \beta_{m-2} \tag{99}$$

and:

$$y_n(x_{m-1}) = \alpha_{m-1} y_n(x_m) + \beta_{m-1} \tag{100}$$

From (98)–(100) we get:

$$y_n(x_m) = \frac{\beta_{m-2} + \beta_{m-1}(2i\lambda\delta x + \alpha_{m-2})}{(1 - \alpha_{m-1}[2i\lambda\delta x + \alpha_{m-2}])} \tag{101}$$

which is our desired result.

In practice, we have found that the above integration scheme works for all reasonable cases, provided $\delta x$ is sufficiently small. We have obtained better than 1% accuracy by choosing:

$$\delta x = \left(\text{minimum value of } \frac{2\pi}{\left[\frac{1}{h_n}\left(\kappa H + \frac{dH}{dx}\right) - \frac{1}{4}\right]^{1/2}}\right) \cdot 10^{-2} \tag{102}$$

However, to the best of my knowledge, the only rigorous proof of the validity of the above procedure is for the situation where:

$$\left[\frac{1}{h_n}\left(\kappa H + \frac{dH}{dx}\right) - \frac{1}{4}\right] < 0 \tag{103}$$

Clearly, this is a sufficient but not a necessary proof. The above procedure can also be extended to a variety of higher order ordinary differential equations and partial differential equations (Lindzen and Kuo, 1969); but, as with the above procedure, necessary conditions have not yet been proven.

## 3.3. Explicit calculation of atmospheric tides

### 3.3.1. Sources of excitation

As mentioned in section 1, a large part of the development of atmospheric tidal theory has been concerned with the specification of the gravitational and thermal sources of excitation. However, time does not permit a detailed derivation of these sources here. A discussion of both gravitational and thermal excitations may be found in Chapman and Lindzen (1970), where additional references are included. Here, we will briefly describe, in approximate form, the most important thermal sources of excitation.

It has already been mentioned that thermal excitation is of dominant importance for the atmosphere, and that thermal excitation is mainly due to absorption of solar radiation by water vapor and ozone. Thermal excitation is commonly expressed by a function $\tau$, rather than $J$, where:

$$\tau = \frac{\kappa J}{i\sigma R} \tag{104}$$

$\tau$ is, approximately, the amplitude of the temperature oscillation that would be produced by $J$ in the absence of motion and dissipation. For a given absorber, $G$, the excitation at frequency $\sigma$, and zonal wavenumber $s$ can usually be written with sufficient accuracy as:

$$\tau_G^{\sigma,s} = f_G^{\sigma,s}(z)g_G^{\sigma,s}(\theta) \tag{105}$$

For the diurnal excitation $\sigma = 2\pi/1$-solar day and $s = 1$; for the semidiurnal excitation $\sigma = 4\pi/1$-solar day and $s = 2$. The $f_G^{\sigma,s}$s for both the diurnal and semidiurnal excitations can be taken to be the same:

$$f_{H_2O}(z) = e^{-z/22.8 \text{ km}} \tag{106}$$

and:

$$f_{O_3}(z) = e^{0.00116(z-z_1)} \sin\frac{\pi}{60}(z - z_1) \text{ for } z_1 < z < z_2$$

$$= 0 \text{ elsewhere;}$$

$$z_1 = 18.0 \text{ km}$$

$$z_2 = 78.0 \text{ km}$$

$$\tag{107}$$

The distributions of $g_{H_2O}$ and $g_{O_3}$ are also taken to be the same for both the diurnal and semidiurnal excitations—with the exception of their overall amplitudes. Fig. 21 shows both the vertical and latitudinal distributions of excitation.

For the diurnal excitation, the phase is such that $\tau$ has a maximum at 1800 LT; for the semidiurnal excitation $\tau$ has maxima at 0300 and 1500 LT.[*]

For tides in the thermosphere, thermal excitation due to Schumann–Runge band absorption by $O_2$ is of finite importance, and EUV absorption between 120 and 160 km can be of major importance. We shall return to this matter in the fourth section.

---

[*] There is some reason to doubt the phase of the semidiurnal excitation; viz., Lindzen and Blake (1971).

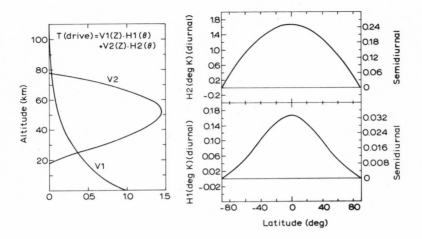

Fig. 21. Vertical distributions of thermal excitation due to water vapor (V1) and ozone (V2); latitude distributions for water vapor (H1) and ozone (H2). (After Lindzen, 1968a.)

For tides below 100 km, $H_2O$ and $O_3$ absorption are indeed of primary importance.

### 3.3.2. Solar semidiurnal thermal tide

The first three symmetric Hough Functions for this tide are shown in Fig. 22. Also shown are the corresponding expansion functions for the northerly and westerly velocity, in Fig. 23 and 24, respectively. Associated with these three modes are the following equivalent depths: $h_2^{2\omega,2} = 7.8519$ km, $h_4^{2\omega,2} = 2.1098$ km and $h_6^{2\omega,2} = 0.9565$ km (Flattery, 1967). Expanding the latitude distributions of the excitation we get:

$$g_{O_3}^{2\omega,2} = 0.249°K\Theta_2^{2\omega,2} + 0.0645°K\Theta_4^{2\omega,2} + 0.0365°K\Theta_6^{2\omega,2} + \cdots \quad (108)$$

$$g_{H_2O}^{2\omega,2} = 0.0307°K\Theta_2^{2\omega,2} + 0.00796°K\Theta_4^{2\omega,2} + 0.00447°K\Theta_6^{2\omega,2} + \cdots \quad (109)$$

We see from Fig. 22 and 21 that the distributions of excitation are similar to $\Theta_2^{2\omega,2}$. This is reflected in eq. 108 and 109 by the fact that $\Theta_2^{2\omega,2}$ receives the bulk of the excitation. Moreover, the equivalent depth of this mode, 7.852 km, is such that $\lambda^2 = [1/h_2^{2\omega,2}(\kappa H + dH/dx) - 1/4]$ is everywhere close to zero; i.e., the vertical wavelength of this mode is very large (ca. 200 km). Thus, not only does most semidiurnal excitation go into this mode, but $\Theta_2^{2\omega,2}$ responds with particular efficiency, since all excitation below 100 km acts "in phase". These facts explain the two most striking observational features of the solar semidiurnal surface pressure oscillation; namely, its strength and regularity. The latter results from the fact that the excitation going into $\Theta_2^{2\omega,2}$ is determined by the overall global distribution of excitation, which doesn't change much; the larger local variations excite the higher order, less efficient Hough Modes.

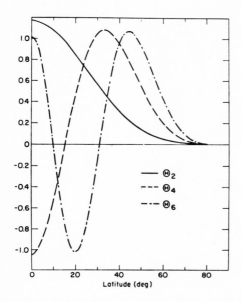

Fig. 22. Latitude distribution for the first three symmetric solar semidiurnal migrating Hough Functions.

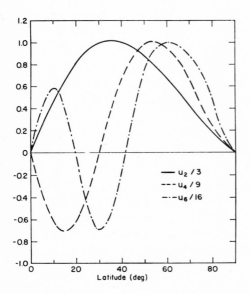

Fig. 23. The expansion functions for the latitude dependence of the solar semidiurnal component of $u$, the northerly velocity. The functions have been divided by the amounts shown.

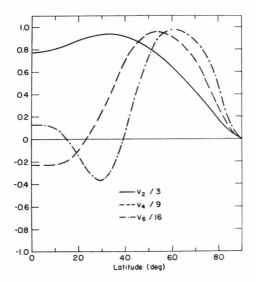

Fig. 24. The expansion functions for the latitude dependence of the solar semidiurnal component of $v$, the westerly velocity. The functions have been divided by the amounts shown.

The surface pressure oscillation resulting from the above excitation is relatively insensitive to the precise choice of basic temperature distribution. For the ARDC, equatorial, and isothermal profiles shown in Fig. 25, the amplitudes of the semidiurnal surface pressure oscillation at the equator are $1.18 \times 10^{-3}$ mbar, $1.27 \times 10^{-3}$ mbar and $1.05 \times 10^{-3}$ mbar respectively; the phases correspond to maxima at 0862 LT, 0844 LT, and 0817 LT. The particular choice of temperature profile does, however, make a difference at higher altitudes. In Fig. 26 we see the amplitude of the westerly velocity over the equator for the various choices to $T_0$. For realistic $T_0$'s, the $\Theta_2^{2\omega,2}$ mode is evanescent between about 50 and 80 km. The variation of semidiurnal tidal fields with latitude is indicated in Fig. 27 and 28, where the amplitude and phase distributions of the northerly velocity at different latitudes are shown. Note the increasing importance of higher order modes at high altitudes and latitudes. These modes do not become evanescent in the mesosphere.

### 3.4. Solar diurnal thermal tide

The first five symmetric Hough Functions (as well as the main antisymmetric Hough Function) for the migrating solar diurnal tide are shown in Fig. 29. The corresponding expansion functions for the northerly and westerly velocity are shown in Fig. 30 and 31. The equivalent depths are $h_1^{\omega,1} = 0.6909$ km, $h_3^{\omega,1} = 0.1203$ km, $h_5^{\omega,1} = 0.0484$ km, $\ldots$, $h_{-2}^{\omega,1} = -12.2703$ km, $h_{-4}^{\omega,1} = -1.7581$ km; for the main antisymmetric mode $h \approx \infty$ (Flattery, 1967). The existence of negative equivalent

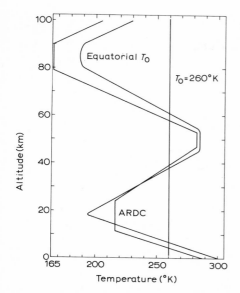

Fig. 25. Different basic temperature profiles used in examining the semidiurnal thermal tide. (After Lindzen, 1968a; Minzner et al., 1959.)

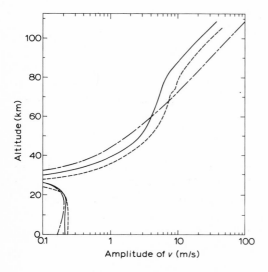

Fig. 26. Amplitude of the solar semidiurnal component of $v$ over the equator for different basic temperature profiles; continuous line = ARDC; dotted line = equatorial; dash-dotted line = isothermal ($T_0 = 260°K$). (After Lindzen, 1968.)

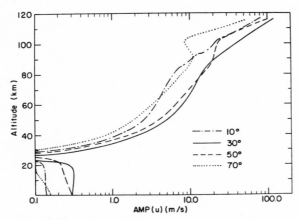

Fig. 27. Amplitude of the solar semidiurnal component of $u$ at various latitudes; equatorial $T_0(z)$ assumed.

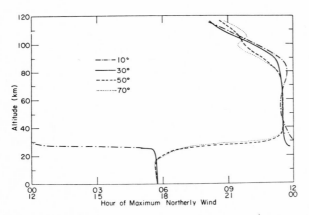

Fig. 28. Phase (hour of maximum) of the solar semidiurnal component of $u$ at various latitudes; equatorial $T_0(z)$ assumed.

depths has already been discussed in section 2. As we see in Fig. 29 and 31, modes with negative equivalent depths have their amplitudes mostly polewards of $\theta = (60°, 120°)$, while modes with positive equivalent depths are confined primarily equatorwards of these colatitudes.

The expansions of the thermal excitations latitude dependences in terms of diurnal Hough Functions are

$$g_{O_3}^{\omega,1} = 1.6308°K\Theta_{-2}^{\omega,1} - 0.5128°K\Theta_{-4}^{\omega,1} + \cdots$$
$$+ 0.5447°K\Theta_1^{\omega,1} - 0.1411°K\Theta_3^{\omega,1} + 0.0723°K\Theta_5^{\omega,1} + \ldots \qquad (110)$$

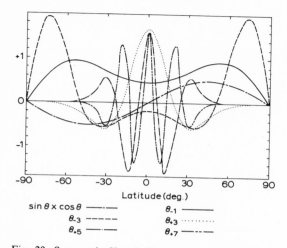

Fig. 29. Symmetric Hough Functions for the migrating solar diurnal thermal tide. Also shown is $\sin \theta \cos \theta$, the most important odd mode. (After Lindzen, 1967.)

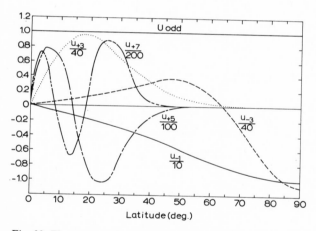

Fig. 30. The expansion functions for the latitude dependence of the solar diurnal component of $u$, the northerly velocity. The functions have been divided by the amounts shown. (After Lindzen, 1967.)

$$g_{H_2O}^{\omega,1} = 0.157°K\Theta_{-2}^{\omega,1} - 0.055°K\Theta_{-4}^{\omega,1} + \cdots$$

$$+ 0.062°K\Theta_{1}^{\omega,1} - 0.016°K\Theta_{3}^{\omega,1} + 0.008°K\Theta_{5}^{\omega,1} \ldots \tag{111}$$

Comparing Fig. 29 and 21 we see that no diurnal Hough Function is particularly like the latitude distribution of excitation—though $\Theta_{-2}^{\omega,1}$ comes the closest. Indeed, we see from eq. 110 and 111 that the $\Theta_{-2}^{\omega,1}$ mode does receive most of the excitation though not to so relatively great an extent as does the $\Theta_{2}^{2\omega,2}$ semidiurnal mode.

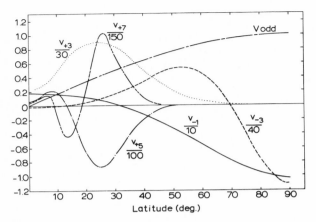

Fig. 31. The expansion functions for the latitude dependence of the solar diurnal component of $v$, the westerly velocity. The functions have been divided by the amounts shown. (After Lindzen, 1967.)

Moreover, since the $\Theta_{-2}^{\omega,1}$ mode cannot propagate vertically, it responds in a relatively poor fashion with respect to surface pressure. This is seen in eq. 112 and 113:

$$\delta p_{H_2O}(O) = \{137\Theta_{-2}^{\omega,1} - 68.2\Theta_{-4}^{\omega,1} + \cdots$$

$$+ 117e^{56°i}\Theta_1^{\omega,1} - 13.0e^{73.3°i}\Theta_3^{\omega,1} + 4.11e^{80.5°i}\Theta_5^{\omega,1} \cdots\}e^{i(\omega t+\varphi)}\mu bar \qquad (112)$$

$$\delta p_{O_3}(O) = \{44.1\Theta_{-2}^{\omega,1} - 3.4\Theta_{-4}^{\omega,1} + \cdots$$

$$+ 94.1e^{12.75°i}\Theta_1^{\omega,1} - 3.75e^{16.1°i}\Theta_3^{\omega,1} + 0.754e^{-6.6°i}\Theta_5^{\omega,1} \cdots\}e^{i(\omega t+\varphi)}\mu bar \qquad (113)$$

where we see the calculated surface pressure oscillations due to water vapor and ozone excitation (Lindzen, 1967). We see that $\Theta_{-2}^{\omega,1}$ no longer dominates; $\Theta_1^{\omega,1}$ is just as important. It is interesting to note, however, that even for the propagating modes, $H_2O$ is the largest contributor to the surface pressure oscillation. This is because the relatively thick $O_3$ source region, together with the relatively short wavelengths of the propagating modes (approximately 25 km, 12 km and 7 km for the modes considered here), leads to some destructive interference.

The above discussion explains why the diurnal surface pressure oscillation is both weak and irregular. The former is due to inefficiency of the diurnal modes, because of either trapping of interference; the latter is due to the fact that the response consists in several modes, each of which is relatively sensitive to local variations in excitation, temperature, etc.

The dependence of diurnal oscillations on latitude is particularly interesting. In Fig. 32 and 33 we can see calculated distributions for the amplitude and phase of the diurnal oscillation in northerly wind at different latitudes. Within regions of excitation, there are comparable amplitudes at all latitudes, but above the excitation there is continued $e^{x/2}$ growth at low latitudes and decay at high latitudes. Similarly, there is phase propagation at low latitudes, and almost none at high latitudes. Of particular importance to upper atmosphere dynamics at low latitudes is the fact that the diurnal propagating modes are not subject to trapping in the mesosphere.

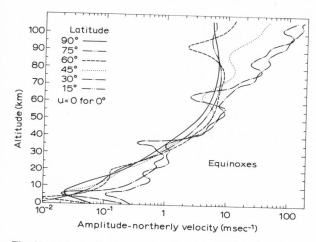

Fig. 32. Altitude distribution of the amplitude of the solar diurnal component of $u$ at 15° intervals of latitude; isothermal basic state assumed. (After Lindzen, 1967.)

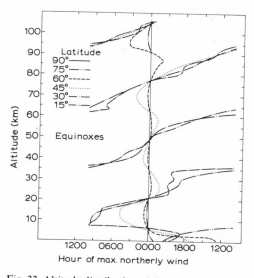

Fig. 33. Altitude distribution of the phase of the solar diurnal component of $u$ at 15° intervals of latitude; isothermal basic state assumed. (After Lindzen, 1967.)

The above results were calculated for an isothermal basic temperature, but they do not change in the overall aspect for more realistic atmospheres (Lindzen, 1968a).

## 3.5. Lunar semidiurnal tide

The Hough Functions for the lunar semidiurnal tide are very similar to those for the solar semidiurnal tide; the equivalent depth of the main lunar mode is 7.07 km. Lunar excitation is gravitational; it is given by:

$$
\Omega = (-23,662\Theta_2 - 5,615\Theta_4 - 2,603\Theta_6 \cdots)
$$
$$
\cdot \cos[2(\sigma_2^L t + \varphi)] \ cm^2 \ sec^{-2} \tag{114}
$$

As we have already mentioned, lunar atmospheric tides are much smaller than solar tides. However, their theoretical behavior is, at least, pedagogically interesting. As may be recalled from section 2, the amplitude of the lunar semidiurnal surface pressure oscillation is about 60–70 $\mu$bar, with a maximum occurring about an hour after transit. If we calculate the response of an isothermal atmosphere to (114), we get:

$$
\delta p_2^{2\sigma_2^L,2}(O) \approx \frac{34.2 \ \mu bar \ e^{i[2(\sigma_2^L t+\varphi)+90°]}}{\left[\left(\dfrac{H}{h_2} - \dfrac{1}{2}\right) + i\sqrt{\dfrac{\kappa H}{h_2} - \dfrac{1}{4}}\right]} \tag{115}
$$

(It will prove a simple but useful exercise for the reader to derive (115).) For $H \sim 7$ km (115) gives results close to the observed ones. However, what happens when one considers more realistic temperature profiles is somewhat surprising. In Fig. 34 we see a number of temperature profiles, differing only in their stratopause temperatures. Sawada (1956) computed the lunar semidiurnal oscillation for each of these profiles. His results are shown in Fig. 35. We see that the amplitude and phase vary significantly with small changes in the basic temperature—in distinct contrast to the solar semidiurnal oscillation. The Hough Functions for these two oscillations are so similar that the different behavior can only result from the different natures of the excitations. The gravitational excitation behaves as though it were a coherent source at the ground, while the thermal excitation is distributed throughout the atmosphere. Apparently, small changes in the distribution of $T_0$ change both the effective height of levels, where semidiurnal tides are partially reflected, and the reflectivities. Repeated reflections produce significant interference for coherent gravitational excitation; for distributed sources these effects tend to cancel each other.

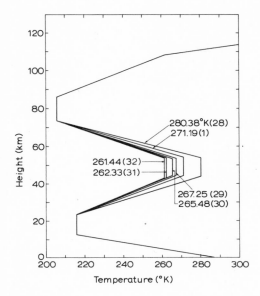

Fig. 34. Various temperature profiles used in calculating the lunar semidiurnal surface pressure oscillation. The maximum temperature of the ozonosphere and a profile number are shown for each of the profiles. (After Sawada, 1956.)

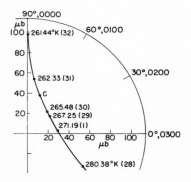

Fig. 35. A harmonic dial for the lunar semidiurnal surface pressure oscillation. Amplitude and phase are shown as functions of the basic temperature profile. (After Sawada, 1956.)

## 3.6. Comparisons with data

Agreement between the above calculations and observations appears to be fairly good for the solar diurnal and semidiurnal surface pressure oscillations. The comparison for the former is shown in Fig. 36. Regions of significant disagreement are also associated with low station density. For the semidiurnal surface pressure oscillation, there is a consistent phase error of about one hour.

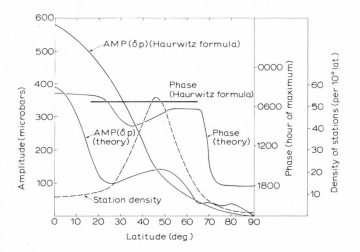

Fig. 36. Calculated amplitude and phase of the solar diurnal surface pressure oscillation. These quantities are also shown as derived from the Haurwitz (1965) empirical formula. The distribution of stations on which his formula is based are shown. (After Lindzen, 1967.)

There is a good agreement between observations of the lunar semidiurnal surface pressure oscillation and calculations for an isothermal atmosphere. However, the calculated variability of the surface oscillation, with small changes in $T_0$, does not appear in the data.

It is difficult to compare theory and observations between the ground and 30 km because of orographic effects, but between 40 and 90 km there is again good agreement for solar oscillations. Lunar tides tend to be buried in noise.

Above 90 km what little data there is suggests significant disagreement with the above theory.

## 4. FURTHER DEVELOPMENTS

In the preceding section we have described atmospheric tides as observed, as well as a mathematical model for atmospheric tides whose consequences have served to explain many aspects of the observed phenomena. The mathematical model incorporated a number of assumptions and approximations, whose consequences I would like to briefly explore. These considerations will naturally lead to the question of the nature of tides in the thermosphere, which I will touch on. I will also try to give some results concerning the effects of mean winds on internal gravity waves— a subject which not only has some bearing on tides, but is also of intrinsic interest for aeronomy.

## 4.1. Neglect of mean winds and horizontal temperature gradients

As we saw in section 3 our equations become immensely complicated when this approximation is dropped. Even if we were to restrict ourselves to steady zonal winds and north–south temperature gradients, we would already lose separability in $z$ and $\theta$ dependence. In general, no comprehensive study of the effects of mean winds and temperature gradients has been made. As a rule, both must be studied together since mean winds and temperatures in the atmosphere are related (to a fair degree of approximation) by the thermal wind equation:

$$\left(\frac{\partial}{\partial z} + \frac{1}{H}\right)(\rho u) = -\frac{g\rho}{afT}\frac{\partial T}{\partial \theta} \tag{116}$$

Limited aspects of the problem have been investigated (Haurwitz, 1957), but the only useful result appears to be that these approximations will not significantly change the resonance properties of the atmosphere (Dikii, 1967). The behavior of internal gravity waves in shear flows for both non-rotating (Booker and Bretherton, 1967) and rotating atmospheres (Jones, 1967) has been investigated in substantial detail. I shall review some of these results soon. They show that mean flows affect internal waves to the degree that they doppler shift the wave's frequency. This must undoubtedly apply to tides as well. Indeed, the recognition that tides are simply rotationally modified internal gravity waves permits us, without further mathematical analysis, to prescribe two necessary (though not necessarily sufficient) conditions for the neglect of mean winds and latitude variations of mean temperature:

(1) Latitude variations in temperature must be sufficiently small so that the variation of vertical wave number (i.e., $1/H^2[1/h_n(\kappa H + \mathrm{d}H\mathrm{d}X) - 1/4]$; viz, eq. 49) be small compared to the vertical wavenumber itself for the mode in question.

(2) Mean zonal winds must be small compared to the zonal phase speed of a tide. If $U$ is a characteristic zonal speed, then:

$$U \ll C = \frac{2\pi \times \text{earth's radius}}{1 \text{ solar day}} \tag{117}$$

where $C \sim 450$ m sec$^{-1}$.

Condition (2) is certainly satisfied on earth. Condition (1) is also fairly well observed. That it is not perfectly satisfied is indicated by the observation of seasonal variations in tides (viz., section 2), which cannot be accounted for by seasonal variations in excitation alone.[*]

---

[*] Calculations of the effects of seasonal variations in excitation may be found in Butler and Small (1963) and Lindzen (1967).

### 4.1.1. Internal gravity waves in shear flow

A study of the behavior of internal gravity waves in shear flow provides a convenient opportunity to look into the mechanisms whereby gravity waves act to transport mean energy and momentum through the atmosphere.

Let us begin by writing down the equations for two dimensional linear wave perturbations on a zonal wind, $U$, varying only with height, $z$. Let us, moreover, ignore both sphericity and rotation:

$$\rho_o\left(\frac{\partial u'}{\partial t} + U\frac{\partial u'}{\partial x} + U_z w'\right) + p'_x = 0 \tag{118}$$

$$\rho_o\left(\frac{\partial w'}{\partial t} + U\frac{\partial w'}{\partial x}\right) + \rho'g + p'_z = 0 \tag{119}$$

$$\frac{\partial u'}{\partial x} + \frac{\partial w'}{\partial z} + \frac{1}{\rho_o}\left(\frac{\partial \rho'}{\partial t} + U\frac{\partial \rho'}{\partial x} + w\frac{d\rho_o}{dz}\right) = 0 \tag{120}$$

$$\left(\frac{\partial p'}{\partial t} + U\frac{\partial p'}{\partial x} + w'\frac{d p_o}{dz}\right) = \gamma gH\left(\frac{\partial \rho'}{\partial t} + U\frac{\partial \rho'}{\partial x} + w'\frac{d\rho_o}{dz}\right) \tag{121}$$

Basic fields are defined as in section 3. Without rotation, one may have shear without horizontal temperature gradients. Let us consider solutions with $x$ and $t$ dependence of the following form:

$$e^{ik(x-ct)}$$

where $k$ is an east-west wavenumber, and $c$ is the wave's westerly phase speed. Eq. 118–121 become:

$$\rho_o\left((U-c)\frac{\partial u'}{\partial x} + U_z w'\right) + p'_x = 0 \tag{122}$$

$$\rho_o(U-c)\frac{\partial w'}{\partial x} + \rho'g + p'_z = 0 \tag{123}$$

$$\frac{\partial u'}{\partial x} + \frac{\partial w'}{\partial z} + \frac{1}{\rho_o}\left((U-c)\frac{\partial \rho'}{\partial x} + w'\frac{d\rho_o}{dz}\right) = 0 \tag{124}$$

$$(U-c)\frac{\partial p'}{\partial x} + \gamma H\rho_o \omega_o^2 w = \gamma gH(U-c)\frac{\partial \rho'}{\partial x} \tag{125}$$

where we have used the fact that:

$$\gamma H\rho_o \omega_o^2 = \frac{d\rho_o}{dz} - \gamma gH\frac{d p_o}{dz}$$

$\omega_o^2$ being the square of the Brunt–Vaisala frequency. From (122) we obtain after averaging over $x$, and assuming either periodicity or boundedness:

$$\overline{pw} = -\rho_o(U-c)\overline{uw} \tag{126}$$

When $U$ is a constant, then $\overline{pw}$ is the upward flux of energy due to the wave. Regardless of $U$, $\rho_0\overline{uw}$ is the upward flux of zonal momentum. Eq. 126 says that an upward propagating wave carries easterly momentum, if its phase speed is easterly relative to $U$, and it carries westerly momentum if $c$ is westerly relative to $U$. This is also true when $U = U(z)$; however, $\overline{pw}$ is no longer the complete expression for the vertical flux of energy. Another term, $\rho_0 U\overline{uw}$, must be added (Hines and Reddy, 1967) to account for the advection of the fluid's energy. The difficulty with this term is that it is undetermined to a constant, depending on one's frame of reference. Hence, the direction of the energy flux is no longer related uniquely to the direction in which a wave packet is traveling.

If we multiply (122) by $u'$, (123) by $w'$ and (124) by $p'$, then eliminate $\rho'$ in (124) using (125), and add the resulting equations we get:

$$\frac{\partial}{\partial x}\left\{\frac{1}{2}\rho_0(U-c)u'^2 + \frac{1}{2}\rho_0(U-c)w'^2 + \frac{1}{2}\rho_0\omega_0^2(U-c)\xi^2\right.$$

$$\left. + \frac{1}{2}\frac{(U-c)}{\rho_0\gamma gH}p'^2 + p'u'\right\} + \frac{\partial}{\partial z}(wp) = -\rho_0 U_z uw \tag{127}$$

where $w = (U-c)\xi x$, and $\xi$ is the vertical displacement. Averaging over $x$, (127) becomes:

$$\frac{d}{dz}\overline{pw} = -U_z\rho_0\overline{uw} \tag{128}$$

which, combined with (126), yields the important result:

$$\frac{d}{dz}(\rho_0\overline{uw}) = 0 \text{ if } (U-c) \neq 0 \tag{129}$$

which states that in the absence of a critical level where $U = c$, there will be no exchange of mean momentum between internal waves and the mean field. Results (126) and (129) were first derived by Eliassen and Palm (1961). If, however, waves are damped by radiation, viscosity, etc., then their mean momentum flux will be deposited in the mean flow, causing the flow speed to approach the wave's phase speed. We will say more about this later. Eq. 129 has counterparts for a rotating system. In such a system, the vertical flux of zonal momentum is given by:

$$F_m = \rho_0\overline{uw} - \rho_0\overline{\eta w}f \tag{130}$$

(Jones, 1967; Bretherton, 1969), where $f$ = twice the vertical component of the rotation rate, and $\eta$ = northward displacement of a fluid element. On a rotating plane where $f$ = constant and $U = U(z)$:

$$\frac{dF_m}{dz} = 0 \quad \text{(Jones, 1967)}. \tag{131}$$

It has also been shown that for at least certain cases on a rotating sphere:

$$\frac{d}{dz}\langle F_m \rangle = 0 \qquad \text{(Lindzen, 1971d)} \qquad (132)$$

where $\langle \ \rangle$ refers to an integral over all latitudes.

An important question is what happens when $U - c$ goes through zero. If the wave goes through such a level, $F_m$ must change sign, implying an infinite flux divergence of momentum. If the wave does not get through, is it absorbed or reflected? This question was answered by Booker and Bretherton (1967) who showed that the wave must be absorbed. Absorption of momentum at a level can lead to the production of sharp shear zones (Lindzen, 1968c; Jones and Houghton, 1971). Moreover, as mentioned before, there can be a more gradual exchange of momentum with the mean flow due to the routine damping of waves. The latter could possibly be the main source of friction for the mesosphere and lower thermosphere (since the average phase speed for all internal waves is probably close to zero), causing such things as the pole to pole temperature reversal at the mesopause (Lindzen, 1968b). While time does not permit further discussion, it should be evident that momentum transport and exchange by internal waves have important ramifications for the structure of wind, temperature, and even ionization in the upper atmosphere. In the lower stratosphere, the process has successfully accounted for the quasi-biennial oscillation (Lindzen and Holton, 1968).

### 4.2. Neglect of dissipation

The consideration of this approximation is not only of intrinsic importance, but is also a necessary prerequisite to the consideration of nonlinearity. As we see from eq. 49–58, the amplitudes of invescid, adiabatic wave solutions grow as $e^{x/2}$. Thus we might expect eventual nonlinearity. However, as we shall show, the inclusion of viscosity and thermal conductivity in the linear problem suppresses this growth at sufficiently great heights. Hence, the consideration of nonlinear effects must be deferred until the linear dissipative solutions have been obtained.

It proves useful to begin with a qualitative discussion of an arbitrary dissipative mechanism whose time scale, $\tau_{diss}$, may be specified. If a given tidal mode has a period, $\tau_{tide}$, then two distinct situations may exist with respect to dissipation: (1) if $\tau_{diss} < \tau_{tide}/2\pi$, then the tidal dynamics are fundamentally altered since dissipation is more important than inertia; (2) the presence of any dissipation will tend to reduce the $e^{x/2}$ growth of propagating modes. If, moreover:

$$\frac{\tau_{tide}}{2\pi} < \tau_{diss} < \frac{\tau_{tide}}{\left(\dfrac{L_v}{2H}\right)} \qquad (133)$$

(where $L_v$ is the vertical wavelength of the tidal mode), then the vertical wavelength of the tide will be relatively unaffected, but the $e^{x/2}$ growth above the region of excitation will be replaced by decay of amplitude.

Condition (1) arises in the atmosphere from molecular viscosity and conductivity whose effectiveness increases as $1/\rho$, and possibly from hydromagnetic drag. For the most important tidal modes, molecular effects assume importance above 100 km, whole hydromagnetic drag becomes important primarily above 200 km. Below 100 km, infrared cooling is of moderate importance—leading to small reductions of amplitude ($\lesssim 20\%$) but not to decay with height.

Having discussed dissipation qualitatively, the problem remains as to how to deal with it quantitatively. If we return to section 3, we will see that separation of variables resulted from the possibility of expressing the horizontal divergence of velocity as an operator on $p/\rho_0$, which depended only on latitude; and on the fact that the equations of state, hydrostatic pressure, continuity and energy depended on $\theta$ only through their dependence on velocity divergence. The inclusion of almost any meaningful model for friction will invalidate the first condition and eliminate the possibility of separating variables. On the other hand, any model of thermal dissipation which may be expressed as an operator on temperature involving only altitude will simply complicate the vertical structure equation (Lindzen and McKenzie, 1967; Dickinson and Geller, 1968); Hough Functions will remain the appropriate latitude expansion functions. Basically, the presence of rotation leads to this state of affairs; friction leads to modified coriolis torques, with the consequence that the horizontal structure of modes changes with height. This does not necessarily happen with thermal dissipation. Thus, reasonable calculations of the effect of infrared cooling have been made (Lindzen, 1968a); however, the proper study of the effects of viscosity on tides has not yet been completed. In general, we are unable to analytically solve non-separable equations. The numerical solution which is currently being performed is exceptionally cumbersome (with the inclusion of conductivity, viscosity and anisotropic hydromagnetic drag we have an eighth order equation, and we need very high resolution). However, the numerical approach permits the inclusion of arbitrary distributions of mean wind and temperature. Only preliminary results are so far available. These will be mentioned later.

Here, however, we can begin to develop some intuition by studying internal gravity waves in a planar non-rotating fluid. Certainly, the alteration of the horizontal structure is not the only effect of viscosity. In addition, as we saw in section 3, there always exist internal gravity waves in planar non-rotating fluid whose frequency and inviscid vertical wavenumber are identical to those of tidal modes (negative equivalent depths are modelled by imaginary north–south wavenumbers). Before discussing studies of the effects of dissipation on internal gravity waves, I would like to point out that there are two important points at which inviscid, adiabatic wave theory may be affected by dissipation.

(1) In the upper atmosphere where molecular viscosity and conductivity assume dominant importance our solutions will be *locally* very different.

(2) In adopting the radiation condition it was assumed that dissipation in the upper atmosphere would absorb all upcoming energy; it will be shown that for very large vertical wavelengths, dissipation increasing in effectiveness as $1/\rho$, can actually cause partial reflection—thus affecting the solution at all levels.

Perturbation studies of the effects of dissipation had been made by Pitteway and Hines (1963). However, Yanowitch (1967) was the first to make a full study of the effects of viscosity increasing as $1/\rho$ in a very difficult analytic study wherein thermal conductivity and compressibility (but no stratification) were neglected. Lindzen (1970, 1971e) and Lindzen and Blake (1971) extended these studies numerically to situations wherein eddy and molecular viscosity and conductivity, ion drag, radiative cooling, and realistic temperature variation and composition were all included. With regard to the most important of the dissipative processes, molecular viscosity and conductivity, all the above studies have basically supported the conclusion of Lindzen (1968d): namely, that any dissipative process whose effectiveness increases as $1/\rho$ tends to affect internal gravity waves in the same way. What happens to the wave depends on two parameters. First, there is the ratio of the time scale for the wave ($1/\omega$) to the time scale for dissipation ($\rho/a$):

$$\chi = \frac{a}{\omega\rho} \tag{134}$$

What $a$ is, depends on the dissipative process. If we are considering viscosity:

$$a = \frac{4\pi^2 \mu}{L_v^2}$$

where $\mu$ = molecular viscosity. Since $\chi \propto 1/\rho$, it is an increasing function with altitude. The second important parameter is a ratio of atmospheric scale height to vertical wavelength:

$$\beta = 2\pi H/L_v \tag{135}$$

If we are dealing with a non-isothermal atmosphere, then $H$ is essentially the local scale height at the altitude where $\chi \sim 1$.

In terms of $\chi$ and $\beta$, the main effects of molecular viscosity and thermal conductivity are:

(1) The increasing dissipation with height serves as an inhomogeneity in the medium, which can cause downward reflection. The magnitude of the reflection is given by:

$$|R| = e^{-\pi\beta} \tag{136}$$

$L_v$ must be $\gtrsim 120$ km for this effect to be significant (assuming $H \sim 6$ km).

(2) For $\beta \lesssim 1$, wave amplitudes increase roughly as $e^{x/2}$ up to the vicinity of $\chi \sim 1$, asymptotically approaching a constant above this level—with little or no decrease of amplitude.

(3) For $\beta > 1$, wave amplitudes increase roughly as $e^{x/2}$ up to the vicinity of $\chi \sim 1$—but then decrease considerably before asymptoting to a constant.

(4) The effects of dissipation become important when $\chi \gtrsim 1$. Hence, for a given $\beta$, the greater $\omega$ is, the greater the height at which dissipative effects set in.

(5) The dominance of dissipation is associated with the constancy of both amplitude and phase with height.

(6) Most of the above results apply to waves which, in the absence of dissipation, propagate vertically. For evanescent waves, increasing dissipation causes wave amplitudes to cease decaying with height, approaching a constant instead. This is accompanied by a change of phase.

Finally, a comment is in order on the effects on ion drag. Ion drag is centered in the F region (ca. 250 km), where it has a time scale of the order of 1 h. A priori, one might expect ion drag to dominate the physics within the F region of waves with periods in excess of 3 h. However, the time scales for viscous diffusion are even shorter than those for ion drag, and the effect of the latter amounts to a reduction of no more than 40% in amplitudes. From the results in section 3 we may compute the vertical wavelengths of the main tidal modes. These and the tidal periods may be used to calculate the height at which $\chi \sim 1$, and the value of $\beta$ at that height. The results are shown in Table II.

TABLE II

All values apply where $\chi \simeq 1$. $\overline{T}$ corresponds to ARDC standard atmosphere

|  | Mode | Altitude (km) | $L_v$ (km) | $H$ (km) | $\beta$ |
|---|---|---|---|---|---|
| S.D. | 2,2 | 145 | 120 | 16 | 0.84 |
|  | 2,4 | 125 | 45 | 11 | 1.6 |
| D. | 1,3 | 105 | 20 | 7.5 | 2.4 |
|  | 1,-1 | 200 | 34* | 24 | — |

*Scale for vertical decay of amplitude.

It is apparent from Table II that none of the diurnal modes is effective in penetrating the thermosphere. In addition, it is shown in Lindzen and Blake (1971) that the main diurnal propagating mode suffers some measure of unstable breakdown near 90 km. I shall discuss this further in the next section. Thus, there is no way in which diurnal oscillations excited by ozone and water vapor absorption can significantly penetrate the thermosphere. The diurnal oscillations of the thermosphere must be excited in situ. The situation for semidiurnal tides is profoundly different. We see that the main semidiurnal mode continues to grow (approximately as $e^{x/2}$, where $x$ is the height in scale heights) to very considerable altitudes, and its amplitudes asymptote to constants with no decay. Thus, it appears that in addition to a diurnal oscillation, excited in situ, the thermosphere should also have a strong semidiurnal oscillation excited in the lower atmosphere. These results are displayed graphically in Fig. 37 and 38 where we show results from numerical calculations for internal gravity waves with the same period and inviscid vertical structure as the main tidal modes. (A standard atmosphere with an 800°K exosphere was assumed.)

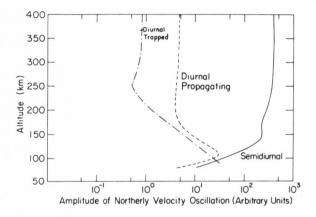

Fig. 37. Theoretical altitude distribution of the amplitude of $u$ for the main semidiurnal (continuous line), the main propagating diurnal (dashed line), and the main trapped diurnal (dash-dotted line) modes on earth. Molecular viscosity and thermal conductivity have been taken into account.

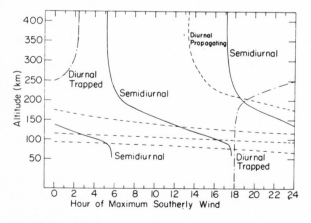

Fig. 38. Same as Fig. 37 except for the phases of the various modes.

Clearly, the above results demand further discussion of thermospheric tides. However, we must first consider the possibility of nonlinear effects.

## 4.3. Nonlinear effects

Had we approached this approximation by means of amplitude expansions for inviscid, adiabatic solution, we would have found that successive terms would have grown as $e^{x/2}$, $e^x$, $e^{3x/2}$, etc.—clearly an untenable situation. This difficulty can, in principle, be eliminated by expanding in terms of viscous, conducting solutions. The difficulty of this approach should be evident from the discussion of 4.2. A qualitative examination of the nature of likely nonlinear effects serves, moreover, to suggest that an amplitude expansion may often be fundamentally inappropriate.

Like internal gravity waves, tides are basically transverse oscillations for which advective (i.e., $\vec{V} \cdot \vec{\nabla}$) terms are very small. Thus, the most important nonlinear terms will result from $\rho/\rho_0$ approaching (or exceeding) unity. For the main semi-diurnal mode, this quantity approaches 0.5 before viscosity and conductivity put an end to amplitude growth. For the diurnal propagating mode, another amplitude-dependent effect proves more important; namely, the instability of the wave. It appears possible for still linearizable wave fields to have shears and/or temperature gradients, which are unstable to perturbations that grow rapidly compared to the wave frequency. This seems likely to be true for most Richardson Number instabilities; i.e., if we write:

$$T = T_0(z) + \delta T$$

then:

$$Ri = \frac{\dfrac{g}{T}\left(\dfrac{\partial T}{\partial z} + \dfrac{g}{c_p}\right)}{\left(\dfrac{\partial u'}{\partial z}\right)^2 + \left(\dfrac{\partial v'}{\partial z}\right)^2}$$

and when $Ri < 1/4$ we have instability (Hodges, 1967; see also Lindzen, 1968a). This occurs for the propagating diurnal mode at 88 km over the equator and must continue until 108 km above which height dissipative processes cause amplitude decay. Amplitude expansions would be useless for this case.

## 4.4. Tides in the thermosphere

### 4.4.1. Diurnal oscillation in the thermosphere

As indicated in section 4.2 the diurnal oscillation in the thermosphere must be excited in situ—primarily, it turns out, by EUV absorbed in the neighborhood of 150 km by $N_2$, O, and $O_2$. I have, therefore, calculated the response of tidal type gravity modes to such excitation—including within the numerical calculations the effects of viscosity, thermal conductivity and ion drag. The details are given in Lindzen (1971e). I considered the excitation due to EUV radiation, for which: $\overline{\varepsilon F_\infty} = 1.0$ erg cm$^{-2}$ sec$^{-1}$ at noon. $F_\infty$ is the incident solar flux, $\varepsilon$ is the thermal efficiency of the absorbing processes, and the overbar indicates that the quantities have been integrated over the wavelength range 220–900 Å. The diurnal component of the excitation was partitioned between the 1, 3 and the 1, -1 modes. Some of the results are shown in Fig. 39 and 40. Fig. 39 shows the distributions with height of the amplitudes of the temperature oscillations associated with the 1, 3 and 1, -1 modes. Fig. 40 shows the phases associated with these modes. The amplitudes are appropriate to the equator. The total diurnal temperature oscillation results from the sum of the two modes. The oscillation at other latitudes may be calculated using Fig. 29. Roughly, we calculate a diurnal temperature oscillation with an amplitude of 200°K, with maximum temperature occurring at 1400 h over the equator, and between 1500 and 1600 h at midlatitudes. While the phases are roughly in agreement with thermospheric observations (largely satellite drag data), the amplitude is substantially larger than the one appropriate to recent Jacchia models. This is consistent with Hinteregger's reduced estimate of solar EUV flux and Hays' calculations of thermal efficiencies (personal communications), which lead to an estimated noontime value of: $\overline{\varepsilon F_\infty} \approx 0.6$ erg cm$^{-2}$ sec$^{-1}$ for moderate solar activity. For such heating, we would obtain an amplitude for the temperature oscillation of 120°K, which is substantially closer to Jacchia's models. Unfortunately, the lower value of $\overline{\varepsilon F_\infty}$ is totally inadequate to maintain the observed mean temperature of the upper thermosphere. A consideration of the semidiurnal tide helps us out of the problem of maintaining mean thermospheric temperatures, but leads to other difficulties.

### 4.4.2. Semidiurnal oscillation in the thermosphere

The results of section 4.2 suggest that semidiurnal tidal oscillations at about 90 km can effectively penetrate the upper thermosphere. Moreover, observations cited in section 2 suggest reasonable agreement between theory and observation at 90 km. Indeed, more recent data by Revah et al. (1967) suggests agreement as high as 100 km over Garchy (47°N); and observations at night from gun-launched vapor canisters (Murphy et al., 1966) suggest agreement up to 120 km. Our remaining task is to calculate the upward extension of the semidiurnal tide into the thermosphere. The results are shown in Fig. 41 and 42, where I show the height distributions of the amplitude and phase at the equator of the temperature oscillation associated with the main semidiurnal mode. Results are shown for Schumann–Runge absorption, Schumann–Runge plus EUV absorption, and for all absorption including UV and near infrared absorption by $O_3$ and $H_2O$. Clearly $O_3$ and $H_2O$ absorption is most important, and EUV is more important than

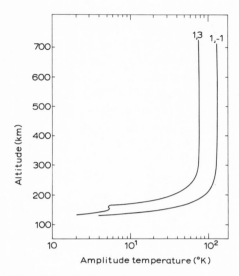

Fig. 39. Amplitudes of the diurnal temperature oscillations at the equator for the 1,3 and 1,−1 modes as functions of height. Excitation is due to EUV absorption in the lower thermosphere. (From Lindzen, 1971e.)

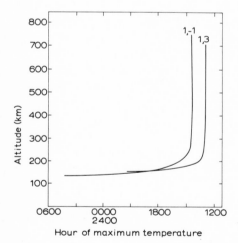

Fig. 40. Same as Fig. 39, but for the phases of the temperature oscillations at the equator.

Schumann–Runge absorption. Also shown are results for different thermospheric basic temperature distributions. The phase of the thermospheric semidiurnal temperature oscillation depends markedly on the basic temperature, though the amplitude does not. Our calculations suggest the presence of a semidiurnal temperature oscillation with an amplitude of 200° K. Since observations, thus far, do not suggest such a strong semidiurnal oscillation, we are left with another quandary. However, the presence of the semidiurnal tide does provide us with a convenient solution to an earlier quandary. The upward propagation of the main semidiurnal mode is associated with a mean upward flux of energy amounting to about 0.3–0.4 erg cm$^{-2}$ sec$^{-1}$. If such a flux were due to solar radiation, a noontime value of $\overline{\varepsilon F_\infty}$ of about 1 erg cm$^{-2}$ sec$^{-1}$ would be necessary. Such fluxes are small compared to UV fluxes. However, in order to heat the thermosphere, the flux must be deposited above about 120 km. As we see in Fig. 43 this is, in fact, the case for the mean energy flux carried by the semidiurnal tide. For comparison, we also show the heating due to the deposition of a *mean* $\overline{\varepsilon F_\infty}$ of 1 erg cm$^{-2}$ sec$^{-1}$ which is currently believed to be inordinately high. In Fig. 44, the mean temperature that would result from the deposition of mean semidiurnal energy flux is shown. Clearly, the semidiurnal tide goes a long way toward relaxing the need for EUV heating. (Details of these calculations are given in Lindzen and Blake, 1971.) However, the question remains as to why the semidiurnal oscillation has not been prominent in the data.

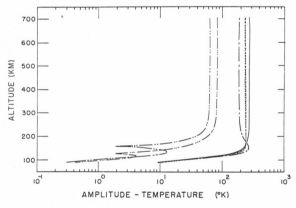

Fig. 41. Amplitude of the solar semidiurnal temperature oscillation over the equator as a function of height. The different curves correspond to different combinations of excitation and to different assumed basic temperature distributions. The curves — - - —, — - - - —, and · · · · · · all pertain to an ARDC Standard Atmosphere with an 800°K exosphere; — corresponds to an atmosphere with an 800°K exosphere, but with a somewhat different temperature profile between 80 km and 200 km; - - - corresponds to an atmosphere with a 1000°K exosphere; and — - ——corresponds to an atmosphere with a 1400°K exosphere. — - - - — is due to excitation by EUV absorption; — - · · - — is due to excitation by Schumann–Runge and EUV absorption, while the remaining curves involve absorption by $O_3$ and $H_2O$ as well as Schumann–Runge and EUV absorption. (From Lindzen, 1971e.)

### 4.4.3. Remaining problems and possible solutions

The above described calculations lead to what is, on the whole, a rather self-consistent picture of the upper thermosphere. The mean temperature is satisfactorily accounted for by the sum of EUV heating, and heating due to the molecular

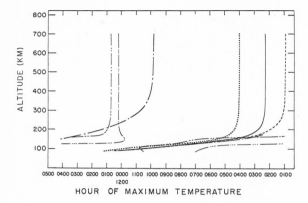

HOUR OF MAXIMUM TEMPERATURE

Fig. 42. Same as Fig. 41, but for the phases of the semidiurnal oscillations.

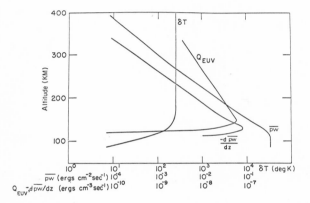

Fig. 43. Distributions with altitude of the amplitude of the solar semidiurnal temperature oscillation as excited by ozone and water vapor insolation absorption ($\delta T$), the mean upward flux of energy due to the solar semidiurnal tide $\overline{pw}$, the mean energy deposited by the solar semidiurnal tide $[d(\overline{pw})dz]$ and the mean energy deposited by absorption of EUV radiation—assuming $\epsilon F_\infty = 1.4$ erg cm$^{-2}$ sec$^{-1}$ at noon ($Q_{EUV}$). The values shown are appropriate to the equator. (From Lindzen and Blake, 1970.)

dissipation of the mean energy carried by the semidiurnal tide. A relatively small EUV heating is sufficient to account for the observed diurnal oscillation in the thermosphere. However, a serious discrepancy remains between theory and observation for the semidiurnal oscillation. We have predicted a semidiurnal temperature oscillation of about 200°K amplitude; Jacchia's models suggest no more than about 40°K. There are two obvious possibilities: the data are wrong or the theory is wrong. Both, in some measure, are true. Satellite drag data implicitly involve smoothing over 3–6 h, depending on orbit characteristics. Hence, such data do tend to suppress semidiurnal oscillations. In effect, satellite drag analyses attribute all

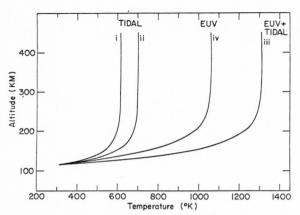

Fig. 44. Thermospheric temperature distributions resulting from the balance of thermal conduction with (i) the tidal heating shown in Fig. 43, (ii) 4/3 times the tidal heating shown in Fig. 43, (iii) the tidal heating use for curve ii plus $Q_{EUV}$ shown in Fig. 43, and (iv) $Q_{EUV}$ alone. From Lindzen and Blake (1970). N.B., Curves i and ii are shown because the tidal energy flux is dependent on the details of the mean mesospheric temperature structure.

drag to the perigee height; however, with a strong semidiurnal oscillation, much of the drag will come from regions away from perigee. Such variations are smoothed and perigee densities are overestimated. The exact extent of such errors has not yet been calculated. Recent incoherent backscatter data (McClure, 1971) can be interpreted as supporting the existence of a substantial semidiurnal oscillation. Similarly, there are errors in the theory. An extensive discussion of such errors is given in Lindzen (1971e). Briefly, such approximations as linearity and the use of "equivalent" gravity waves for the tidal modes can lead to errors of as much as 50% in the amplitude of the semidiurnal oscillation in the upper thermosphere[*]. (N.B. Observations support our calculations up to about 100 km.) Even so, we would be left with an amplitude (i.e., half the range) of 100°K, which is larger than current data analyses suggest.

At present, numerical calculations are under way for the semidiurnal tide in a rotating, spherical, viscous, conducting atmosphere with realistic distributions of ion drag, wind and temperature. Our very preliminary findings suggest semidiurnal amplitudes in the thermosphere over the equator are less than we predicted with the equivalent gravity mode formalism. Otherwise, the vertical structure over the equator is similar to the EGM calculations. Away from the equator, in the thermosphere, amplitudes decrease more slowly with increasing latitude than they do in inviscid regions. It is also found that temperature oscillations in the thermosphere at high latitudes lag those at the equator by 90°, in the absence of the mean wind. With mean wind, the phase lag may increase to 180° depending on

---

[*] Recent calculations suggest ion drag is more important than suggested in the EGM calculations.

season. The last effect, if real, would significantly complicate the detection of the semidiurnal oscillation in satellite drag data. Even polar orbits would average out the effect.

## ACKNOWLEDGEMENTS

The preparation of these notes was supported by Grant NSF-GA 25904 from the National Science Foundation and contract NGR 14-001-193 from the National Aeronautics and Space Administration.

REFERENCES

Bartels, J., Gezeitenschwingungen der Atmosphäre. *Handb. Exp. Phys.*, **25** (*Geophys.* 1): 163–210, 1928.
Beyers, N. J., Miers, B. T. and Reed, R. J., Diurnal tidal motions near the stratopause during 48 hours at White Sands Missile Range. *J.Atmos.Sci.*, **23**:325–333, 1966.
Booker, J. R. and Bretherton, F. P., The critical layer for internal gravity waves in shear flow. *J. Fluid Mech.*, **27**:513–539, 1967.
Bretherton, F. P., Momentum transport by gravity waves. *Q. J. R. Meteorol. Soc.*, **95**:213–243, 1969.
Butler, S. T. and Small, K. A., The excitation of atmospheric oscillations. *Proc. R. Soc. Lond. Ser.* A, **274**:91–121, 1963.
Carrier, G. F. and Pearson, C. E., *Ordinary Differential Equations*. Blaisdell, Waltham, Mass., 1968.
Chapman, S., The semi-diurnal oscillation of the atmosphere. *Q. J. R. Meteorol. Soc.*, **50**:165–195, 1924.
Chapman, S., Atmospheric tides and oscillations. In: *Compendium of Meteorology*. Am. Meteorol. Soc., Boston, Mass., pp. 262–274, 1951.
Chapman, S. and Lindzen, R. S., *Atmospheric Tides*. Reidel, Dordrecht; Gordon and Breach, New York, N.Y., 200 pp., 1970.
Chapman, S. and Westfold, K. C., A comparison of the annual mean solar and lunar atmospheric tides in barometric pressure as regards their world-wide distribution of amplitude and phase. *J.Atmos.Terr.Phys.*, **8**:1–23, 1956.
Dickinson, R. E. and Geller, M. A., A generalization of "tidal theory with Newtonian cooling". *J. Atmos. Sc.*, **25**:932–933, 1968.
Dikii, L. A., Allowance for mean wind in calculating the frequencies of free atmospheric oscillations. *Izv. Atmos. Ocean. Phys.*, **4**:583–584, 1967 (English edition).
Elford W. G., A study of winds between 80 and 100 km in medium latitudes. *Planet Sp. Sci.*, **1**:94–101, 1959.
Eliassen, A. and Palm, E., On the transfer of energy in stationary mountain waves. *Geofys. Publ.*, **22**:1–23, 1961.
Flattery, T. W., Hough functions. *Tech. Rep.* 21. Dep. Geophys. Sci., Univ. Chicago, Chicago, Illinois, 175 pp., 1967.
Goldstein, S., *Modern Developments in Fluid Dynamics* (2 vols.). Oxford University Press, Oxford, 1938.
Golitsyn, G. S. and Dikii, L. A., Oscillations of planetary atmospheres as a function of the rotational speed of the planet. *Izv. Atmos. Ocean. Phys.* **2**:137–142, 1966 (English edition).
Greenhow, J. G. and Neufeld, E. L., Winds in the upper atmosphere. *Q. J. R. Meteorol. Soc.*, **87**:472–489, 1961.
Harris, I. and Priester, W., On the dynamical variation of the upper atmosphere. *J. Atmos. Sci.*, **22**:3–10, 1965.
Harris, M. F., Finger, F. G. and Teweles, S., Diurnal variations of wind, pressure and temperature in the troposphere and stratosphere over the Azores. *J. Atmos. Sci.*, **19**:136–149, 1962.

Haurwitz, B., The geographical distribution of the solar semidiurnal pressure oscillation. *Meteorol. Pap.* 2. New York University, New York, 1956.

Haurwitz, B., Atmospheric oscillations and meridional temperature gradient. *Beitr. Phys. Atmos.*, **30**:46–54, 1957.

Haurwitz, B., Tidal phenomena in the upper atmosphere. *W.M.O. Rep.* 146, 1964.

Haurwitz, B., The diurnal surface pressure oscillation. *Archiv. Meteorol. Geophys. Bioklimatol.* A,**14**:361–369, 1965.

Hines, C. O., Internal gravity waves at ionospheric heights. *Can. J. Phys.*, **38**:1441–1481, 1960.

Hines, C. O., The upper atmosphere in motion. *Q. J. Meteorol. Soc.*, **89**:1–42, 1963.

Hines, C. O., Diurnal tide in the upper atmosphere. *J. Geophys. Res.*, **71**:1453–1459, 1966.

Hines, C. O. and Reddy, C. A., On the propagation of atmospheric gravity waves through regions of wind shear. *J. Geophys. Res.*, **72**:1015–1034, 1967.

Hodges, Jr., R. R., Generation of turbulence in the upper atmosphere by internal gravity waves. *J. Geophys. Res.*, **72**:3455–3458, 1967.

Hough, S. S., The application of harmonic analysis to the dynamical theory of the tides, Part II. On the general integration of Laplace's dynamical equations. *Philos. Trans. Soc. Lond. Ser.* A **191**:139–185, 1898.

Jacchia, L. and Kopal, Z., Atmospheric oscillations and the temperature profile of the upper atmosphere. *J. Meteorol.*, **9**:13–23, 1954.

Jones, W. L., Propagation of internal gravity waves in fluids with shear flow and rotation. *J. Fluid Mech.*, **30**:439, 1967.

Jones, W. L., Internal gravity waves. *Handb. Phys.*, 1971.

Jones, W. L. and Houghton, D. D., Gravity wave propagation with a time-dependent critical level. *J. Atmos. Sci.*, **28**:604–608, 1971.

Kato, S., Diurnal atmospheric oscillation, 1: Eigenvalues and Hough Functions. *J. Geophys. Res.*, **71**:3201–3209, 1966.

Kelvin, The Lord (W. Thomson), On the thermodynamic acceleration of the Earth's rotation. *Proc. Roy. Soc., Edinb.*, **11**:396–405, 1882.

Kertz, W., Components of the semidiurnal pressure oscillation. *Sci. Rep.* 4. New York Univ., Dept. Meteorol. and Ocean., 1956.

Lamb, H., On atmospheric oscillations. *Proc. Roy. Soc., Lond. Ser.* A, **84**:551–572, 1910.

Lamb, H., *Hydrodynamics*. University Press, Cambridge, 4th edition, 1916.

Laplace, S. L., *Mécanique Céleste*. Duprat, Paris, 1799.

Lindzen, R. S., On the theory of the diurnal tide. *Mon. Weather Rev.*, **94**:295–301, 1966.

Lindzen, R. S., Thermally driven diurnal tide in the atmosphere. *Q. J. R. Meteorol. Soc.*, **93**:18–42, 1967.

Lindzen, R. S., The application of classical atmospheric tidal theory. *Proc. Roy. Soc. Lond. Ser.* A, **303**:299–316, 1968a.

Lindzen, R. S., Lower atmospheric energy sources for the upper atmosphere. *Meteorol. Monogr.*, **9**:37–46, 1968b.

Lindzen, R. S., Some speculations on the roles of critical level interactions between internal gravity waves and mean flows. *Proc. ESSA/ARPA Symp., Acoustic-gravity Waves in the Atmosphere, Boulder, Colorado, July, 1968*. U.S. Gov. Print. Off., Washington, D.C., 1968c.

Lindzen, R. S., Vertically propagating waves in an atmosphere with Newtonian cooling inversely proportional to density. *Can. J. Phys.*, **46**:1835–1840, 1968d.

Lindzen, R. S., Internal gravity waves in atmospheres with realistic dissipation and temperature, 1: Mathematical development and propagation of waves into the thermosphere. *Geophys. Fluid Dyn.*, **1**:303–355, 1970.

Lindzen, R. S., Lecture notes prepared in connection with the Summer Seminar on Mathematical Problems in the Geophysical Sciences. *Lect. Appl. Math.*, **14**:293–362, 1971a.

Lindzen, R. S., Tides and gravity waves in the upper atmosphere. In: *Mesospheric Models and Related Experiments* (G. Fiocco, Editor). Reidel, Dordrecht, Holland, 1971b.

Lindzen, R. S., Some aspects of atmospheric waves in realistic atmospheres, In: *Physics of the Space Environment* (R. E. Smith and S. T. Onu, Editors). *NASA Rep.* SP-305, 1971c.

Lindzen, R. S., Equatorial planetary waves in shear, 1. *J.Atmos.Sci.*, **28**:609–622, 1971d.

Lindzen, R. S., Internal gravity waves in atmospheres with realistic dissipation and temperature, 3: Daily variations in the thermosphere. *Geophys.Fluid Dyn.*, **2**:89–121, 1971e.

Lindzen, R. S. and Blake, D., Mean heating of the thermosphere by tides. *J. Geophys. Res.*, **33**:6868–6871, 1970.

Lindzen, R. S. and Blake, D., Internal gravity waves in atmospheres with realistic dissipation and temperature, 2: Thermal tides excited below the mesopause. *Geophys. Fluid Dyn.*, **2**:31–61, 1971.

Lindzen, R. S. and Holton, J. R., A theory of the quasi-biennial oscillation. *J. Atmos. Sci.*, **25**:1095–1107, 1968.

Lindzen, R. S. and Kuo, H. L., A reliable method for the numerical integration of a large class of ordinary and partial differential equations. *Mon. Weather Rev.*, **97**:732–734, 1969.

Lindzen, R. S. and McKenzie, D. J., Tidal theory with Newtonian cooling. *Pure Appl. Geophys.*, **66**:90–96, 1967.

Longuet-Higgins, M. S., The eigenfunctions of Laplaces's tidal equations over a sphere. *Philos. Trans. R. Soc. Lond. Ser.* A, **269**:511–607, 1967.

Margules, M., Über die Schwingungen periodisch erwarmter Luft. *Sitzber.Akad.Wiss.Wien, Abt.* IIa, **99**:204–227, 1890.

Margules, M., Luftbewegungen in einer rotierenden Spharoidschale. *Sitzber.Akad.Wiss.Wien, Abt.* IIa, **101**:597–626, 1892.

Margules, M., *Sitzber.Akad.Wiss.Wien, Abt.* IIA: **102**:11–56; 1369–1421, 1893.

Martyn, D. F. and Pulley, O. O., The temperatures and constituents of the upper atmosphere. *Proc. R. Soc. Lond. Ser.* A, **154**:455–486, 1936.

McClure, J. P. Thermospheric temperature variations inferred from incoherent scatter observations. *J. Geophys. Res.*, **76**:3106–3115, 1971.

Minzner, R. A., Champion, K. S. W. and Pond, H. L., The ARDC Model Atmosphere, 1959. *Air Force Cambridge Res. Cent. Rep.*, AFCRC-TR-59-267 (*Air Force Surv. Geophys.* 115), 1959.

Murphy, C. H., Bull, G. V. and Edwards, H. D., Ionospheric winds measured by gun-launched projectiles. *J.Geophys. Res.*, **71**:4535–4544, 1966.

Pekeris, C. L., Atmospheric oscillations. *Proc. R. Soc. Lond. Ser.* A: **158**:650–671, 1937.

Phillips, N. A., The equations of motion for a shallow rotation atmosphere and the "traditional approximation". *J.Atmos.Sci.*, **23**:626–628, 1966.

Phillips, N. A., Reply to "comments on Phillips" simplification of the equations of motion, by G. Veronis. *J.Atmos.Sci.*, **25**:1155–1157, 1968.

Pitteway, M. L. V. and Hines, C. O., The viscous damping of atmospheric gravity waves, *Can. J. Phys.*, **41**:1935, 1963.

Rayleigh, The Baron (J. W. Strutt). On the vibrations of an atmosphere. *Philos. Mag.*, **29**:173–180, 1890.

Reed, R. J., Semidiurnal tidal motions between 30 and 60 km. *J. Atmos. Sci.*, **24**:315–317, 1967.

Reed, R. J., Oard, M. J. and Sieminski, M., A comparison of observed and theoretical diurnal tidal motions between 30 and 60 km. *Mon. Weather Rev.*, **97**:456–459, 1969.

Revah, I., Spizzichino, A. and Massebeuf, Mme., Marée semi-diurnelle et vents dominants zonaux mesurés a Garchy (France) de Novembre 1965 à Avril 1966. *Note Tech.* GRI/NTP/27. Cent. Nat. Étud. Telécommun., Issy-des-Moulineaux, France, 1967.

Sawada, R., The atmospheric lunar tides and the temperature profile in the upper atmosphere. *Geophys. Mag.*, **27**:213–236, 1956.

Siebert, M., Atmospheric tides. *Adv. Geophys.*, **7**:105–182, 1961.

Taylor, G. I., The resonance theory of semidiurnal atmospheric oscillations. *Mem. R. Meteorol. Soc.*, **4**:41–52, 1932.

Taylor, G. I., The oscillations of the atmosphere. *Proc. R. Soc. Lond. Ser.* A, **156**:318–326, 1936.

Wallace, J. M. and Hartranft, F. R., Diurnal wind variations; surface to 30 km. *Mon. Weather Rev.*, **96**:446–455, 1969.

Whittaker, E. T. and Watson, G. N., *A Course of Modern Analysis*. Cambridge Univ. Press, London, 4th ed., 608 pp., 1927.

Wilkes, M. V., *Oscillations of the Earth's Atmosphere*. Cambridge Univ. Press, New York, 74 pp. 1949.

Yanowitch. M., A remark on the hydrostatic approximation. *Pure Appl. Geophys.*, **64**:169–172, 1966.

Yanowitch, M., Effect of viscosity on gravity waves and the upper boundary condition. *J. Fluid Mech.*, **29**:209–231, 1967.

# ELECTRICAL STRUCTURE OF THE LOWER IONOSPHERE

WILLIS L. WEBB

*Atmospheric Sciences Laboratory, U.S. Army Electronics Command, White Sands Missile Range, N.M. (U.S.A.)*

## SUMMARY

Synoptic rocket exploration of the stratospheric circulation has revealed the presence of hemispheric tidal circulations which seem to be in part characterized by systematic vertical motions in low latitudes of the sunlit hemisphere. These vertical motions are powered by meridional oscillations in the stratospheric circulation produced by solar heating of the stratopause region, and they serve as the energy source of electrical current systems which are postulated to result from an impressed electromotive force produced by charged particle mobility differences in the lower ionosphere as the tidal circulations tend to force these particles across the earth's magnetic field. These dynamo currents are variable with geometry and time variabilities of the tidal circulations, as well as with variability in the solar-induced conductivity of the E region. The semiconducting lower atmosphere and highly conducting earth's surface occupy the near field of the lower side of this current system, with a resulting complex tropospheric electrical structure. Low-impedance electric current paths along magnetic field lines result in the development of currents in the exosphere that are driven and controlled by the electrical structure of the primary dynamo circuit and that exert a control of their own through interaction with the solar wind. The basic physical process which provides the required electromotive force for maintenance of the electrical structure of the earth's atmospheric environment is thus indicated to center in thermally driven tidal motions in the lower ionosphere, with locally observed structures such as the fair-weather electric field, thunderstorms, lightning discharges, aurora, airglow, electrojets and radiation belts, playing supporting roles. A sectional model illustrating the low-latitude global D- and E-region electrical structure is presented. Model conductivity structures are derived from mean electron density and collision frequency profiles and are applied to delineate atmospheric sources and sinks of electromotive energy and the resulting electric current and potential fields. These models indicate that the lower ionosphere is the site of important electrical structure which must be joined with the observed lower atmosphere and exospheric structures to obtain the complete global electrical structure. The lower ionospheric dynamo current is indicated by these considerations to dominate the global electrical structure, with tropospheric and exospheric electrical phenomena resulting principally from relaxation currents below and above the lower ionospheric region.

## 1. INTRODUCTION

The electrical structure of the earth remains today one of the most intriguing problems faced by the geo-scientist. The earth's permanent magnetic field is generally believed to result from electrical phenomena in the earth's interior, and certain small variations in that magnetic field have been demonstrated to be related to direct electromagnetic interaction of the earth system with the local solar environment. In addition, some systematic variations are believed to originate in motions of the weak ionospheric plasma through the permanent magnetic field. Electrical manifestations of this comprehensive magneto-electrodynamic system are commonly noted in thunderstorm electrification, the fair-weather electric field and current, aurora, airglow and numerous other facets of an obviously complex system. The certainty that these puzzling phenomena (as viewed individually) must fit into a satisfying global electrical structure has led to inspection of new information on the physical structure of the atmosphere in a search for clues which will serve to bind the currently fragmented picture together.

In October, 1959, a new system for synoptic exploration of the earth's upper atmosphere using small rocket vehicles was initiated to extend the region of meteorological study to higher altitudes (Webb, 1966a; Webb et al., 1966). This Meteorological Rocket Network (MRN) has expanded the atmospheric volume which is currently subject to meteorological scrutiny from limitations of the order of 30 km peak altitude to a current synoptic data ceiling of the order of 80 km. A number of very important findings have resulted from MRN synoptic exploration of the upper atmosphere, the most notable of which (for our current purposes) is the discovery of large diurnal variations in the temperature (Beyers and Miers, 1965), wind (Miers, 1965) and ozone (Randhawa, 1967) fields of the stratopause region. The data indicate that solar ultraviolet heating of the ozonosphere is concentrated in middle and low latitudes in a relatively thin layer in the 45 to 50 km altitude region. The distribution of this diurnal heat pulse downward and upward from the stratopause level is then accomplished by secondary physical processes. Consideration of this new information has led to the realization that circulation systems exist in the atmosphere above 40 km altitude that exert a profound influence on the upper atmospheric structure by dynamic altering of physical processes and by mixing and transport of atmospheric constituents of the upper atmosphere.

An important result of these synoptic meteorological studies of the upper atmosphere concerns the fact that vertical motions of considerable magnitude in specific locales are indicated by the data. Upward motions in the summer high-latitude night-time sky, indicated by the data, imply the presence of downward motions in polar regions of the opposite hemisphere and in middle and low latitudes of both hemispheres on a diurnal basis (Webb, 1966b). Downward motions are indicated to be located in the sunlit hemisphere from early morning until early afternoon with maximum intensity at low latitudes. Such vertical motions exert pronounced influences on the electrical structure of the atmosphere because they transport charged particles of the atmosphere vertically through regions where differing mobilities of the positive and negative charge carriers produce electromotive forces. Charge separation caused by this basic mechanism is indicated to occur in a predictable fashion in the mesosphere and lower ionosphere, and such a separated charge can be shown to result in electric fields and currents which modify the "normal" ionosphere (Fejer, 1953; Beynon and Brown, 1959; Martyn, 1959; Chandra and Rangaswamy, 1967; Yuen and Roelofs, 1967) in accord with the observed global electromagnetic structure. Some of the more relevant observations are listed below.

Short-term variations in surface observations of the earth's magnetic field have been analyzed extensively (Chapman and Bartels, 1940; Matsushita, 1965). These studies indicate the following general features: (1) occurrence of magnetic disturbances is directly correlated to sunspot activity; (2) occurrence of magnetic disturbances is maximum at equinox and minimum at solstice times; (3) occurrence of magnetic disturbances is minimum in midmorning and maximum just after midnight local time; and (4) a systematic diurnal variation occurs in the magnetic field intensity with an amplitude of several tens of gammas.

It is generally assumed that these short-term variations in the magnetic field are caused by electric currents which flow in the global shell centered near the 100-km level, and are the result of atmospheric tidal motions (Stewart, 1883; Chapman and Bartels, 1940; Chapman, 1954). Lack of adequate environmental data has prevented verification of these "dynamo" theories, and in general there is question about the adequacy of the tidal motions used in these initial estimates. Dynamo currents clearly do exist in the upper atmosphere, but their structure and origin remain open to question (Kato, 1956).

Available data indicate the existence of a characteristic electric field over the surface of the earth, except in storm areas (Chalmers, 1957; Imyanitov and Shifrin, 1962). Observational data concerning the earth's fair-weather electrical structure provide the following general information for the tropospheric situation: (1) the fair-weather electric field is negative (electrical potential increases with height); (2) intensity is variable with location and time in the 100 V/m range at the earth's surface; (3) an electric current of approximately 1400 A flows to the earth (Kraakevik, 1961); and (4) a net negative charge of approximately 450,000 C appears to reside in the earth's surface.

Integration of the known troposphere field gradient data indicates that the ionosphere is at a potential difference of the order of 300,000 V above the potential of the earth's surface.

A most popular theory relative to the atmospheric fair-weather electric field pictures thunderstorms as the generators which, through the energy available in gravitational separation of precipitation, provide a steady overall potential difference between the ionosphere and the earth's surface (Wilson, 1920). The fair-weather electric current is then pictured as a uniform leak between highly conducting ionospheric and earth's surface spherical condenser plates. The current flowing in the fair-weather electric field is such that in a matter of a few tens of minutes the stored charges on the earth's surface would be effectively eliminated, so it is necessary to show that the generating mechanism is in continuous operation. Thunderstorm activity on the global scale has been shown to be at least generally capable of providing the return current required to keep the fair-weather electric field current flowing (Brooks, 1925; Gish and Wait, 1950; Stergis et al., 1957). Thunderstorms have not been shown, however, to provide the organized electrification required to produce the earth's fair-weather electric field, and there is even the suspicion that they might not produce any electrification at all if the fair-weather electric field did not already exist. These considerations lead to the view that the observational evidence of electrification in the earth's troposphere allows the possibility that these phenomena are directly developed by some basic electrification mechanism, and that local electrical effects associated with thunderstorms, duststorms, air pollution, etc., simply represent local modifications of the general electrical structure of the earth's atmospheric system.

Electric currents flow in the surface layers of the earth (Barlow, 1849; Gish and Rooney, 1937; Chapman and Bartels, 1940; Redding, 1967). Complex conductivities in land surface areas assure complications in the global field of telluric currents, but the following generalizations appear to be characteristic of average conditions: (1) telluric currents of $10^{-10}$ A/m$^2$ are representative of surface currents through land areas under undisturbed conditions; (2) a strong diurnal variation is evidenced, with the current flow toward the equator in the daytime; (3) certain earth current variations are coincident with certain magnetic field variations; and (4) the range of diurnal variation is greatest in summer. The question of whether these currents represent a transport of charge through the earth's surface to some external circuit or are simply induction currents caused by ionospheric currents has been considered at some length in the literature without resolution.

Auroral activity is observed in a nearly circular band of approximately 23 degrees magnetic colatitude in both hemispheres (Chapman and Bartels, 1940; Parker, 1959; Chamberlain, 1961; Hines and Reid, 1965; Barcus and Brown, 1966; Barcus and Rosenberg, 1966; Bates, 1966; Feldstein and Starkov, 1967; Piddington, 1967).

Auroral phenomena appear to be manifestations of electrical activity which are in part controlled by the earth's magnetic field. Principal characteristics are: (1) the base of auroral activity exhibits a maximum occurrence at approximately 100 km altitude; (2) many magnetic field disturbances are directly related to auroral occurrences and fluctuations (Nichols, 1959; Feldstein and Starkov, 1967); (3) auroral activity in both hemispheres varies similarly with the sunspot cycle (Chivers and Hargreaves, 1966); (4) auroral activity is maximum at equinox times and minimum at solstice times; and (5) auroral activity generally exhibits a maximum just before midnight and a minimum just before noon local time.

Auroras are observed visually and through scattering of electromagnetic energy, with the two observational systems giving differences in details but general agreement (Blevis and Collins, 1965; Montalbetti, 1965) regarding the physical characteristics of the phenomena. The visual observations appear to deal principally with phenomena occurring along the magnetic field, while the radio observations appear to deal principally with a layer of enhanced electron concentration centered in the 100 km altitude region and frequently extending significantly above and below that level. Various concepts of the physical processes associated with auroral activity have been published (Wulf, 1953; Vestine, 1954; Cole, 1960; Chamberlain, 1965). Auroral activity is located in the correct position to be the precipitation ground for Van Allen (1959) outer belt particles, but the energy in the trapped particles is apparently inadequate for continued auroral generation.

Airglow represents characteristic optical line emissions from the ambient constituents of the upper atmosphere that result from their being in an excited state (Chamberlain, 1961; Greenspan and Woodman, 1967). The nightglow ·is quite striking when viewed under the proper conditions (Carpenter et al., 1962; Glenn, 1962; Hennes and Dunkelman, 1966). Principal features of the airglow include: (1) it is global in distribution, although the daytime intensity is not known with certainty because of interference by scattered sunlight; (2) it is inhomogeneous, with detail structure which appears to drift with the wind; and (3) it is centered near the 100-km level, and generally extends well below and above that level.

Airglow at night is generally regarded as residue from daytime solar ultraviolet excitation, although the nocturnal distribution does not necessarily support such a concept.

Thunderstorms generally occupy less that 1% of the earth's surface areas and exhibit gross spatial and temporal variations (Chalmers, 1957; Smith, 1958; Coroniti, 1965). Their electrical characteristics are complex, with the most distinctive feature centering around the lightning flash. Thunderstorms are characterized very generally by: (1) globally, approximately 2000 thunderstorms are in continuous operation with an estimated lightning frequency of approximately 10 cloud-to-ground strokes per second (Vorpahl, 1967); (2) maximum occurrence is in early evening; (3) an average of the order of $10^2$ C of charge is transferred vertically over an average altitude range from the surface to 10 km by each system of strokes; and (4) a majority of lightning discharges produce a transfer of negative charge to the earth.

Severe modifications of the tropospheric electric structure are observed in the vicinity of thunderstorms, but an organized transport of charge is not obvious except in the lightning discharge path, and even there the situation is highly variable.

Whistlers are electromagnetic signals of a few thousand cycles per second frequency, which have been identified as being of lightning origin and are known to travel between hemispheres along magnetic field lines (Helliwell and Morgan, 1959; Gendrin, 1961; Liemohn and Scarf, 1964; Angerami and Carpenter, 1966; Carpenter, 1966; Gurnett and Shawhan, 1966). Whistlers are probably a special extreme case of a general class of events which involve propagation and transport along magnetic field lines (Gallet, 1959; Gringauz et al., 1961; Krasovskii et al., 1961). Whistlers have been observed to: (1) travel from one hemisphere to the other in times of the order of one second; (2) oscillate between hemispheres as many as twenty times; and (3) occur most frequently at night, possibly as a result of D-region absorption in the daytime.

In all cases the whistler mode of propagation implies tubular enhancements of electron densities along field lines to guide the waves, and, in the lower frequency case studies of Gallet, the flow of discrete packets of electrons along the field lines is indicated.

Using the information which has been briefly reviewed above, the following pages will be devoted to the development of a comprehensive picture of the electrical structure of the atmosphere, from the earth's surface layers to distances of roughly ten earth's radii.

## 2. DIURNAL CIRCULATIONS

Diurnal variations in the temperature (Beyers and Miers, 1965) and wind (Miers, 1965) fields of the stratospheric circulation of the order of 15°C and 30 m/sec respectively were discovered through application of the sensitive measuring systems of the Meteorological Rocket Network. The diurnal variation of the wind field has been shown to be a general characteristic of MRN data (Reed et al., 1966). These measurements came as a surprise since theoretical considerations of solar ultraviolet absorption by ozone had led to the conclusion that the diurnal temperature variation would be smaller, of the order of 4°C (Craig, 1950; Leovy, 1964). These new data would then indicate that solar heating of the ozonosphere is confined to a much thinner region of the upper stratosphere than had previously been assumed, or that other heat transport processes exist, with a resulting need for an experimental reexamination of ozone structure and dynamic equilibrium in the upper stratosphere and lower mesosphere. This was done for the ozone case by Randhawa (1967), and significant differences from theoretical expectations were found, with a greater concentration and a pronounced diurnal variation in ozone concentration in the stratopause region where a maximum ozone concentration occurs during night-time hours.

The above information has been combined with geometric and theoretical considerations to provide a global picture of the tidal circulations of the stratospheric circulations (Webb, 1966b). Principal features of this model are a heat wave oriented latitudinally, with a trough near sunrise and a crest near 2 P.M. Winds at the stratopause level are accelerated by the resulting ridge of high pressure, propagating away from the equator in the increasing temperature sector and back toward the equator during the late afternoon and night-time. At equinox times the tidal circulation winds are zonal westerly around the high latitude ends of the heated ridge in the sunlit hemisphere, in both the northern and southern hemispheres.

The situation is materially altered as the solar aspect angle changes toward solstice time. The ridge of high pressure (heated, expanded air) quickly includes the polar region of the developing summer season so that the zonal portion of the tidal circulation of that hemisphere can no longer be westerly on the sunlit side of the summer pole, but must be easterly in the night-time sky and forms the Stratospheric Tidal Jet (STJ) (Webb, 1966b) indicated in Fig.1. In the winter hemisphere, on the other hand, the westerly zonal portion of the tidal circulation can be expected to move to lower latitudes. It is evident from stratospheric circulation data acquired by synoptic sounding that these events have a marked impact on the general circulation (Webb, 1966a).

Consideration of the physics of ozonospheric absorption of solar ultraviolet radiation indicates that observed diurnal variations should have an altitude dependence with solar aspect. Deposit of heat from a uniform source into a near-exponential atmosphere of spherical form should be at lowest altitudes at the subsolar point and should be found at higher altitudes at the limb of this interaction. Thus, the observed altitude of the tidal circulation at the stratopause (45–50 km) in low latitudes would indicate that the local tidal effects in polar regions should be found at higher altitudes. These considerations are based on the assumption of a uniform absorbing medium, which is obviously in error, but should at least give a first-order approximation to the actual geographic distribution of the tidal circulations. The stratospheric circulation in the mesopause region of high latitudes in the summer solstice period (the STJ) has been investigated by Morris (1967), who discovered that maximum winds of the summer easterly stratospheric circulation are to be found at that point in space and time. Further, the upper mesosphere of high latitudes ($>60°$) has been shown by these data to have an easterly thermal wind.

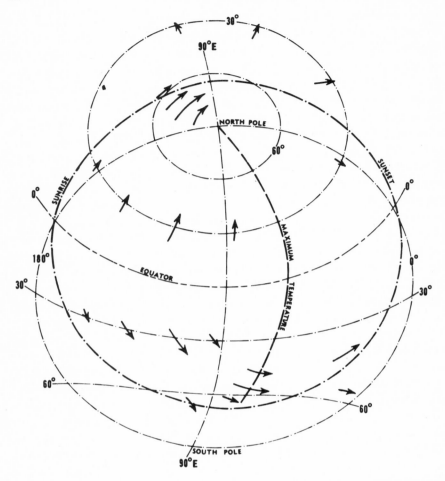

Fig. 1. Diurnal tidal circulation of the summer solstice northern hemisphere mesosphere projected on the stratopause which is represented by an equidistant projection centered on 42°N latitude and 103°W longitude.

This information indicates that the cold mesopause of summer high latitudes has the form of a ring around the summer pole in the region where noctilucent clouds and the STJ are located.

The global tidal circulations discussed above imply vertical motions of gross dimensions in the mesosphere and ionosphere. The import of such motions on the physical, chemical and electrical structures of the upper atmosphere may be significant, so their spatial and temporal characteristics will be considered in more detail here. The implication of vertical upward motions in the night-time sky of the summer high-latitude mesopause region has long been available from noctilucent cloud observations (Vestine, 1934; Ludlam, 1957; Webb, 1965; Fogle and Haurwitz, 1966). Consideration of the known physical characteristics of these clouds leads to the conclusion that upward motions of the order of tens of centimeters per second are required in the summer high-latitude night-time sky, with speeds in the m sec$^{-1}$ range compatible with available data. Mesospheric air is mixed by these tidal circulations, and a portion of the air involved in those circulations is injected across the mesopause into the lower ionosphere in high latitudes of the summer hemisphere. Using order of magnitude values of 1 m sec$^{-1}$ vertical speed, 0.01 of the mesopause hemispheric area involved in the STJ and a mesopause density of 2 · 10$^{-5}$ kg m$^{-3}$, the mesospheric mass transported across the mesopause boundary by this tidal circulation is evaluated to be of the order of 10$^6$ kg sec$^{-1}$. Obviously, this mass must return downward across the mesopause surface at some other location.

It is well known that the mesopause region of the winter polar zones is quite warm compared to the temperatures expected from radiational processes. Various mechanisms have been suggested to account for this anomalous thermal distribution, with only compressional heating from downward motions possessing the requisite power capability. A downward flow of the order of cm sec$^{-1}$ would be adequate to provide the observed heating if adiabatic conditions prevail. The area of the winter hemisphere in which this source of heat appears to occur is of the order of one tenth of the total hemispheric area, and a simple calculation of the mass of that downward transport across the mesopause indicates that it must be of the order of one tenth of the 10$^6$ kg sec$^{-1}$ which is postulated to move upward across the mesopause in the STJ. In view of the gross approximations which we are forced to make in these estimates, the downward mass transport over the winter pole could be a more significant fraction of the upward mass transport of the STJ at the mesopause level.

Consideration of earth rotational effects on the stratospheric tidal circulation indicates that special features of these circulations should be expected in equatorial regions. Reversal in direction of the horizontal component of the Coriolis force means that the hemispheric tidal circulations will separate over the rotational equator. Divergence associated with this rotational separation results in transport of stratospheric mass out of equatorial regions along the leading edge of the heat wave from sunrise to 2 P.M., with a maximum in late morning. Observations indicate that tidal motions near 30° latitude are constrained to roughly the 40–60 km altitude range, and rough estimates of the mass outflow across 30° latitude north and south boundaries indicate that about 4% of the mass of that sector is removed during the 8-h heating period. As has been pointed out (Webb, 1966b), the mass deficit of stratopause equatorial regions can be replaced in three principal ways: (1) convergence of the horizontal wind field as reflected by a decrease in zonal wind velocity during the heating period. The data indicate that such a mechanism is operating; (2) subsidence of the entire upper atmosphere, in which case the 60 km level would lower by approximately 3 km; and (3) development of vertical downward circulation which would transport the required mass into the region.

The 4% mass deficit mentioned above is of the order of $10^{14}$ kg, which, over the 8-h period, yields a mass transport of approximately $10^{10}$ kg sec$^{-1}$. Clearly, then, the mass injected into the lower ionosphere by the STJ is very small compared to the mass involved in this equatorial oscillation. It seems probable that all of the mechanisms mentioned above are involved in relaxation of equatorial mass losses during the morning hours of the heat wave. The downward circulation hypothesized and presented above could take the subsidence form in one region and the convective form in another. In particular, eddy motions could mask the circulation in the mesosphere, while the great stability of the lower thermosphere could foster the development of a localized downward wind system to provide the required mass transport. Downward-directed winds are then to be expected in the lower ionosphere at low latitudes on the sunlit side of the globe on each side of the equator with maxima as illustrated in Fig.2, while an upward-directed circulation will occur between 2 P.M. and sunrise. It is important to note that these tidal circulations do not reach below the stratopause level and that they should evidence an exponential increase with height, with a scale height of roughly 8 km, except as modified by convergence and divergence.

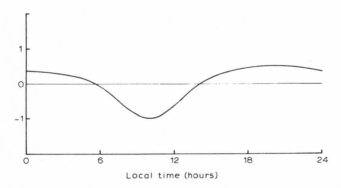

Fig. 2. Diurnal character of vertical motions across the mesopause surface at 15° latitude. The ordinate has been normalized to a nondimensional unity at the peak of the daytime downward motion.

Anomalies in latitudinal distributions of electron densities of the ionosphere in equatorial regions have been reported in the F region from data obtained by ground-based ionosounders (Rawer, 1952; Hanson, 1965) and above the peak of electron density with topside satellite probing equipment (Krishnamurthy, 1966). These data have the general character of enhanced electron density (the parameter measured) in magnetic equatorial latitudes (Appleton, 1946), centrally over the equator to 15° north and south in topside altitudes of 300 to 1000 km, and separately in each hemisphere at about 15° latitude at approximately 140 km altitude in the F region (Brace et al., 1967; Chandra and Rangaswamy, 1967). The temporal variations are diurnal in nature with maximum effects during daylight hours at midafternoon in low latitudes. In addition, marked reductions in electron density are noted in the topside data poleward from approximately 20° magnetic latitudes. These anomalous electron densities are associated with the equatorial and polar electrojets (Chapman, 1951, 1954) which operate near the 100-km level within a few degrees of the magnetic equator and in the auroral zone, respectively.

While these variations are generally considered to be tidal in origin because of their diurnal occurrence, the details of the physical processes involved are complex (Dougherty, 1961; Axford, 1963; Hanson and Moffett, 1966). The excess electrons are usually postulated to be held in equatorial regions by electric fields which have their origin in the D and E regions. Ionospheric plasma fountains with downward motions in the F region of greater than 10 m sec$^{-1}$ have been postulated to explain these diurnal electron density variations (Martyn, 1945).

It must be noted here that the very smooth picture of the tidal circulation presented above has imposed on it a large amount of detail structure. These small-scale features appear to be generated by a variety of physical processes, ranging from synoptic-scale disturbances to turbulent eddies, as well as the full spectrum of mechanical body wave perturbations. We should emphasize that a great amount of variability is a characteristic feature of all upper atmospheric data of sensitivity, and these considerations of uniformity which have been used here for clarity must not be allowed to mislead the investigator as to the true state of the medium.

## 3. ELECTRICAL PROCESSES

Motion of a gas containing a mixture of electrons and ions through a magnetic field will result in production of an electromotive force oriented in the direction of flow, as a result of differing mobilities of the charge carriers. Fejer (1965) has stated the general relation:

$$q(\vec{E} + \vec{w} \times \vec{B}) + m\vec{g} + mv\vec{v} - \frac{\vec{\nabla} p}{n} = 0 \tag{1}$$

where $\vec{E}$ is the electric field, $\vec{w}$ the velocity of the charged ($q = 1.6 \cdot 10^{-19}$C) particles, $\vec{B}$ the magnetic field, $m$ the mass of the particles, $\vec{g}$ the gravity, $v$ the collision frequency, $\vec{v}$ the velocity of the charged particles relative to the neutral gas and $\vec{\nabla} p/n$ is the pressure gradient of the charged molecules of $n$ particles m$^{-3}$. He derived the vertical component solution for the case of a vertical wind exerting a force ($F$) through collisions between neutral and charged molecules to obtain the

charged particle speed ($v$) relative to the neutral gas:

$$v = -\frac{F}{m}\left(\frac{\omega}{v^2 + \omega^2}\right) \tag{2}$$

where $\omega$ is the gyrofrequency for the particular charged particles involved. The gyrofrequency is given by $\omega = qB/m$, which indicates that the gyrofrequency of electrons will be significantly greater than that for the ions as a result of the difference in mass ($9.1 \cdot 10^{-31}$ kg for electrons and greater than $1.67 \cdot 10^{-27}$ kg for ions).

Collision frequencies are significantly different for ions and electrons. An expression for the collision frequency of the simple electron population has been derived by Nicolet (1953) to be:

$$v_e = 5.4 \cdot 10^{-10} N T_e^{1/2} \tag{3}$$

where $N$ is the number density of the gas and $T_e$ is the temperature of the electron gas. Chapman (1956) has derived the expression:

$$v_i = 2.6 \cdot 10^{-9} N m^{1/2} \tag{4}$$

for the collision frequency of ions. In both cases the collision frequency decreases with increasing altitude in accord with the usual density scale height. Collision frequency dependence of electrons on their possible non-gaussian temperature distribution in the ionosphere and ions on the mass of the parent molecule results in complex and relatively uncertain profiles in the upper atmosphere. Estimates of representative profiles are illustrated in Fig. 3; the arrows indicate the points at which gyro and collision frequencies are equal according to Fejer (1965).

These are important altitudes for our purposes since they indicate the regions in which control is shifted from the magnetic field mode above to the dynamic mode below. Thus at higher altitudes the individual charged particles would essentially be held in place by the magnetic field if a wind tended to transport them across the field; whereas at lower altitudes, collisions would disrupt the gyro motions, and the charged particles would be carried along with the wind. These considerations would then indicate that in the 70–150 km altitude region particularly, electrons would be restrained in their vertical motion with the neutral gas of the tidal circulation, while their partner positive ions would move with the neutral gas. The force producing this separation will increase with decreasing height as a result of an increasing collision frequency until the collision frequency becomes high enough to carry the electrons along with the wind also. The current which would result from this impressed electromotive force, if no limiting impedance were involved, can be determined by evaluating the difference in relative speeds ($\vec{v}$) of the electrons and an equal number of positive ions from eq. 2, and using their difference as a measure

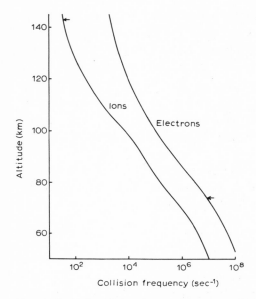

Fig. 3. Typical collision frequencies versus altitude for ions and electrons. Arrows indicate the approximate gyrofrequencies for electrons and oxygen molecules of unit charge.

of the rate of charge separation. The gravitational and pressure gradient terms of eq. 1 and 2 can be considered negligible in the 50–150 km region. In the special open circuit case where the electrons, as a result of magnetic field constraints, have a speed of $-\vec{v}$ relative to the flow and the motion of positive ions is controlled by collisions, the vertical electric field ($\vec{E}$) required to prevent further charge separation (that is, to force the positive ions also to move upstream with a speed of $v$) can be calculated approximately from the relation:

$$q\vec{E} = mv\,\vec{v} \tag{5}$$

where the gravitational and pressure gradient terms are considered negligible and the effects of charge motions in the horizontal are neglected for the moment. If a return current path is available, such a velocity of separation ($\vec{w}$) of the positive ions and electrons yields the current density ($\vec{J}$):

$$\vec{J} = nq\vec{w} \tag{6}$$

In general, in the atmosphere a circuit is neither open nor closed. The presence of mobile charge carriers affords the atmosphere a certain conductivity through a unit cross section which is defined by the field form of Ohm's law:

$$\vec{J} = \sigma\vec{E} \tag{7}$$

where $\vec{J}$ is the current density produced by an electric field $\vec{E}$, and $\sigma$ is the conductivity in mho m$^{-1}$ through a column of 1 m$^2$ cross section. The resistivity (ohm m) is thus the reciprocal of the conductivity. The specific conductivity for a direct current in an ionized gas such as the atmosphere is given by (Hanson, 1965):

$$\sigma = \frac{nq^2}{m_e \nu_e} + \frac{nq^2}{m_i \nu_i} \tag{8}$$

This equation yields the value to be used if there is no magnetic field or along the field lines.

Hanson (1965) has given eq. 9, relating the electrical parameters of a gaseous medium, such as the ionosphere, to obtain the conductivity in a direction perpendicular to a magnetic field for the direct current case (the Pederson conductivity):

$$\sigma' = \frac{nq^2 \nu_e}{m_e(\nu_e^2 + \omega_e^2)} + \frac{nq^2 \nu_i}{m_i(\nu_i^2 + \omega_i^2)} \tag{9}$$

In the presence of combined magnetic and electric fields an apparent force is exerted on ambient charged particles which is normal to the plane of the field vectors in the direction of $\vec{E} \times \vec{B}$ (the Hall conductivity). In the absence of a difference in mobilities of the charged particles, the positive and negative particles will simply be displaced together, but in general in the lower ionosphere the ions are effectively immobilized while the electrons are highly mobile, and an electric current is thus generated in the $-\vec{E} \times \vec{B}$ direction. The conductivity for such a current is given by the relation (Hanson, 1965):

$$\sigma'' = \frac{nq^2 \omega_e}{m_e(\nu_e^2 + \omega_e^2)} + \frac{nq^2 \omega_i}{m_i(\nu_i^2 + \omega_i^2)} \tag{10}$$

The above considerations form the principal basis for interaction between the neutral diurnal circulation systems of the upper atmosphere and the earth's electrical structure. There are other modes of interaction indicated by eq.1, and in specific cases they may play a dominant role. For our purposes here we will center our attention on the electrodynamic processes of interaction characterized by asymmetries in the Pederson and Hall conductivities.

## 4. ELECTRICAL STRUCTURE

The electrical structure of the atmosphere varies with time and space in a complex fashion, and thus can be represented in only a limited fashion by a particular model or set of data. While the general features of electron structure are adequately portrayed by the theory of Chapman layer formation (Chapman, 1931), it is apparent that important detail features can only be explained by significant deviations from this assumed radiational equilibrium under static conditions (Rawer, 1952; Appleton, 1959; Gibbons and Waynick, 1959; Maeda and Sato, 1959; Martyn, 1959; Nichols, 1959; Belrose, 1965a, b; Heikkila and Axford, 1965; Herman, 1966). It is thus desirable to make gross generalizations relative to the general electrical structure in order to obtain a first look at the nature of important physical processes which may occur, and then to compare the expected results of these processes with modifications of the static electrical structure which are known to exist. This procedure is reasonable as long as it is clearly remembered that a uniformity has been assumed which does not exist.

The vertical distributions of specific, Pederson and Hall conductivities for a mean midlatitude noontime atmospheric model are illustrated in Fig.4. Sea water has a conductivity of a few mho m$^{-1}$ (variable with temperature and salinity) which, as a result of the large oceanic area, determines the general electrical characteristics of the earth at sea level. Topsoil conductivity falls in the range $10^0$ –$10^{-2}$ marked "A" in Fig. 4, and thus land areas present a greater and more variable resistance to the flow of an electric current. It should be noted, however, that a smaller columnar resistance may be obtained over a particular earth-atmosphere columnar path over land due to low resistance surface elevations (mountains), the effect of which is to short some of the very high resistance paths of the lower atmosphere.

The very low conductivity of the lower atmosphere ($\sim 2 \cdot 10^{-14}$ mho m$^{-1}$) results from the high collision frequency (and thus low mobility) produced by high air density and the small number of charged particles. The conductivity increases rapidly with altitude, starting off from surface values to the order of $10^{-9}$ mho m$^{-1}$

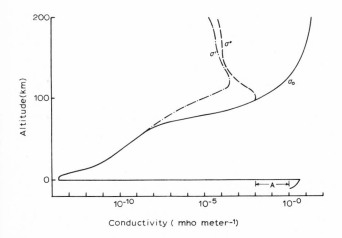

Fig. 4. Vertical distribution of the specific ($\sigma_0$), Pederson ($\sigma'$) and Hall ($\sigma''$) conductivities for typical midlatitude noon conditions after Cole and Pierce (1965) from the surface of the ocean to 100 km and Hanson (1965) from 100 km upward. The symbol $A$ refers to the range of variable conductivity of the earth's surface layers.

at the base of the D region, partly as a result of an increase in number density of charged particles. Above about 50 km, the presence of free electrons becomes important as a result of gross differences in mass and collision cross sections of the charge carriers, so that the conductivity becomes nonisotropic, with a component structure illustrated in the upper portions of Fig.4. At low latitudes the specific conductivity ($\sigma_0$) is most applicable meridionally; the Pederson conductivity ($\sigma'$) is most applicable vertically, and the Hall conductivity ($\sigma''$) is most applicable zonally.

Vertical motions produced by the tidal circulations described in section 2 will, at low latitudes of the morning and early afternoon sunlit hemisphere, result in electrical charge separation in the vertical direction, with positive charges forced downward by collison between those ions and molecules of the neutral flow when the flow is downward. Using the open circuit approximation of eq. 5 (in the absence of an effective return circuit path), with an assumed exponential distribution of speeds tied to 1 m sec$^{-1}$ at 80 km and 10 m sec$^{-1}$ at 100 km acting on ions of molecular weight 30, a vertical upward-directed equilibrium electrical field of the order of 0.066 V m$^{-1}$ at the 80-km level is obtained, decreasing in strength below and above that level with a value of 0.026 at the 100-km level as is illustrated in Fig.5. The charge separation process will thus produce an opposing vertical electric field which will rigidly maintain a positive space charge in the stratopause region and a negative space charge in the ionosphere during the morning and early afternoon. This separated charge region will exhibit a diurnal structure under control of the tidal circulation which continuously rotates around the earth at a speed of approximately 460 m sec$^{-1}$ at 15° latitude.

The same reasoning may be applied to evaluate the electrical structure that will

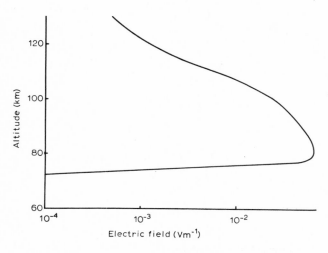

Fig. 5. Vertical distribution of the negative vertical electric field generated by downward motions of the tidal circulations.

result from the tidal upward flow which is indicated to occur (Fig.2) as the heat wave recedes locally from 2 P.M. until sunrise. Using a mean vertical circulation value of 0.4 of the morning values (from mass continuity considerations) for the late afternoon period from 2 P.M. to sunset, mean values of $E_v$ will be approximately one fourth the noontime values illustrated in Fig.5, with the principal additional difference being that the direction of the resulting electric field will be reversed. The efficiency of this source of electric field generation will decrease during the evening as a result of the diurnal decrease in electron density.

The general circulation will also produce electrical charge separation through horizontal application of the process indicated in eq.5. This mode will be globally asymmetric in that it will be most effective in the sunlit hemisphere as a result of the enhanced concentrations of electrons and positive ions. This time, assuming that the electrons of the 80-km region will remain essentially with the magnetic field while the ions will be carried along with the wind (a short circuit), nominal values of $10^9$ electron-positive ion pairs $m^{-3}$ and a wind speed of 100 m sec$^{-1}$ normal to the magnetic field, a current density ($\vec{J}$) is obtained from eq.6 to be of the order of $1.6 \cdot 10^{-8}$ A m$^{-2}$ in the direction of the wind. Using the same wind speed, a positive ion-electron concentration of $10^{10}$ at 90 km yields $1.6 \cdot 10^{-6}$ A m$^{-2}$. The latter value would remain essentially unchanged with height up to approximately the 140-km level, where the collision frequency is inadequate to carry the positive ions across the magnetic field, and the charge separation process would stop. In the case of a west wind, which corresponds to the winter season in the monsoonal circulation, the circulation-produced current will be in the same direction as the tidally produced current, so the current produced by the general circulation of winter will enhance the dynamo current in the morning circuit and decrease the intensity of the evening circuit. An east wind, typical of the summer season, will tend to move positive ions westward, and thus the current produced by the summer general circulation will oppose that produced by the morning tidal circulation and enhance the evening current. This mode of electrical structure generation will then introduce an annual variation into the intensity of the dynamo current circuits, and will tend to shift the time of day in which the maximum electrical intensity is reached.

A final mode of generation of electrical structure concerns the cross product of the general circulation and the magnetic field (the first term of eq.1). This force will be vertical so that the circuit will be largely open and the equilibrium electric field will be given approximately by:

$$\vec{E} = \vec{v} \times \vec{B} \tag{11}$$

Using 100 m sec$^{-1}$ and $0.34 \cdot 10^{-4}$ Wb m$^{-2}$ an equilibrium electric field of 0.0034 V m$^{-1}$ is obtained. This electric field will have the same orientation as the tidally produced field during morning hours at low latitudes when the circulation is easterly, but will diminish that tidally produced electric field in the case of winter westerlies.

Juxtaposition of electric and magnetic fields in an ionized gas will result in imposition of apparent forces on charged particles in a direction normal to the plane of the field vectors in the direction of the cross product of the electric and the magnetic vectors (Fejer, 1965). In the D and E regions the ions are immobilized (relative to motions through the gas) by collisions so that the electrons are the principal carriers of this "Hall" current. The velocity of electron motion (unhampered by collisions) constituting such a current is given by eq.12:

$$\vec{v} = \frac{\vec{E} \times \vec{B}}{B^2} \tag{12}$$

Using values of $\vec{B} = 0.34 \cdot 10^{-4}$ Wb m$^{-2}$ directed northward, and $\vec{E} = 0.066$ V m$^{-1}$ (Fig.5) directed upward as obtained during the morning hours (Fig.2), a westward velocity of electron motion of approximately $2 \cdot 10^3$ m sec$^{-1}$ is obtained for the 80-km altitude.

The current density carried by such a system can be calculated by use of eq.6 which yields an 80-km value of $2.6 \cdot 10^{-7}$ A m$^{-2}$, when a concentration of $10^9$ electrons m$^{-3}$ is used with the motion derived above. Similar calculations at other altitudes indicate that the vertical distribution of the midday Hall current of low latitudes is of the type illustrated in Fig.6.

The motion of electrons constituting the Hall current will not be free of impedance in the lower ionosphere. Collisions with neutral molecules will transfer heat from the current to the environment, and the collisions plus interaction with

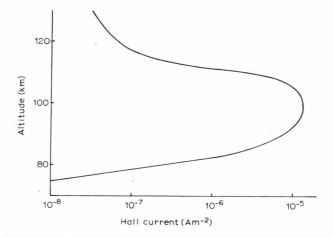

Fig. 6. Hall current produced at low latitudes by an average tidal downward motion of 1 m sec$^{-1}$ at 80 km altitude as the heat wave approaches. The direction of current flow is to the east. (Reproduced from Webb, 1968.)

the magnetic field will produce a drop in electrical potential along the current path. As is indicated by the Hall conductivity curve of Fig.4, the resistivity to this zonal dynamo current will vary from a gross value of $1.7 \cdot 10^5$ ohm m at 80 km to a minimum of $10^2$ at 100 km, and increases to another high value of $10^4$ at 130 km altitude. Under these circumstances, the maximum specific electrical potential drops resulting from the Hall current will be found near the base and top of the current region, while the potential drop along the centerline of the current (100 km) will be minimum.

Now the electric field associated with the Hall current can be calculated from eq.7, using a value of $6 \cdot 10^{-6}$ mho m$^{-1}$ obtained from Fig.4. This calculation yields an eastward-directed horizontal electric field at 80 km of 0.043 V m$^{-1}$ which, over a longitudinal span of $1.5 \cdot 10^7$ m over which the average downward flow is assumed to occur, indicates a potential difference of 650,000 V between sunrise and early afternoon along the longitude circle at 15° latitude. Calculations for other altitudes indicate the vertical distribution of horizontal electric field illustrated in Fig.7. The principal region of electric potential field is thus indicated to be near the 80-km level where the electromotive force is being generated, with potentials of the E region a full order of magnitude less. These potential gradients in the horizontal mean that the ambient electric fields which produce the Hall current are no longer vertical (as were generated by the tidal motions), and are directed so that the transport of charges abets the fundamental charge separation process produced by the vertical tidal motions. The circulation-generated portion of the dynamo current will reverse direction seasonally with the monsoon and should have a small influence on the upper E region electrical structure where the tidally produced structure is small and

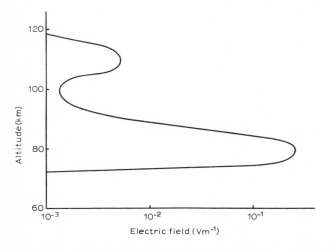

Fig. 7. Horizontal electric fields in the lower ionosphere derived from a downward tidal motion of 1 m sec$^{-1}$ at 80 km altitude. The field is toward the east.

at other locations where the dominant tidally produced electrical structure is reduced. These considerations lead to the conclusion that in a longitudinal belt at low latitudes ($\sim$ 15°) there will be a region of electromotive force (e.m.f.) generation, with a potential difference of the order of $6 \cdot 10^5$ V in the D region between the early morning sector and the early afternoon sector for each meter per second of average vertical downwind of the tidal circulation at 80 km altitude. The direction of this potential gradient is such that the early morning sector ($B$) is depressed in potential relative to the early afternoon sector ($A$). (See Fig.8.)

In the afternoon and evening, the tidally generated electric fields which generate the Hall currents will reverse direction and be oriented downward. The Hall current will then flow westward after 2 P.M. The electron density of the E region decreases throughout this period to make the Hall current generating process ineffective during the late night-time (smaller than at noontime by two orders of magnitude). It is clear, then, that the sunset dynamo circuit is weaker than the morning circuit.

Flow of these dynamo currents in the D and E regions (75–140 km) of the atmosphere will result in development of a diurnal pattern of electrical potential distribution in that region. The above analysis indicates that the maximum electrical potential will be located near 80 km altitude, geographically near the point marked $A$ in Fig.8 (near 2 P.M. at low latitudes). Minimum potential should be located near the point of Fig.8 marked $B$. Significant diurnal differences will be induced in the hemispheric dynamo currents even at equinox times as a result of asymmetries between the rotational and magnetic axes, and as a result of equatorial separation of the vertical tidal circulations over the rotational equator we can expect the existence of a reduced potential between the low latitude e.m.f. regions of the two hemispheres. More substantial hemispheric differences will develop as the subsolar point moves away from the equator, with the summer hemisphere being favored with a stronger tidal circulation and enhanced electron concentrations, and thus with a stronger dynamo current and greater potential differences. All of these factors indicate that potentials in equatorial regions will be different between hemispheres for similar geomagnetic latitudes, and make it probable that trans-equatorial currents will flow between the hemispheric e.m.f. zones, probably from the summer toward the winter hemisphere.

Such currents will flow principally along magnetic field lines, so the specific conductivity (solid line) of Fig.4 would be applicable. These currents would start near the 100-km level of one hemisphere on a magnetic field line, progress upward in a divergent field to cross the magnetic equator at a few hundred (and/or thousand) kilometers altitude, and converge back to the 100-km level at the same low latitudes of the other magnetic hemisphere. This low latitude interhemispheric dynamo current circuit is in the location of the inner Van Allen radiation belt (Van Allen, 1959) and presumably provides the basic energy for that phenomenon. The inner radiation belt electric current path then serves to smooth inequalities in the

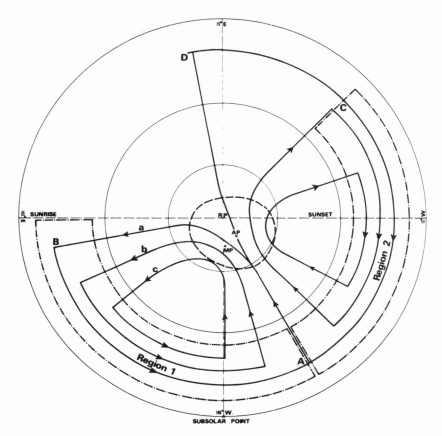

Fig. 8. Equatorial projection of the dynamo current system. Regions 1 and 2 (enclosed by dash-dot curves) represent regions of electromotive force generated by vertical tidal motions. Positions of the rotational (*RP*), magnetic (*MP*) and auroral (*AP*) poles for the Northern Hemisphere are indicated. The point of highest electric potential is at *A*. The dashed circular curve represents the centerline of the auroral zone. (Reproduced from Webb, 1968.)

hemispheric tidally produced e.m.f. regions. There should be marked diurnal variations in the basic currents flowing in the region of the inner radiation belt. Some of the electrons and ions taking part in these exospheric currents will, as a result of their favored initial trajectory angles and thermal energies, be trapped in the magnetic field for short or long periods. These trapped particles will drift longitudinally (electrons drifting to the east and positive ions to the west) along L-shells (Heikkila and Axford, 1965) to envelop the entire low latitude global region in a few tens of minutes, in the case of the higher energy particles. Thus, the daytime dynamo regions supply the inner Van Allen radiation belt with ions and electrons, which then interact with the night-time upper atmosphere of middle and low latitudes as they move from one hemisphere to the other or as they drift around the globe, resulting in enhanced ionization and heating in low and middle latitudes of the night-time D and E regions, and forming a ring current in the upper ionosphere which serves to reduce the meridional component of the earth's magnetic field.

In order for these hemispheric generators of dynamo currents in the lower ionosphere to be effective, there must be return current paths in the lower ionosphere. Such paths will be minimum resistance paths from the high to low potential regions. The very low conductivity which results from a high collision frequency of ions in the lower D region will limit electrical currents which could constitute return flows at altitudes below the e.m.f. zone, and rigidity of the magnetic field will preclude an effective zonal electron current flow above. An important return current path is available in the meridional direction, where at least part of the path can be traversed at an enhanced conductivity (the specific conductivity) along the magnetic field lines. At the magnetic equator the meridional conductivity would be essentially the one indicated by the solid curve of Fig.4, decreasing with increasing magnetic latitude until in polar regions the meridional conductivity will be reduced to essentially the Pederson values, which are representative of those regions.

TABLE I

Columnar resistivities $(R)$ and fractional parts $(F)$ for paths a, b and c of Fig. 8; columnar resistance units are $10^{10}$ ohm m$^{-2}$

|             | a   |      | b   |      | c   |     |
|-------------|-----|------|-----|------|-----|-----|
|             | $R$ | $F$  | $R$ | $F$  | $R$ | $F$ |
| Poleward    | 1.2 | 0.31 | 1.2 | 0.33 | 1.0 | 0.5 |
| Auroral     | 0.7 | 0.25 | 0.5 | 0.20 |     |     |
| Equatorward | 0.9 | 0.44 | 0.9 | 0.47 | 1.0 | 0.5 |
| Total       | 2.8 |      | 2.6 |      | 2.0 |     |

Estimated columnar resistivities along three legs of three selected paths in Region 1 illustrated in Fig.8 are tabulated in Table I. Also listed are the fractional parts $(F)$ of the total path resistance which are contained in the separate path segments. These values indicate that gross lateral differences in electric potential exist in the lower ionosphere on a global scale. Assuming that the dynamo current is conservative (no convergence or divergence) along the path "$a$", the potential will fall from its highest value at point $A$ (say $10^6$ V) to 0.69 (see Table I) of that value ($0.69 \cdot 10^6$ V) at the auroral zone. Across the auroral zone the current will not be confined to the 100 km region, but in any case the IR drop will result in a potential fall of an additional 0.25 of the total to $0.44 \cdot 10^6$ V, and over the equatorward leg of path "$a$" the potential would fall to its lowest value at $B$. The sunlit hemisphere of the E region is thus always at an elevated potential relative to the night-time region, with a maximum near 80 km altitude at low latitudes near 2 P.M., while the hemispheric minimum is near 4 A.M. at low latitudes. It is clear that such a structure in the potential field of the base of the ionosphere will have important implications for the boundary regions above and below the D and E regions in which this basic phenomenon is located.

The basic mode of interaction between the ionospheric dynamo current and the earth's magnetosphere is illustrated in Fig.9 and 10. The situation is illustrated for the equinox case, with the sun's rays roughly normal to the plane containing the rotational and magnetic axes. In Fig.9 the subsolar point is located over 160° W, and in Fig.10 it is over 20° E, just twelve hours later. The projections of the dynamo current circuits of each hemisphere are idealistically represented by dashed curves. It is obvious that the electric potential of the lower ionosphere considered above will not generally be equal at each end of any high latitude geomagnetic equipotential

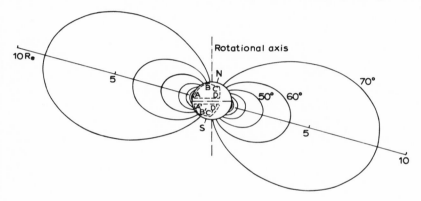

Fig. 9. Equinoctial cross section of the earth's geomagnetic field when the sun is over 160°W longitude. The dashed curves represent projections of hemispheric circuit elements of the tidally generated dynamo current. (Reproduced from Webb, 1968.)

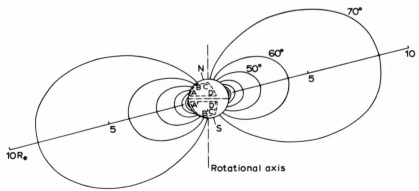

Fig. 10. The same situation illustrated in Fig. 9 except that the sun is over 20°E longitude. (Reproduced from Webb, 1968.)

line, even if the interhemispheric inner Van Allen radiation belt currents do maintain the hemispheric e.m.f. zones at essentially equal potential. In the case illustrated in Fig.9 the northern hemisphere magnetic field lines at sunrise (on the left) will be at a higher electric potential (in auroral zones of the order of $10^4$ V for each meter per second of average downward tidal wind) than will the southern hemisphere ends of those magnetic field lines (for example, compare $B$ and $B'$). These voltages are quite high when considered in the light of the impedance of the magnetic field line paths, which might have a total columnar resistance of the order of $10^5$ ohm m$^{-2}$ from the 100-km levels between hemispheres at 70° geomagnetic latitude. Large currents of the order of $10^{-1}$ A m$^{-2}$ can be expected to flow from one hemisphere to the other along the magnetic field lines (Somayajulu, 1964; Blake et al., 1966; Mozer and Bruston, 1966) limited principally by the reduced conductivity of the field lines in the upper atmosphere near the dynamo level and the mechanics of individual charge motion in the exosphere along the magnetic field lines.

It is informative to look at the situation twelve hours later, as is illustrated in Fig.10. In this case the northern hemisphere ends of the magnetic field line ending at $B$ are at a lower electric potential then those same field lines at $B'$ in the southern hemisphere. These considerations indicate that there must be a diurnal variation in all of the phenomena that result from this interhemispheric electrical difference which basically is a result of asymmetries between the rotational and magnetic axes. That is, the exospheric current will flow from $B$ to $B'$ in Fig.9, and it will flow from $B'$ to $B$ in the case illustrated in Fig.10.

In addition, it is clear that at the time a current is flowing from $B$ to $B'$ in the case illustrated in Fig.9, a current will also be flowing from $C'$ to $C$ in the early afternoon sector. A possible current circuit is then along the 70° magnetic latitude field line from the northern hemisphere auroral zone ($B$) out through the earth's near space and back to the southern hemisphere auroral zone ($B'$), eastward along that auroral zone to the point $C'$, and then back to the northern hemisphere auroral zone ($C$) along the 70° magnetic latitude field line which projects toward the sun in the early afternoon sector. Since we are here considering only an extreme simplification of the actual dynamo current distribution, it is clear that the longitudinal high latitude currents termed auroral electrojets (Davis and Sugiura, 1966; Gottlieb and Fejer, 1967) in the high conductivity E region of auroral zones will play important parts in the global electrical current circuits.

The exospheric current path discussed above is representative of the principal exospheric current paths which must exist. That is, in the sunlit hemisphere over the primary dynamo circuits the potential differences between hemispheres will be maximum, and the conductivities required to transport the currents will be maximum. This does not preclude the flow of currents in the night-time exosphere, but surely should result in gross diurnal changes in character of the currents. Drifts of electrons (eastward) and positive ions (westward) which become trapped (a small fraction of the particles participating in the interhemispheric currents) would populate the night-time outer Van Allen belt, and the precipitation of both types of particles in the night-time auroral zones would provide enhanced conductivity so that longitudinal currents would flow in the night-time auroral zones, completing additional exospheric current circuits. These processes, along with the known diurnal expansion and contraction of the exosphere through the permanent magnetic field (Harris and Priester, 1962; Jacchia and Slowey, 1964) and variations in the dynamo electric potentials, could be the mechanisms which result in buildup of a stored component of hard radiation (Vestine, 1954) and subsequent preferred dumping in the night-time region. The increased conductivity in night-time auroral zones caused by precipitation of these particles would then provide additional current paths through high latitudes from the nearest high conductivity region of the daytime sector. The division of the high latitude portion of the current circuit marked "a" in Fig.8 between the two dynamo current systems of early morning and late afternoon would then become more complex in the night-time auroral zone, with the switch between control by the two dynamo systems occurring in the early morning hours near 2 A.M. Observational data on auroral characteristics during this period do indeed indicate a dramatic change (Davis, 1960).

Many of the observable electromagnetic phenomena exhibit semiannual variations with maximums at equinox times. An understanding of the reason for this pronounced cyclic variation is essential for any general concept of electrical and magnetic structure. The point is well illustrated in Fig.11. A cross section is presented here which roughly includes the rotational axis, the magnetic axis and the sun at solstice time in the northern hemisphere. The most obvious fact is that all of the high latitude magnetic pole in the southern hemisphere is denied the ionizing solar ultraviolet radiation during a part of the diurnal cycle, and only the lower portion of the auroral zone magnetic field lines are connected when maximum opportunity is afforded twelve hours later than the case illustrated in Fig.11. In general, lack of solar ultraviolet radiation means that the D and E regions of the ionosphere will have roughly two orders of magnitude less in electron concentration. This, coupled with the failure of the dynamo current to reach high latitudes in the winter hemisphere, limits the hemispheric potential differences and the flow of current into the higher latitude magnetic field lines of the outer Van Allen belt and thus limits the occurrence of phenomena which depends on this mode of operation. This does not necessarily mean that less total interhemispheric current flows, but simply that the current must follow other paths, or accomplish it in a way which is less efficient in generating observable auroral and high-latitude magnetic phenomena. For instance, it is well known that ionospheric electron densities are noticeably high in middle latitudes of the winter hemisphere (the winter anomaly) when compared to the calculated values obtained from static solar-atmospheric interactions.

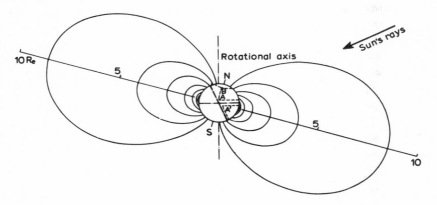

Fig. 11. Cross section of the geomagnetic field in the plane of the sun's rays when the sun is over 70°W longitude. The heavy dashed line marks sunrise, and the light dashed line indicates an element of the dynamo circulations and the northern auroral zone projected on the above plane.

The high geomagnetic latitude mode of hemispheric interaction between the dynamo current circuits occupies the exospheric location of the outer Van Allen (1959, 1961) radiation belt. Thus, the basic control and power for the outer radiation belt appears to center in high latitude potential differences and variations of the primary dynamo current circuits between hemispheres. Interaction of the earth's magnetosphere with the solar wind will complicate the current system involving the outer radiation belt. These exospheric currents may, on occasion, result in gross modifications of the potential distributions along the primary dynamo circuit. For instance, if the inner radiation belt current should produce a net flow from low latitudes of one hemisphere to the other, the outer radiation belt is very likely to constitute the return path of that current. Such a current path could be illustrated by a flow from point $A$ (Fig.9) of the northern hemisphere through the inner Van Allen belt to point $A'$, meridionally to the auroral zone at $B'$, exospherically to $B$ and meridionally back to $A$. A large number of similar parallel circuits are immediately obvious.

The known presence of auroral activity in high magnetic latitudes of both hemispheres will result in significant modification of the simplified picture of the dynamo currents presented in Fig.8. Measurements of precipitation electrons (Gledhill et al., 1967) and protons (Sharp et al., 1967) in auroral zones indicate the presence of currents along the magnetic field lines of the order ($10^{-5}$ A m$^{-2}$) of those calculated for the dynamo current. Much larger fluxes of energetic electrons have been noted in middle latitudes (Paulikas et al., 1966), and there is reason to expect the precipitation of protons in these regions (Prag et al., 1966). Brace et al.

(1967) have suggested such a source of precipitating particles for maintenance of observed high electron (and possibly ion) temperatures in the ionosphere at night, and the presence of airglow throughout the night would indicate some such energetic flux. In addition, Brace et al. (1967) have shown the thermal structure at the 1000-km level which requires special sources and sinks of heat. All of these observations can be considered to be in general agreement with heating and ionization effects of important interhemispheric currents flowing along the magnetic field lines.

The electrons and positive ions in these interhemispheric current flows, particularly in the outer belt circuit, will attain high speeds in the electric potential drop (of the order of $10^4$ V for each meter per second of the average tidal vertical motion) as they traverse the earth's near space along the magnetic field lines. This energy will be dissipated in collision processes as the individual charged particle mode of current flow is interchanged for the usual conduction mode. Hot electrons, ionization and photoemission will be expected results in the E and F regions along the paths near terminal points of these exospheric current circuits. Examples of observed ionospheric thermal and kinetic structure which can now possibly be ascribed to joule heating by these electric currents have been published by Schilling and Sterne (1959), Harris and Priester (1962), Jacchia and Slowey (1964), Banks (1967), Becker, (1967), Cook (1967), Cummings and Dessler (1967), Dalgarno et al. (1967), Gleeson and Axford (1967), Jacchia et al. (1967) and Nagy and Walker (1967).

Superimposed on the hypothetical current pattern of the dynamo circuit which is illustrated in Fig.8, there will be the above described regions of convergence and divergence in the dynamo current system which will involve a comparatively large exospheric current flow into and out of the horizontal primary dynamo circuit from above. The geometry and intensity of these currents will be determined by interaction between the earth's magnetosphere and the solar wind, solar-induced variations in the electrical conductivity of the dynamo region and through variations in the tidal motions which serve as the basic source of power for the dynamo current systems. The first two controls represent the well-known solar influence through particulate and wave emissions incident on the electrical and magnetic structure of the earth, while the last source of control represents a new quasi-independent neutral atmosphere mode of establishing the electrical and magnetic structure of the atmosphere. Observational data point very clearly to the fact that the smooth character of the curves of Fig.8 is a gross simplification, and that a principal feature of the stratospheric tidal circulation and ionospheric conductivity is a strong variability. The variability of the earth's electrodynamic structure, which is known through observations of particular events, is then a composite result of variable inputs from the atmospheric circulation and the solar impact on this dynamic medium.

The above described production of a Hall current system in each of the hemispheres through vertical tidal motions at low latitudes can be expected to result in certain complications in equatorial regions. This is true because the rotational and magnetic equators do not coincide and further because the magnetic intensity along the magnetic equator is variable. Thus, when the magnetic equator is at low latitudes of the northern hemisphere, such as in the Indian Ocean, the electrical structure of the northern hemisphere should be intensified, because the Hall effect will be most effective where the tidal downflow of low latitudes is most nearly normal to the magnetic field. By the same token, maximum electrical intensity should shift to the southern hemisphere when the magnetic equator is south of the rotational equator, as will be the case in the South American region. There should be, then, a marked longitudinal variation in the intensity of the interhemispheric currents.

Splitting of the tidal circulations on each side of the rotational equator may well provide a structure in which the vertical winds over the equator are quite small at the 80-km level. The tidally produced electrification would then be low in that region, and the other modes of electrification might become significant. In particular, advective transport of positive ions across the magnetic field by the general circulation should first be noted. If such were the case, certain results should be clearly observed; namely, an easterly circulation over the rotational equator should serve to diminish the characteristic eastward-directed dynamo current, while a westerly circulation should enhance that current.

Just such effects have been observed by Cahill (1959) in measurements of the equatorial electrojet (Chapman, 1951; Balsley, 1965; Maynard and Cahill, 1965; Gassmann and Wagner, 1966; Sugiura and Cain, 1966; Ogbuehi et al., 1967) with a rocket-borne magnetometer. On 17 August 1957, at 1359 165th meridian time, at 159° W longitude and 3° N latitude, a weak westward-flowing current was observed with the base at 104 km. On 18 October 1957, at 1356 near the same location, a strong eastward-flowing current was observed at approximately 97 km, and on 19 October 1957, at 0907, a much stronger eastward-flowing electrojet was noted, based just above 90 km. These data would indicate that the monsoonal circulation is indeed significant in this region, since it is known that the upper atmospheric circulation has easterly winds at that location in August and westerly winds in October (Webb, 1966a, fig.4.21). The equatorial high altitude circulation exhibits a semiannual variation, with easterly winds during the period from mid-May to mid-August and mid-November to mid-February, and westerly winds during the intervening periods. Cahill's results would then indicate that the circulation transport mode of charge separation is effective, but is smaller than the tidal mode except immediately above the equator, so that the circulation mode simply modulates the total electrical structure in other regions, making the total intensity of the dynamo current circuit greatest where the circulation is westerly. The strength of the tidal mode may vary in some special way, so the electrical structure resulting from these two processes may well vary in a more complex fashion than is indicated here.

The role of thunderstorms in atmospheric electrification has been the subject of considerable discussion over the years. They have been assigned roles which range from the original source of electromagnetic force for all electrification effects in the lower atmosphere, to a strictly passive role in the earth's electrical structure. There is no firm evidence to support differentiation between these two extremes. It is true that, whatever the cause or purpose of thunderstorm electrification, the physical processes and particularly the lightning discharges associated with these storms provide a low impedance path across the lower atmosphere, and thus provide a partial short across the lower troposphere where maximum atmospheric impedance to vertical flow of electric current exists. Assuming continuous lightning discharge paths across the lower 10 km of the atmosphere, the columnar resistance between the earth's surface and the ionosphere is reduced from a nominal value of $5 \cdot 10^{16}$ ohm m$^{-2}$ to the order of $5 \cdot 10^{15}$ ohm m$^{-2}$, and may well be further reduced at higher altitudes (Cole et al., 1966). Thunderstorms thus provide a process which results in the earth's surface assuming an electrical potential almost equal to that of the ionospheric dynamo current region above the thunderstorms and providing an additional current path for return flow of the dynamo currents.

Thunderstorms exhibit a maximum occurrence at low latitudes, at about 1900 local time over land, and shortly after midnight over the oceans. The thunderstorm short-circuits are then postulated to tie the earth's surface in those regions to the dynamo circuit potentials above those regions. Negative charge is known to be transported downward by lightning discharges in an amount approximately equal to the positive charge which is constantly transported downward by diffusion to the earth's surface by the fair-weather electric field. This particular leg of the dynamo electrical circuit carries a total current of approximately 1500 A, which is of the order of 1% of the current carried by the primary dynamo circuit. Tropospheric electrification is thus indicated to be a result of the dynamo electrification, possibly locally in particular cases, supporting or opposing the operation of the primary dynamo e.m.f., but always responsive to it.

The picture of tropospheric electrification drawn above attributes the observed effects to the dynamics of a current system, which means that the current collected by the earth in fair-weather regions must flow through the crust of the earth to the thunderstorm locations in which the current flows back upward into the ionosphere. Measurements of telluric currents in the earth's crust do indeed indicate a systematic flow of electric current from the poles toward the equator during the afternoon and from the equator toward the poles at night (Chapman and Bartels, 1940). This would indicate that the early evening maximum in thunderstorm occurrence is of prime importance in establishing the difference in potential between the E region and the earth's surface.

The thermal structure of a medium is modified by the flow of an electric current (Cole, 1962; Kato, 1962). In the direct current case, the power dissipated by the current is given by:

$$H = I^2 R \tag{13}$$

in J m$^{-3}$ sec$^{-1}$. In the primary dynamo circuit where the Hall current is generated, the peak current of $10^{-5}$ A m$^{-2}$ (Fig.6) with an impedance of $10^2$ ohm m (Fig.4) will yield a heat deposit of approximately $10^{-7}$, or 1 erg m$^{-3}$ sec$^{-1}$.

Exospheric electric currents will be characterized by two modes of electric energy dissipation. The first will be standard $I^2 R$ conversion of eq. 13 to thermal energy in the upper atmosphere immediately above the 80-km level, although the sharp reduction of resistivity with altitude along the magnetic field lines will significantly reduce the heat deposited by these currents. The exponential decrease in heat capacity of the upper atmosphere with height will far exceed the divergence of the magnetic field lines, however, so that as long as the charged particles experience sufficient collisions to exchange their energy to the neutral molecules as thermal energy the temperature of the upper atmosphere should increase with height. When the trajectories of the charged particles are free of collisions, they will accumulate a speed consistent with the difference in potential between the ends of the magnetic field lines along which they are traveling. It is important to note that the dynamo current will not be a steady current, but will be characterized by a high degree of variability. These variations in electric field strength can provide for acceleration of these charged particles, which enter the exosphere with sufficient thermal energy, in addition to their static acceleration in the electric field, to become a part of the hard trapped radiation.

## 5. D AND E REGIONS

The earth's atmosphere may be characterized electrically as a semiconducting medium permeated by a relatively strong magnetic field. The ions and electrons which represent the electrical composition of the atmosphere are influenced in their electrical role through collision interaction with neutral constituents of the atmosphere. A heterogeneous atmospheric electrical structure results, with high impedance electrical phenomena most obvious in the dense lower atmosphere and large current flows a principal characteristic of the upper atmosphere. The D and E regions of the lower ionosphere represent the zone of interaction between these markedly different electrical processes. This interaction zone then provides that element which forms the electrical structure of the earth's atmosphere into a unified global electrical system.

A model low-latitude electrical conductivity structure of the earth's atmosphere is illustrated in Fig.12. The earth exhibits a wide range of specific ($\sigma$) electrical

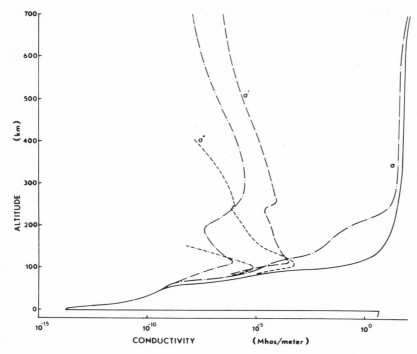

Fig. 12. Model vertical specific ($\sigma$), Pederson ($\sigma'$) and Hall ($\sigma''$) conductivity profiles for low latitudes. The labeled curves are for noon-time and companion dotted curves for night-time. These curves are based on Cole and Pierce (1965) below 100 km and Hanson (1965) in the ionosphere (at sunspot maximum).

conductivity, with highest values found in low-latitude ocean areas. These rather good conducting surfaces at the base of the atmosphere are matched above in the F region of the ionosphere at about 300 km altitude by another highly conducting layer. The semiconducting lower atmosphere is sandwiched between these conductors. The high impedance of the lower atmosphere results from high collision rates with neutral particles. A reduction.in conductivity is observed in the lower ionosphere in those directions normal to the ambient magnetic field as is indicated by the Pederson ($\sigma'$) and Hall ($\sigma''$) conductivity profiles.

Early models of the earth's electrical structure generally employed the idea that an acceptable model could be characterized by a spherical condenser system, with a conductor at the earth's surface and another concentric conductor in the lower ionosphere. The spacing between conducting surfaces in these models has varied with time, generally decreasing as our knowledge of the electrical structure of the atmosphere has increased, until today the upper surface is usually placed somewhere in the 30–70 km altitude range. A principal difficulty associated with these models concerns the fact that the atmosphere is not, in accordance with their assumptions, devoid of sources and sinks of electrical energy at any level.

Early modeling of exospheric electrical structure generally involved the same basic assumptions; in this case it is assumed that ambient electrical fields were negligible. Impedance to particle motion between hemispheres along the earth's magnetic fields was generally assumed to be zero, and thus the potential fields of the ionosphere of the two hemispheres were essentially symmetric. Such assumptions have a certain merit for problems in the outer exosphere, where Coulomb interactions are less pronounced, but these assumptions lead to gross errors in the lower ionospheric altitudes where regions of high charged particle concentration and greater collision rates are involved (Ossakow, 1968).

The D and E regions of the lower ionosphere represent that interaction zone where weak electric currents and high potential fields of the lower atmosphere must be converted to the large currents and low potential fields of the ionosphere and exosphere. The electrical structure of this important region of the atmosphere is of prime interest today, since it is now apparent that the arbitrary separation of the lower and upper atmospheres from an electrical standpoint may lead to gross misunderstanding of the overall global electrical structure.

The electrical structure of the D and E regions is extremely complex and difficult to observe. In the latter case this is so because commonly available electrical techniques for sounding the atmosphere, which operate well at high altitudes, are rather difficult to apply in this intermediate zone. The same difficulty applies for commonly employed meteorological sensors used to explore the lower atmosphere. Adequate exploration of this intermediate zone requires application of experimental techniques which are both difficult and expensive, and thus the amount of data available for analyses is apt to be meager for the immediate future. This lack of experimental data would not be so crucial if some simple theory of electrical structure could be applied to this intermediate region. Interactions between a moving semiconducting atmosphere and the earth's magnetic field produce observed ionospheric currents, and assure that the electrical structure of this intermediate region will be highly dynamic in nature. Such dynamic processes have only recently been envisioned, and today a great deal of clarification remains to be accomplished.

The observational problem has improved in recent years through the upward expansion of synoptic meteorological information provided by the Meteorological Rocket Network and through downward extension of electrical probing systems. The data thus far obtained clearly show that this intermediate zone of the atmosphere is the most active and interesting region from an atmospheric physical point of view. With these first understandings of the electrical processes that shape the structure of this region, a new concept of a unified electrical structure of the entire earth's atmosphere is already beginning to emerge (Webb, 1968). The relative significance of the diverse electrical phenomena which have been observed and studied in the past is becoming clear as their contributions toward a unified global electrical structure are now being realized.

## 6. ATMOSPHERE CONDUCTIVITY STRUCTURE

Under the impulse of the earth's net captive negative electrical charge, the atmosphere in fair weather is characterized by a geocentric electrical potential field. The atmospheric electric currents motivated by this electric potential are thus directed downward in a radial fashion. This simple structure is highly modified in the vicinity of thunderstorms, but generally quite accurately represents the atmospheric electrical structure of fair weather which has been observed to date.

As is obviously the case in the vicinity of thunderstorms, the presence of any sources or sinks of atmospheric electrification will generally result in complication of the symmetrical electrical structure which is observed in the lower atmosphere. It is well-known from magnetic field measurements (Chapman and Bartels, 1940) that sources and sinks of atmospheric electrification exist in the upper atmosphere, and thus it is to be expected that the simple fair-weather electrical structure of the lower atmosphere will become more complex with increasing altitude.

It is informative to inspect the earth's atmospheric spherical electrical condenser system for paths of least resistance along which electric currents will flow if special potential differences are imposed upon the atmosphere. A model conductivity structure for this purpose over the first 140 km of the atmosphere is illustrated in Fig. 13. Very low conductivity of the order of $10^{-14}$ mho m$^{-1}$ are characteristic of the fair-weather surface layer over the entire globe, while at the 100 km altitude region more than two orders of magnitude variation are observed diurnally, with conductivity values in the $10^{-2}$ to $10^{-5}$ range. Thus, bulk resistivities of the atmosphere range from approximately $10^{14}$ ohm m near the surface to low values of 100 ohm m near noon at the 100 km altitude level.

If some physical process results in separation of electric charge such that the positive charges are at a particular altitude, say at the northern hemisphere auroral zone and the negative charges are at that same altitude at the opposite auroral zone, it is possible to estimate the effective path resistances over the various available paths for relaxation of this separated charge. One extreme case is that the current could flow down to the earth's surface, through the earth and up to the conjugate point. It is immediately obvious that for only a few kilometers of altitude this path will involve total path resistances of the order of $10^{17}$ ohm.

Another path of interest would involve flow around the globe on the geocentric shell on which the charges are located. Approximate calculations of these path resistances yield values indicated in Table II. It is clear that above 40 km altitude path resistances around the globe will be less than those down through the higher

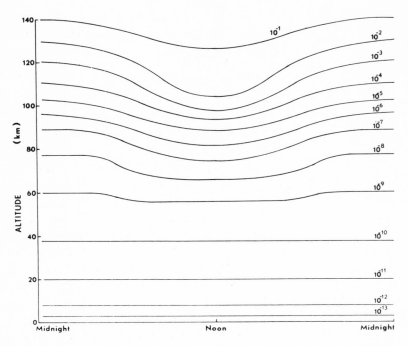

Fig. 13. Zonal distribution of specific conductivity ($\sigma$) in mho m$^{-1}$ at low latitudes.

resistance lower atmosphere and up on the other side. In fact, the lowest resistance path above a few kilometers altitude will be upward to the 100-km level, around the globe and down on the other side.

Clearly, then, potential distributions in the upper atmosphere may result in extensive electric current flow patterns which do not involve the lower atmosphere to an appreciable extent.

TABLE II

Approximate path resistances between auroral zones and specified altitudes

| Altitude (km) | $\sim R$ (ohms) |
|---|---|
| 10 | $5 \cdot 10^{18}$ |
| 20 | $1 \cdot 10^{18}$ |
| 30 | $3 \cdot 10^{17}$ |
| 40 | $1 \cdot 10^{17}$ |
| 50 | $2 \cdot 10^{16}$ |
| 60 | $1 \cdot 10^{16}$ |
| 70 | $1 \cdot 10^{15}$ |

## 7. THE DYNAMO CURRENTS

The dynamo currents of the lower ionosphere have been studied extensively through analysis of surface diurnal variations ($Sq$) in the earth's magnetic field (Chapman and Bartels, 1940; Matsushita, 1965). While this observational technique has been most informative on the general global structure of the dynamo currents, it has a serious shortcoming in that it sums all currents above the earth's surface. This technique does not provide for resolution of complexities in upper atmospheric current systems, and yields a net value which may be very misleading as far as the actual electrical structure of the upper atmosphere is concerned.

A model vertical zonal structure of current density in the northern hemispheric electromotive-force-generating zone at 15° latitude (see Fig. 6 and 8) is presented by the curves in Fig.14. It is important to note that the current is flowing up the potential gradient in the "dynamo" region above approximately 75 km altitude.

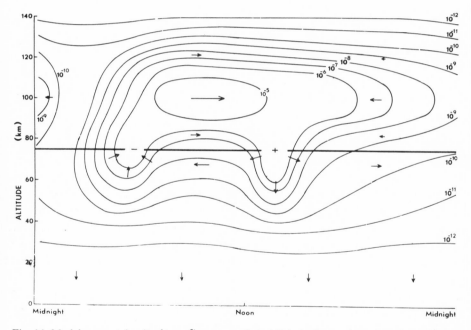

Fig. 14. Model current density (A m⁻²) structure in a 15° latitude zonal cross section. The heavy line denotes a transition region separating the dynamo region above and the relaxation currents below. Arrows indicate direction of current flow.

While the Hall effect imposes this rigidity of electrical structure in this 75 to 140 km altitude region, those regions below the dynamo levels are not so constrained. That is, in addition to the lateral return current paths (Fig. 8) which are observed in the 100-km region, return currents may be expected to flow westward underneath the dynamo region. The general structure of this return current system is illustrated by the curves in Fig. 14. The structure of the transition region between the dynamo-generating region and the relaxation current region is unknown, so it is simply indicated here with a heavy line.

The dynamo current magnetic field variation at the earth's surface will then be reduced below the actual value generated in the dynamo region by the amount of this return current system. This reduction in $Sq$ is of small consequence for most purposes, but because of the greater resistances of the lower atmosphere, is of great import for the general electrical structure.

In particular, those return currents ($\sim 10^{-8}$ A m$^{-2}$) beneath the dynamo currents ($\sim 10^{-5}$ A m$^{-2}$) are large relative to the tropospheric current system ($\sim 10^{-12}$ A m$^{-2}$). Based on these model structures of conductivity (Fig. 13) and current density (Fig. 14), the electrical potential structure may be derived for this zonal vertical plane, and this structure is found to be compatible with available measurements.

## 8. CONCLUSIONS

The earth's general electric structure is found to have its basic origin in charge separation in the lower ionosphere, which is produced principally by differential transport of charge carriers by low latitude vertical transport of the stratospheric tidal circulations. Vertical motion of less than 1 m sec$^{-1}$ at 80 km appear to be adequate to account for all observed phenomena. Vertical electric fields are generated in the E region in low latitudes at each hemisphere which, in conjunction with the earth's permanent magnetic field, power Hall currents that flow longitudinally toward the early afternoon sector and have their principal lower ionosphere return paths through high latitudes. These primary dynamo currents produce gross structure in the horizontal electrical potential distribution of the D and E regions that is not symmetrical between the magnetic hemispheres, with a resulting development of interhemispheric currents along magnetic field lines that provide the basic particle fields and power sources which form the radiation belts. The inner radiation belt currents serve to equalize hemispheric differences between the e.m.f. regions, and the outer radiation belt currents serve to equalize differences in potential between auroral zones. A third electrical circuit is formed between the lower ionosphere and the earth's surface by a relative short circuit effected by thunderstorms and their lightning discharges that connect the earth's surface to the negative potential of the night-time dynamo circuit so that a small part of the dynamo currents flows in the fair-weather electric field circuit.

This electrical structure generating mechanism is then postulated to serve as the source of the following atmospheric phenomena:

(1) Airglow, which apparently is powered by electrical phenomena associated with flow of the primary dynamo currents and their exospheric components

and interaction of these systems with the neutral atmosphere.

(2) Auroral activity, which in part represents current flows between hemi-spheres along high latitude magnetic field lines as a result of differences in potential developed by the E-region dynamo currents in polar regions. Charged particles collected from the solar wind are also sorted and precipitated as well as accelerated under the special influence of these geo-atmospheric internal electric fields.

(3) Thunderstorm electrification, where the lightning discharge paths provide an effective short circuit connecting the earth's surface with the primary dynamo circuit in its low potential region.

(4) Fair-weather electric field, which represents the difference between the average potential of the atmosphere and the surface potential established as indicated in (3) above.

(5) Fair-weather electric current, which conducts approximately 1500 A to the earth continuously as an auxiliary return path of the primary dynamo circuit.

(6) Telluric currents, which flow in the earth's surface to complete in part the global circuit which includes the fair-weather electric current and current flow through thunderstorm lightning paths.

(7) Magnetic disturbances at the earth's surface, which are the result of variations in current intensities in the primary dynamo and exospheric circuits caused by combined variations in the tidal circulation, E-region conductivity and solar particle collection. In addition, modifications of the exospheric magnetic fields must be expected as a result of the currents flowing in circuits located in vertical meridional planes.

The electron density of the upper ionosphere and the exosphere is thus controlled by the electrical structure of the lower ionosphere. The charged particles of the exosphere and night-time ionosphere are generally in transit under the direction of the electric fields of the primary dynamo circuits. It is necessary to this concept that these exospheric currents exhibit a diurnal reversal in direction. Such reversals will not modify the trapped components of these currents other than possibly to accelerate them as a result of short-term variations in the electric fields associated with variations in the dynamo e.m.f.'s. These internal variations in the dynamo circuits will, in addition, be modulated by solar interaction through charged particle capture by the outer belt. However, the tender point for solar control of this system is in the lower ionosphere, through solar-radiation-induced variations in the conductivity.

This initial look at the comprehensive electrical structure of the earth's atmo-sphere has clarified some aspects of atmospheric physical processes. It has pointed the way toward investigations that should illuminate some of the points which, for experimental or theoretical reasons, remain as difficulties in the clear understanding of the earth's electrical structure. Progress has come from the application of synoptic principles, emphasizing the fact that in a complex system, such as the earth's atmosphere, isolated investigations depend on sheer luck for success.

Inspection of upper atmospheric electric current flows which may be expected with reasonable models of dynamo current and conductivity structure indicates that a complex electrical structure is to be expected. In addition, it has been shown that this complex structure can be modeled in a manner which is compatible with available observational data.

Variations in electric current density and potential fields of the upper atmosphere will in large part be eliminated by counter currents and fields in the atmosphere below the dynamo region. Thus, electrical measurements in the lower atmosphere cannot necessarily detect the presence of this complex electrical structure in the upper atmosphere.

Similarly, magnetometer observations of $Sq$ variations at the earth's surface detect only the net effect of electrical currents above that station and cannot delineate details of electrical structure. Since the return currents beneath the dynamo level are small compared to the dynamo currents, they will have small effect on the magnetic field generated by the dynamos.

In the absence of sufficient data in the 30–80 km range, the models presented here must be considered very tentative. They, however, do point out the fact that assumptions concerning the electrical neutrality of the upper atmosphere must be tested, and they indicate that a complex structure probably exists. Clearly, electrical measurements in the 30–80 km region are required.

## ACKNOWLEDGEMENT

Some parts of this paper were reproduced from the article: Webb, W.L., Source of atmospheric electrification. *J.Geophys.Res.*, **73**:5061–5071, 1968, with the permission of the American Geophysical Union.

## REFERENCES

Angerami, J. J. and Carpenter, D. L., Whistler studies of the plasma pause in the magnetosphere, 2: Electron density and total tube electron content near the knee in magnetospheric ionization. *J. Geophys. Res.*, **71**:711–725, 1966.

Appleton, E. V., Two anomalies in the ionosphere, *Nature*, **157**:691–692, 1946.

Appleton, E. V., The normal E region of the ionosphere. *Proc. Inst. Radio Eng.*, **47**:155–159, 1959.

Axford, W. I., 1963. The formation and vertical movement of dense ionized layers in the ionosphere due to neutral wind shears. *J. Geophys. Res.*, **68**:769–779, 1963.

Balsley, B. B., Some additional features of radar returns from the equatorial electrojet. *J. Geophys. Res.*, **70**:3175–3182, 1965.

Banks, P. M., The temperature coupling of ions in the ionosphere. *Planet. Space Sci.*, **15**:77–93, 1967.

Barcus, J. R. and Brown, R. R., Energy spectrum for auroral-zone X rays, 2. Spectral variability and auroral absorption. *J. Geophys. Res.*, **71**:825–834, 1966.

Barcus, J. R. and Rosenberg, T. J., Energy spectrum for auroral-zone X rays, 1: Diurnal and tide effects. *J. Geophys. Res.*, **71**:803–823, 1966.

Barlow, W. H., On the spontaneous electrical currents observed in the wires of the electric telegraph. *Philos. Trans. R. Soc.*, 61, 1849.

Bates, H. F., Latitude of the dayside aurora. *J. Geophys. Res.*, **71**:3629–3633, 1966.

Becker, W., The temperature of the F region deduced from electron number-density profiles. *J. Geophys. Res.*, **72**:2001–2006, 1967.

Belrose, J. S., The lower ionospheric regions. In: *Physics of the Earth's Upper Atmosphere* (C. O. Hines, I. Paghis, T. R. Hartz and J. A. Fejer, Editors). Prentice-Hall, Englewood Cliffs, N. J., 434 pp., 1965a.

Belrose, J. S., The ionospheric F region. In: *Physics of the Earth's Upper Atmosphere* (C. O. Hines, I. Paghis, T. R. Hartz and J. A. Fejer, Editors). Prentice-Hall, Englewood Cliffs, N. J., 434 pp., 1965b.

Beyers, N. J. and Miers, B. T., Diurnal temperature in the atmosphere between 30 and 60 km over White Sands Missile Range. *J. Atmos. Sci.*, **22**:262–266, 1965.

Beynon, W. J. G. and Brown, G. M., Geomagnetic distortion of region-E. *J. Atmos. Terr. Phys.*, **14**:138–166, 1959.

Blake, J. B., Freden, S. C. and Paulikas, G. A., Precipitation of 400 keV electrons in the auroral zone. *J. Geophys. Res.*, **71**:5129–5134, 1966.

Blevis, B. C. and Collins, W. C., Radio aurora. In: *Physics of the Earth's Upper Atmosphere*. (C. O. Hines, I. Paghis, T. R. Hartz and J. A. Fejer, Editors). Prentice-Hall, Englewood Cliffs, N. J., 434 pp., 1965.

Brace, L. H., Reddy, B. M. and Mayr, H. G., Global behavior of the ionosphere at 1000 km altitude. *J. Geophys. Res.*, **72**:265–283, 1967.

Brooks, C. E. P., The distribution of thunderstorms over the globe. *Geophys. Mem.*, **3**:145–164, 1925.

Cahill, Jr., L. J., Investigation of the equatorial electroject by rocket magnetometer. *J. Geophys. Res.*, **64**:489–503, 1959.

Carpenter, D. L., Whistler studies of the plasmapause in the magnetosphere, 1: Temporal variations in the positions of the knee and some evidence of plasma motions near the knee. *J. Geophys. Res.*, **71**:693–709, 1966.

Carpenter, M. S., O'Keefe, J. A. and Dunkelman, L., Visual observations of the airglow layer. *Science*, **138**:978–980, 1962.

Chalmers, J. A., *Atmospheric Electricity*. Pergamon, London, 327 pp., 1957.

Chamberlain, J. W., *Physics of the Aurora and Airglow*. Academic Press, New York and London, 704 pp., 1961.

Chamberlain, J. W., Discharge theory of auroral rays. In: *The Airglow and the Aurora* (E. B. Armstrong and A. Dalgarno, Editiors). Pergamon Press, London, pp. 206–221, 1965.

Chandra, S. and Rangaswamy, S., Geomagnetic and solar control of ionization at 1000 km. *J. Atmos. Terr. Phys.*, **29**:259–265, 1967.

Chapman, S., The absorption and dissociative or ionizing effect of monochromatic radiation in an atmosphere on a rotating earth. *Proc. Phys. Soc.*, **43**:483–496, 1931.

Chapman, S., The equatorial electrojet as detected from the abnormal electric current distribution above Huanayo, Peru and elsewhere. *Arch. Meteorol. Geophys. Bioklimat. A.*, **44**:368–390, 1951.

Chapman, S., Rockets and the magnetic exploration of the ionosphere. In: *Rocket Exploration of the Upper Atmosphere* (R. L. Boyd and M. J. Seaton, Editors). Pergamon, London, pp. 292–305, 1954.

Chapman, S., The electrical conductivity of the ionosphere: a review. *Nuovo Cimento*, **4**(suppl.): 1385–1412, 1956.

Chapman, S. and Bartels, J., *Geomagnetism*, I,II. Clarendon Press, Oxford, 1940.

Chivers, H. J. A. and Hargreaves, J. K., Slow fluctuations between conjugate points in the auroral absorption of cosmic noise. *J. Atmos. Terr. Phys.*, **28**:337–342, 1966.

Cole, K. D., A dynamo theory of the aurora and magnetic disturbance. *Aust. J. Phys.*, **13**:484–497, 1960.

Cole, K. D., Joule heating of the upper atmosphere. *Aust. J. Phys.*, **15**:223–235, 1962.

Cole, Jr., R. K. and Pierce, E. T., Electrification in the earth's atmosphere for altitudes between 0 and 100 kilometers. *J. Geophys. Res.*, **70**:2735–2749, 1965.

Cole, Jr., R. K., Hill, R. D. and Pierce, E. T., Ionized columns between thunderstorms and the ionosphere. *J. Geophys. Res.*, 71:959–964, 1966.

Cook, G. E., The large semi-annual variation in exospheric density: A possible explanation. *Planet. Space Sci.*, 15:627–632, 1967.

Coroniti, S. C., *Problems of Atmospheric and Space Electricity*. Elsevier, Amsterdam, 616 pp., 1965.

Craig, R. A., The observations and photochemistry of atmospheric ozone and their meteorological significance. *Meteorol. Monogr., Am. Meteorol. Soc.*, Boston, 1 (2):50 pp., 1950.

Cummings, W. D. and Dessler, A. J., Ionospheric heating associated with the main-phase ring current. *J. Geophys. Res.*, 72:257–263, 1967.

Dalgarno, A., McElroy, M. B. and Walker, J. C. G., The diurnal variation of ionospheric temperatures. *Planet. Space Sci.*, 15:331–345, 1967.

Davis, T. N., The morphology of polar aurora, *J. Geophys. Res.*, 65:3497–3515, 1960.

Davis, T. N. and Sugiura, M., Auroral electrojet activity index AE and its universal time variations. *J. Geophys. Res.*, 71:785–801, 1966.

Dougherty, J. P., On the influence of horizontal motion of the neutral air on the diffusion equation of the F-region. *J. Atmos. Terr. Phys.*, 20:167–176, 1961.

Fejer, J. A., Semidiurnal currents and electron drifts in the ionosphere. *J. Atmos. Terr. Phys.*, 4:184 –203, 1953.

Fejer, J. A., Motions of ionization. In: *Physics of the Earth's Upper Atmosphere* (C. O. Hines, I. Paghis, T. R. Hartz and J. A. Fejer, Editors). Prentice-Hall, Englewood Cliffs, N. J., 434 pp., 1965.

Feldstein, Y. I. and Starkov, G. V., Dynamics of auroral belt and polar geomagnetic disturbances. *Planet. Space Sci.*, 15:209–229, 1967.

Fogle, B. and Haurwitz, B., Noctilucent clouds. *Space Sci. Rev.*, 6:279–340, 1966.

Gallet, R. M., The very low-frequency emissions generated in the earth's exosphere, *Proc. Inst. Radio Eng.*, 47:211–231, 1959.

Gassmann, G. J. and Wagner, R. A., On the equatorial electrojet. *J. Geophys. Res.*, 71:1879–1890, 1966.

Gendrin, R., Le guidage des whistlers par ler champ magnetique. *Planet. Space Sci.*, 5:274–282, 1961.

Gibbons, J. J. and Waynick, A. H., The normal D region of the ionosphere. *Proc. Inst. Radio Eng.*, 47:160 –161, 1959.

Gish, O. H. and Rooney, W. J., New aspects of earth-current circulation revealed by polar-year data. *Int. Union Geod. Geophys. Terr. Magn. Electr., Bull.* 10 (*Edinburgh Meeting*), pp.378–382, 1937.

Gish, O. H. and Wait, G. R., Thunderstorms and the earth's general electrification. *J. Geophys. Res.*, 55:473–484, 1950.

Gledhill, J. A., Torr, D. G. and Torr, M. R., Ionospheric disturbance and electron precipitation from the outer radiation belt. *J. Geophys. Res.*, 72:209–214, 1967.

Gleeson, L. J. and Axford, W. I., Electron and ion temperature variations in temperate zone sporadic-E layers. *Planet. Space Sci.*, 15:123–136, 1967.

Glenn, J. H., The Mercury-Atlas 6 space flight. *Science*, 136:1093–1095, 1962.

Gottlieb, B. and Fejer, J. A., Critical examination of a theory of auroral electrojets. *J. Geophys. Res.*, 72:239–244, 1967.

Greenspan, J. A. and Woodman, J. H., Synoptic description of the 5577 Å nightglow near 78°W longitude. *J. Atmos Terr. Phys.*, 29:239–250, 1967.

Gringauz, K. I., Kurt, V. G., Moroz, V. I. and Shklovskii, I. S., Ionized gas and fast electrons in the vicinity of the earth and in interplanetary space. *Iskusstvennye Sputniki Zemli*, 6:108–112, 1961.

Gurnett, D. A. and Shawhan, S. D., Determination of hydrogen ion concentration, electron density, and proton gyrofrequency from the dispersion of proton whistlers. *J. Geophys. Res.*, 71:741–754, 1966.

Hanson, W. B., Structure of the ionosphere. In: *Satellite Environment Handbook* (F. S. Johnson, Editor). Stanford University Press, Stanford, Calif., 187 pp., 1965.

Hanson, W. B. and Moffett, R. J., Ionization transport effects in the equatorial F region. *J. Geophys. Res.*, 71:5559–5572, 1966.

Harris, I. and Priester, W., Time-dependent structure of the upper atmosphere. *J. Atmos. Sci.*, 19:286 –301, 1962.

Heikkila, W. J. and Axford, W. I., The outer ionospheric regions. In: *Physics of the Earth's Upper*

*Atmosphere*. (C. O. Hines, I. Paghis, T. R. Hartz and J. A. Fejer, Editors). Prentice-Hall, Englewood Cliffs, N. J., 434 pp., 1965.

Helliwell, R. A. and Morgan, M. G., Atmospheric whistlers. *Proc. Inst. Radio Eng.*, **47**:200–208, 1959.

Hennes, J. and Dunkelman, L., Photographic observations of nightglow from rockets. *J. Geophys. Res.*, **71**:755–762, 1966.

Herman, J. R., Spread F and ionospheric F-region irregularities. *Rev. Geophys.*, **4**:255–299, 1966.

Hines, C. O. and Reid, G. C., Theory of geomagnetic and auroral storms. In: *Physics of the Earth's Upper Atmosphere*. (C. O. Hines, I. Paghis, T. R. Hartz and J. A. Fejer, Jr., Editors). Prentice-Hall, Englewood Cliffs, N. J., 434 pp., 1965.

Imyanitov, I. M. and Shifrin, K. S., Present state of research on atmospheric electricity. *Uspekhi Fiz. Nauk*, **76**:593–642, 1962.

Jacchia, L. G. and Slowey, J., Atmospheric heating in the auroral zones: a preliminary analysis of the atmospheric drag of the Injun 3 satellite. *J. Geophys. Res.*, **69**:905–910, 1964.

Jacchia, L. G., Slowey, J. and Verniani, F., Geomagnetic perturbations and upper-atmosphere heating. *J. Geophys. Res.*, **72**:1423–1434, 1967.

Kato, S., Horizontal wind systems in the ionospheric E region deduced from the dynamo theory of the geomagnetic $S_q$ variation, Part II. Rotating earth. *J. Geomagn. Geoelectr.*, **8**:24–37, 1956.

Kato, S., Joule heating and temperature in the upper atmosphere. *Planet. Space Sci.*, **9**:939–946, 1962.

Kraakevik, J. H., Measurements of current density in the fair weather atmosphere. *J. Geophys. Res.*, **66**:3735–3748, 1961.

Krasovskii, V. I., Shklovskii, I. S., Gal'perin, Yu. I., Svetlitskii, E. M., Kushnir, Yu. M. and Bordovskii, G. A., The detection of electrons with energies of approximately 10 keV in the upper atmosphere. *Iskusstvennye Sputniki Zemli*, **6**:113–128, 1961.

Krishnamurthy, B. V., Behavior of topside and bottomside spread F at equatorial latitudes. *J. Geophys. Res.*, **71**:4527–4533, 1966.

Leovy, C., Radiative equilibrium of the mesosphere. *J. Atmos. Sci.*, **21**:238–248, 1964.

Liemohn, H. B. and Scarf, F. L., Whistler determination of electron energy and density distributions in the magnetosphere. *J. Geophys. Res.*, **69**:883–904, 1964.

Ludlam, F. H., Noctilucent clouds. *Tellus*, **9**:341–364, 1957.

Maeda, K. I. and Sato, T., The F region during magnetic storms. *Proc. Inst. Radio Eng.*, **47**:232–239, 1959.

Martyn, D. F., Anomalous behavior of the F2 region of the ionosphere. *Nature*, **155**:363–364, 1945.

Martyn, D. F., The normal F region of the ionosphere. *Proc. Inst. Radio Eng.*, **47**:147–155, 1959.

Matsushita, S., Global presentation of the external $S_q$ and $L$ current systems. *J. Geophys. Res.*, **70**:4395–4398, 1965.

Maynard, N. C. and Cahill, Jr., L. J., Measurement of the equatorial electrojet over India. *J. Geophys. Res.*, **70**:5923–5936, 1965.

Miers, B. T., Wind oscillations between 30 and 80 km over White Sands Missile Range, New Mexico. *J. Atmos. Sci.*, **22**:382–387, 1965.

Montalbetti, R., Optical aurora. In: *Physics of the Earth's Upper Atmosphere* (C. O. Hines, I. Paghis, T. R. Hartz and J. A. Fejer, Editors). Prentice-Hall, Englewood Cliffs, N. J., 434 pp., 1965.

Morris, J. E., Wind measurements in the subpolar mesosphere region, *J. Geophys. Res.*, **72**:6123–6129, 1967.

Mozer, F. S. and Bruston, P., Observation of the low-altitude acceleration of auroral protons. *J. Geophys. Res.*, **71**:2201–2206, 1966.

Nagy, A. F. and Walker, J. C. G., Direct measurements bearing on the question of the nighttime heating mechanism in the ionosphere. *Planet. Space Sci.*, **15**:95–101, 1967.

Nichols, B., Auroral ionization and magnetic disturbances. *Proc. Inst. Radio Eng.*, **47**:245–254, 1959.

Nicolet, M., The collision frequency of electrons in the ionosphere. *J. Atmos. Terr. Phys.*, **3**:200–211, 1953.

Nicolet, M., Structure of the thermosphere. *Planet. Space Sci.*, **5**:1–32, 1961.

Ogbuehi, P. O., Onwumechilli, A. and Ifedili, S. O., The equatorial electroject and the world-wide $S_q$ currents. *J. Atmos. Terr. Phys.*, **29**:149–160, 1967

Ossakow, S. L., Anomalous resistivity along lines of force in the magnetosphere. *J. Geophys. Res.*, **73**:6366–6369, 1968.

Parker, E. N., Auroral phenomenon. *Proc. Inst. Radio Eng.*, **47**:239–244, 1959.

Paulikas, G. A., Blake, J. B. and Freden, S. C., Precipitation of energetic electrons at middle latitudes. *J. Geophys. Res.*, **71**:3165–3172, 1966.

Piddington, J. H., A theory of auroras and the ring current. *J. Atmos. Terr. Phys.*, **29**:87–105, 1967.

Prag, A. B., Morse, F. A. and McNeal, R. J., Nightglow excitation and maintenance of the nighttime ionosphere by low-energy protons. *J. Geophys. Res.*, **17**:3141–3154, 1966.

Randhawa, J. S., Ozonesonde for rocket flight. *Nature*, **213**:53–54, 1967.

Rawer, K., *The Ionosphere*. Frederick Ungar Publishing Co., New York, N. Y., 202 pp., 1952.

Redding, J. L., Diurnal variation of telluric currents near the magnetic equator. *J. Atmos. Terr. Phys.*, **29**:297–305, 1967.

Reed, R. J., McKenzie, D. J. and Vyrerberg, J. C., Diurnal tidal motions between 30 and 60 kilometers in summer. *J. Atmos. Sci.*, **23**:416–423, 1966.

Schilling, G. F. and Sterne, T. E., Densities and temperatures of the upper atmosphere inferred from satellite observations. *J. Geophys. Res.*, **64**:1–4, 1959.

Sharp, R. D., Johnson, R. G., Shea, M. F. and Shook, G. B., Satellite measurements of precipitating protons in the auroral zone. *J. Geophys. Res.*, **72**:227–237, 1967.

Smith, L. G., *Recent Advances in Atmospheric Electricity*. Pergamon, Oxford, 631 pp., 1958.

Somayajulu, Y. V., Evidence on the horizontal diffusion of F region ionization along the magnetic lines of force in equatorial latitudes. *J. Geophys. Res.*, **69**:561–563, 1964.

Stergis, C. G., Rein, G. C. and Kangas, T., Electric field measurements above thunderstorms. *J. Atmos. Terr. Phys.*, **11**:83–90, 1957.

Stewart, B., Terrestrial Magnetism. *Encyclopaedia Britanica*, 9th ed., **16**:159–162, 1883.

Sugiura, M. and Cain, J. C., A model equatorial electrojet. *J. Geophys. Res.*, **71**:1869–1877, 1966.

Van Allen, J. A., Radiation observations with satellite 1958 Epsilon. *J. Geophys. Res.*, **64**:271–286, 1959.

Van Allen, J. A., The geomagnetically trapped corpuscular radiation. In: *Science in Space* (L. V. Berkner and H. Odishaw, Editors). McGraw-Hill, New York, N. Y., 458 pp., 1961.

Vestine, E. H., Noctilucent clouds, *J. R. Astronom. Soc. Can.*, **28**:249–272 and 303–317, 1934.

Vestine, E. H., Winds in the upper atmosphere deduced from the dynamo theory of geomagnetic disturbance, *J. Geophys. Res.*, **59**:93–128, 1954.

Vorpahl, J. A., *The Frequency and Intensity of Lightning within 30° of the Equator*. Thesis, University of Minnesota, Physics Dept., Minneapolis, Minn., 66 pp., 1967.

Webb, W. L., Morphology of noctilucent clouds. *J. Geophys. Res.*, **70**:4463–4475, 1965.

Webb, W. L., *Structure of the Stratosphere and Mesosphere*. Academic Press, New York and London, 382 pp., 1966a.

Webb, W. L., Stratospheric tidal circulations. *Rev. Geophys.*, **4**:363–375, 1966b.

Webb, W. L., Source of atmospheric electrification. *J. Geophys. Res.*, **73**:5061–5071, 1968.

Webb, W. L., Giraytys, J., Tolefson, H. B., Forsberg, R. C., Vick, R. I., Daniel, O. H. and Tucker, L. R., Meteorological Rocket Network probing of the stratosphere and lower mesosphere. *Bull. Am. Meteorol. Soc.*, **47**:788–799, 1966.

Wilson, C. T. R., Investigations on lightning discharges and on the electric field of thunderstorms. *Philos. Trans. R. Soc. Lond. Ser. A*, **221**:73–115, 1920.

Wulf, O. R., On the production of glow discharges in the ionosphere by winds. *J. Geophys. Res.*, **58**:531–539, 1953.

Yuen, P. C. and Roelofs, T. H., Seasonal variations in ionospheric total electron content. *J. Atmos. Terr. Phys.*, **29**:321–326, 1967.

# OBSERVATIONS OF THE GLOBAL STRUCTURE OF THE STRATO-SPHERE AND MESOSPHERE WITH SOUNDING ROCKETS AND WITH REMOTE SENSING TECHNIQUES FROM SATELLITES*

DONALD F. HEATH, ERNEST HILSENRATH, ARLIN J. KRUEGER, WILLIAM NORDBERG, CUDDAPAH PRABHAKARA AND JOHN S. THEON

*NASA Goddard Space Flight Center, Greenbelt, Md. (U.S.A.)*

## SUMMARY

Rockets and satellites have observed and led to the understanding of several aspects of the physical state and composition of the stratosphere and mesosphere on a global scale. The results obtained from programs conducted or sponsored by NASA are presented.

Rocket soundings have provided in situ measurements of temperature, density, winds and composition ranging from balloon altitudes to the mesopause. Temperatures were measured with the acoustic grenade technique and thermistor dropsondes, and vertical density distributions were derived from pitot probes and falling spheres. Winds were derived from radar tracking of chaff as well as from the grenade soundings. The distribution of ozone has been derived from measurements of absorption of solar radiation and of chemiluminescence. The concentration of atomic oxygen has been sampled in the upper mesosphere with an oxidation sensor.

Satellite platforms have expanded the scope of temperature and compositions monitoring to a truly global scale, with the development of remote sensing techniques. Stratospheric temperatures were inferred from measurement of infrared radiant energy in the 15-$\mu$m $CO_2$ band. Initially broad-band, spatially scanning radiometers have revealed the horizontal temperature field at one single level, and more recently, infrared spectrometers have produced vertical temperature profiles to stratopause levels. The global distribution of total ozone was deduced from measurements of the upwelling radiation in the 9.6-$\mu$m ozone band and from solar ultraviolet radiation in the 2500–3100 Å region which is scattered by the atmosphere to space. Such measurements were used to infer the vertical ozone profile to 55 km.

From the rocket observations it was found that the mean tropical temperature distribution in the stratosphere and mesosphere is invariant with season but that

---

* Presented by Dr. William Nordberg, Chief, Laboratory for Meteorology and Earth Sciences.

significant diurnal variations, attributable to thermal tide effects, are present. At middle and high latitudes, marked seasonal changes exist. In the polar summer the stratopause is warmer than 270°K while the mesopause is sometimes colder than 150°K. These low temperatures were found to be sufficient to explain the formation of noctilucent clouds. This substantial temperature lapse rate in the polar mesosphere is absent in winter, when extreme variability, suggestive of internal gravity waves, is found.

The diurnal change in mesospheric ozone, measured by rockets at low latitudes, is found to be smaller than expected from the classical photochemical theory but is consistent with a chemistry including the presence of water. Soundings during the polar night show little vertical structure contrary to what might be expected from the highly structured temperature distributions.

The horizontal temperature fields observed with satellite radiometers and spectrometers indicate that in the lower stratosphere there is a monotonic decrease in temperature from summer to winter pole. This temperature field is significantly coupled with the troposphere and surface characteristics. The manner in which synoptic scale events, such as midwinter stratospheric warmings, develop in time and space has been revealed in the infrared satellite data.

The total ozone measurements from space have been used to construct daily and monthly mean global maps of the ozone field. Analysis of these maps has shown persistent features at high latitudes and a seasonal modulation of the Hadley cells in the tropics. A strong relationship between jet streams and total ozone amount was found. The high level ozone distributions derived from the UV measurements show strong seasonal differences and an apparent temperature dependence in the upper stratosphere where the ozone appears to be in an equilibrium state. In the tropical middle stratosphere, however, the effects of vertical motions are exhibited as small-scale modulations in the horizontal field.

## 1. INTRODUCTION

Until about eight years ago, analyses of the structure and behavior of the upper stratosphere and of the mesosphere were based on rather indirect ground based observations and on infrequent rocket soundings from isolated sites. Ground-based observations resulted in the description and understanding of only the most basic physical processes of that region of the atmosphere: the absorption of near-ultraviolet solar radiation resulting in the temperature maximum at the stratopause; the photochemistry and transport phenomena responsible for the distribution of ozone and of atomic oxygen in the upper stratosphere and mesosphere; the predominant radiative cooling of the mesosphere resulting in very low temperatures near the mesopause; the accumulation of particles near the mesopause at high latitudes in summer which relate to the noctilucent cloud phenomenon; and the generally easterly winds during summer, alternating with predominantly westerly flow in winter throughout the upper stratosphere and lower mesosphere.

From occasional rocket soundings during the IGY and during subsequent years we have learned: (1) that the temperature structure varies with latitude and season in such a way that at high latitudes during summer, maximum temperatures near 50 km and minimum temperatures near 80 km are extremely pronounced, while in winter the temperature extremes at these levels are much less distinct; (2) that in winter the temperature structure in the mesosphere displays large, short-term local variations; (3) that the circulation of the atmosphere up to 70 km follows patterns of high and low pressure systems which may bring about the onset of sudden stratospheric warmings; and (4) that the general concept of the meridional circulation in the mesosphere, with upward motion over the summer pole and downward motion over the winter pole, which had been derived theoretically, is basically correct.

However, until recently, observations were lacking to describe the structure and phenomena of this region in greater detail and on a truly global scale. Systematic exploration of the atmosphere up to about 100 km began in the mid-1960's with simultaneous, seasonal rocket soundings distributed along latitude circles, almost around the globe or along meridians from the equator to high latitudes. This became possible because of the evolution of relatively simple and economical rocket instruments for the direct measurement of such parameters as pressure, temperature, wind, and ozone, and because of the strong interest in exploring this part of the atmosphere by many nations around the world. Also, during these recent years, meteorological satellites have carried "remote sensors", such as spectrometers and image-forming radiometers, which have permitted the observation of temperature patterns in the stratosphere, and the distribution of ozone, as well as the variation of solar radiation absorbed in the mesosphere on a daily basis over the entire globe.

The systematic rocket soundings and global surveys by satellites have resulted in considerably better descriptions of the structure of, and in a better understanding of processes in the stratosphere and mesosphere than was possible previously. For example, from continuous observations with Nimbus III and IV since April, 1969, we have been able to determine the correlation of variations in the intensity of ultraviolet solar radiation between 1200 and 1800 Å, with the 27-day rotation period of the sun and with other solar activity, while no such correlation was observed at longer wavelengths. The genesis and morphology of stratospheric warmings has been observed with Tiros and Nimbus satellites on numerous occasions in both hemispheres and on a global scale since 1963. Ozone distributions have been measured with Nimbus IV up to 50 km, daily and globally, since April, 1970. Systematic rocket soundings established a relationship between the occurrence of noctilucent clouds and the mesospheric temperature profile. They also demonstrated the vertical propagation of wavelike phenomena and of tides through the mesosphere, and measured the changes in the distribution of ozone between day and night, as well as during the polar night up to 70 km.

In the following discussion, a brief description of the techniques involved in these observations and a survey of the results, many of which are still preliminary, will be given. The discussion will be limited to those techniques and results dealing with rocket and satellite programs conducted or sponsored by the National Aeronautics and Space Administration (NASA) of the U.S.A.

## 2. TECHNIQUES FOR MEASURING THE COMPOSITION, THERMODYNAMIC STRUCTURE AND MOTION FIELD OF THE NEUTRAL ATMOSPHERE UP TO 80 KM

Observations of the upper atmosphere are carried out in a complementary fashion: rocket soundings provide measurements with detailed resolution and accuracy along the vertical scale, but they cannot be deployed in sufficient numbers to provide more than a coarse picture on the horizontal and time scales. Conversely, as many as several thousand satellite observations of temperature or ozone distribution with height are being made daily over the entire globe, but in the vertical scale, each measurement is usually averaged over an interval of 5–10 km. Thus, satellite observations now provide for an unprecedented number of measurements (especially of temperature and ozone in the stratosphere) on a global scale, while rocket soundings complement and supplement these observations with detailed measurements of the vertical structure and of parameters, such as winds which are not yet observable with satellites.

### 2.1. Rocket techniques

Generally, instruments for measuring temperature, density, pressure, composition and winds in the upper stratosphere and mesosphere are carried on two-stage sounding rockets of the Nike–Cajun type. Some observations are based on direct measurements, such as those of temperature and pressure with thermistor and electrical gauges, respectively, and the measurement of ozone by means of chemical reactions with substances carried aloft by rockets. Other observations are derived from indirect measurements, such as temperature from the measured speed of sound propagation, density from the measured acceleration of spheres dropped from rockets, and ozone from the attenuation of solar radiation. Winds are usually measured by tracking the horizontal drift of objects such as smoke, chemicals, chaff, or parachutes released from rockets, or of sound waves propagating through the atmosphere. The following techniques have produced significant measurements of the structure of the upper stratosphere and mesosphere during the past several years. About 50 of the larger, two-stage rockets are launched every year to heights of about 100 km from several sites between the equator and Alaska to measure temperature, pressure, density, wind, and ozone. Many more smaller rockets (several per week from about 14 sites, and one or two per month from a few additional sites in the Americas) provide temperature and wind soundings up to about 60 km.

### 2.1.1. Radar tracking of chaff

A vertical profile of the horizontal winds and wind shears can be obtained by tracking the horizontal drift of falling chaff with a radar (Warner and Bowen, 1953). Chaff consists of a large number ($\sim 10^6$) of dipole reflectors, usually fine wire or electrically conductive thin ribbon which is cut to half the wavelength of the tracking radar to improve its reflectivity, and is most often deployed in the 60–100 km altitude region by Loki-class rockets. Considerations of desired fall rate and altitude determine the type of chaff used, but all chaff inherently begins to disperse after falling 20–30 km, so that the radar no longer has a single target and tracking becomes unreliable (Beyers, 1969).

### 2.1.2 Thermistor dropsondes

Dropsonde instruments containing immersion temperature sensors are ejected from Arcas-class rockets near apogee to measure (and telemeter) the ambient temperature of the atmosphere as they descend on parachutes or other drag devices from about 60 to 20 km (Ballard and Rofe, 1969). The pressure profile can then be obtained by the upward integration of the hydrostatic equation using the measured temperature profile and a balloonsonde pressure measurement; the density profile can be calculated from the equation of state (Nordberg and Rasool, 1968). Radar tracking of the metallized parachute provides a wind profile as the sonde descends. The temperature sensor most commonly employed is a 10-mil bead thermistor which has a rapid time response, but at altitudes above 45 km, it is subject to errors caused by radiation, aerodynamic heating, heat conduction from the payload, etc. When the parachute undergoes high fall velocities at the upper altitudes, it does not respond to wind shears effectively.

### 2.1.3. Falling spheres

Falling spheres are used as sensors to measure density profiles in the upper atmosphere. Although several variations of the sphere technique are used, i.e., rigid or inflatable, the passive type deployed from a small meteorological sounding rocket near apogee is the most common (Wright, 1969). This sphere is tracked by radar as it falls through the 100 to 35 km altitude range, yielding density and, below 70 km, winds also. Changes in the fall rate of the sphere are measured by the radar and, with the use of aerodynamic theory, are interpreted in terms of an atmospheric density profile. Using the density profile derived from the measured accelerations of the falling sphere, the hydrostatic equation can be integrated downward to produce a temperature profile. A pressure profile can be calculated from the density and temperature profiles, using the equation of state. Active spheres contain accelerometers and telemetry instrumentation, and they rely on radars for altitude discrimination and wind determination only. However, these systems require the use of larger, Nike–Cajun class rockets to deliver the payload to the desired altitude.

### 2.1.4. The pitot probe technique

The pitot probe technique utilizes pressure sensors mounted in the forward tip of the payload to measure the impact pressure as the Nike–Cajun class rocket ascends through the 30–120 km region of the atmosphere (Ainsworth et al., 1961). The pressure measurements are telemetered to the ground. The impact pressure profile is used, with appropriate aerodynamic theory, to derive a profile of atmospheric density. The temperature profile is calculated by integrating the hydrostatic equation downward using the derived density profile, and the pressure profile is obtained from the equation of state (Horvath et al., 1962). This technique provides data with 0.5-km vertical resolution, but no wind measurement.

### 2.1.5. The acoustic grenade technique

In the grenade technique, explosive charges (grenades) varying in weight from 0.1 to 1.8 kg are carried aloft in the nose cone of a Nike–Cajun rocket. Up to 31 grenades are ejected and detonated at 2–4 km intervals as the rocket traverses the 35–95 km region of the atmosphere (Nordberg and Smith, 1964).

The position of the rocket, and therefore of each explosion, is determined by a Doppler tracking system or a high precision radar such as the FPQ-6, or both. The time of each explosion is detected by sensors in the payload and telemetered to the ground. A ground-based array of hot-wire microphones, capable of responding to frequencies between 1 to 20 Hz, is used to detect and record the arrivals of the sound waves generated by the exploding grenades. The times and positions of the grenade explosions, and the arrival times of the sound waves at the ground-based microphones are measured.

The elevation and azimuth angles of the normal to each arriving spherical sound wave front are computed by applying a least-squares fit to the arrival times at the various microphones. Each wave front is then analytically traced back along its path of propagation through the atmosphere by means of Snell's Law. Wind and temperature data from balloon sondes and rocket sondes below the grenade explosions are used to trace the path of the sound wave from the ground to the level of the first explosion; above this altitude, the results of the grenade sounding itself are used for each succeeding explosion. The origin of the sound wave as determined by ray tracing is compared with the known position of the explosion, and the horizontal difference by which the sound wave has been displaced from one explosion to the next is a measure of the average wind velocity in the layer bounded by the two explosions. The average speed of sound, and hence the average temperature of the atmosphere between adjacent explosions, may also be determined. The temperature and wind profiles consist of discrete points, each representing the average temperature and average wind, respectively, of the vertically stacked horizontal layers between consecutive explosions. The pressure profile is derived from the temperature profile, using the pressure measured by an accompanying balloon sonde as a reference value. Pressure is calculated as a function of altitude from the hydrostatic equation, by integrating the pressure upward over the temperature profile. The density is then calculated as a function of altitude from the temperature and pressure, using the equation of state.

### 2.1.6. Ozone sondes

*2.1.6.1 Optical.* The most common ozone sensing technique measures the intensity of sunlight or moonlight as a function of height, with photometers at one or more wavelength bands between 2500 and 3500 Å. The method is characterized by its absolute accuracy; in a dropsonde configuration, the ozone concentration can be measured with 10% accuracy, with 1-km resolution throughout the stratosphere. Early experiments used spectrographs (Johnson et al., 1952), but filter photometers are used now, since narrow band interference filters have been developed for that spectral region. Ozone concentrations are computed by iteration from the attenuation of radiation intensity with height in each spectral channel (Paetzold, 1961a). For soundings through the height region of 20 to 60 km, a minimum of two optical filters is required. This technique has been developed for routine daytime use on small rockets by several investigators. The dropsonde approach, in which the instrument is deployed on a parachute at apogee, has been pursued by Krueger and McBride (1968) in the United States, and by Sissons (1968) and Beach (1970) in Australia. More than 20 soundings have been made by the American group using Arcas rockets. Soundings have been conducted to survey the latitudinal and temporal changes in the ozone distribution, and to provide reference data for calibration or verification of satellite ozone measurements.

An alternate approach has been taken by Nagata et al. (1971) in Japan, where they made measurements from the rocket during ascent. Five soundings have been reported to date. Weeks and Smith (1968) have used a single channel radiometer at 2550 Å on larger sounding rockets to monitor ozone above 50 km in conjunction with other experiments.

Night-time ozone has been measured optically by Reed (1968) in the U.S., and by Carver et al. (1966) in Australia. Reed's experiment was unusual in that the UV airglow was used as a light source.

*2.1.6.2. Chemical.* In contrast to optical techniques, where natural light sources are required, chemical ozone probes can be used routinely at night. Nocturnal ozone soundings of the mesosphere provide information on the increase of ozone concentration due to "three body" collisions of O with $O_2$, and the absence of dissociating radiation. This technique has also been used to observe ozone variations in the mesosphere during the polar winter night.

A chemiluminescent sonde dropped from a rocket on a parachute (Hilsenrath et al., 1969) makes measurements of ozone from 70 to 15 km. It works on a principle that was first discussed by Beranose and Rene (1959). Exposure of a chemiluminescent material (Rhodamine B) to ozone causes luminescence which is proportional to the ozone concentration and the sampling rate. The atmosphere is sampled by self pumping as the sonde descends on a specially designed parachute. A ballast chamber is connected to the ambient atmosphere through an inlet pipe and is initially in pressure equilibrium with the atmosphere. The sensor is released at apogee and as it descends through the atmosphere on the parachute, the pressure inside the ballast chamber is lower than the increasing external pressure, resulting in a net flow of gas through the inlet pipe. The chemiluminescent detector and a photometer are positioned along the inlet pipe; thus, the ambient atmosphere is continuously sampled. Photometer, pressure, and flow rate data are telemetered to the ground during the entire parachute descent. Ozone mixing ratios vs. height are derived from the measured light intensity. Calibration of the sensor is performed by exposing the instrument to known concentrations of ozone at pressures and flow rates that are expected to occur during the flight. Numerous laboratory studies have been performed to establish the validity of this method and to verify the assumptions upon which it is based. These studies have also demonstrated that under flight conditions, the sonde will not be affected by atomic oxygen.

### 2.1.7. Atomic oxygen probes

A sensor for measuring the height profile of oxygen atoms from approximately 70 to 100 km has been developed by Henderson and Schiff (1970). This sensor consists of a thin film of silver deposited on a pyrex rod which is sufficiently small to permit free molecular flow of atomic oxygen onto its surface at the measurement altitude range. As the sensor is exposed to the ambient atmosphere during the rocket ascent, the atom flux is determined from the rate of change of the electrical resistance of the silver film. This information is telemetered to the ground during the rocket ascent. Laboratory measurements indicate that this technique is accurate to at least 25%.

### 2.1.8. Water vapor probes

A promising technique for water vapor measurements in the mesosphere is a rocket-borne aluminum oxide hygrometer (Chleck, 1966). The hygrometer consists of a thin aluminum foil strip that has been anodized. A coating of gold is vacuum-deposited over this strip, to produce an aluminum oxide capacitor. Changes in the ambient water vapor pressure result in corresponding changes in the electrical impedance of the sensor. The measurement range is from approximately 300°K to 150°K dew/frost point temperatures, with an error of about ± 2°K at the lower temperature.

This hygrometer, which has been flown routinely on a jet aircraft, has produced continuous water vapor data from the ground to just above the tropopause (Hilsenrath and Coley, 1971). Balloon flights have yielded data in the lower stratosphere. Rocket flights to measure water vapor concentrations in the stratosphere are planned for 1972.

## 2.2. Satellite techniques

All observations of the atmosphere below 100 km from satellites are necessarily based on indirect measurements. They permit the derivation of atmospheric composition or temperature from radiometric measurements of the spectral attenuation of either direct solar radiation or of solar radiation reflected by the earth, and from spectral measurements of infra-red and microwave radiation emitted by the upper atmosphere. Both methods have been employed successfully on meteorological satellites for the United States (Tiros and Nimbus), to make observations of temperature and ozone up to about 50 km. In addition, the spectral intensity of incident solar radiation between 1200 and 2600 Å was measured with such satellites. These techniques are described below.

### 2.2.1. Broad-band infra-red radiometers for mapping stratospheric temperatures

The first satellite observations of global stratospheric temperature fields were based on measurements of the radiant emittance of the atmosphere in the spectral interval 14.8–15.5 $\mu$m (Kennedy and Nordberg, 1967). They were obtained by Tiros VII during June, 1963 to November, 1964. The radiometer consisted of a thermistor bolometer, filters, light gathering mirrors and lenses, and associated electronics. The instantaneous field of view was approximately $5° \times 5°$ which was scanned across the surface of the earth by the spinning motion of the satellite. A description of this instrument is contained in the Tiros VII Radiation Data Catalog and User's Manual (1964). During each 12-h period which corresponded to 14 consecutive orbits, the instrument observed the entire zone of the earth between 65°N and 65°S. Thus, radiation patterns over this quasi-global zone could be mapped by the satellite every 12 h. Practically, however, the time interval during which full coverage was achieved ranged over several days because measurements could not be transmitted to the ground for each and every full orbit. Also, the spatial resolution of the measurements was considerably coarser than the ideally possible one: although each instantaneous radiation measurement corresponded to an area of $50 \times 50$ km$^2$ on the surface of the earth, the radiometric accuracy of such a measurement when converted to equivalent black-body temperatures was generally poorer than ±5°C, and required the averaging of many hundred measurements over larger areas and longer time periods. This resulted in nearly full coverage of the "quasi globe", with a relative accuracy of better than $\pm 1°C$ in periods of about 10 days.

The radiant emittances measured by the satellite sensor in this spectral interval (14.8–15.5 $\mu m$) were primarily due to thermal emission in the vibration-rotation band of carbon dioxide. It was assumed that carbon dioxide is distributed uniformly throughout the upper troposphere and stratosphere. Also, assuming a typical temperature profile for the troposphere and stratosphere as given by the U.S. Standard Atmosphere (COESA, 1962), one may compute, for any given height interval in the atmosphere, the radiant emittance which is transmitted to the satellite within this spectral interval. The result of such a computation shows that more than 90% of the total radiation sensed by the satellite is emitted by the atmosphere above 10 km. More than 70% of that total radiation is emitted at heights between 10 and 30 km. Thus, if the measured radiant emittances are converted to equivalent black-body temperatures, they can generally be interpreted as atmospheric temperatures averaged over the height range from 10 to 30 km. Equivalent black-body temperature is defined here as the temperature of an isothermal black-body filling the field of view of the sensor which would cause the same response from the radiometer as does the radiation emerging from the top of the atmosphere in the direction of the satellite.

A similar technique was also used with Nimbus II between April and July, 1966. In this case, the spectral interval was somewhat wider than in the previous experiment on Tiros VII. This resulted in considerably more accurate measurements, but decreased the height range of the temperature measurements to somewhat lower altitudes than with Tiros VII. Maximum radiation was received from near the 16-km level, and about 75% of the radiation emanated from above 10 km. This permitted inferences of average temperatures in the upper troposphere and lower stratosphere (Nordberg et al., 1966). In contrast to Tiros VII, Nimbus satellites were fully stabilized so that the radiometer viewed the earth at all times. Scanning across the earth was therefore achieved with a rotating mirror in front of the radiometer optics (Nimbus II Data User's Guide, 1966).

*2.2.2. Infra-red spectrometers for measurements of the temperature and ozone distributions in the stratosphere*

Nimbus III carried two and Nimbus IV carried three spectrometers which measured, among other quantities, spectral radiances in the 14.5–15 $\mu m$ wavelength range from which temperatures in the stratosphere could be derived. In contrast to Tiros VII and Nimbus II, these instruments provided a very high spectral resolution (5 cm$^{-1}$ or better) and measurements were made in more than one spectral channel in the 14.5–15.0 $\mu m$ interval, which allowed derivations of temperatures at different heights in the stratosphere.

The satellite infra-red spectrometer (SIRS) of Wark and Hilleary (1969) obtained measurements relating to stratospheric temperatures in two discrete intervals. The temperature profile in the troposphere was derived from radiances in several additional wavelengths shorter than 14.5 $\mu$m that were also measured. The SIRS instrument is a conventional diffraction grating spectrometer, and has been described in detail in the Nimbus III (1969) and Nimbus IV (1970) User's Guides. On Nimbus III, it measured radiances from the atmosphere in a narrow strip about 200 km wide along the sub-satellite track. On Nimbus IV, the 11° field of view was scanned to 38° on each side of the satellite track. Two observations per day were made at every latitude from 81°N to 81°S: one near noon, and one at night near midnight. The instrumental accuracy was such that each measurement yielded temperatures of a $200 \times 200$ km$^2$ area up to an altitude of about 40 km within about $\pm 1°$C. Measurements with the SIRS have been made continuously since April, 1969.

The infra-red interferometer spectrometer (IRIS) by Conrath et al. (1970) obtained measurements of spectral radiances over the entire range from 6 to 20 $\mu$m, which included not only the wavelengths from 14.5 to 15 $\mu$m from which stratospheric temperature could be derived, but also the emission from ozone between 9 and 10 $\mu$m. The instrument was a Michelson interferometer with an 8° field-of-view which pointed 2° ahead along the sub-satellite track at the beginning of an interferogram. At the end of the interferogram, or 10 sec later, the field-of-view was directed about 2° behind the sub-satellite point. Therefore, the radiation contained in a single interferogram originated from a circular area, about 150 km in diameter. Spectra were obtained by Fourier analyses of the interferograms. About 4000 spectra, each containing information on stratospheric temperatures and total ozone amounts, were obtained daily. Measurements were made with Nimbus III continuously from April to July, 1969, and with Nimbus IV since April, 1970.

The third instrument on Nimbus IV to measure stratospheric temperatures was a selective chopper radiometer (SCR), by Ellis et al. (1970). The object of the selective chopper radiometer was to determine the temperature of the atmosphere from the surface of the earth or cloud top level to a height of 60 km. Temperature soundings were achieved by observing the emitted infrared radiation in the 15-$\mu$m band from atmospheric carbon dioxide. Height resolution was obtained by a combination of optical multi-layer filters, and selective absorption of radiation using carbon dioxide-filled cells within the experiment. The four lower channels consisted of a cantilever-mounted blade shutter which oscillated at 10 Hz and successively chopped the field of view between earth and a cold reference source (space). The chopped radiation was then passed through a 10 cm path length of carbon dioxide, the pressure being set for each channel to define the viewing depth in the atmosphere. Behind the $CO_2$ path there was a narrow band filter, the centers of which were different for each channel, and a light pipe which converged the radiation on a thermistor bolometer detector. In order to obtain adequate height resolution in the upper layers of the atmosphere, the upper two channels operated as double cell channels, switching the radiation between two half-cells, semicircular in shape and with a 1 cm path length, containing different pressures of carbon dioxide. The oscillating shutter used in the four lower channels was replaced by a vibrating 45° mirror. During one half-period, earth radiation passed through one half-cell and space radiation through the other; the situation was reversed during the other half-period. This system assumed that, apart from the $CO_2$ pressures, both halves of a cell had equal optical transmissions. A special in-flight calibration procedure, known as "imbalance calibration" was required to verify this assumption.

Radiances measured by any of these instruments could be converted to temperatures of stratospheric layers, each corresponding to a given spectral interval, similar to the broad-band measurements made earlier with Tiros VII and Nimbus II. However, it was also possible to "invert" simultaneously measured radiances in different spectral intervals to continuous profiles of atmospheric temperatures with height. This was based on the fact that radiation near the center of the carbon dioxide absorption band (15 $\mu$m) is emitted at higher altitudes (30–60 km) than radiation in the weaker absorption portion of the $CO_2$ spectrum, such as near 14.4 $\mu$m, which is emitted from an altitude range from about 7 to 20 km.

It was first pointed out by King (1958) and Kaplan (1959) that vertical temperature profiles could be obtained from remote infrared measurements. Since then, analytical techniques have been developed by which estimates of temperature profiles could be obtained from actual radiance measurements (Rodgers, 1966; Wark and Fleming, 1966; Chahine, 1968; Conrath, 1968; Strand and Westwater, 1968). These retrieval techniques can be divided into two general categories: techniques which utilize only the radiances measured from the satellite, and statistical techniques which, in addition to the satellite measurements, employ statistical information on the behavior of the atmospheric temperature profiles. Measurements with the SIRS and SCR were "inverted" to temperature profiles by means of statistical analyses (Smith, 1969; Ellis et al., 1970, respectively). A nonstatistical method, based on an iterative commutational technique developed by Chahine (1968) was used for the inversion of IRIS measurements.

The emission of radiation by ozone in the 9–10 $\mu$m band which was measured with the IRIS, was used to extract information on the total amount and the vertical distribution of ozone (Prabhakara, 1969). A representation containing free parameters was assumed for the vertical distribution of ozone, and the free parameters were evaluated by making a least squares fit of the theoretically calculated radiances to the measured radiances in the ozone band. The vertical ozone distribution was represented in terms of empirical orthogonal functions or characteristic patterns (Obukov, 1960; Mateer, 1965). The empirical orthogonal functions were calculated from ensembles of historical ozone profiles obtained with ozone sondes. From the spectral resolutions and accuracies of the radiation measurements obtained in the IRIS experiment, essentially only one parameter of the ozone distribution could be derived. This was the expansion coefficient of the first empirical orthogonal function. While such a representation provided only relatively crude information on the vertical distribution of ozone, it did, however, permit a meaningful estimate of the total ozone.

### 2.2.3. Spectrophotometers to measure backscattered ultra-violet (BUV) solar radiation for determination of ozone and its vertical distribution

An experiment for obtaining the spatial distribution of atmospheric ozone on a global scale, by inversion of measurements of ultraviolet radiation backscattered by the atmosphere, completed one year of continuous operation aboard Nimbus IV in April, 1971.

The instrument consists of a double (tandem) Ebert–Fastie spectrophotometer in conjunction with a narrow band interference filter photometer. Both instruments view along the nadir direction of the spacecraft. The spectrophotometer measures spectral intensities at 12 wavelengths from 2555 Å, with a 10-Å band pass. The interference filter photometer measures a 50-Å band centered at 3800 Å. A depolarizer is inserted in front of the entrance slit of the spectrophotometer to eliminate any effects due to its polarization properties. The double monochromator is composed of two 25 cm focal length Ebert–Fastie monochromators. Light entering the entrance slit is rendered parallel by a spherical collimation mirror, and is then diffracted by a $52 \times 52$ mm grating of 2400 grooves mm$^{-1}$ (solid angle is 0.043 steradians). The diffracted light returns to the spherical collimating mirror, passes through a roof prism, and is imaged onto an intermediate slit. The light passing through the intermediate slit is dispersed again by a second monochromator. A field lens at the exit slit is used to image the grating onto the photocathode of a photomultiplier. Both gratings are mounted onto a common rigid shaft so that no wavelength tracking error can occur between the two monochromators. A roof prism is used to invert the image in the direction of dispersion at the intermediate slit. This is necessary if one is to double the dispersion in passing through the second monochromator. When passing over the polar regions, diffuser plates are deployed in front of the spectrophotometer and the filter photometer in order to view the sun for calibration.

The primary reason for using a double monochromator is to obtain the desired spectral purity, i.e., the elimination of scattered light, which permits one to use a high quantum efficiency ($>20\%$ for the 12 wavelengths) photomultiplier tube with a dark current of the order of $10^{-11}$ amp at a gain of $10^{6}$. A secondary reason is that the aberrations at the exit slit are much less than those produced by a single monochromator. In addition, the use of a double dispersion system permits one to double the slit width while maintaining the band pass of a single dispersion system.

Since the radiation energy throughout increases with the square of the slit width, the transmission losses in passing through the second half of the double monochromator are almost offset by the doubled dispersion.

The problem of measuring the earth radiance in terms of a laboratory standard of spectral irradiance is illustrated in Fig. 1. The standard of spectral irradiance (1000 W, quartz-iodine lamp) and the solar irradiance decrease by only an order of magnitude from 3200 to 2500 Å, while the earth radiance decreases by more than $10^3$ over this wavelength interval. In addition, one must either know the polarization characteristics of the measurement equipment or make it insensitive to the polarization of the incident radiation, since for single scattering the polarization of the earth radiation may be as much as 60% at 60° from the sun. The BUV double monochromator uses a Lyot type depolarizer to make it insensitive to the state of polarization.

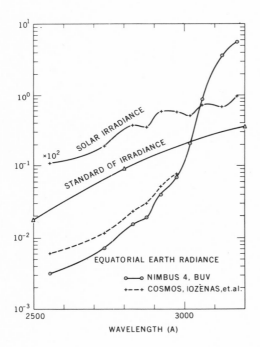

Fig. 1. Wavelength dependence of solar irradiance, laboratory standard of irradiance, and radiance backscattered from the earth at the equator. The units of irradiance are erg/cm$^2$ sec Å and radiance, erg/cm$^2$ sec Å sterad. The ordinate should be multiplied by $10^2$ to read solar irradiance. The earth radiance measured by Iozenas et al. (1969) on the Cosmos satellites is shown for comparison with the Nimbus BUV measurements. Both measurements are from the equatorial regions where the polarization is essentially zero. The agreement is quite good at the long-wavelength end; however, at 2555 Å the U.S.S.R. measurement of the earth radiance is larger by about a factor of two. We believe that this discrepancy could be due to stray light originating in the U.S.S.R. instrument.

Because ozone absorption is very strongly wavelength dependent through much of the middle ultraviolet, observations of backscattered energy at several wavelength bands in this spectral region provide a mechanism for scanning through the atmosphere in the vertical. This wavelength effect is illustrated in Fig. 2 which shows the relative amount of energy backscattered at various levels in the atmosphere for 8 of the 12 wavelengths measured by the Nimbus IV BUV instrument. As wavelength increases, the absorption coefficient decreases, and there is an increase in the depth of penetration of the solar radiation into the atmosphere. Each curve has been normalized to unity at the level of maximum contribution, and the area between the curve and the y-axis is directly proportional to the backscattered energy observed at the spacecraft. The curves are based on a mid-latitude ozone distribution having a total ozone of 0.336 atm cm as measured by Hering and Borden (1965).

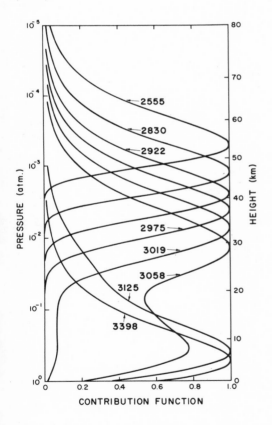

Fig. 2. Contribution to the backscattered radiation from various levels in the atmosphere. Each curve is normalized to unity at level of maximum. (Surface reflectivity = 0, solar zenith angle = 60°, nadir angle of observation = 0°, total ozone = 0.336 atm cm.)

According to Fig. 2, the radiance measurements contain information about vertical ozone distribution from the earth's surface up to about 60 km. However, because of the rather large half-width of the curves, of the random errors introduced by the instrument, and of random deviations in the atmosphere during the period of measurement, a fundamental limitation is placed on the vertical resolution that can be obtained by the inversion of these measurements (Wei, 1962; Mateer, 1965; Twomey, 1965, 1966), especially at lower altitudes where the resolution is particularly poor because of the bimodal nature of the weighting curves (e.g., 3058 Å). Therefore an iterative scheme (Mateer, 1972) is used in the derivation of the total amount of ozone assuming certain statistics for its vertical distribution, and using the measured ratios of earth radiance to solar irradiance at each wavelength. Total amounts of ozone can be determined within an accuracy of about 5%.

The distribution of ozone with height above 30 km is determined by means of an iterative inversion of the measured radiance ratios at wavelengths shorter than 2976 Å. For altitudes below 30 km, statistical methods (Strand and Westwater, 1968) may be used to infer the ozone distribution with height.

### 2.2.4. Photometers to monitor ultraviolet solar energy

A monitor of ultraviolet solar energy (MUSE) experiment was flown on Nimbus III and Nimbus IV. A detailed description of these observations has been given by Heath (1971). The absolute solar flux in the 1100–3000 Å region has been measured continuously since April, 1969. Each of the five wavelength intervals was several hundred Ångstroms wide. Solar energy in this region represents the major radiative energy input into the lower thermosphere, mesosphere, and upper stratosphere. This wavelength range also covers the transition from photospheric to chromospheric radiation which passes through the region of the solar temperature minimum. The instrument consists of five vacuum photodiodes, each with a nominal 90° field of view. The sensors are fully illuminated by the sun for about 20 min of each orbit, and they view the sun at near normal incidence on every crossing of the terminator in the north polar regions. The angle of solar illumination to the normal of the face of the sensors is measured precisely with a solar aspect sensor. The short wavelength response is determined by a suitable radiation resistant optical filter. The long wavelength cutoff is achieved through the use of photocathode materials of varying degrees of "solar blindness."

## 3. RESULTS OF OBSERVATIONS

### 3.1. The mean thermodynamic structure and circulation of the mesosphere from rocket soundings

The mean profiles of temperature and pressure were computed from 227 soundings carried out from five sites which covered a wide range of geographical latitudes during all seasons. The launch sites included: Natal, Brazil (6°S) and Ascension Island (8°S) which were combined to represent a tropical regime; Wallops Island, Virginia, U.S.A. (38°N) representing a temperate regime; Churchill, Manitoba, Canada (59°N) representing a subarctic regime; and Point Barrow, Alaska, U.S.A. (71°N) representing an arctic regime. (See Smith et al., 1964, 1966, 1967, 1968a, 1969, 1970, 1971.)

Soundings conducted in December, January, and February were averaged to produce mean winter profiles; those conducted in June, July, and August were averaged to produce a mean for summer; and those in March, April, May, September, October, and November were averaged to produce mean profiles for the transition seasons at the Wallops Island, Churchill, and Barrow sites. Since only a very small seasonal dependence was detected in the Natal–Ascension soundings, the data for all months were averaged together to produce a mean annual profile for the tropics. The numbers of soundings and the months included in each model are listed in Table I. Profiles are shown in Fig. 3.

TABLE I

Number of soundings included in each mean model atmosphere

| Site | Winter | Summer | Spring/Fall | Annual |
|------|--------|--------|-------------|--------|
|      | Dec. Jan. Feb. | Jun. July Aug. | Mar. Apr. May Sep. Oct. Nov. | Jan. to Dec. |
| Natal–Ascension (6–8°S) | – | – | – | 34 |
| Wallops (38°N) | 28 | 24 | 41 | – |
| Churchill (59°N) | 29 | 12 | 13 | – |
| Barrow (71°N) | 19 | 17 | 10 | – |

The mean temperature profile for the tropical sites, indicates a mean stratopause temperature of about 270°K, and an indistinct mesopause of about 200°K at 80 km. The mesospheric lapse rate is small and no substantial seasonal effect is observed. The mean seasonal temperature profiles for the middle latitudes (Wallops), indicate that the stratopause temperature is higher than 270°K during all seasons. The mean mesopause temperature is 180°K in summer, and about 200°K in winter. Note that the structure in the mesosphere is more disturbed than in the tropics during all

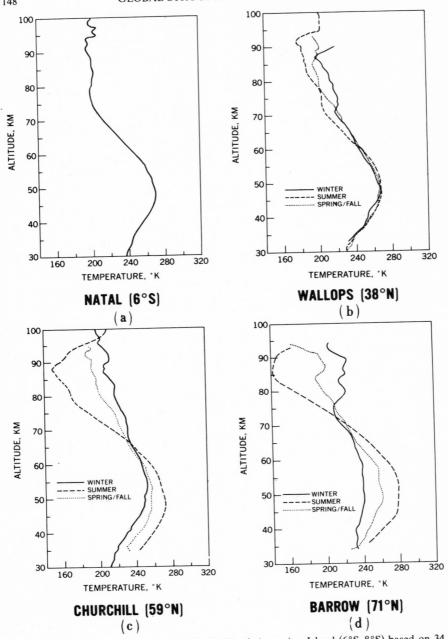

Fig. 3. Mean annual temperature profile for Natal–Ascension Island (6°S–8°S) based on 34 soundings (a); mean seasonal temperature profiles for Wallops Island (38°N) based on 28 winter, 24 summer and 41 transition soundings (b); for Churchill (59°N) based on 29 winter, 12 summer and 13 transition soundings (c); and for Barrow (71°N) based on 19 winter, 17 summer and 10 transition soundings (d). (After Theon et al., 1971.)

seasons, and that distinct differences occur with the change of season. The mean seasonal profiles for the subarctic site (Churchill) show an even more pronounced seasonal variation of temperature. The mean summer stratopause temperature exceeds 270°K, while the mean winter stratopause temperature is only about 250°K. The mean mesopause temperature is as low as 150°K in summer, while in winter, the mesopause is undiscernable; during winter, the mean temperature above the stratopause is never lower than 195°K. The mean profiles for the arctic site (Barrow) show a 40°C variation of the stratopause temperature, and a 80°C variation in the mesopause temperature with season. Note the smooth, steep lapse rate in summer, and the very cold (140°K) mean mesopause temperature. The winter profile, on the other hand, shows a very shallow lapse rate, and a warm upper mesosphere.

Comparisons of these data with the U.S. Standard Atmospheres Supplements (COESA, 1966) demonstrate that some substantial differences exist between the statistical mean of these observations and the "Standard" profiles for the corresponding latitude and season (Theon et al., 1970). Of course, the argument can be made that the observations are valid for only one station while the standard model is an attempt to give a representative value for a given latitude (i.e., all longitudes). However, the standard profiles were drawn from earlier observations which were sparser than the data reported here, and, as is evident from recent observations, conditions at any one latitude vary greatly with longitude.

Quasi-meridional cross-sections along a diagonal path traced across the North American continent from Barrow southeastward through Churchill, Wallops Island and across the western Atlantic Ocean to Natal have been drawn from seasonal mean profiles of Fig. 3. The resulting temperature cross-section, shown in Fig. 4, forms an organized pattern dominated by the warm stratopause and cold mesopause of the high latitudes in summer, and the almost isothermal structure of the high latitudes in winter. These features are, in general terms, similar to the earlier models of Murgatroyd (1957) except that the high latitude summer stratopause and mesopause are colder than in Murgatroyd's model. Fig. 4 is also similar to the cross section given in the U.S. Standard Atmospheres Supplements (COESA, 1966), except that the winter mesosphere shown here is colder.

Combining the mean seasonal pressure profiles into the same quasi-meridional cross-section results in Fig. 5. Here the values are shown in percent difference from the COESA U.S. Standard Atmosphere, (1962). Note that the zero percent difference line (i.e., exact agreement with the standard) is most nearly approximated by a low latitude pressure profile in winter, while a well developed low pressure region dominates the winter mesosphere at higher latitudes, and a high pressure region dominates the summer mesosphere. These latitudinal pressure differences drive the mean circulation in the mesosphere, and are consistent with the observed winds. The low pressure (cyclonic) region in the winter mesosphere underlies a high pressure (anticyclonic) region, and the high pressure region in the summer meso-

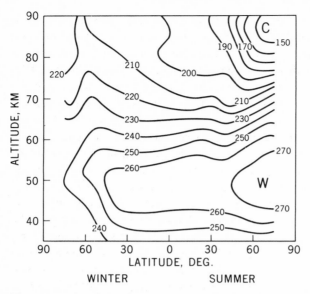

Fig. 4. A quasi-meridional cross-section of temperature in °K derived from mean profiles in Fig. 3. (After Theon and Smith, 1970.)

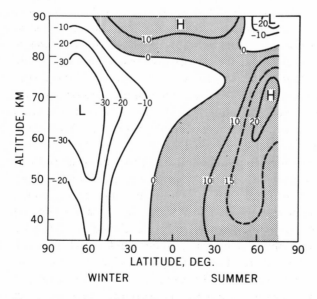

Fig. 5. A quasi-meridional cross-section of pressure (as percent difference from the COESA U. S. Standard Atmosphere, 1962), derived from mean profiles in Fig. 3. (After Theon and Smith, 1970.)

sphere underlies a low pressure region. This vertical alternation of pressure systems closely resembles the patterns observed in the troposphere, but the vertical scale sizes are much larger in the upper atmosphere. The patterns shown in Fig. 5 are not necessarily an exact description since we expect that the latitudinal and vertical extent of these systems varies with longitude. Such variation cannot be determined from the observations shown here, since they were made only over a very limited longitudinal segment. Nevertheless, the basic nature of the atmospheric mass distribution in the mesosphere is quite evident. For example, latitude zones where the tightest horizontal pressure gradients are shown in Fig. 5, namely, near 60 km at 45° in winter, are also zones where the most intense zonal winds occur. This is consistent with the observed wind fields. An analysis of the circulation of the mesosphere was made by Theon and Smith (1971) based on the monthly and seasonal mean values of pressure and winds observed from three sites over North America. These observations provide a considerable improvement over earlier circulation estimates for this region, which were based on observations at only one or two sites.

Fig. 6 presents mean seasonal "weather" maps over the North American continent for the 60- and 80-km levels. These maps are polar stereographic projections with the north pole indicated by the + at the top center of the figure. Longitudes radiate from that point, and latitudes are concentric circles, the center of which is the pole. The maps were analyzed by plotting the mean wind and mean pressure for the appropriate level for each of the three sites previously mentioned. In addition, the mean wind data from the Meteorological Rocket Network (MRN) in January and July were plotted in the winter and summer respectively, to aid in the analyses. As the altitude increases, horizontal pressure gradients become weaker, necessitating the choice of smaller intervals of pressure to describe the flow at higher levels. These analyses are geostrophic, which means that the curvature of the isobars, friction, and all short term effects have been neglected. Fig. 6a, which gives the mean winter circulation at 60 km, indicates that the flow is dominated by a vortex centered over north central Canada, far from the geographic pole. This circulation produces strong westerly winds over most of the continent, and a predominately northerly component over Alaska. By extrapolating the analysis over the Atlantic (broken lines), one may infer strong southerly winds over Greenland. The polar asymmetry of the flow implies the transport of atmospheric properties across latitude circles. The mean summer circulation at 60 km (Fig. 6b) is dominated by an anticyclone whose center cannot be determined from the available data. This pattern produces easterly winds over most of the continent with the strongest zonal components occurring along the southern portion of the United States. The pressure gradients are smaller in magnitude in summer than in winter, producing generally lighter winds. Fig. 6c shows the mean winter circulation at 80

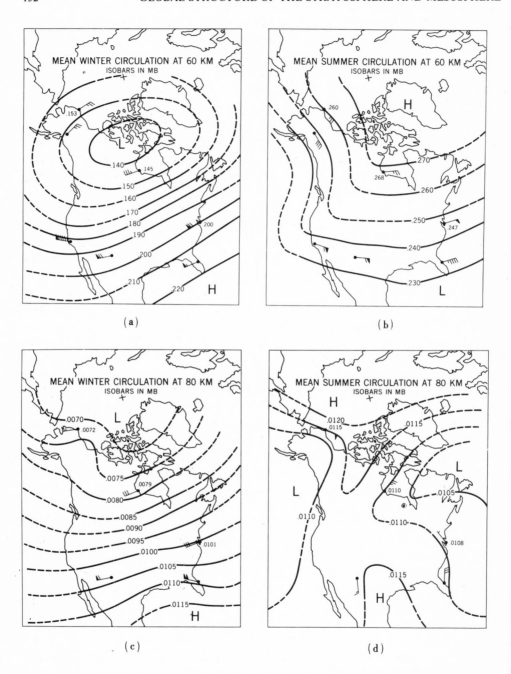

Fig. 6. Mean winter circulation at 60 km (a) and 80 km (c), and mean summer circulation at 60 and 80 km (b and d respectively). Isobars are given in mbar, winds in m sec$^{-1}$. (After Theon and Smith, 1970.)

km. This coherent circulation is somewhat unexpected since the data from individual soundings fluctuate widely. The averaging process appears to filter out most of the ageostrophic components which appear to be quite large in individual soundings. The prevailing west winds, which are generated by the strong vortex at 60 km, remain essentially intact up to 80 km. But, two disturbances stand out in the mean zonal flow: the ridging over eastern Alaska, which produces a southerly wind over Barrow, and the divergent flow between Wallops Island and Cape Kennedy. The mean circulation in summer at 80 km, shown in Fig. 6d, is vastly different from that at 60 km. The high pressure region to the north of the continent still exists in summer, but ridging appears across the center of the continent in a N–S direction producing a seemingly chaotic circulation. Low pressure regions extend onshore from both the Atlantic and Pacific Oceans, and the flow is generally light (except for Barrow). The appearance of this map may result from the breakdown of the geostrophic assumption at these altitudes. If tides and/or gravity waves dominate the flow, as discussed in the next section, then large accelerations in the flow occur, and the geostrophic balance no longer applies. Averaging a sample of this small size may not adequately remove short-term influences, since the variability of the individual soundings about the mean is quite large. Thus, Fig. 6d may not represent a true mean circulation.

## 3.2. Small-scale and short-term variations in the vertical structure of the mesosphere

Since 1967, we have conducted soundings of the atmosphere above 50 km, primarily utilizing the acoustic grenade and pitot-static tube techniques to observe short-term variations in the temperature, pressure, and density profiles to altitudes as high as 120 km. Soundings closely spaced in time have demonstrated that the structure of the atmosphere above the stratopause is subject to rapid and dynamic changes. Temperature variations are used here to demonstrate clearly the variability of the upper atmosphere.

Superimposed upon the gross seasonal differences in temperature presented in section 3.1, there are temperature changes which have been observed to occur over short periods of time, e.g., days or hours. Examination of these observations reveals that the variations in the tropics are quite different from those which occur at middle and high latitudes (Theon, 1968). Several different mechanisms appear to be responsible for these short-term fluctuations in the upper atmosphere, and several theories have been advanced to explain the observations. Hines (1960) proposed internal gravity waves to explain the irregular motions observed in the lower thermosphere, and he suggests that this gravity wave mechanism may also be the cause of the wave-like structure in the temperature profiles lower in the atmosphere. Lindzen (1967) proposed thermally driven diurnal tides to explain the periodic motions observed in the stratosphere, mesosphere, and lower thermosphere. In both cases, the propagation of energy from lower levels upward disturbs the temperature and wind profiles of the upper atmosphere. Examples of observed, short-term disturbances in the temperature profiles, and comparisons of these with the appropriate theory follow.

### 3.2.1. Thermal tides

Large variations in individually observed tropical temperature profiles occur, despite the absence of seasonal changes in the mean profiles. An example of the variability of the tropical upper atmosphere is given in Fig. 7, which shows three temperature profiles observed with the acoustic grenade technique at Natal, Brazil (6°S) in October, 1966. This series of soundings was initiated at sunrise, and the second and third soundings were launched approximately 12 and 24 h later, at the following sunset and sunrise, respectively. Temperature changes of up to 50°C (at 91 km) occurred during the two 12-h periods, and these far outweighed the small changes which occurred during the 24-h period. Thus the diurnal variation was far greater than other variations. In another instance, a 100°C temperature change was observed in 12 h at 105 km, with a pair of pitot-static tube soundings conducted at the equator in March, 1965. In Fig. 7, the temperature changes vary considerably with altitude so that heating occurred at the 68- and 90-km levels and cooling occurred at the 80-km level during the first 12-h period.

When the profiles in Fig. 7 are paired and the temperatures at sunrise are subtracted from the temperatures at each sunset, the two profiles of temperature

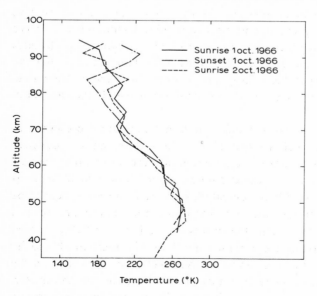

Fig. 7. Temperature profiles at Natal (6°S). The soundings were separated in time by twelve hours. (After Smith et al., 1968b.)

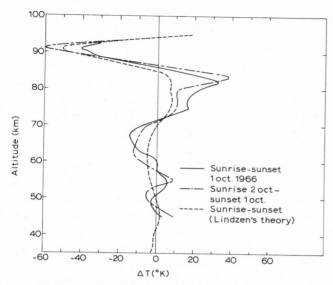

Fig. 8. Twelve hour temperature changes as a function of altitude at Natal compared with theoretical change. (After Smith et al., 1968b.)

differences, shown in Fig. 8, result. These temperature difference profiles change sign and increase in amplitude with altitude. The dashed curve in Fig. 8 gives the temperature changes which were derived by Lindzen from theoretical consideration of the thermal tides. The agreement between observation and theory is good above 70 km, although the theory underestimated the amplitude of the cooling above 90 km. The agreement is remarkable considering that the theoretical profile is based on assumed vertical distributions of temperature, water vapor, and ozone.

According to theory, these temperature variations are due to solar heating of water vapor in the troposphere and ozone in the stratosphere; these then generate periodic disturbances that propagate upward into the mesosphere and lower thermosphere. The amplitudes of these disturbances are small in the region where they are generated, but they increase inversely as the square root of the density. The phase of the disturbance varies continuously and in an irregular manner with altitude. The distinguishing characteristic of the short-term temperature variations produced by the thermal tides is their regular, periodic recurrence. In Lindzen's model, the amplitude of the thermal tide diminishes rapidly with increasing latitude. Thus, the diurnal temperature variations produced by this mechanism become small at middle and high latitudes, and are masked by other effects.

### 3.2.2. Gravity waves

Wave-like features in the temperature structure of the mesosphere and lower thermosphere, which are superimposed upon the average seasonal profiles, have been observed at both middle and high latitudes in winter, and at middle latitudes in summer. Fig. 9 gives typical temperature profiles obtained with the acoustic grenade technique at a mid-latitude site, namely Wallops Island, Virginia (38°N). These profiles were observed over a period of several years and grouped according to season. The three upper plots are for consecutive winters, and the two lower plots are for consecutive summers. For two of the three winters, temperature fluctuations were much greater than in summer; while during the winter of 1965, a stratospheric warming was observed, and fluctuations in the mesosphere were relatively small. Typical temperature profiles obtained with acoustic grenades for the winters and summers of several years at Churchill, Canada (59°N) are shown in Fig. 10. The average amplitude for the winter temperature fluctuations is about 30°C, but in summer, there is no consistent wave-like structure in the mesosphere. If the profiles in Fig. 10 are compared with the seasonal averages given in section 3.1, the summer profiles are seen to be well-described by a single average temperature profile, but the winter mesosphere changes so dynamically from day to day that the average winter profile is not representative of any given time. At even higher latitudes, namely, at Point Barrow, Alaska (71°N) this contrast is still more pronounced. The summertime mesosphere temperature profile is extremely uniform, as can be seen in Fig. 11. The seven acoustic grenade soundings were made over periods spanning two summers; yet, the temperature spread is no greater than 10–15°C at any given altitude below the mesopause, and there is no trace of the wave-like structure. Compare these results with the 6 temperature soundings given in Fig. 11b. The latter were observed with the acoustic grenade technique during a 15-h period at Barrow on 31 January–1 February, 1967. Temperature changes of up to 80°C were observed to occur in less than 3 h. The wavelengths of the structure averaged about 10–15 km, and the amplitudes increased with altitude (Smith et al., 1968b). The following analysis was performed on the soundings shown in Fig. 11b; average temperatures were computed at each height level from the six soundings, and were subtracted from the temperatures of each of the individual soundings. These differences were then plotted in Fig. 11c as a function of altitude in a time cross section for the 15-h period. Isolines for zero temperature change were drawn to separate the zones of positive difference (heating) from the negative difference (cooling). The phase of the waves was assumed to propagate downward. An average wave period of approximately 200 min is obtained from Fig. 11c, with minimum periods of 85 min and maximum periods of 330 min occurring at the higher and lower altitudes, respectively.

Temperature variations observed at middle and high latitudes thus have much shorter periods than the diurnal waves observed in the tropics. As at low latitudes, there is a definite periodicity, and amplitudes increase with height. There is some selective mechanism which allows these disturbances to propagate upward in winter, but not in summer at high latitudes. Perhaps this dependence on season is a result

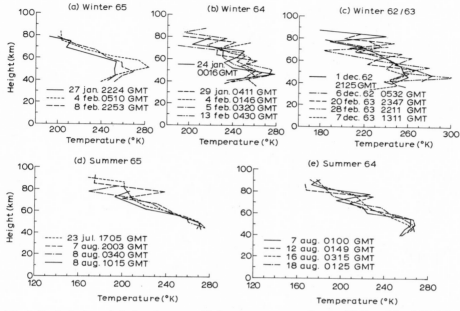

Fig. 9. Typical winter (a,b,c) and summer (d,e) temperature soundings at Wallops Island (38°N).

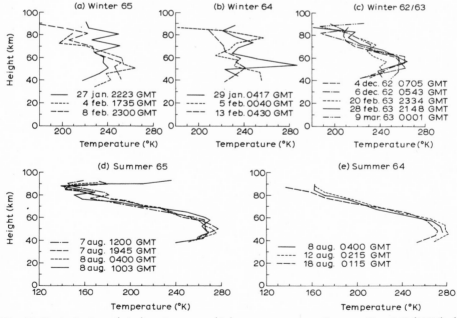

Fig. 10. Typical winter (a,b,c) and summer (d,e) temperature soundings at Churchill (59°N). (After Theon, 1968.)

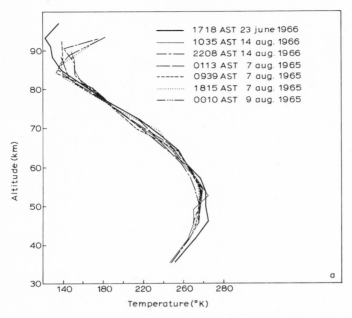

Fig. 11a. Summer temperature profiles at Barrow (71°N). (After Theon, 1968.)

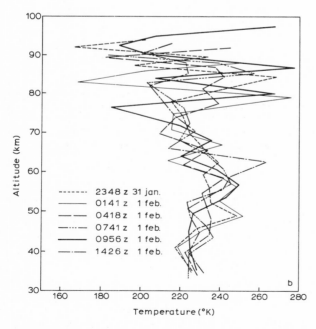

Fig. 11b. Winter temperature profiles at Barrow in 1967. (After Smith et al., 1968b.)

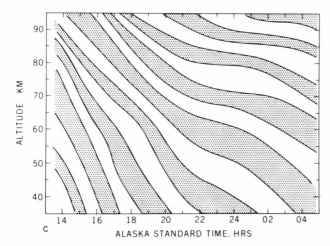

Fig. 11c. Sign of temperature variations for 31 January–1 February 1967 at Barrow, Alaska (71°N). Phase propagation is downward.

of the generating mechanism at low levels, or of the vertical wind shears suggested by Hines and Reddy (1967). The regular vertical wavelength of 10–15 km, and the relatively short period suggest that a mechanism such as the internal gravity waves proposed by Hines are responsible for the variations, which amount to as much as 80°C in 3 h near the mesopause. At higher altitudes (120 km), temperature changes of over 150°C were observed at Churchill, but these large temperature variations may have been caused by intense auroral activity which was observed at the same time.

### 3.3. Temperature structure in noctilucent clouds

Noctilucent clouds have been a subject of interest for many years because of the great heights above the earth at which they occur. These clouds have generally been observed at altitudes ranging from 78 to 90 km: the majority of observations indicates a cloud height of about 83 km (Fogle, 1966). They have been sighted only at high latitudes, and most frequently during the 4- to 6-week period following the summer solstice. Noctilucent clouds are visible only during the time that the sun is below the observer's horizon, and they are directly illuminated by sunlight against a darkened sky background. High latitudes provide favorable geometry for observation of the clouds for considerably longer periods each day than middle or low latitudes do, but the fact that clouds exist only at high latitudes and only during summer can be explained in terms of the temperature structure of the mesosphere at these latitudes.

The acoustic grenade technique was employed to provide measured profiles of atmospheric temperature, pressure, density, and wind up to an altitude of 95 km during displays of noctilucent clouds. A total of ten grenade soundings were carried out from Kronogard, Sweden (66°N) (Witt et al., 1965), and Barrow, Alaska (71°N), during the summers of 1963 through 1965 to determine the relation between the mesospheric temperature structure and the occurrence of noctilucent clouds. Results of soundings at Barrow (August, 1965) and Kronogard (1963) are given in Fig. 12. As discussed in section 3.2, these profiles display remarkable uniformity. The mesopause temperatures ranged from 130 to 148°K, and the steep uniform lapse rate, which is typical of the high latitude summer mesosphere, was observed in all the soundings. An error analysis has been performed on these data, and the errors associated with the temperatures mentioned above are 1 to 3°C. Very little change in the temperature structure can be seen between soundings made in the presence and absence of noctilucent clouds. The first sounding at each site was conducted during a display of noctilucent clouds, and the second served as a control sounding, having been carried out in confirmed absence of the clouds. At Barrow, the minimum temperature of the profile made in the presence of clouds was 139°K, which is 3°C warmer than the mesopause temperature of the control sounding. At Kronogard, however, the minimum temperature of the profile made during the cloud display was 130°K, or 18°C colder than the mesopause temperature of the control.

The profile of the average temperature for five soundings conducted during displays of noctilucent clouds (Fig. 13), and the average profile for three soundings conducted during the confirmed absence of these clouds do not differ by more than 5°C at any point between 45 and 90 km. The warm stratopause, the steep uniform

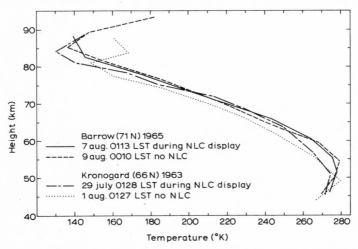

Fig. 12. Temperature profiles above Kronogard, Sweden (66°N), and Barrow (71°N) during the summers of 1963 and 1965, respectively. Note that the Barrow mesopause was colder in the absence of noctilucent clouds than during the cloud display, but the reverse was observed at Kronogard. (LST = local standard time; NLC = noctilucent cloud.) (After Theon et al., 1967.)

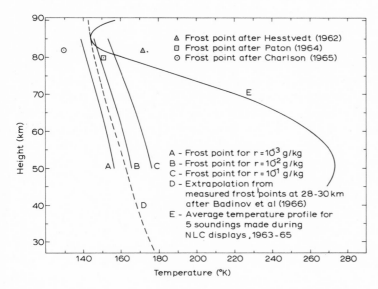

Fig. 13. Comparison of estimates of frost points by several authors, with the average temperature profile for the five soundings conducted during displays of noctilucent clouds. (After Theon et al., 1967.)

lapse rate, and the extremely cold mesopause are essentially identical for both average profiles. For five soundings conducted during displays of noctilucent clouds, mesopause temperatures varied from 130 to 147°K. Correspondingly, for soundings conducted in the absence of clouds, the mesopause temperatures ranged from 129 to 149°K. Thus the coldest temperatures did not necessarily produce noctilucent clouds, but the clouds were always accompanied by mesopause temperatures less than 150°K.

There are two schools of thought concerning the composition of noctilucent cloud particles. One theory assumes the presence of sufficient water vapor at the mesopause to form ice particles by a process of saturation and condensation (Humphreys, 1933). The other does not accept the presence of water vapor and ice, but explains the noctilucent clouds in terms of the light-scattering properties of the dust alone (Ludlam, 1957), which is believed to originate from the vaporization of incoming meteors or from the surface of the earth. Sampling experiments have been conducted to resolve the question of the composition of cloud particles, and traces of a volatile substance believed to be ice were found surrounding many of the larger particles obtained from a noctilucent cloud (Hemenway et al., 1964).

In view of the temperature data reported above, it appears that the occurrence of noctilucent clouds depends largely on the amount of water vapor present at the mesopause. Khvostikov (1966) has postulated that "Noctilucent clouds appear in the atmosphere at the place and the time where and when the temperature of the air turns out to be low enough". However, the results shown in Fig. 12 indicate that a given low temperature alone is not sufficient to produce noctilucent clouds, unless the low temperature occurs in conjunction with sufficient water vapor. Thus the water vapor content of the high atmosphere must be considered.

Seasonal variations of the mesopause temperature not only produce sufficiently cold conditions for the formation of noctilucent clouds, but also provide a circulation consistent with the transport of water vapor to the mesopause at high latitudes in summer. According to section 3.1, the high-latitude summertime mesopause is about 80°C colder than the wintertime mesopause. These seasonal variations of temperature are not consistent with considerations of radiation alone, since at high latitudes, the summer mesosphere is heated almost 24 h a day, and the winter mesosphere is dark almost 24 h a day. Leovy (1964) demonstrated that a meridional circulation superimposed on an atmosphere in radiative equilibrium produced good qualitative agreement with observed seasonal variations of temperature. This meridional circulation caused ascending motion at the summer pole and descending motion at the winter pole, thereby transferring heat from the radiatively heated upper atmosphere of summer to the heat-deficient upper atmosphere of winter.

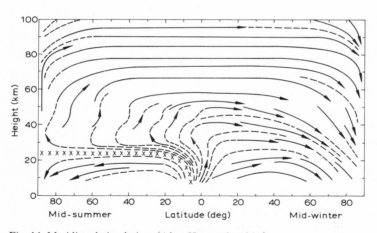

Fig. 14. Meridional circulation. (After Hesstvedt, 1964.)

Hesstvedt (1964) used a meridional circulation similar to that proposed earlier by Murgatroyd and Singleton (1961) to explain the presence of water vapor in the upper atmosphere. This model is also consistent with the observed variations of temperature. As can be seen in Fig. 14, Hesstvedt's model shows that the source of water vapor is the tropical troposphere, and that water vapor enters the stratosphere through the gap in the tropopause. From this relatively narrow latitudinal band near the equator, air rises to an altitude of 25 km, moves meridionally toward the summer pole, and then ascends rapidly at high latitudes. Poleward of 60° latitude in the summer hemisphere, air at 80 km is seen to originate from the equatorial troposphere. Such a model qualitatively explains the mechanism both for transporting water vapor to the summer mesopause and for transferring heat from the summer mesosphere to the winter mesosphere, thereby accounting for the observed seasonal variations in the temperature structure. There is no more reason to believe that water vapor is homogeneously distributed in the high-latitude mesosphere than to believe that such a situation exists in the troposphere. The meridional circulation shown in Fig. 14 represents the average motion in the stratosphere and mesosphere, but this circulation is subject to frequent and dynamic changes, and therefore, any assumptions of steady-state flow or homogeneous composition are unrealistic. Thus the low temperatures at the mesopause may not produce noctilucent clouds if there is insufficient water vapor available. Questions of the amount of water vapor available and the magnitude of the vertical velocity necessary to transport water vapor to the mesopause must remain unanswered for the present, since no in situ measurements have been made to confirm or to refute the estimates and extrapolations that have been published. Fig. 13 shows that Hesstvedt (1962) estimated the mixing ratio at the mesopause to be $1 \text{ g kg}^{-1}$, which corresponds to a frost point of 172°K at 82 km. Hesstvedt (1964) later revised this value downward to the order of $10^{-2} \text{ g kg}^{-1}$, which is consistent with Paton's (1964) estimate of $6 \cdot 10^{-3} \text{ g kg}^{-1}$ ($6 \cdot 10^{-3} \text{ g kg}^{-1}$ corresponds to a frost point of 150°K at 80 km). Charlson (1965) used the mesopause temperature measured at Kronogard (130°K) as a conservative approximation to develop a steady-state model for noctilucent clouds. Thus, estimates of the water vapor content at the mesopause have grown smaller with the observations of lower mesopause temperatures. Fig. 13 also gives a comparison of the frost points for various mixing ratios, and a curve which has been extrapolated from measurements at altitudes reached by balloon-borne instrumentation, with the average temperature profile for the five soundings made during displays of noctilucent clouds. Curves $A$, $B$, and $C$ were computed for constant mixing ratios, with the use of the pressure profile derived from the average of 15 measured temperature profiles at high latitudes in summer. Curve $A$ gives the frost points for a mixing ratio of $10^{-3} \text{ g kg}^{-1}$, curve $B$ for $10^{-2} \text{ g kg}^{-1}$, and curve $C$ for $10^{-1} \text{ g kg}^{-1}$. Badinov et al., (1966), have extrapolated values for frost points from measurements made by various techniques at balloon altitudes of 28 to 30 km, and these are given by curve $D$. The average temperature profile for the five soundings of noctilucent clouds is given by curve $E$. The mesopause temperature of 143°K corresponds to a saturation mixing ratio of approximately $1.3 \cdot 10^{-3} \text{ g kg}^{-1}$. It must be remembered, however, that this is an average value of the mixing ratio and that both cooler and warmer temperatures, corresponding to lower and higher saturation mixing ratios, have been measured during the cloud displays.

Since no significant difference in the observed temperature was noted between the soundings conducted in the presence of noctilucent clouds and those conducted in the absence of the clouds, it appears reasonable to conclude that a mesopause temperature of less than 150°K is a necessary, but not sufficient, condition for the existence of noctilucent clouds. Variability of the water vapor content at the mesopause is also believed to be a key factor in the occurrence of these clouds. In view of the circulation that is implied by the seasonal differences in temperature in the mesosphere, small amounts of water vapor may be transported by this circulation into the mesosphere during the summer at high latitudes. It is at these latitudes that saturation or super-saturation occurs in the narrow layer at the mesopause where these extremely cold temperatures occur. Dust particles, probably originating from incoming meteors, serve as sublimation nuclei in this saturated region, and grow to sufficient size to scatter sunlight, thus producing noctilucent clouds.

### 3.4. Global stratospheric temperature fields observed with satellites

The first global patterns of stratospheric temperatures observed with satellites were presented by Kennedy and Nordberg (1967). They dealt with the results from Tiros VII which produced measurements of temperatures "weighted" over a broad slab of most of the lower stratosphere. Nevertheless, these patterns provided a synoptic insight into the seasonal variations of that region which was not possible with previous conventional observations. A total of about 70 "quasi-global" maps, from 65°N to 65°S were analyzed by Kennedy (1966). Two of these maps are shown here in Figs. 15 and 16. The temperature patterns are indicative of the summer polar anticyclones and the winter polar vortices. During June, 1963, the warm core was centered over the north polar cap, and the cold core over the south polar cap. A significant deviation in the symmetry of the rather closely spaced isotherms around the south pole occurs over the central South Pacific. A ridge of warm air (high pressure) extends to high latitudes southeast of Australia and over New Zealand while cold air (low pressure) penetrates to low latitudes over the eastern Pacific Ocean, suggesting anticyclonic circulation over Australia. Differences of about 6°C in equivalent black-body temperatures along latitude circles are observed between the western and eastern South Pacific. This asymmetry was observed during both the 1963 and 1964 southern hemisphere winters, and was observed again with greater detail in the south polar region during 1966 with the Nimbus II satellite (see Fig. 19, below). During September 1963, the warm air ridge spread considerably westward so that the equivalent black-body temperature difference between the Pacific and the Indian Oceans along 60°S was almost 20°C.

Northern hemisphere isotherms display a pattern which is well correlated with the circulation features already known from radiosonde observations in this region (Fig. 16). This correlation lends confidence to the application of satellite observations to infer circulation in areas where no conventional observations exist.

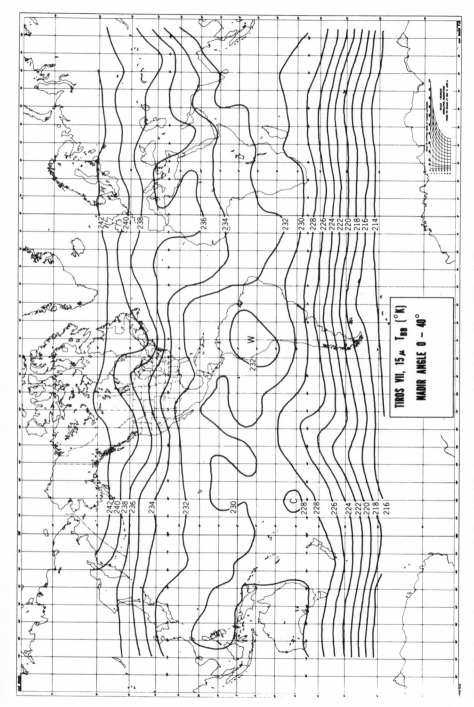

Fig. 15. Isotherms for average equivalent black-body temperatures derived from Tiros VII radiation observations during the period 19 June–28 June 1963. Numbers along isotherms refer to °K. Radiation observations were restricted to nadir angles 0–40°. (After Kennedy and Nordberg, 1967.)

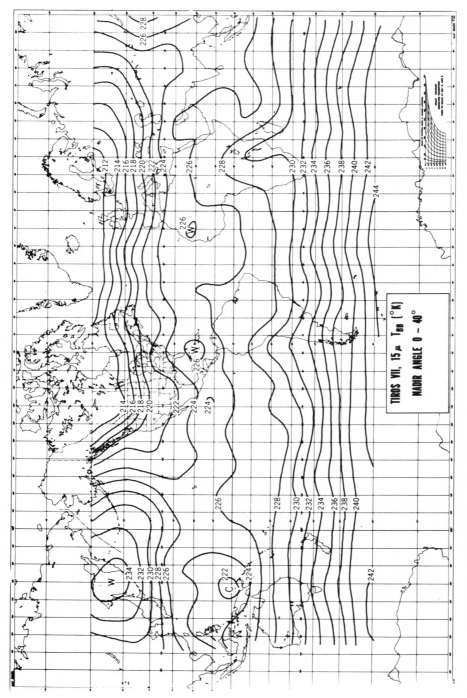

Fig. 16. Same as Fig. 15 for period 10–19 December 1963.

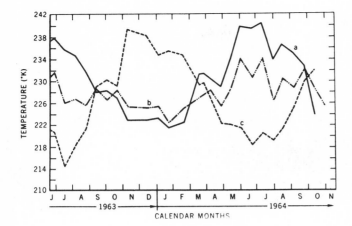

Fig. 17. Variations of average black-body temperatures derived from Tiros VII radiation observations with time for: (a) latitudes 40°–65°N; (b) latitudes 30°N–30°S; and (c) 40°–65°S. Averages are for typical 10-day periods for each respective season. (After Kennedy and Nordberg, 1967.)

Temperatures in the tropics are considerably lower during December–January than they are during June–July (Fig. 17). There is an almost perfect phase relationship between the high latitude northern hemisphere and tropical tempera- ture curves (a and b) and a 180° phase lag between the tropical and high latitude southern hemisphere temperature cycle (curves b and c). Note the almost identical zonal average temperatures at high northern and southern latitudes, and in the tropics during the equinoxes. The difference in the amplitudes of the high latitude curves in the two hemispheres in winter indicates that the magnitude of the temperatures in the stratosphere are strongly influenced by probably both radiative exchange with the surface and lower atmosphere, and eddy transfer of energy within the stratosphere. The warm anti-cyclone over the North Pacific is obviously responsible for the fact that zonal average temperatures between 40–65°N are 4–6°C higher in December–January than corresponding temperatures between 40–65°S during June–July. The almost equal amplitudes of the two high latitude curves in Fig. 17 during summer indicate that such effects are of lesser consequence in summer.

The Tiros VII data also permitted, for the first time, the analysis of the development of the final stratospheric warming in the northern and southern hemispheres during March, 1964 and September, 1963, respectively. The essence of this analysis is illustrated in Fig. 18a and b.

Similar measurements were made with Nimbus II (Warnecke and McCulloch, 1967). In this case, because of the higher inclination of the satellite orbit, these measurements included both polar regions. The position of the southern polar vortex, with coldest temperatures located over the South pole is shown in Fig. 19 as

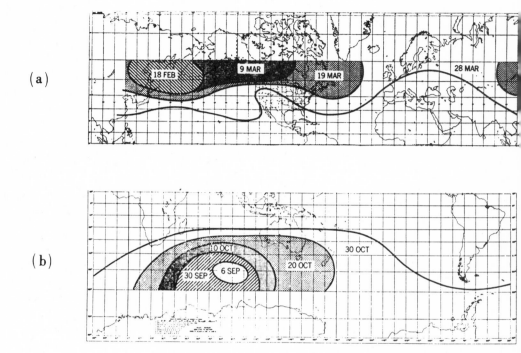

Fig. 18. Progression of the final warming in the northern hemisphere stratosphere in 1964 (a) and for the southern hemisphere in 1963 (b). Solid lines indicate the 230°K equivalent black-body isotherms for the dates shown. (After Kennedy and Nordberg, 1967.)

derived from Nimbus II observations on June 10, 1966. The temperature pattern is extremely asymmetric. Warmer mid-latitudes and stronger temperature gradients than elsewhere exist toward the Australian and western Pacific side of the hemisphere. This asymmetry, which has been observed consistently in the Antarctic winter with the instruments on Tiros VII and Nimbus II, III, and IV, seems to indicate that dynamic processes strongly counteract the radiative heat loss which is responsible for the cold air cyclone in the southern hemisphere winter stratosphere.

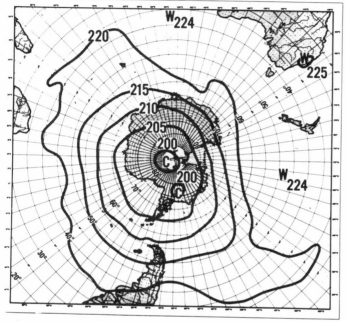

Fig. 19. Stratospheric temperature distribution (°K) from Nimbus II measurements on June 10, 1966.

Global stratospheric temperature analyses were made by Prabhakara et al. (1971) based on the "inversion" of 14 to 15 $\mu$m radiances measured with the IRIS. In that analysis, global temperature maps for both the 10-mbar and 50-mbar levels were drawn for the period 25 to 29 April 1969. Fig. 20 shows one set of these maps for the 50- and 10-mbar levels. The 50-mbar (approximately 21 km) map reflects the lower stratospheric regime: the middle and high latitudes of the northern hemisphere display a series of waves of wave number 4 and 5 in the zonal temperature pattern around the globe; the tropical regions between 20°N and 20°S are characterized by a belt of very cold air separating warmer air on both sides. The temperatures, in general, increase from the equator (approximately 208°K) to the North pole (approximately 228°K), while toward the southern hemisphere warm air (228°K over the Indian Ocean and 218°K over the eastern Pacific) is sandwiched

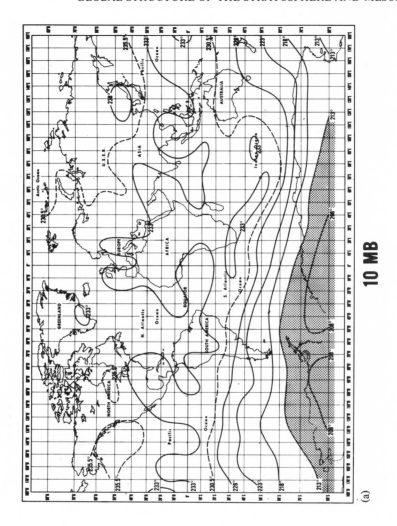

**10 MB**

(a)

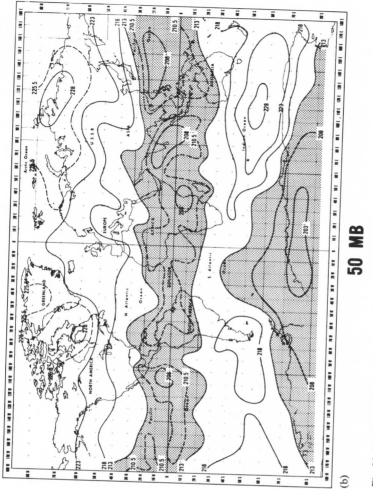

**50 MB**

(b)

Fig. 20. Nimbus III IRIS temperatures (°K) on April 29, 1969 for 10-mbar level (a) and for 50-mbar level (b). (After Prabhakara et al., 1971.)

between the cold tropics and the cold vortex over the Antarctic. This is in accord with the earlier Tiros and Nimbus II observations. The 10-mbar (approximately 32 km) level maps show a steady increase of temperature from the South pole (208°K) to the North pole (235°K) indicating that the cooling of the southern hemisphere has progressed considerably further at this level than at 50 mbar. The characteristic asymmetry in the south polar vortex mentioned earlier is still evident at the 10-mbar level.

The most complete analysis to date of stratospheric temperatures observed by satellites was made by Fritz and Soules (1970). Radiances measured by the SIRS instrument on Nimbus III in a narrow 5 cm$^{-1}$ wide spectral interval near 15 $\mu$m were analyzed globally for the period April, 1969 to April, 1970. No attempt was made to convert the radiances to atmospheric temperatures, but the radiance patterns in themselves are extremely indicative of the temperature structure of the "middle stratosphere". All radiation in this spectral channel is received from the stratosphere and more than half of it is received from altitudes higher than 30 mbar. The annual variations at different latitudes analyzed by Fritz and Soules (Fig. 21) are in essential agreement with those derived from Tiros VII (Fig. 17), but they were obtained with a much higher resolution of both time and latitude zones. Once again, the remarkable uniformity of stratospheric temperatures over the entire globe during the equinoxes can be seen. The warming in the northern winter from January to March is also extremely pronounced. The minimum temperature in the tropics occurs again in December/January, but maxima at the equator are observed at the equinoxes, which is in contrast to the Tiros VII results that showed the maximum temperatures in the tropics in June/July. Apparently, the Tiros VII observations were averaged over too wide a latitude band (30°S–30°N) to show the behavior of the equatorial zone. An extremely significant result of Fritz and Soules is that all stratospheric warmings in the winter hemisphere are accompanied by simultaneous coolings in the stratosphere of the tropics and the summer hemisphere. Fig. 22 shows this out-of-phase relationship in the radiances averaged around latitude circles. However, these changes did not occur at all longitudes. When the warmings reached their maxima, the higher radiances occurred only in one part of a latitude zone, while widespread cooling took place in the tropics and in the summer hemisphere. Fritz and Soules believe that these out-of-phase changes of stratospheric temperature may be explained by heat transfer changes which result from variations in the meridional circulation and large-scale eddies.

Radiances measured by the SCR instrument on Nimbus IV have been inverted to temperature profiles up to 60 km, and analyzed in terms of daily variations of the global temperature structure since April, 1970 by E. J. Williamson and J. T. Houghton of Oxford University, England. Fig. 23 and 24 are examples of these analyses, and are shown here through the courtesy of Williamson and Houghton. Fig. 23 displays meridional temperature cross-sections for four days (near the

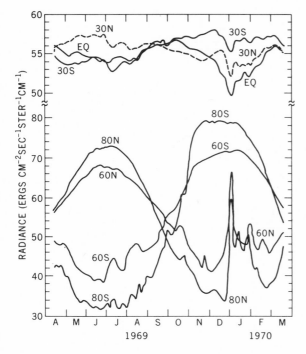

Fig. 21. Annual march of radiances measured with the SIRS at 15.0 μm (669.3 cm⁻¹) for selected latitudes showing the seasonal warming and cooling of the stratosphere. For each latitude indicated, measurements were averaged daily over a 4° zone around the latitude circle. (After Fritz and Soules, 1970.)

solstices and equinoxes) from the surface to heights of about 60 km. The upper portions of Fig. 23 are comparable to the lower portion of Fig. 4. The SCR measurements for 21 January 1971 correspond indeed to the mean cross-section shown in Fig. 7. However, the 16 July SCR data show a much warmer and higher stratopause for the winter (south) polar region than would be expected from Fig. 7. It should be noted that it took some 200 rocket soundings over a period of nearly a decade, and the assumption of complete seasonal symmetry between northern and southern hemispheres to interpret Fig. 7 on a truly global scale, while one global cross-section of Fig. 23 was observed in one single day. The asymmetry in the seasonal temperature extremes between the north and south polar caps is quite interesting. During summer, both regions show the well-known weak temperature lapse rate in the troposphere, a shallow tropopause and a very steep temperature increase throughout the stratosphere. In winter, however, the north polar region shows the expected, nearly isothermal troposphere and stratosphere with a very shallow tropopause and stratopause, while the south polar region displays a remarkably cold tropopause (180°K) at the extremely high altitude equivalent to 30

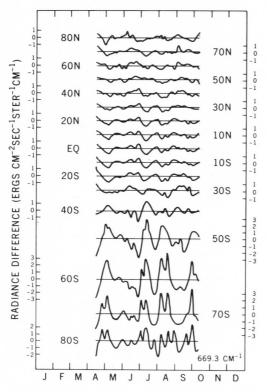

Fig. 22. Deviation of averaged latitudinal radiances measured with SIRS at 15 μm from a least-squares fit for 80°N to 80°S from April 14–October 10, 1969 showing close relationship between the southern hemisphere warming and cooling periods in the high latitudes and corresponding cooling and warming periods in the tropical and northern hemisphere latitudes. (After Fritz and Soules, 1970.)

mbar; indications of an unusually warm (270°K) and high (0.5 mbar) stratopause at the South pole are also found in Fig. 23 for 16 July 1970. The dynamic structure of the stratosphere is illustrated by Fig. 24, where vertical cross-sections across longitudes of the temperature deviations from a mean for the latitude belt of 55–66°S are shown for three days during September, 1970. On each of the three days, the westward slope of the temperature deviations with height is clearly evident. For example, on 4 September, at a height of about 1 mbar, the characteristic warm sector of the southern hemisphere (+20°C deviation) is located near 30°E, while at 200 mbar a deviation of equal amplitude is found near 90°E. Five days later, on 9 September, this warm sector has intensified to +25°C deviation and moved further eastward in the lower stratosphere, while at the 1-mbar level it has split into two centers one at 50°E, the other at 120°W. In general, it is observed that cold and warm regions in the lower stratosphere are overlain by warm and cold regions, respectively in the upper stratosphere.

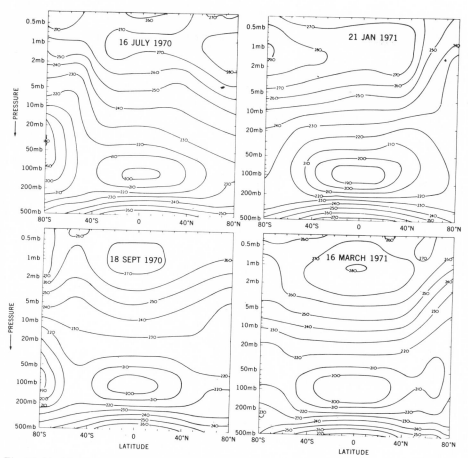

Fig. 23. Meridional temperature cross-sections from surface to the 0.5 mbar pressure level derived from SCR measurements on 16 July 1970, 18 September 1970, 21 January 1971 and 16 March 1971. Data from about twelve satellite passes at longitudes progressing successively westward by about 30° were averaged for each latitude for each 24-h period. (Courtesy Williamson and Houghton.)

## 3.5. Global ozone fields from satellite observations

The principal impetus for observing atmospheric ozone on a *global* scale arises from the fact that the ozone concentration at altitudes below 30 km serves as a tracer of the general circulation of the atmosphere, while at higher altitudes, ozone relates to the heat budget and the photochemistry of the atmosphere.

Ozone is produced in the middle stratosphere (30–40 km) at low and mid-latitudes by photochemical processes, and then transported to other heights and latitudes by the vertical and meridional motions of the atmosphere. About two-thirds of the ozone is found in the lower stratosphere (below 30 km), where it is protected from

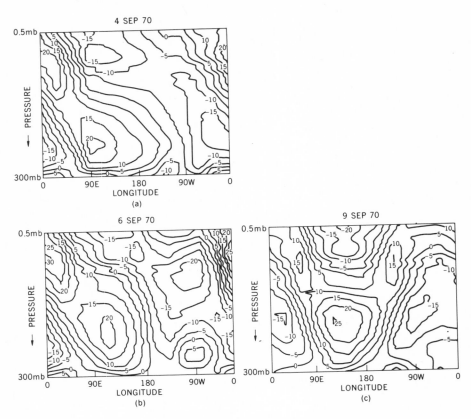

Fig. 24. Cross-section of temperature deviation between the 300- and 0.5-mbar levels at longitudes around the 55° to 60°S latitude zone, from the mean temperatures for this latitude zone on 4 September (a), 6 September (b), and 9 September (c) 1970. Temperature deviations are shown in °C. (Courtesy Williamson and Houghton.)

photochemical modification by the absorption of dissociating radiation at higher altitudes. The lower stratosphere ozone is moved by winds and eddy motions until it is either destroyed by oxidation reactions or lost in the troposphere. Dobson spectrophotometer measurements of the total columnar ozone content have given systematic data on the fluctuations of the lower stratosphere ozone. It has become apparent that the total ozone field is more complex than can be determined from the existing network of about 80 stations, located primarily in the northern hemisphere. The general latitudinal gradient of total ozone is modulated by longitudinal patterns which seem to be related to tropospheric pressure patterns.

Balloon ozone soundings have been useful in demonstrating the relationships between total ozone amounts and the vertical ozone distribution. They have also served in characterizing the quasi-horizontal transport processes near the base of the stratosphere. The density of balloon ozone sounding stations, however, is inadequate for a study of the general circulation mechanisms.

The ozone source region in the middle stratosphere is above balloon altitudes. The amount of ozone in that region is a small fraction of the total ozone amount which is dominated by the lower altitude ozone. The only reliable source of data for this region has been from sounding rockets. The number of flights of instruments with well-defined absolute accuracy has only been sufficient to provide a general understanding of the photochemical processes and of static-equilibrium ozone distributions, and to hint that transport processes might actively modify the distributions. The systematic, global observations of both the total amount of ozone and its vertical distribution, especially above 30 km, from Nimbus afford an excellent opportunity to analyze the circulation of the lower stratosphere, and to study the combination of photochemistry and transport processes in the middle stratosphere.

### 3.5.1 Total ozone

Figure 25 shows a latitudinal profile of total ozone observed during one single orbit of Nimbus IV on 15 April, 1970. Simultaneous observations by the BUV and

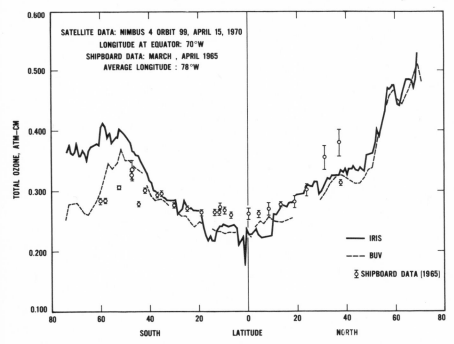

Fig. 25. Comparison of total ozone amounts derived from the IRIS and BUV instruments on Nimbus IV for orbit 99 on April 15, 1970. Points with error bars are results from an earlier latitudinal survey with a Dobson spectrophotometer on board a ship.

IRIS instruments are compared to results from earlier observations from shipboard by the Dobson spectrophotometer (White and Krueger, 1968). It is interesting to compare the time scale of the observations. It took 54 days to acquire the shipboard data, while the satellite data were taken in 54 minutes!

The shipboard Dobson data are consistent with climatological mean values where available from fixed stations. The differences between the recent satellite and earlier shipboard observations are representative of the short-term variability at mid latitudes. The IRIS and BUV results are in reasonably good agreement between 70°N and 45°S, but total ozone measured with the BUV is about 7% lower than that measured with IRIS. According to Mateer et al. (1971), the BUV measurements systematically underestimate total ozone by about 0.020 atm cm, after simultaneous comparisons with Dobson observations. The same major features of the variation of total ozone amount with latitude appear in both sets of data, but the fine scale structure differs between the two observing techniques. This is probably due to the differences in the field-of-view of the two instruments (5° × 5° for IRIS and 11° × 11° for the BUV). Between 45°S and 74°S the measurements diverge considerably, although similar structural features appear in both profiles. The BUV values seem in better agreement with the shipboard observations and with climatological data for Antarctica. The reason for this discrepancy appears to be that the near-isothermal temperature structure of the stratosphere over Antarctica introduces errors in the retrieval of ozone amounts from the IRIS observations. The IRIS measurements depend on the thermal emission from the atmosphere, and therefore on a strong temperature gradient with height. Thirteen profiles such as the one given in Fig. 25 may be obtained from Nimbus IV daily, with each of the two instruments. Daily global maps of total ozone were constructed from the BUV observations which are shown in Fig. 26a and 26b. Longitudinal variations in the total amount of ozone are strongly related to upper tropospheric and lower stratospheric pressure patterns. For example, high amounts of ozone near Greenland and over Siberia appear to be related to persistent low pressure systems in these regions. In the southern hemisphere autumn, the latitudinal gradients of the ozone amount are less steep than gradients in the northern hemisphere during spring.

Mean monthly global maps of total ozone were prepared from the IRIS observations with Nimbus III for April, May, June, and July, 1969. They are shown in Fig. 27. The minimum in the total ozone over the equatorial region is produced by the transport of ozone away from the equator to higher latitudes by the Hadley circulation in the stratosphere. However in July, the total ozone values in this equatorial minimum increase to about 270 Dobson units, suggesting the probable weakening of the Hadley cell at that time.

### 3.5.2. High altitude ozone distributions

A parameter which is particularly useful for analysis of the middle and upper stratosphere is the ozone-to-air mixing ratio. Within a given parcel of air, the mixing ratio is unaltered by vertical or horizontal displacements; only photochemical or chemical reactions can change its value. Latitudinal cross-sections of mixing ratios in the stratosphere from 10 mbar to 0.2 mbar have been prepared from the BUV observations over the latitude range 70°S to 80°N. One such cross-section for an equatorial crossing of 56°E on July 5, 1970, is shown in Fig. 29. The contours are values of constant $O_3$ to air mass mixing ratio in $\mu g\ g^{-1}$. At pressure levels greater than 4 mbar (lower than 37 km), mixing ratios are practically symmetrical with latitude about the equator in April and 10°N in early July. Relatively high mixing ratios of ozone are centered in the tropics, but they decrease abruptly toward the poles at about 20° latitude from the center. A cellular structure can be observed within the tropical region. Mixing ratios in the polar winter region (South pole) are 15–20% lower than those measured in the polar summer (North pole). At pressures lower than 4 mbar (heights greater than 37 km), distributions are not symmetric about the equator. Mixing ratios at those altitudes increase monotonically from north to south. The winter hemisphere between 40°–60°S exhibits a striking maximum at the 1–3-mbar levels.

The global nature of these characteristics is shown in Fig. 28. In Fig. 28a and b the total ozone above 10 mbar has been plotted for both hemispheres. They show symmetry about the equator, with zonal gradients circling about the minima at both poles. The total amounts are approximately equal in the two hemispheres with highest values at low latitudes and minima at both poles. Longitudinal variation features appear to bear little relationship to the lower stratosphere ozone. Fig. 28c and d show the total ozone above the 2.8-mbar level. The two hemispheres are remarkably dissimilar in this case. The northern hemisphere has weak gradients and irregular contours. The southern hemisphere has strong gradients with contours centered at the pole. Total ozone amounts in the southern hemisphere are about double those in the northern hemisphere. The highest values are found, not at the pole, but near 50°S.

Calculations of the ozone distribution assuming a static atmosphere model have shown the mixing ratio maximum near the equator (Dütsch, 1969), and correspond approximately to our observations in the northern hemisphere (summer). The southern hemisphere (winter) maximum, however, has not been reproduced by these calculations.

## 3.6. Nocturnal structure of $O_3$ in the mesosphere

Ozone measurements up to 67 km with the chemiluminescent rocket sounding technique, described in 2.1.6, were first obtained at Wallops Island, Va., in 1968, and were compared with simultaneous optical rocket and chemical balloon soundings. Good agreement among all techniques was demonstrated in the regions where the measurements overlap, as shown by Fig. 30.

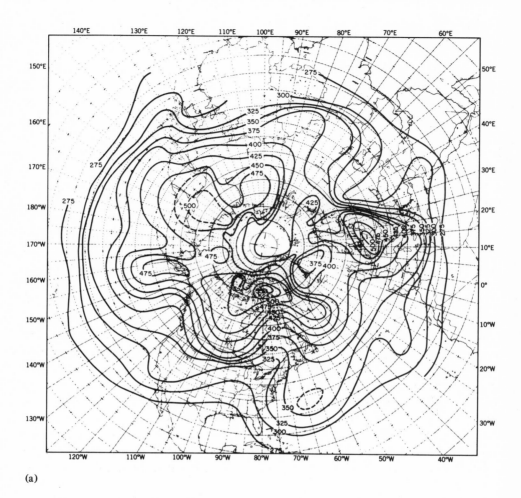

(a)

In March, 1970, and March, 1971, pairs of chemiluminescent ozone soundings were made at Wallops Island, Va. (38°N), and at the Guiana Space Center (5°N) to determine the diurnal variation of ozone concentrations at various latitudes in the mesosphere. The results of these experiments are shown in Fig. 31. At both locations, the day/night variations in $O_3$ concentration are smaller than those calculated on the basis of even a "wet" model of the atmosphere where the hydrogen compounds play a major role in the photochemistry of the mesosphere. Water vapor or other constituents responsible for the decomposition of ozone may be present at these heights in greater amounts than previously expected. In the pure oxygen or "dry" model only the following reactions are considered:

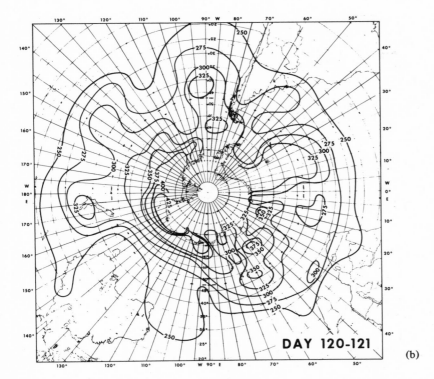

Fig. 26. Total ozone contours for the northern and southern hemispheres derived from BUV measurements on April 30 and May 1, 1970, orbits 294–312. Units are milli atm cm.

$$O_2 + h\nu \to O + O \qquad O + O_3 \to 2O_2$$

$$O + O + M \to O_2 + M \qquad O_3 + h\nu \to O_2 + O_2$$

$$O + O_2 + M \to O_3 + M$$

In the "wet" model, described in detail by Hunt (1966), the decomposition of $H_2O$ by oxygen atoms starts a chain of reactions which, by their reaction rates, are more effective in controlling O and $O_3$ in the mesosphere than if only oxygen were considered. The following equations demonstrate these additional sinks for O and $O_3$, due to the presence of water vapor:

$$O + H_2O \to OH + OH \qquad OH + O_3 \to HO_2 + O_2$$

$$H_2O + h\nu \to H + OH \qquad HO_2 + O_3 \to OH + 2O_2$$

$$H + O_3 \to OH + O_2 \qquad H + OH + M \to H_2O + M$$

$$O + OH \to O_2 + H$$

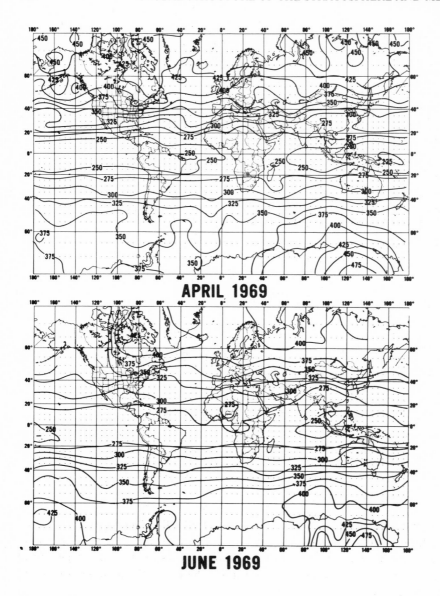

**APRIL 1969**

**JUNE 1969**

Significantly, the rate coefficients of these reactions have no known temperature dependence, whereas those in the pure oxygen atmosphere are temperature dependent.

In January, 1969, two chemiluminescent ozone soundings were made from Point Barrow, Alaska (71°N). The objective of this experiment was not only to measure the ozone distribution in the polar winter night, but also to measure this distribution

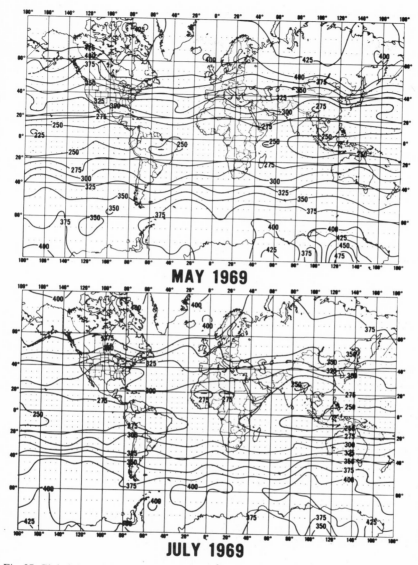

Fig. 27. Global distribution of total ozone ($10^{-3}$ cm STP) derived from Nimbus III IRIS for April (18–30) May, June and July (1–22) 1969.

before and after a stratospheric disturbance (Hilsenrath, 1971). Fig. 32 shows the results of the two ozone soundings performed on January 11 and 30. The solar zenith angle was 125° for both flights, while the minimum zenith angle during the days of the launch was 93° and 80° for the 11th and 30th of January, respectively.

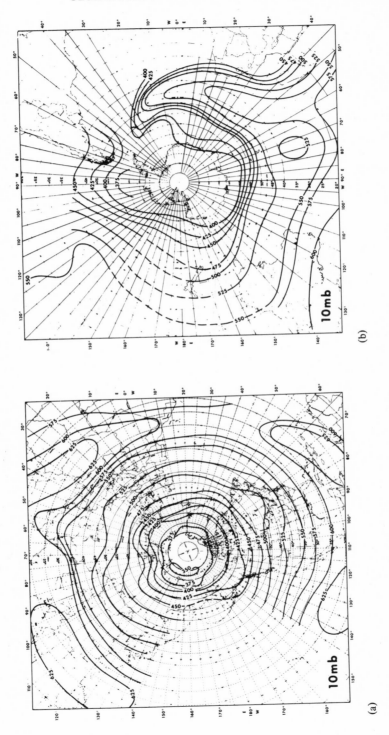

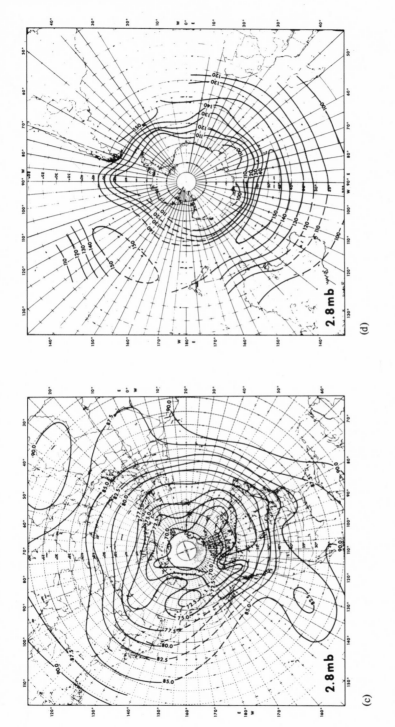

Fig. 28. Contours of total ozone above 10 mbar (top) and total ozone above 2.8 mbar (bottom) for northern and southern hemispheres on July 5, 1970 derived from BUV high altitude data. Ozone amounts are in $10^{-4}$ atm cm.

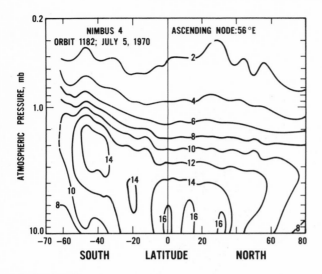

Fig. 29. Meridional cross section of ozone mass mixing ratio ($\mu$g g$^{-1}$) in the atmospheric pressure region from 10 to 0.2 mbar derived from BUV earth radiance data at wavelengths from 2555 to 2976 Å from Nimbus IV orbit 1182 on July 5, 1970.

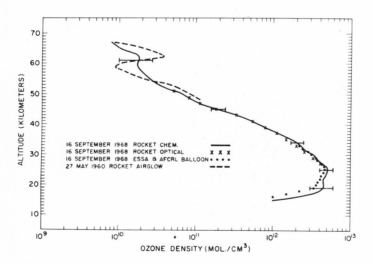

Fig. 30. Ozone density profile measured with chemiluminescent probe on 16 September 1968 over Wallops Island, Va. and compared with simultaneous balloon and optical sondes, and airglow experiment by Reed (1968).

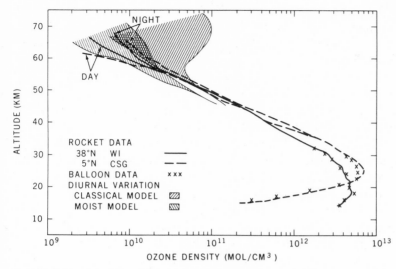

Fig. 31. Diurnal variation of ozone. Measurements at two latitudes 5°N and 38°N, compared with photochemical models.

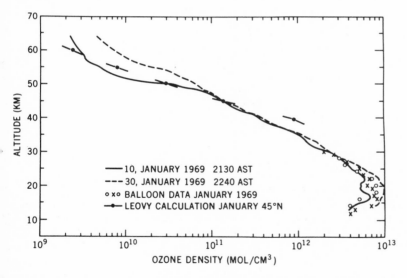

Fig. 32. Ozone density profile measured over Point Barrow Alaska 71°N on 11 and 30 January of 1969. (After Hilsenrath, 1971.)

In the mesosphere, the ozone profiles show an essentially monotonic decrease of ozone concentration with height, in contrast to the distribution observed at lower altitudes. It was thought that in the absence of ultraviolet radiation responsible for the production of O and loss of $O_3$, some structure would appear in the profiles even at higher altitudes, especially in light of the highly disturbed temperature and wind fields that occur at high latitudes during the winter. Since this structure did not appear, one might conclude that the chemical restoration time constants are short enough, or mixing is rapid enough to yield the measured smooth profiles. Another significant feature of this experiment is the variation of ozone near 60 km. Temperature soundings at this altitude conducted within 3 h of the ozone soundings, indicated a temperature decrease of approximately 50°C near 60 km, for the same period during which the ozone concentration increased by almost a factor of four. This result is consistent with the pure oxygen model of the atmosphere in photochemical equilibrium, where ozone destruction is favored by the warmer temperature. The colder temperature enhances the formation of ozone by recombination of available oxygen atoms with molecular oxygen. This conclusion is based on the assumption of a constant source of atomic oxygen at these altitudes. The latitudinal differences in the diurnal variation of the ozone distributions, as well as the results from the polar night measurements, imply a fairly strong horizontal and seasonal variation of atomic oxygen and water vapor in the mesosphere.

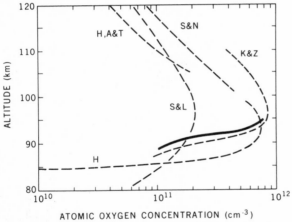

Fig. 33. Atomic oxygen in the mesosphere. Silver film measurements are presented as heavy solid line and compared to calculations of Shimazaki and Laird (1970) (*S&L*), Hesstvedt (1968) (*H*), Keneshea and Zimmerman (1970) (*K&Z*) and rocket mass spectrometer measurements of Hedin et al. (1964). (After Henderson, 1971.)

### 3.7. Results of atomic oxygen sounding

The first flight of the atomic oxygen experiment discussed here was performed at Wallops Island, Va., on a Nike Cajun sounding rocket in October, 1970. Fig. 33 shows the results of this flight compared to theoretical models. Though this first flight yielded data from only 88 to 95 km, future flights will extend measurements to a height range of 75–120 km. Measurements of oxygen atoms in this altitude region are important to determine the state of the mesosphere and lower thermosphere, because of the wide range of characteristic time constants (or residence times) for oxygen atoms, i.e., from less than one day at the lower altitudes to months at altitudes above 110 km. These measurements would then yield information on the photochemical processes as well as the dynamics of this part of the atmosphere. Since only one set of data has been obtained, it would be premature to make any definitive conclusions on the results of this experiment, except that the results are consistent with models that include molecular and eddy diffusion and turbulent mixing, but not with those that assume steady state conditions. Therefore inferences concerning the large-scale dynamics in this altitude range, such as the nature of meridional flow from the summer pole to winter pole near the mesopause, will be difficult. Oxygen atom data will also be applicable to the description of phenomena in the ionosphere such as D and E region anomalies, since they both most likely involve neutralization of atomic oxygen atoms.

### 3.8. Long-term variations of solar UV energy inputs to the mesosphere

Variations of solar UV radiation have been derived from the three UV sensors which were common to a rocket flight in 1966, and the Nimbus III and IV MUSE experiments during 1969, and 1970, respectively. The MUSE sensor response distribution functions are shown in Fig. 34. The response distribution function at any wavelength represents the fraction of the sensor output which comes from shorter wavelengths.

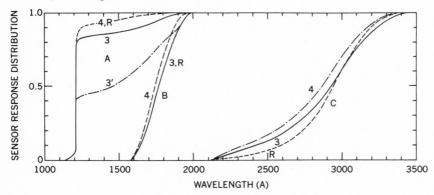

Fig. 34. Sensor response distribution is the fraction of the total signal coming from wavelengths shorter than λ. *A*, *B*, *C* designates the three sensors used on the rocket flight (*R*), on Nimbus III (3) and on Nimbus IV (4); and 3′ illustrates the computed response distribution for sensor *A* after degradation in orbit. (After Heath, 1971.)

There are three types of solar UV variations which have been observed: those related to the long term or solar cycle variation; those related to the solar rotation period; and those associated with flare activity. They are listed in order of decreasing magnitude.

### 3.8.1. Solar cycle variations

Significant variations of solar irradiance were observed in accordance with the 11-year solar cycle (minimum in 1964, maximum in late 1968). The best demonstration of this variation is found in the 1750 Å radiation, which originates near the region of solar temperature minimum. In August, 1966, the solar irradiance observed by the sensor at 1750 Å was only 41% of that observed in April, 1969; by April, 1970, the irradiance had decreased to 76% of the 1969 value. Such variation should produce significant changes in the production rate of atomic oxygen in the lower thermosphere–upper mesosphere region. A discussion of the importance of dissociation in the tail of the Schumann–Runge continuum and pre-dissociation of the vibrational levels at wavelengths shorter than 1972 Å was given in detail by Hudson et al. (1969).

An increase in the production rate of atomic oxygen in the mesosphere–lower thermosphere region, without a corresponding change in the photodissociation rate of $O_3$, leads to an increase in the amount of ozone. There is some experimental evidence that this may be taking place. Gattinger and Jones (1966) observed a fourfold decrease in the twilight brightness of the O, 1 band of the $O_3$, $^1\Delta_g - {}^3\Sigma_g$ system, between 1960 and 1964. One of the theories for the production of $O_2(^1\Delta_g)$ is via the photodissociation of $O_3$ in the Hartley continuum. Additional experimental evidence has been given by Paetzold (1961b), who reported a small enhancement in the amount of ozone above 35 km, which was greater in 1958 than in 1952, and which showed a small positive correlation with sunspot numbers and the decimeter solar radio flux.

### 3.8.2. 27-day variability

The most unambiguously observed type of UV variability is the one associated with the 27-day solar rotational period. These variations are easily observed since the amount of sensor degradation which occurs in the 27-day period is practically negligible.

If the slowly varying exponential decay factors are removed from the MUSE photometer measurements, and corrections are made for the variable earth–sun distance, one obtains the curves of Fig. 35 showing the 27-day solar UV variability. The curves labeled A, B, and C refer to sensors whose response functions are shown in Fig. 34. Each data point represents a daily average of one to eight observations per day at the terminator. In this figure, the long-term trends of the curves over a period of a year are due to the way in which the exponential decay factors were removed. However, the following significant factors inherent in the solar UV flux are clearly evident in Fig. 35: (1) two UV flux maxima per solar rotation are observed; (2) the magnitude of the UV flux variation decreases with increasing wavelengths; (3) the UV flux variation per rotation is decreasing with time; (4) the UV flux variations correlate with other indicators of solar activity.

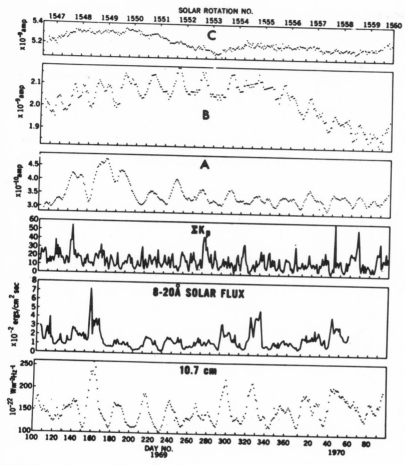

Fig. 35. Time plot of MUSE sensor currents, with exponential decay factors removed, compared with other indicators of solar activity. (After Heath, 1971.)

Three indicators of solar activity are shown below the MUSE measurements. The 8–20 Å X-ray flux background (flare enhancements removed) is from the experiment on Explorer 37 of R. Kreplin of the U.S. Naval Research Laboratory. The values for the daily $\Sigma Kp$, and the 10.7 cm solar flux are taken from "Solar-Geophysical Data" of the NOAA Environmental Data Service. Active solar regions producing enhancements in the 8–20 Å X-ray flux and in the 10.7 cm radio flux are definitely related to the UV enhancements. In addition, it appears that some relationship does exist between fluctuations in the UV radiation and perturbations of the geomagnetic field $Kp$ which are related to the solar wind. Increases to values greater than 40 frequently follow 0–3 days after a UV maximum, which sometimes does not correlate well with the other solar activity indicators.

The UV maxima and minima correlate better with the 10.7 cm radiation than with the calcium plage area, or the Zurich provisional mean sunspot numbers. Similarly, Timothy and Timothy (1970) have observed that the correlation between the EUV helium II Lyman-α line at 304 Å and the standard indicators of solar activity is poor. For the MUSE experiment, the correlation is much worse for the longer wavelength sensors (B) and (C).

The percentage variation of solar irradiance versus wave number (cm$^{-1}$), is shown in Fig. 36 for the spectral region spanned by sensors A and C. No scientific justification is proposed for the logarithmic relation between percentage variation and wave number.

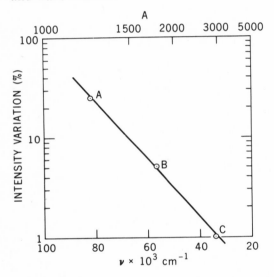

Fig. 36. Percentage solar UV flux variations typically observed with the MUSE sensors versus the wave number of the 0.5 point of the response distribution given in Fig. 34. (After Heath, 1971.)

It can be seen in Fig. 35 that two UV maxima per solar rotation are frequently observed from sensors A and B. A graph of the Carrington longitude of the central meridian, on the days of the UV flux maxima, is presented in Fig. 37 which covers a period of 28 solar rotations. The UV maxima can be determined with an accuracy of about ±1 day, or ±13° in longitude. Two distinct phenomena are observed. The UV maxima appear to originate from two UV active regions, as seen by the clustering of the points about two straight lines. The regions have persisted through 28 solar rotations, and they have been observed by the MUSE experiments on Nimbus III and IV. The longitude of the strongest or primary feature is indicated by (×), whereas the secondary region is described by (⊙). At the time of launch of Nimbus III, the two regions were separated in longitude by 225°, and 28 rotations later, the separation had decreased to 190°. If one uses $13°.199$ day$^{-1}$ as the rotation rate of the Carrington longitude, then the rotation rate of the primary UV active region is $13°.14$ day$^{-1}$, while that of the secondary one is $13°.11$ day$^{-1}$.

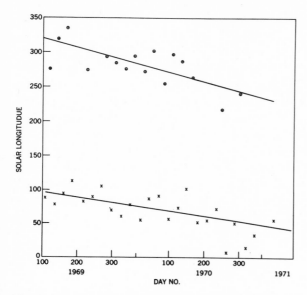

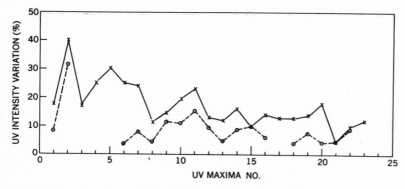

Fig. 37. Carrington longitude of the central meridian at the time of observed UV flux maxima. The stronger region is indicated by ×, while the secondary region is shown by ⊘. (After Heath, 1971.)

The time dependence of the variations of the UV irradiances (sensor $A$) per solar rotation are given in Fig. 38 for the two regions shown in Fig. 37. The primary region peaked during solar rotation 1548, while the secondary one appears to have reached a maximum around rotation number 1557.

Fig. 38. Time dependence of the UV flux variation observed for the two active regions. (After Heath, 1971.)

Evidence for the existence of persistent zones of activity on opposite sides of the sun has been presented by Dodson (1970) and Bumba (1970). In general, the active regions appear to be better defined in the UV. Moreover, there is an approximate agreement as to strength and longitude of the active regions, as determined from ground based and satellite UV measurements.

### 3.8.3. Flare UV variability

On April 21, 1959, at 1959 UT, a major optical flare of optical importance 3B, small class X was recorded. The response of the solar UV flux measured with the MUSE experiment to this flare is shown in Fig. 39. The ordinate is the sensor current normalized to the value at the terminator on orbit 101. The function $f(\phi)/\cos \phi$ represents the sensor response function to changing angles of illumination of the photocathode. The flare occurred during orbit 102, 15 frames or 12 min prior to the sun appearing in the field-of-view. Only the shortest wavelength channel showed any flare enhancement. If an exponential decay for the flare radiation is assumed, then this leads to an enhancement of 16% above the pre- and post-flare values of solar radiation, which produces the signals in the shortest wavelength channels. Based on the sensor response functions given in Fig. 34, the enhancement shown for sensor $A$ should be principally due to H, Ly-$\alpha$.

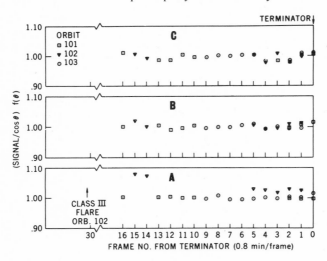

Fig. 39. Enhancement of UV flux associated with major optical flare. (After Heath, 1971.)

REFERENCES

Ainsworth, J. E., Fox, D. F. and LaGow, H. E., Upper atmosphere structure measurement made with the pitot-static tube. *J. Geophys. Res.*, **66**:3191–3211, 1961.
Badinov, I. Y., Andreyev, S. D. and Lipatov, V. B., Measurements of humidity in the upper atmosphere. *Izdatel'stvo Nauka (Moscow)*, pp. 66–79, (*NASA Tech. Transl.* F10, 303), 1966.

Ballard, H. N. and Rofe, B., Thermistor measurement of temperature in the 30–65 km atmospheric region. In: *Stratospheric Circulation* (W. L. Webb, Editor). Academic Press, New York, N.Y., pp. 141–166, 1969.

Beach, C. J., A rocket-borne photometric ozone and temperature measuring sonde. *Aust. Def. Sci. Serv., WRE, Salisbury, S. A., Tech. Note HSA* 179, 1970.

Beranose, H. J. and Rene, M. G., Oxyluminescence of a few fluorescent compounds of ozone. In: *Ozone Chemistry and Technology. Adv. Chem. Sci.,* 21:7–12, 1959.

Beyers, N. J., Radar chaff as a wind sensor. In: *Stratospheric Circulation* (W. L. Webb, Editor). Academic Press, New York, 89–96, 1969.

Bumba, V., Large-scale magnetic field and activity patterns on the sun. In: *Solar Terrestrial Physics. Proceedings International Symposium, Leningrad, May, 1970* (E. R. Dyer, Editor). D. Reidel Dordrecht, pp. 21-37, 1970.

Carver, J. H., Horton, B. H. and Burger, F. G., Nocturnal ozone distribution in the upper atmosphere. *J. Geophys. Res.,* 71:4189–4191, 1966.

Chahine, M. T., Determination of the temperature profile in an atmosphere from its outgoing radiance. *J. Opt. Soc. Am.,* 58:1634, 1968.

Charlson, R. J., Noctilucent clouds: A steady-state model. *Q. J. R. Meteorol. Soc.,* 91:517–523, 1965.

Chleck, X., Aluminum oxide hygrometer, laboratory performance. *J. Appl. Meteorol.,* 5:878–885, 1966.

COESA, (Committee on Extension of the Standard Atmosphere), *U.S. Standard Atmosphere, 1962.* U.S. Government Printing Office, Washington, D.C., 278 pp., 1962.

COESA, *U.S. Standard Atmosphere Supplement, 1966.* U.S. Government Printing Office, Washington, D.C., 289 pp., 1966.

Conrath, B. J., Inverse problems in radiative transfer: A review. In: *Proceedings of the XVIIIth International Astronautical Congress, Belgrade, 1968* (M. Junc, Editor). Academic Press, New York, N.Y., p. 339, 1968.

Conrath, B. J., Hanel, R. A., Kunde, V. G. and Prabhakara, C., The infrared interferometer experiment on Nimbus III. *J. Geophys. Res.,* 75:5831–5857, 1970.

Dodson, H. W., Comments on large scale organization of solar activity in time and space. *AIAA Bull.,* 7:563, 1970.

Dütsch, H. U., Atmospheric ozone and ultraviolet radiation. In: *Climate of the Free Atmosphere* (D. F. Rex, Editor). Elsevier, Amsterdam, pp. 383–432, 1969.

Ellis, P. J., Peckham, G., Smith, S. D., Houghton, J. T., Morgan, C. G., Rodger, C. D. and Williamson, E. J., First results from the selective chopper radiometer on Nimbus IV, *Nature,* 228:139–143, 1970.

Fogle, B., Noctilucent clouds. *Rep.* UAG R-177, Geophysical Inst., Univ. of Alaska, College, pp. 34–66, 1966.

Fritz, S. and Soules, S. D., Large-scale temperature changes in the stratosphere observed from Nimbus III. *J. Atmos. Sci.,* 27:1091–1097, 1970.

Gattinger, R. L. and Jones, A. V., The $^1\Delta_g - {}^3\Sigma_g^+$ $O_2$ bands in the twilight and day airglow. *Planet. Space Sci.,* 14:1, 1966.

Heath, D. F., Observations of solar long term variability and irradiance in the near and far ultraviolet. *GSFC Rep.* X-651-71-116. NASA Goddard Space Flight Center, Greenbelt, Md., 1971.

Hedin, A. E., Avery, C. P. and Tschetter, C. D., An analysis of spin modulation effects on data obtained with rocket-borne mass spectrometer. *J. Geophys. Res.,* 69:4637, 1964.

Hemenway, C. L., Fullum, E. F., Skrivanek, R. A., Soberman, R. K. and Witt, G., Electron microscope studies of noctilucent cloud particles. *Tellus,* 9:96–102, 1964.

Henderson, W. R., D-region atomic oxygen measurement. *J. Geophys. Res.,* 76:3166–3167, 1971.

Henderson, W. R. and Schiff, H. I., A simple sensor for the measurement of atomic oxygen height profiles in the upper atmosphere. *Planet. Space Sci.,* 18:1527–1534, 1970.

Hering, W. S. and Borden, Jr., T. R., Mean distributions of ozone density over North America, 1963–1964. *U.S.A.F., Off. Aerosp. Res., Environ. Res. Pap.* 162. Air Force Cambridge Res. Lab., Bedford, Mass., 19 pp., 1965.

Hesstvedt, E., On the possibility of ice cloud formation at the mesopause. *Tellus,* 14:290–296, 1962.

Hesstvedt, E., On the water vapor content of the high atmosphere. *Geofys. Publ. Oslo,* 25:1–18, 1964.

Hesstvedt, E., On the effect of vertical eddy transport on atmospheric composition in the mesosphere and lower thermosphere. *Geofys. Publ. Oslo*, **27**:1–35, 1968.

Hilsenrath, E., Ozone measurements in the mesosphere and stratosphere during two significant geophysical events. *J. Atmos. Sci.*, **28**:295–297, 1971.

Hilsenrath, E. and Coley, R., Performance of an aluminum oxide hygrometer on the NASA CV-990 aircraft meteorological observatory. *GSFC Document* X-651-71-37. NASA Goddard Space Flight Center, Greenbelt, Md., 1971.

Hilsenrath, E., Seiden, L., and Goodman, P., An ozone measurement in the mesosphere and stratosphere by means of rocket sonde. *J. Geophys. Res.*, **74**:6873–6879, 1969.

Hines, C. O., Internal atmospheric gravity waves at ionospheric heights. *Can. J. Phys.*, **38**:1441–1481, 1960.

Hines, C. O. and Reddy, C. A., On the propagation of atmospheric gravity waves through regions of wind shear. *J. Geophys. Res.*, **72**:1015–1034, 1967.

Horvath, J. J., Simmons, R. W., and Brace, L. H., Theory and implementation of the pitot-static technique for upper atmospheric measurements. *Univ. Mich. Sci. Rep.* NS-1. Univ. Mich., Ann Arbor, Mich., 1962.

Hudson, R. D., Carter, V. L. and Breig, E. L., Predissociation in the Schumman–Runge band system of $O_2$: laboratory measurements and atmospheric effects. *J. Geophys. Res.*, **74**:4079, 1969.

Humphreys, W. J., Nacreous and noctilucent clouds. *Mon. Weather Rev.*, **61**:228, 1933.

Hunt, B. G., Photochemistry in a moist atmosphere. *J. Geophys. Res.*, **71**:1385–1398, 1966.

Iozenas, V. A., Krasnopol'skiy, V. A., Kuznetsov, A. P. and Lebedinskiy, A. I., Studies of the earth's ozonosphere from satellites. *Izv., Atmos. Oceanic Phys.*, **5**:77–82, 1969.

Johnson, F. S., Purcell, J. D., Tousey, R. and Watanabe, K., Direct measurements of the vertical distribution of atmospheric ozone to 70 km altitude. *J. Geophys. Res.*, **57**:157–176, 1952.

Kaplan, L. D., Inference of atmospheric structure from remote radiation measurements. *J. Opt. Soc. Am.*, **49**:1004, 1959.

Keneshea, T. J. and Zimmerman, S. P., The effect of mixing upon atomic and molecular oxygen in the 70–170 km region of the atmosphere. *J. Atmos. Sci.*, **27**:831, 1970.

Kennedy, J. S., An atlas of stratospheric mean-isotherms derived from Tiros VII observations. *NASA Rep.* X-622-66-307. Goddard Space Flight Center, Greenbelt, Md., 1966.

Kennedy, J. S. and Nordberg, W., Circulation features of the stratosphere derived from radiometric temperature measurements with the Tiros VII satellite. *J. Atmos. Sci.*, **24**:711–719, 1967.

Khvostikov, I. A., Present state of the problem of noctilucent clouds. *Izdatel'stvo Nauka (Moscow)*, pp. 5–10, 1966 (*NASA Tech. Transl.* F10, 301).

King, J. I. F., The radiative heat transfer of planet earth. In: *Scientific Uses of Earth Satellites* (J. A. van Allen, Editor). Univ. of Mich. Press, Ann Arbor, Mich., 2nd ed., 316 pp., 1958.

Krueger, A. J. and McBride, W. R., Sounding rocket—OGO-4 satellite ozone experiment: rocket ozonesonde measurements. *Naval Weapons Cent. Rep.* TP4667. Naval Weapons Center, China Lake, Calif., 1968.

Leovy, C., Simple models of thermally driven mesospheric circulation. *J. Atmos. Sci.*, **21**:327–341, 1964.

Lindzen, R. S., Thermally driven diurnal tides in the atmosphere. *Q. J. R. Meteorol. Soc.*, **93**:18–42, 1967.

Ludlam, F. H., Noctilucent clouds. *Tellus*, **9**:341–364, 1957.

Mateer, C. L., On the information content of Umkehr observations. *J. Atmos. Sci.*, **22**:370–381, 1965.

Mateer, C. L., A review of some aspects of inferring the ozone profile by inversions of ultraviolet radiance measurement. In: *The Mathematics of Profile Inversions* (L. Colin, Editor). *NASA TMX 62/50*, pp. 2–25, 1972.

Mateer, C. L., Heath, D. F. and Krueger, A. J., Estimation of total ozone from satellite measurements of backscattered ultraviolet earth radiance. *J. Atmos. Sci.*, **28**:1307–1311, 1971.

Murgatroyd, R. J., Winds and temperatures between 20 km and 100 km—A review. *Q. J. R. Meteorol. Soc.*, **83**:417–458, 1957.

Murgatroyd, R. J. and Singleton, F., Possible meridional circulations in the stratosphere and mesosphere. *Q. J. R. Meteorol. Soc.*, **87**:125, 1961.

Nagata, T., Tohmatsu, T. and Ogawa, T., Sounding rocket measurement of atmospheric ozone density, 1965–70. In: *Space Research, XI* (K. Ya. Komdratyev, M. J. Rycroft and C. Sagan, Editors). Akademie Verlag, Berlin, pp. 849–855, 1971.

*Nimbus II Data User's Guide.* National Space Science Data Center, Goddard Space Flight Center, Greenbelt, Md., 73–91, 1966.

*Nimbus III User's Guide.* National Space Science Data Center, Goddard Space Flight Center, Greenbelt, Md., 1969.

*Nimbus IV User's Guide.* National Space Science Data Center, Goddard Space Flight Center, Greenbelt, Md., March 1970.

Nordberg, W., McCulloch, A. W., Foshee, L. L. and Bandeen, W. R., Preliminary results from Nimbus II. *Bull. Am. Meteorol. Soc.*, **47**:857–872, 1966.

Nordberg, S. and Rasool, S. I., *Introduction to Space Science* (W. N. Hess and G. D. Mead, Editors). Gordon and Breach, New York, N.Y., pp. 265–313, 1968.

Nordberg, W. and Smith, W. S., The rocket grenade experiment. *NASA Tech. Note TND-2107*, 1964.

Obukhov, A. M., The statistically orthogonal expansion of empirical functions. *Izv. Ser. Geofic., Akad. Nauk S.S.R.*, **3**:432, 1960.

Paetzold, H. K., Messungen des Atmospherischen Ozone. In: *Handbook der Aerologie* (W. Hesse, Editor). Akademische Verlagsgesellschaft, Leipzig, pp. 458–531, 1961a.

Paetzold, H. K., The photochemistry of the atmospheric ozone layer. In: *Chemical Reactions in the Lower and Upper Atmosphere*. Interscience, Chichester, England, p. 181, 1961b.

Paton, J., Noctilucent clouds. *Meteorol. Mag.*, **93**:161–179, 1964.

Prabkahara, C., Feasibility of determining atmospheric ozone from outgoing infrared energy. *Mon. Weather Rev.*, **97**:307–314, 1969.

Prabhakara, J., Conrath, B. J., Allison, L. J. and Steranka, J., Seasonal and geographic variation of atmospheric ozone derived from Nimbus III. *GSFC Document* X-651-71-38, p. 75. Goddard Space Flight Center, Greenbelt, Md., 1971.

Reed, E. I., A night measurement of mesospheric ozone by observations of ultraviolet airglow. *J. Geophys. Res.*, **73**:2951, 1968.

Rodgers, C. D., Satellite infrared radiometer; A discussion of inversion methods. Univ. of Oxford, *Clarendon Lab. Mem.* **66.13**:25 pp., 1966.

Shimazaki, T. and Laird, A. R., A model calculation of the diurnal variation in minor neutral constituents in the mesosphere and lower thermosphere including transport effects. *J. Geophys. Res.*, **75**:3221, 1970.

Sissons, N. V., Ozone measuring techniques and their assessment for W. R. E. Dropsonde Use. *Aust. Defense Sci. Ser., WRE Note SAD* 196, 1968.

Smith, W. L., Statistical estimation of the atmosphere's geopotential height distribution from satellite radiation measurements. *ESSA Tech. Rep. NESC* 48, 1969.

Smith, W. S., Katchen, L. B., Sacher, P., Swartz, P. C. and Theon, J. S., Temperature, pressure, density, and wind measurements with the rocket grenade experiment, 1960–63. *NASA Tech. Rep.* TR R-211, 1964.

Smith, W. S., Theon, J. S., Katchen, L. B. and Swartz, P. C., Temperature, pressure, density and wind measurements in the upper stratosphere and mesosphere, 1964. *NASA Tech. Rep.* TR R-245, 1966.

Smith, W. S., Theon, J. S., Swartz, P. C., Katchen, L. B. and Horvath, J. J., Temperature, pressure, density, and wind measurements in the upper stratosphere and mesosphere, 1965. *NASA Tech. Rep.* TR R-263, 1967.

Smith, W. S., Theon, J. S., Swartz, P. C., Katchen, L. B. and Horvath, J. J., Temperature, pressure, density, and wind measurements in the stratosphere and mesosphere, 1966. *NASA Tech. Rep.* TR R-288, 1968a.

Smith, W. S., Katchen, L. B. and Theon, J. S., Grenade experiments in a program of synoptic meteorological measurements. In: *Meteorological Investigations of the Upper Atmosphere* (R. S. Quiroz, Editor). *Meteorol. Monogr., Am. Meteorol. Soc.*, **9**:170–175, 1968b.

Smith, W. S., Theon, J. S., Swartz, P. C., Casey, J. F. and Horvath, J. J., Temperature, pressure, density, and wind measurements in the stratosphere and mesosphere, 1967. *NASA Tech. Rep.* TR R-316, 1969.

Smith, W. S., Theon, J. S., Casey, J. F. and Horvath, J. J., Temperature, pressure, density, and wind measurements in the stratosphere and mesosphere, 1968. *NASA Tech. Rep.* TR R-340, 1970.

Smith, W. S., Theon, J. S., Casey, J. F., Azcarraga, A. and Horvath, J. J., Temperature, pressure, density, and wind measurements in the stratosphere and mesosphere, 1969. *NASA Tech. Rep.* TR R-360, 1971.

Strand, O. N. and Westwater, E. R., Minimum rms estimation of the numerical solution of a fredholm integral equation of the first kind. *SIAM J. Numerical Analysis*, 5:287, 1968.

Theon, J. S., Short term temperature variations in the atmosphere below 120 km. *Proc. Nat. Conf. Aerosp. Meteorol., 3rd, New Orleans, May 1968. Am Meteorol. Soc. Publ.,* pp. 449–456.

Theon, J. S. and Smith, W. S., The meteorological structure of the mesosphere including seasonal and latitudinal variations. In: *Mesospheric Models and Related Experiments* (G. Fiocco, Editor). D. Reidel, Dordrecht, pp. 131–146, 1971.

Theon, J. S., Nordberg, W. and Smith, W. S., Temperature measurements in noctilucent clouds. *Science*, 157:419–421, 1967.

Theon, J. S., Smith, W. S. and Casey, J. F., Statistical models of the density and wind profiles in the mesosphere based on 208 soundings. *Proc. Am. Meteorol. Soc. Nat. Conf. Aerosp. Meteorol., 4:h, Las Vegas, Nev., May 1970. Am. Meteorol. Soc. Publ.,* pp. 306–311, 1970.

Theon, J. S., Smith, W. S., Casey, J. F., and Kirkwood, B. R., The mean observed structure and circulation of the stratosphere and mesosphere. *NASA Tech. Rep.* TR R-375, 1972.

Timothy, A. F. and Timothy, J. G., Long-term intensity variations in the solar helium II Lyman alpha line. *J. Geophys. Res.*, 75:6950, 1970.

*Tiros VII Radiation Data Catalog and User's Manual*, Volume 1. Goddard Space Flight Center, Greenbelt, Md., 256 pp., 1964.

Twomey, S., The application of numerical filtering to the solution of integral equations encountered in indirect sensing measurements. *J. Franklin Inst.*, 279:95–109, 1965.

Twomey, S., Indirect measurements of atmospheric temperature profiles from satellites, II. Mathematical aspects of the inversion problem. *Mon. Weather Rev.*, 94:363–366, 1966.

Wark, D. Q. and Fleming, H. E., Indirect measurements of atmospheric temperature profiles from satellites. *Monthly Weather Rev.*, 94:351, 1966.

Wark, D. Q. and Hilleary, D. T., Atmospheric temperature: successful test of remote probing. *Science*, 165:1256, 1969.

Warnecke, G. and McCulloch, A. W., Stratospheric temperature patterns derived from Nimbus II measurements. *GSFC Document* X-622-67-436. Goddard Space Flight Center, Greenbelt, Md., 1967.

Warner, J. and Bowen, E. G., A new method of studying the fine structure of air movements in the free atmosphere. *Tellus*, 5:36–41, 1953.

Weeks, L. H. and Smith, L. G., A rocket measurement of ozone near sunrise. *Planet. Space Sci.*, 16:1189–1195, 1968.

Wei, Ding-wen, The non-uniqueness of the solution for the vertical distributions of ozone from the calculation by the Umkehr method B. *Acta Geophys. Sinica*, 11:123–135 (in Chinese), 1962. English summary in: *Scientia Sinica*, 12:1065–1070, 1966.

White, W. C. and Krueger, A. J., Shipboard observations of total ozone from 38°N to 60°S. *J. Atmos. Terr. Phys.*, 30:1615–1622, 1968.

Witt, G., Martin-Lof, J., Wilhelm, N. and Smith, W. S., High latitude summer mesospheric temperatures and winds with particular regard to noctilucent clouds. In: *Space Research, V.* (D. G. King-Hele, P. Muller and G. Righini, Editors). North-Holland, Amsterdam, p. 820, 1965.

Wright, J. B., The robin falling sphere. In: *Stratospheric Circulation* (W. L. Webb, Editor). Academic Press, New York, N.Y., 115–139, 1969.

# UPPER ATMOSPHERE ROCKET SOUNDINGS IN POLAND

JACEK WALCZEWSKI

*Hydro-Meteorological Institute, Krakow (Poland)*

## SUMMARY

Upper atmosphere research by means of sounding rockets in Poland is described. The aim of the work is to study the seasonal variation of the stratospheric circulation system. By seeking correlations between the characteristics of the seasonal circulation reversals in the stratosphere, and the seasonal characteristics in the troposphere, we look for a method of applying high-atmosphere data to support the seasonal forecasts of troposheric parameters. After briefly describing the rockets used in the sounding program, the main results so far obtained are reviewed. The dependence of the spring wind reversal on the phase of the quasi-biennial cycle of tropical stratospheric circulation appears to be confirmed.

## 1. INTRODUCTION

Regular rocket soundings of the stratosphere started in Poland in 1965. The research carried out is focused on a study of the seasonal variation of the stratospheric circulation system in spring and in autumn. Such seasonal changes in the stratospheric and mesospheric circulation reflect the basic changes in the energy system of the whole atmosphere and so it seems worthwhile to seek correlations between the characteristics of the seasonal circulation reversals in the stratosphere and mesosphere, and the characteristics of the seasons in the troposphere. This was the basic concept of the research, the purpose of which is to find a method of applying high-atmosphere data to support the seasonal forecasts of tropospheric parameters. On the assumption that a study of seasonal wind reversals is the first task of the sounding system, the characteristics of the rockets and the sounding schedule were established.

## 2. ROCKETS, INSTRUMENTATION AND SOUNDING SCHEDULE

The first Polish meteorological rocket produced in series was the Meteor-1 rocket (Fig. 1). It was used in regular soundings during the years 1965–70. The Meteor-1 is a small rocket intended for wind measurements (Walczewski, 1969). The rocket consists of a motor section and a dart. The overall length of the rocket is 2.5 m and the weight at take-off is 32.5 kg. The peak altitude reached with an 85° launch angle is 37 km. The dart is equipped with chaff for wind measurements. The launcher is transportable and can operate in field conditions.

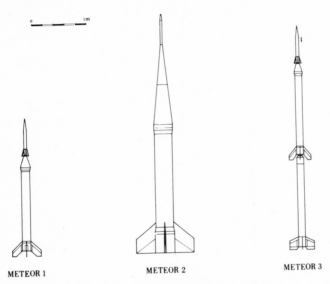

Fig. 1. Schematic representations of the "Meteor" rockets.

An improved model of Meteor-1 is the Meteor-3 rocket with the peak altitude increased up to 65 km by using two rocket motors in tandem (Nowak, 1969). Both motors of this two-stage rocket are identical to the "Meteor-1" rocket motor. The basic type of dart used here is the "Meteor-1" dart. The overall length is 4.5 m, the take-off weight is 65 kg. The first successful launching of "Meteor-3" took place in 1969 (Walczewski, 1971a) and the rocket was used operationally for the first time in 1970.

Radar tracking of chaff is the standard method of wind measurements with Meteor-1 and Meteor-3 rockets. Different types of chaff are in use. A parachute sonde equipped with a thermistor is now being developed for the Meteor-3. The sonde will enable us to measure wind and temperature.

The larger multipurpose meteorological rocket Meteor-2 was successfully launched in 1970 (Harazny, 1971). This is a one-stage, solid-propellant rocket with a 10 kg payload capacity. The simple one-stage version reaches a height of 65 km; the version with boosters climbs to 85-90 km. The length of the rocket is 4.3 m, the weight at take-off is 480 kg, and the boosters are two Meteor-1 motors parallely attached on both sides of the main motor. The experimental series of rockets was equipped with parachute sondes for wind and temperature measurements (Bielak et al., 1971). The rocket has not been used operationally until now and the parachute sonde has been miniaturized and adjusted for use in the dart of the Meteor-3 rocket.

Rocket soundings in the period 1965–1969 were performed in field conditions at Ustka on the Baltic coast. In 1970, a sounding base was made operational at Leba on the Baltic coast, about 40 km from Ustka. This base is equipped with a launcher for Meteor-2 and Meteor-3 rockets; Meteor-1 rockets were launched here, too. The latitude of both sounding points is about 54.5°N.

The launching schedule contained at first (in 1965–1966) double soundings on Quarterly World Days (according to the International Geophysical Calendar) and then, since 1967, two sounding series each year, one in spring and one in autumn. The sounding series is summarized in Table I. Most of the soundings were performed at 1200 GMT ± 1 hour, except for the 1970 autumn series, when rockets were fired at 2000 GMT ± 1 hour.

TABLE I

Rocket soundings, 1965–1970

| Year | Total number of rockets fired | Periods of sounding |
|------|-------------------------------|---------------------|
| 1965 | 6 | Quarterly World Days |
| 1966 | 12 | Quarterly World Days |
| 1967 | 42 | 5 April–13 May and 12 Aug.–26 Aug. |
| 1968 | 45 | 1 April–15 May and 12 Aug.–13 Sept. |
| 1969 | 35 | 11 April– 5 May and 8 Aug.–15 Sept. |
| 1970 | 41 | 12 April– 7 May and 15 Aug.– 3 Oct. |

## 3. RESULTS

As a result of the soundings, the time-height cross-sections of the wind field were obtained for the seasonal wind reversals since 1967. These cross-sections were reaching levels equivalent to 10 and 5 mbar and in autumn 1970 were extended higher, up to 0.2 mbar (containing the 0.4 mbar level used for the high-atmosphere synoptic charts). Initial work was devoted to an evaluation of this material and attempts were made to compare the sounding results with the data from other sounding stations (Walczewski, 1971b). The most important conclusions from the preliminary analyses may be summarized as follows: the synoptic conditions in which the spring wind reversal takes place differ very much from the synoptic conditions of the autumn reversal. The spring reversals over the Polish coast of the Baltic sea, on the 30–35 km levels, demonstrated higher variability in time, changing their dates between the beginning of April and the beginning of May; maximum difference being about one month. The dates of these reversals were determined by the rotational movements of the main high- and low-pressure centers (travelling around the pole), and by the influence of a high pressure center emerging over the Atlantic at that time.

The autumn wind reversals were surprisingly stable in time and took place in the last decade of August. This refers to the 30–35 km level. The reversal at the 60 km level was observed in 1970, about two weeks later than at 30 km.

It seems that the dependence of the spring reversal date on the phase of quasi-biennial cycle of tropical stratospheric circulation (described by Labitzke, 1966) was confirmed at least in 1969 and 1970 when the type of reversal ("early" or "late") was foreseen, although the definition of "early" and "late" reversals can still be discussed.

Comparison of the sounding data from the Ustka and Berlin–Tempelhof stations (located about 330 km apart) led to an important conclusion. The observed discrepancies in measurement results seem to suggest that a meso-structure of the baric field is possible, at least in autumn, even at 30–35 km levels. This mesostructure may create comparatively small (less than 400-km diameter) synoptic systems, and this phenomenon should be taken into account in the interpretation of high-altitude sounding data from a network made up by few stations.

The analysis of both Polish and foreign data describing the seasonal wind reversals led to the conclusion that four basic types of reversals may be distinguished, depending on the course of the zero-isoline of the zonal wind and on the time-height cross-section of the wind field (Walczewski and Wludarska, 1971). The most frequently observed reversal type is type 1, where the new circulation system, characteristic for the coming season, establishes itself first at higher levels and then at the lower ones.

Study of the reversal processes reported above was greatly limited because only the morphology and synoptics of the process were investigated and only one parameter, in a comparatively low height range, was taken into account. As we mentioned in paragraph 2, attempts are being made to increase the ceiling of measurements and to increase the number of parameters measured at the Polish sounding stations. It seems that the introduction of more correlation parameters and more sophisticated methods of analysis would reveal new interesting features of the seasonal reversal process. The latter is not an isolated phenomenon, but is strictly connected with the characteristics of the succession of seasons. Therefore, its study is particularly promising in view of long-range weather forecasting.

REFERENCES

Bielak, A., Ksyk, A. and Walczewski, J., The prototype of the RAMZES rocket sonde for the wind and temperature measurements. *Artif. Satell.*, **6**:3–10, 1971.
Harazny, J., Rakieta Meteor-2. *Skrzydlata Polska*, **1**:9–12, 1971.

Labitzke, K., The nearly two-year cycle of midwinter warmings and of the final Spring warmings in the stratosphere. *Symp. Interactions between Upper and Lower Layers of the Atmosphere, Vienna, May 1966* (presented paper).

Nowak, K., Polska rakieta meteorologiczna "Meteor-3". *Skrzyndlata Polska,* **34/35**, 1969.

Walczewski, J., Polish meteorological rocket system Meteor-1. In: *Stratospheric Circulation* (W. L. Webb, Editor). Academic Press, New York, N.Y., 600 pp., 1969.

Walczewski, J., First measurements of stratospheric and mesospheric winds over the station Leba. *Artif. Satell.,* **6:**21–28, 1971 a.

Walczewski, J., Seasonal reversals of stratospheric winds over the Polish coast of the Baltic sea in the years 1967–1969. *Artif. Satell.,* **6:**29–46, 1971 b.

Walczewski, J. and Wludarska, J., The morphological types of seasonal wind reversals in the stratosphere. *Artif. Satell.,* **6:**11–20, 1971.

# ACTIVE AND PASSIVE OPTICAL DOPPLER TECHNIQUES FOR THE DETERMINATION OF ATMOSPHERIC TEMPERATURE, 1: AN AIRGLOW SPECTROPHOTOMETER WITH INTERNAL LASER REFERENCE[*]

GLAUCO BENEDETTI-MICHELANGELI,[1] GIORGIO CAPPUCCIO[2], FERNANDO CONGEDUTI AND GIORGIO FIOCCO[3]

*European Space Research Institute (ESRIN), Frascati (Italy)*

## SUMMARY

An optical spectrometer for use in airglow observations is described. The instrument utilizes a single frequency highly stable laser for internal reference. Some improvements enhancing the sensitivity at low photon counting levels are described. Accurate measurements of temperature are possible.

## 1. THE AIRGLOW SPECTROPHOTOMETER

For about two years we have been making intermittent observations of the airglow using a Fabry–Perot spectrophotometer with a 10 cm aperture. A few instruments of this type and their use in the spectrophotometry of the aurora and the airglow have been described, among others, by Armstrong (1956,1968), Hernandez (1966,1970), Shepard (1969), Hays and Roble (1971).

The aim of these measurements is to obtain spectral density profiles of airglow emissions. Under thermal equilibrium conditions, these Doppler-broadened profiles give a measurement of the atmospheric temperature. $\Delta\lambda$, the full width at half power of the line is thus related to the temperature $T$:

$$\frac{\Delta\lambda}{\lambda} = 7.16 \cdot 10^{-7}\left(\frac{T}{\mu}\right)^{1/2}$$

where $\lambda$ is the line wavelength and $\mu$ is the mass of the emitting species in a.m.u. Bulk shift of the line is an indication of wind.

A new feature of our instrument consists in the use of a single-frequency highly stable laser as a reference for periodically checking the instrumental response. This

---

[*] Lecture delivered by Dr. Fiocco.
[1] Present address: CNR, c.p. 27, Frascati, Italy.
[2] Present address: Istituto di Fisica dell'Atmosfera, Rome, Italy.
[3] Present address: Università di Firenze, Italy.

makes it possible, in the analysis of the data, to correct unavoidable instrumental drifts and integrate spectra over arbitrarily long times.

Armstrong (1968) has utilized a $\lambda = 5461$-Å line from a $^{198}$Hg source as a reference; the width of this line, although narrow, is comparable to the instrumental response width. The advantage of using the laser is that its equivalent line width (taking into account modulation and stability) is about two orders of magnitude narrower than that of the incoherent source. Thus, by using the laser, the true instrumental response at $\lambda = 6328$ Å can be found directly.

Basically the instrument consists of a 10 cm aperture flat Fabry–Perot interferometer, a single frequency He–Ne laser, interference filters and two photomultipliers used to alternately observe the airglow and the laser reference, the electronics for photon counting, and various controls that make the operation entirely automatic. The characteristics are listed in Table I.

TABLE I

Characteristics of the spectrophotometer

| Fabry–Perot | | Interference filter | |
|---|---|---|---|
| Diameter | 100 mm | Diameter | 50 mm |
| Spacing | 31.5 mm | Half-width | 2.3 Å |
| Flatness before coating | $\lambda/100$ over 94 mm | Peak transmission | 45% |
| Reflectivity at $\lambda = 6328$ Å | 92% | | |
| Reflectivity at $\lambda = 5577$ Å | 89% | | |
| Estimated overall finesse | 8 | | |

In designing the interferometer, care was taken to eliminate the stresses on the plates from the use of fixed spacers. The axis of the interferometer is vertical. Each plate is held by a support that ultimately rests on three adjustable screws with fine reducing gears. The upper plate support includes columns to achieve the necessary spacing. Suitable materials were used to compensate for thermal expansion, and the use of piezoelectric ceramics for control is foreseen. The interferometer is housed in an enclosure for pressure scanning and is kept at a constant temperature by the circulation of water in a jacket. The input window is a field lens ($F = 5$ m) to provide some averaging of the signal over the accepted solid angle. The output window is another lens ($F = 1.5$ m) focusing the fringes on the iris plane.

The internal reference is provided by a single-frequency laser Spectra Physics mod.119. In this instrument the emitted frequency is electronically locked to the Lamb dip of the laser gain curve and the manufacturer claims it is stable within the limit of 1 MHz/day. At periodic intervals the laser light enters the Fabry–Perot along its axis by means of a small prism. Thus, the response around $\lambda = 6328$ Å is obtained: notice, however, that only the central area of the plates, a few mm in diameter, is explored. This permits an accurate determination of the free spectral range and of the location of the peaks with respect to the scanning. The entire optical system is pneumatically suspended in order to reduce vibrations.

To observe the airglow, at first we used an ITT FW 130, and then an EMI 9558 A with equivalent results. Both tubes were cooled to $-20°C$. With the 9558, a magnetic field produced by a coil in the proximity of the cathode reduces its effective area to a few $mm^2$, with a proportional reduction of the dark current. In the electronic chain each photomultiplier is followed by an impedance adapter, a preamplifier, a discriminator and a counter. By means of a programming unit, all data are printed and transferred to paper tape.

To improve detection, in the discriminator a delay line arrangement sets an oscillator free for the duration of the pulse at the threshold level. Thus, each original photomultiplier pulse, variable in amplitude and duration, produces a train of identical pulses proportional to its duration. These pulses are counted, thus enhancing the signal-to-noise discrimination since dark current photomultiplier pulses tend to have a smaller amplitude and duration. The operation cycle of the instrument is as follows: during one scan, usually lasting about 300 sec, the pressure is allowed to rise in the Fabry–Perot chamber within electronically established limits; therefore, more than one free spectral range is explored: every 10 sec, two measurements, sampling the airglow and the reference respectively, are carried out. For 7 sec, light from the sky filtered by the Fabry–Perot and the interference filter is allowed to enter the sky photomultiplier and output is counted. During this time a measurement of the pressure is also carried out electronically. After 7 sec a set of mirrors and shutters is moved. Then, between the 8th and the 9th second the sky-photomultiplier dark current is recorded; the laser light is allowed to enter the Fabry–Perot whose output is observed with a second photomultiplier, and the instrumental response is thus sampled.

The pressure interval, corresponding to one free spectral range at $\lambda_1 = 6328$ Å, produces a wavelength shift $\delta\lambda_2 = 0.0568$ Å at $\lambda_2 = 5577$ Å.

At present, observations have been limited to the OI 5577 Å line. At temperate latitudes, the $\lambda = 5577$ Å emission is essentially produced by OI excited in the $(^1 S)$ state from the reaction:

$$O + O + O \rightarrow O_2 + O(^1 S)$$

in a narrow layer around 100 km. In view of the long lifetime of the excited state $(\sim 0.7$ sec) and the high collision frequency at this level $(\sim 3500$ sec$^{-1})$, the measured temperature can be assumed to be equal to the ambient air temperature.

Data reduction, carried out by a computer, first requires the spectral response at $\lambda = 6328$ Å for each pressure scan. Then the amount of drift can be ascertained and a correction can be introduced for the instrumental response at $\lambda = 5577$ Å. The pressure range, equivalent to one free spectral range at $\lambda = 6328$ Å, is then divided into 100 intervals where data from successive scans at $\lambda = 5577$ Å are accumulated.

Synthetic spectra, taking the instrumental response into account, are then constructed and fitted to the experimental data. The instrumental response is assumed to be a periodically spaced set of Gaussian curves. The fit is carried out first by making use of the properties of the Fourier integrals of a Gaussian curve, and then by an optimization of the coefficients with the $\chi^2$ criterion.

Typical data from a single scan are shown in Fig. 1. The result of an all-night integration for 4 December 1970 is shown in Fig. 2. The points represent the total photoelectron count in each 100th interval, corresponding to an integration time of 134 sec per interval. The result of fitting is shown as a continuous curve: a temperature of $(265 \pm 3.3)°$K is inferred although the error does not include possible systematic effects.

Fig. 3 is a plot of the stability of the Fabry–Perot interferometer versus time, obtained by the analysis of the data of Fig. 2. The ordinate scales represent the

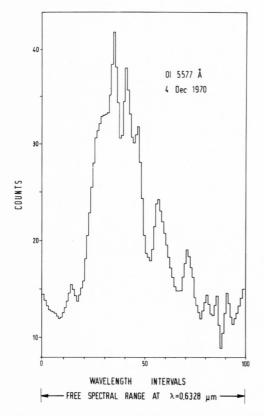

Fig. 1. Typical result of a single scan for the night of 4 December 1970. The total wavelength scanning range ($\delta\lambda_2 = 0.0563$ Å) is divided into 100 intervals: integration time is 2.52 sec per interval.

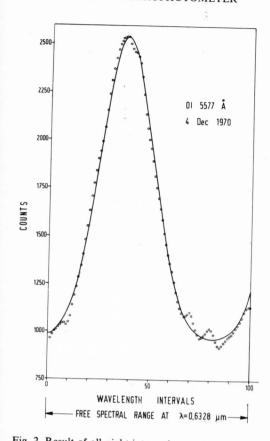

Fig. 2. Result of all night integration on 4 December 1970: integration time is 134 sec per interval. The result of fitting is shown as a continuous curve. Temperature inferred is 265°K.

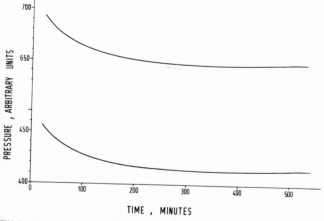

Fig.3. Stability of the interferometer versus time.

pressure values (in arbitrary units) corresponding to the laser peaks.

A total drift of 17% of the free spectral range in about 5 h was observed: its effect was removed in the data reduction, thus demonstrating the advantages of the technique.

## ACKNOWLEDGEMENT

Suggestions by D. Parkinson and I. Taylor regarding data processing are gratefully acknowledged.

## REFERENCES

Armstrong, E. B., The observation of line profiles in the airglow and aurora with a photoelectric Fabry–Perot interferometer. In: *The Airglow and the Aurorae* (E. B. Armstrong and A. Dalgarno, Editors). Pergamon Press, London, 1956.

Armstrong E. B., Variations in the width of OI λ 5577 line in the night airglow *Planet. Space Sci.*, **16**:211–229, 1968.

Hays, P. B., and Roble, R. G., A technique for recovering Doppler line profiles from Fabry–Perot interferometer fringes of very low intensity. *Appl. Opt.*, **10**:193–200, 1971.

Hernandez, H. G., Analytical description of a Fabry–Perot photoelectric spectrometer, *Appl. Opt.*, **5**:1745–1748, 1966.

Hernandez, H. G., Analytical description of a Fabry–Perot photoelectric spectrometer, 2: Numerical results. *Appl. Opt.*, **9**:1591–1596, 1970.

Shepherd, G. G, Airglow spectroscopic temperatures. In: *Atmospheric Emissions* (B. M. McCormac and A. Omholt, Editors). Van Nostrand, Reinhold, New York, N.Y., pp. 411–412, 1969.

# ACTIVE AND PASSIVE OPTICAL DOPPLER TECHNIQUES FOR THE DETERMINATION OF ATMOSPHERIC TEMPERATURE, 2: A HIGHLY COHERENT LASER RADAR

GLAUCO BENEDETTI-MICHELANGELI AND GIORGIO FIOCCO

*European Space Research Institute (ESRIN), Frascati (Italy)*

## SUMMARY

An optical radar system with a high degree of coherence is used for the determination of atmospheric temperature. The instrument utilizes a highly stable, single frequency $Ar^+$ laser for transmitting, and a 50-cm diameter Cassegrain telescope with a piezoelectrically driven Fabry-Perot interferometer for receiving.

The device can emit pulses of monochromatic light of about 0.5 W and can analyze the back-scattered radiation spectrum. Because of different amounts of Doppler shift, it is possible to separate the echoes caused by aerosols from those caused by molecules. The experiments carried out so far have been limited to tropospheric heights because of sensitivity. At these levels the molecular spectrum is broadened by pressure fluctuations (Brillouin scattering).

## 1. CHARACTERISTICS OF THE LASER RADAR

In the preceding paper (Benedetti-Michelangeli et al., 1974) hereafter referred as part 1, an interferometer for use in the determination of upper atmospheric temperature from airglow observations was described. That technique is passive because it utilizes a natural source of radiation. On the other hand, atmospheric motions can be observed in an optical radar experiment where a highly coherent laser is used as a transmitter to illuminate the atmosphere. The receiver includes a telescope and an interferometer similar to the one used for airglow observations. In last year's lectures at this school (Fiocco, 1971), the use of optical radars for atmospheric investigations was considered. In this paper, the instrumentation utilized in recent optical radar experiments (Fiocco et al., 1971) is described.

The optical radar is capable of radiating pulses of highly monochromatic laser light and of resolving the frequency spectrum of the light back-scattered by atmospheric constituents. Because of the different amounts of Doppler broadening affecting the spectra of the echoes from aerosols and molecules, it is possible to distinguish the contributions of the two species to the scattering cross section of a volume of air and infer their relative concentration. Measurements of atmospheric temperature are obtained from the width of the molecular spectrum. By pulsing the outgoing beam and by gating the receiver, these quantities can be measured at different heights. While present measurements are limited to the troposphere, we expect to extend the range of the instrument to the upper atmosphere. We also foresee that with suitable modifications the technique will be able to measure winds and detect turbulence.

If an analysis of the spectrum of the light scattered in an optical radar experiment is performed, a wealth of information related to the atmospheric composition and dynamics can be obtained (Fiocco and De Wolf, 1968).

We do not wish to take into account selective scattering processes related to Raman effects or the matching of resonant transitions here, but we simply mean to consider the Doppler frequency shift that affects echoes from molecules and aerosols in a different manner. In a back-scattering experiment the frequency shift $\Delta f$ associated with a radial component of velocity $v_z$ is given by $\Delta f \simeq 2fv_z/c = 2v_z/\lambda$. Thus, at a wavelength $\lambda = 0.4880\,\mu$, a velocity component $v_z = 10$ m sec$^{-1}$ will give a frequency shift $\Delta f = 41$ MHz. The observations described here were carried out with the optical radar looking at the zenith; no large values of vertical velocities were then expected and the resolution available was not adequate for wind velocity measurements. On the other hand, random velocities associated with thermal molecular motions were much larger and were detectable.

For particles that can be regarded as point scatterers, the differential radar (back-scattering) cross section per unit volume can be written as

$$\frac{d\Sigma_{rad}}{d\omega} = \Sigma_{rad} \cdot \Phi(\omega) \tag{1}$$

The function $\Phi$ contains the dependence of the scattered light on the angular frequency $\omega$.

In a manner similar to the airglow experiments described in part 1, the spectral density $\Phi$ of the molecular echoes is a Gaussian curve as long as the molecules can be regarded as non-interacting and randomly distributed. In this case, the half-power full width of the spectral density curve is given by:

$$\frac{\delta\lambda}{\lambda} = \frac{\delta\omega}{\omega} = 14.32 \cdot 10^{-7} \left(\frac{T}{\mu}\right)^{1/2} \tag{2}$$

where $T$ is the temperature, $\lambda$ the centre wavelength, and $\mu$ is the mass of the molecular species in a.m.u. In order to write this simplified result we have assumed here that the molecules were of one species only ($\mu = 28.9$). In addition, correlations arising because of interactions between molecules (such as Brillouin scattering) were disregarded.

This is possible when the ratio of the wavelength to the collision mean free path is small, which is the case in the upper atmosphere. At higher pressures, pressure fluctuations modify the spectrum by introducing two displaced components (the Brillouin doublet). At tropospheric heights, one does not expect the peaks to be distinguishable, but we should observe a broadening of the spectrum and slight modifications in its shape. Yip and Nelkin (1964) computed spectral profiles for scattering from a monoatomic gas for various values of the parameter $y$, a measure of the ratio of the wavelength to the collision mean free path:

$$y = \frac{5k_B^{3/2} \lambda \rho T^{1/2}}{8\sqrt{2}\,\pi m^{1/2} k_T} \tag{3}$$

where $k_B$ is the Boltzmann gas constant, $\rho$ the atmospheric number density, $m$ the molecular mass and $k_T$ the coefficient of thermal conductivity. The variation of $y$ with height is shown in Fig. 1.

From Yip and Nelkin's paper, we computed the expected spectral curves: these are shown in Fig. 2 as a function of the normalized parameter:

$$x = \frac{\lambda \omega m^{1/2}}{4\pi\sqrt{2}\,k_B^{1/2} T^{1/2}} \tag{4}$$

The curve $\Phi$ tends to a Gaussian shape for $y \to 0$ (see eq. 2).

The spectrum due to the aerosol component would also be affected by their motion. However, in the present experiments, since instrumental broadening is wider than broadening associated with aerosol motion, the latter is disregarded and the contribution of the aerosols appear unshifted. Thus, to conclude this section, the spectrum of the back-scattered light would appear as a broad bell-shaped curve due to contributions from molecules with a superimposed spike caused by the aerosols.

A schematic diagram of the optical radar is given in Fig. 3; Fig. 4 is a more detailed diagram of the laser source. For these experiments the radiation of highly monochromatic light is necessary. Our transmitter utilizes an argon ion laser in continuous operation as a light source. The laser is basically the commercial Coherent-Radiation Mod. 52, modified by us to radiate at one highly stable frequency. Without modifications the laser would operate simultaneously at several wavelengths and longitudinal modes, with a total power output of approximately 3 W. Single wavelength operation ($\lambda = 0.4880\ \mu$) is achieved by introducing a prism assembly into the laser cavity. Then, by replacing the other end-mirror by a 3-mirror-cavity, we are able to select one longitudinal mode only (Smith, 1966). Furthermore, in order to assure high frequency stability, a Spectra-Physics model 119 He-Ne laser (single frequency, stabilized to the Lamb dip to an accuracy of $\pm$ 1 MHz per day) was used as a frequency reference for the 3-mirror cavity of the argon laser. A feedback loop controls the separation of the mirrors with the result that the frequency of the argon laser is linked to the frequency of the He–Ne laser.

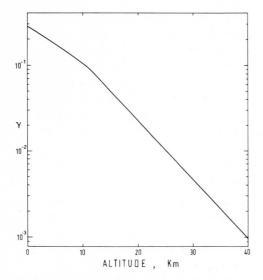

Fig. 1. Variation of the parameter $y$ with altitude.

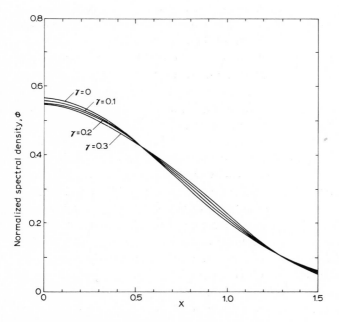

Fig. 2. Normalized spectral density $\Phi$ as a function of $x$ and $y$.

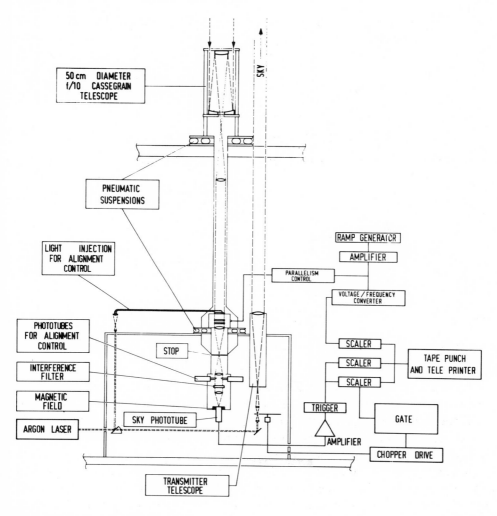

Fig. 3. Block diagram of the experiment.

By controlling the 3-mirror cavity, the stability of the argon laser is better than ± 50 MHz. In order to obtain a high degree of isolation from vibrations, the device is kept on a table resting on an air cushion. The maximum power output at a single longitudinal mode operation is 0.5 W. For a detailed description of the laser, see Maischberger (1971).

In order to determine the range, the laser light is chopped with a mechanical chopping wheel. The light pulses are approximately 12 μsec long and are repeated with a period of 166 μsec, corresponding to a maximum range of 25 km. The light can also be sampled and used for calibrating the receiver. After chopping, the

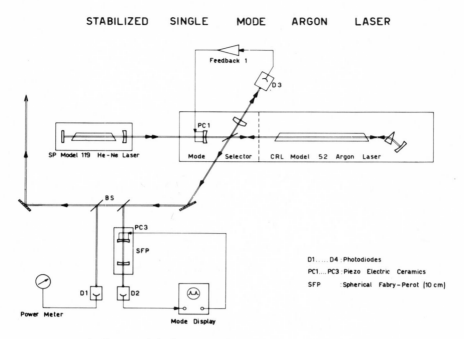

Fig. 4. Schematic diagram of the laser transmitter.

transmitted light is collimated by a telescope with an 8 cm diameter and 60 cm focal length. The telescope is mounted on gimbals to adjust the direction of the output beam and permit overlap with the receiver beam after a minimum height $H_0$.

In the receiver, the light scattered by the atmosphere is collected by a Cassegrain telescope, 50 cm diameter, and is collimated by a lens system. The telescope is located on the roof of the ESRIN building. Through a vertical tunnel the collimated light is sent to a Fabry–Perot interferometer located in the air-conditioned laboratory where stable temperature conditions exist. Telescope and interferometer are connected by a metallic frame and are also resting on low pressure rubber tubes to minimize vibrations. In designing the Fabry–Perot utilized here, care was taken to obtain high resolution and it is similar to the one described in part 1. The axis of the Fabry–Perot is vertical; each plate is independently supported by a frame made of NILO in its essential parts and which includes piezoelectric ceramics for scanning and controlling the parallelism.

The light output of the interferometer is focused by means of an achromatic doublet with a 1 m focal length; on the focal plane a diaphragm with a small adjustable diameter intercepts the on-axis light. Beyond the diaphragm, the light is again collimated. When scattering measurements are made, the light passes through an interference filter and is then focused on the cathode of a photomultiplier. For alignment purposes, in order to optimize the parallelism of the plates, the adjust-ment is carried out in the following way: laser light is intercepted by means of a mirror (a Leitz Visoflex) and brought normal to the plates by three optical pipes at three input points, separated by 120°, close to the periphery of the Fabry–Perot plates. After collimation, the light, at three points corresponding to the three previously mentioned input points, is sampled by three photomultipliers. The three outputs are then simultaneously displayed on an oscilloscope, while a scanning voltage is applied to the piezoelectric ceramic cylinders and to the x-axis of the oscilloscope. Thus, any difference in plate parallelism is easily checked and corrected. The alignment system could be made automatic in a relatively easy fashion. The stability, however, seems good enough to permit only checks. Although the room temperature is stable to about 1°C, circulation of water in the walls of the chamber housing the interferometer keeps its temperature stable to 0.1°C. The Fabry–Perot can also be scanned by changing the pressure of $N_2$ inside its chamber.

For the scattering experiments reported here, a photomultiplier EMI 9558 Å was utilized. In order to reduce its dark current, a magnetic deflection system was mounted outside the tube. The system consists of a coil and an iron structure that permits only those electrons coming from a small central region of the photo-cathode to fall on the diode structure, and excludes all the other electrons emitted. The photomultiplier output pulses, corresponding to amplified photoelectrons, are further amplified and sent to a trigger (Laben FT 130) to establish a threshold: pulses exceeding a minimum level are sent to two counters (Laben PFS 236). One of the counters can be gated so that only pulses from a certain height interval are accepted; the other counter accepts all pulses. Of course, several counters gated at successive height intervals can be added in order to carry out simultaneous measurements. Scanning voltages are converted into variable frequency signals; such signals are also fed to another counter for reference. Periodically, the outputs of all counters are sent through a multiplexer unit (Laben PA 860) to a teleprinter (Olivetti TE 300). Thus both a printed record and a perforated tape can be produced.

Measurements reported in Fig. 5 and 6 were carried out during the night of 6 July, 1970 at Frascati. Each graph is the result of summing several scans. In both Fig. 5 and 6 each point represents the counts obtained in 75 sec of observation; it took about one hour to obtain these curves. Fig. 5 is the spectrum of the echoes received without gating the counter. The echoes are contributed by the volume of air common to the receiver and transmitter beams. The minimum range $H_0$ below which the beams do not intersect is estimated to be a few hundred meters in this case. Laser stray light is negligible. Fig. 6 is the spectrum of the echoes received from a range of approximately 3 to 5 km, obtained by suitably gating the counter. In both cases the contribution of the aerosols can be separated from those brought about by molecules. The noise level is determined by the photomultiplier dark

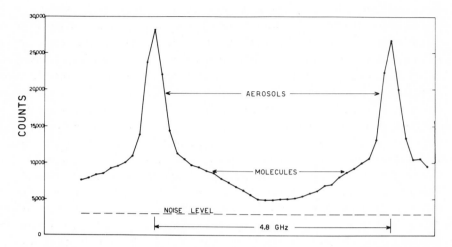

Fig. 5. Spectrum of echoes from air. Contributions of molecules from aerosols can be separated.

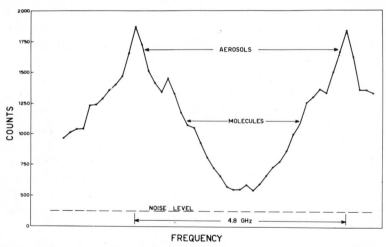

FREQUENCY

Fig. 6. Spectrum of echoes from air; receiver was gated so that only echoes from approximately 3–5 km were received.

current and by sky background. It appears that the echoes from aerosols are relatively more important at closer ranges. The molecular spectra are, after deconvolution, wider than expected from expression 2 because of the Brillouin effect. By taking the latter into account, we estimate a temperature of 303°K from Fig. 5 and a temperature of 276°K from Fig. 6. The difference of 27°K between the two measurements, taken at a difference in height of about 4 km, conforms to expectation.

A balloon sounding made at Fiumicino airport (at a distance of about 30 km) indicated temperatures of 18°C at the ground, of 2°C at 3.12 km height and of −11°C at 5.79 km height. It is difficult at this time to give estimates of the accuracy in our experiment. Aspects requiring further attention are, for instance, scanning linearity, increased resolution and more accurate data reduction, taking the Brillouin spectral shapes into account.

## ACKNOWLEDGEMENT

K. Maischberger substantially contributed to this work by developing the $Ar^+$ laser control system. E. Madonna and H. Orrhammer designed and built some of the electronics. F. Congeduti developed the programs for the data analysis.

## REFERENCES

Benedetti-Michelangeli, F., Cappuccio, G., Congeduti, F. and Fiocco, G., Active and passive optical Doppler techniques for the determination of atmospheric temperature, 1: An airglow spectrophotometer with internal laser reference. In: *The Structure and Dynamics of the Upper Atmosphere* (F. Verniani, Editor), pp. 205–210 (this volume), 1974.

Fiocco, G., Use of optical radars for atmospheric studies. In: *Physics of the Upper Atmosphere* (F. Verniani, Editor). Editrice Compositori, Bologna, pp. 407–439, 1971.

Fiocco, G. and De Wolf, J. B., Frequency spectrum of laser echoes from atmospheric constituents and determination of the aerosol content of air. *J. Atmos. Sci.*, **25**:488–496, 1968.

Fiocco, G., Benedetti-Michelangeli, G., Maischberger, K. and Madonna, E., Measurement of temperature and aerosol to molecule ratio in the troposphere by optical radar. *Nature Phys. Sci.*, **229**: 78–79, 1971.

Maischberger, K., Long-term frequency stabilization of a composite cavity argon laser. *IEEE J. Quantum El.*, **QE-7**: 250–252, 1971.

Smith, P. W., On the stabilization of a high-power single- frequency laser. *IEEE J. Quantum El.*, **QE-2**: 666–668, 1966.

Yip, S. and Nelkin, M., Application of a kinetic model to time-dependent density correlations in fluids. *Phys. Rev.*, **135**:A1241–A1247, 1964.

# MORPHOLOGY OF THE LOWER IONOSPHERE

EWALD HARNISCHMACHER

*Ionospheric Institute, Breisach (Germany)*

## SUMMARY

The aim of this paper is to give a survey of methods of observations and layer parameters of the ionosphere. The observational data show the solar control of the parameters and their dependence on solar and magnetic activity. Global features are shown when possible or at least latitudinal variations for temperate zones are given.

Now, after some ten years of measurements at many ionospheric stations, the data are abundant: a statistical analysis yields results for the mean behaviour (year-to-year, seasonal and diurnal) of the layer parameters. The mean values reveal anomalies in the solar control. Finally, "noise charts" give an impression of how closely solar and layer parameters are correlated.

## 1. THE D REGION

This section discusses first methods of observation and layer parameters. Then, we present a statistical treatment for the noon values of absorption measured over Freiburg at eight different frequencies over a 15-year period. The result of the analysis is a nomogram that gives the mean noon absorption (including selective absorption and winter anomaly) for every month as a function of $f_0 E$ noon vs. $f_0 F1$ noon.

### 1.1. Methods of observation

The D region cannot be described so easily by an electron density profile as is the case of the higher layers. This is due to the fact that at 80 km the electron collision frequency is rather high ($10^6 \sec^{-1}$).

Various methods for investigating the D layer are used and global results are discussed in the following pages.

### 1.1.1. Echo observations

During the day, echoes are obtained from virtual heights between 68 and 100 km. Echoes below 80 km are only seen below 0.5 MHz; they almost disappear in summer. Stronger echoes coming from 85 to 95 km often appear in a nearly horizontal trace. The top frequency is of the order of 1 MHz with practically no retardation. It is not completely certain that these traces are all due to total reflection: whereas for the lower edge of the normal E layer ($h' = 98$–$100$ km) total reflection is always certain, in the D layer partial reflections also often occur.

At night the E-trace is the lowest; it appears regularly, but most often followed by an intermediate layer at roughly 140 km.

The conclusion is that at the lower edge of the E layer the electron density sharply increases. Down in the D region, rather variable thin layers of the sporadic type seem to be quite common. The electron density $N$ below 90 km rarely exceeds some $10^9 \, \mathrm{m}^{-3}$ during the daytime and is quite low in non-disturbed nights.

By using the oblique sounding technique at low frequencies, such low electron densities can be observed. While at higher frequencies the attenuation of a reflected wave is mostly greater at oblique incidence, the inverse is often true at low frequencies; the steepness of the vertical gradient of $N$ is important in this respect for wavelengths which are comparable with the size of the vertical structure. From the total r.f. phase change between night and day, the corresponding height difference can be determined. Values between 7 and 18 km have been found at frequencies comprised between 30 and 65 kHz, but about 15 km at 16 kHz. Absolute height determinations on this latter frequency give about 70 and 85 km for day and night, respectively. A stratification of the lower part of the D region, near 70 km, often seems to be present. It is, however, rather difficult to deduce a reliable electron density profile from echo observations only.

### 1.1.2. In-situ observations

In-situ observations are therefore of considerable importance for studying the D region. Day-time rocket observations with the differential Doppler method sometimes gave a shallow maximum at about 80 km, but often yielded a monotonous decrease of electron density below the E region down to about $10^{10} \, \mathrm{m}^{-3}$ near 86 km (Fig. 1).

An abrupt decrease is often seen at about the 86 km altitude (Fig. 2).

### 1.1.3. Partial reflection observations

Partial reflection observations with penetrating waves at frequencies above 1 MHz allow sharp vertical gradients of the electron density to be detected. Partial echoes appear between 80 and 90 km; such echoes occur especially during those temperate latitude winter days when "excessive absorption" is observed. The corresponding height increases with the increasing solar zenith angle (Fig. 3).

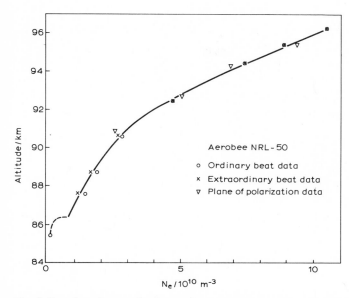

Fig. 1. Day-time electron density profile of lower E and upper D region obtained during a rocket flight with three different methods. "Beat data" means differential Doppler-effect observations; "plane of polarization" means Faraday-effect observations. (Reproduced from Rawer and Suchy, 1967.)

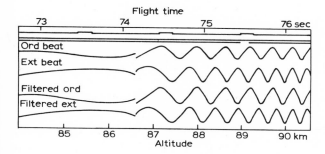

Fig. 2. Differential Doppler record, corresponding to the lower end of Fig. 1. The beat frequency is proportional to the local electron density. Note the abrupt change at 86.5 km altitude. (Reproduced from Rawer and Suchy, 1967.)

In a desert region of extremely low noise, very faint echoes (down to a reflection loss of 120 db) can be observed. Fig. 4 shows noon measurements at 2.3 MHz. Below the normal echo (80 db), and the above-mentioned 90-km echo (60 db), a third echo appears at 70 km (20 db, i.e., 100 db loss).

### 1.1.4. Multifrequency measurements

The multifrequency measurement of the absorption of penetrating waves is a

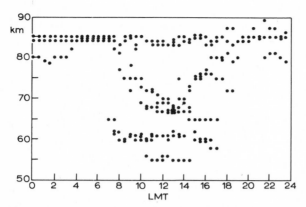

Fig. 3. Lowest reflection heights obtained with the partial reflection technique in New Zealand. (Reproduced from Rawer and Suchy, 1967.)

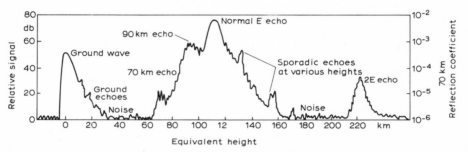

Fig. 4. Relative signal strength (left) and amplitude of the reflection coefficient (right) as a function of equivalent height (abscissa), observed in Australia. (Reproduced from Rawer and Suchy, 1967.)

powerful tool for investigating the D region (Rawer and Suchy, 1967). While reflection methods have a tendency to indicate stratifications, method A1, which uses pulses to distinguish E- and F-layer reflections, gives an integrated picture of the D layer. It is usually applied at frequencies from 1.5 to 8 MHz. The absorption decrement, i.e., the loss $L$ is given by:

$$L = \int_{h_1}^{h} \mathrm{d}h\, \bar{\nu}\, N \tag{1}$$

where $\bar{\nu}$ is the collision frequency between electrons and neutral molecules. For average day-time conditions, $N\bar{\nu}$ has a maximum near 85 km; during perturbed conditions the maximum can be at a considerably lower altitude.

Unfortunately, attenuation by deviative absorption in the E region gives an appreciable contribution to the total absorption loss. Many observations have been interpreted (see below) with only one absorption parameter $A$. Noon observations in Europe (at Freiburg and Slough) in 1949/50 gave $A$-values of the order of 500 db

MHz$^2$. Considerably lower values were found during the minimum of solar activity.

Temperate latitude observations gave rather regular diurnal variations with cos $\chi$, $\chi$ being the solar zenith angle. If only the absorption loss $L$ on a single frequency $f$ is considered, the observed diurnal curve depends mainly on the position of $f$ with respect to $(f_0 E)_{noon}$. If $f < (f_0 E)_{noon}$, then a maximum of deviative absorption is reached twice; i.e., before and after noon, when $f_0 E \simeq f$. Simpler curves with only one maximum (at about noon) are obtained with $f > (f_0 E)_{noon}$. Rather different values of the exponent $p$ in a $(\cos \chi)^p$-law have been reported in the literature (values between 0.6 and 1.8), but these are not physically significant. The rough absorption parameter $A$ can be used to obtain a somewhat more consistent picture. $A$ varies roughly with cos $\chi$.

The simple formula for non-deviative absorption is:

$$L = \frac{A}{(f \pm |f_L|)^2} \tag{2}$$

A typical plot of absorption decrements obtained at noon at different frequencies is shown in Fig. 5.

Generally, $L$ decreases with increasing frequency as a result of non-deviative absorption in lower layers. However, deviative absorption in the E region is selective and causes a "bump" near $f_0 E$; absorption in F1 also causes a small deformation. In cases where blanketing Es layers are present, the bump is cut because the waves do not reach the deviative region; quite generally, blanketing Es has a high reflection coefficient and gives the lowest L-values.

A much better representation of these results can be obtained by adding a term for deviative absorption:

$$L = \frac{B}{(f \pm |f_L|)^2} + C\Delta\left(\frac{f_0 E}{f}; \frac{2q}{H}\right) \tag{3}$$

where $2q$ is the thickness parameter of the E layer and $H$ is the scale height.

Detailed analysis with two parameters $B$ and $C$ gives more reliable information. At a temperate latitude it has been shown that the parameter $B$ characterizing the D region varies with $(\cos \chi)^p$, the exponent $p$ being about 0.8 in summer but only 0.45 in winter. The cos $\chi$-law is valid in similar form in the morning and afternoon when a time delay is allowed for. This delay has been determined as 18 min. The dispersion of individual points on a logarithmic plot of $B$ versus cos $\chi$ is rather great, as shown in Fig. 6.

In fact, the variations of $B$ for individual days are quite important, with sudden increases occurring particularly during solar flare effects, even weak ones.

Compared with $B$, the absorption parameter $C$ in the E layer is nearly independent of $\chi$ (Fig. 7).

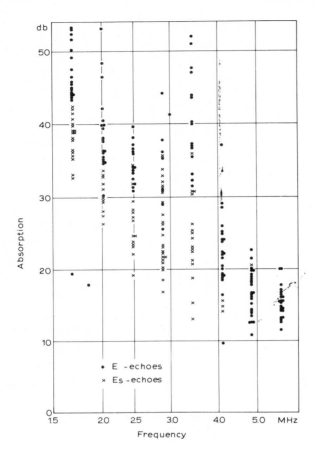

Fig. 5. Absorption decrement as a function of the frequency. Mass-plot of noon measurements during one month. Selective absorption appears clearly near $f_0E = 3.5$ MHz, but only for E-echoes, not for Es-echoes. (Ionosphären-Institut Breisach.) (Reproduced from Rawer and Suchy, 1967.)

The long-term variation is shown in Fig. 8. The correlation is positive for $\overline{\overline{B}}$ and $\overline{\overline{C}}$.

For high sunspot numbers $\overline{\overline{R}} > 160$, we find a saturation effect.

The seasonal variation of $B$ seems to be quite small (see Fig. 9a and b). A deep minimum is regularly observed in November; later in winter the values compare with the summer values as a result of excessive winter absorption. The seasonal effect on $\overline{\overline{C}}$ is inverse, with a maximum in summer, about twice as large as the winter values. A small lunar variation was also found.

The influence of the geographical coordinates is difficult to evaluate for the time being. The non-deviative part does not appear to increase towards lower latitudes, except for the magnetic equator. On the other hand, it is quite certain that the

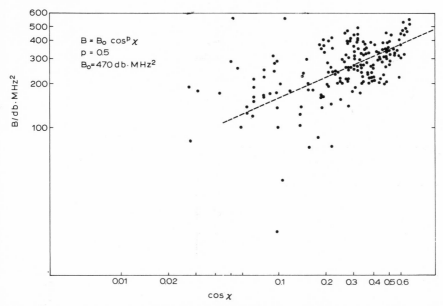

Fig. 6. A logarithmic plot of individually observed absorption values $B$ (non-deviative) versus $\cos \chi$. All observations were made at Freiburg (48°N, 8°E, $\psi = +64°$) in winter 1957/58. (Reproduced from Rawer and Suchy, 1967.)

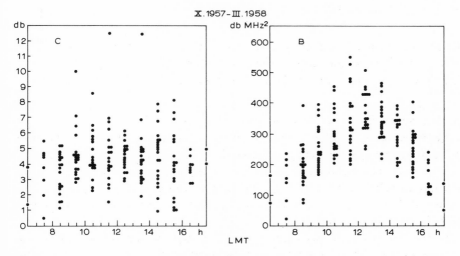

Fig. 7. Diurnal variation of individual observed values of $C$ and $B$ during a winter season. Winter 1957/58 was a period of very high solar activity. (Reproduced from Rawer and Suchy, 1967.)

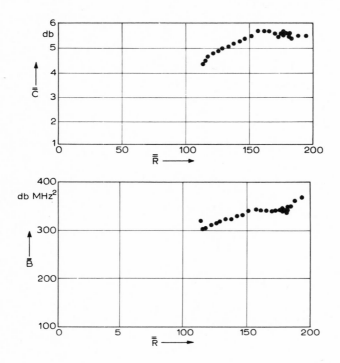

Fig. 8. Correlograms between running means of absorption and solar activity: $\overline{\overline{C}}$, $\overline{\overline{B}}$ = running means of monthly median values characterizing absorption conditions in the E layer ($\overline{\overline{C}}$, above) and the D layer ($\overline{\overline{B}}$, below). Period 1958–1960, characterized by high solar activity. $\overline{\overline{R}}$ is the Zürich sunspot number. (Reproduced from Rawer and Suchy, 1967.)

auroral zone, on the average, has increased absorption because additional low level ionization occurs instantaneously during disturbances. Median values of high latitude stations are considerably influenced by these "polar black-out" phenomena.

### 1.1.5. Oblique incidence field-strength observations "A3"

Continuous wave field-strength observations at oblique incidence over distances of about 300 km with a frequency of about 2.5 MHz (method "A3") have been interpreted. After extrapolation to subsolar conditions ($\chi = 0$), values $A_{max}$ for 1958 (maximum solar activity) and for zero sunspot number are indicated as 560 and 237 db MHz$^2$. $A$ follows the solar cycle and shows solar control from spring through fall, but not in winter.

This winter anomaly is mainly a phenomenon of higher temperate or subauroral zones. It is said that there is no day-to-day correlation with magnetic activity, but the number of anomalous days increases with the average magnetic activity.

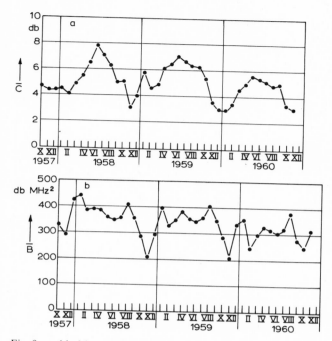

Fig. 9a and b. Monthly median values of absorption parameters $\overline{C}$ (above) and $\overline{B}$ (below) in 1957–1960, a period of decreasing yet rather high solar activity. (Reproduced from Rawer and Suchy, 1967.)

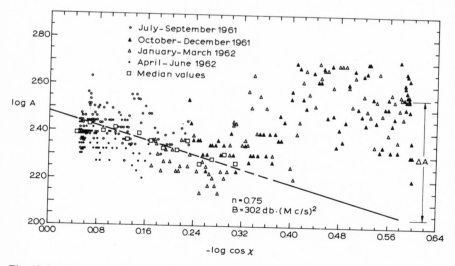

Fig. 10. log $A$ versus ($-$log cos $\chi$) for a period of 1 year, showing very distinctly two groups of points. The anomalous absorption in winter can be seen clearly. (Reproduced from Schwentek, 1963.)

An example for "$A$" is shown in Fig. 10. The values for $A$ are calculated with the formula

$$A = (f + |f_L|)^2 \cos \alpha \; 20 \log\left(\frac{E_n}{E_d} \cdot \frac{p'_n}{p'_d}\right) \tag{4}$$

where $f_L$ is the effective gyro-frequency of the circuit and $\alpha$ is the angle of incidence; $E_n$ and $E_d$ are night-time and day-time field strengths; $p'_n$ and $p'_d$ are night-time and day-time path lengths.

### 1.1.6. Cosmic noise observations

The cosmic noise absorption is particularly useful at high latitudes for polar black-out control. At temperate latitudes, we may control solar flare effects.

### 1.2. Electron density profiles

Electron density profiles of the D region have been tentatively determined by different methods: by multifrequency riometer observations, by partial reflection technique and by cross modulation. Average results (for southern Norway) are reproduced in Fig. 11. It appears that day-time D-region ionization is found between 75 and 95 km with a maximum sometimes near 85–90 km. At night, the tail of the E-region ionization falls off monotonously.

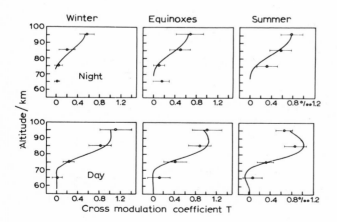

Fig. 11. Typical profiles of D-region ionization expressed by the cross-modulation coefficient (Landmark and Lied, 1960). Night above, day below, three seasons. (Reproduced from Rawer and Suchy, 1967.)

*1.3. Multifrequency absorption over Freiburg "A1"*

Multifrequency absorption measurements at eight frequencies: 1.725 MHz, 2.05 MHz, 2.44 MHz, 2.90 MHz, 3.45 MHz, 4.1 MHz, 4.9 MHz and 5.8 MHz have been made in Freiburg since 1956. We measure at noon during half an hour, and in the same manner at midnight in order to have a standard to use for comparison with noon values. Depending on the frequency, the reflection height may be in the E layer or in the F layer. Different virtual heights are normalized. We set 105 km for the E layer, 325 km for the F layer. The standard midnight absorption is the mean of the last fifteen nights.

Figures 12–19 show the monthly means of the absorption loss, after the above-described normalization, for each frequency and for the different months and years. The values change with frequency, season, and solar cycle.

We know at least five reasons for this great variability in the absorption loss for frequencies between 1.7 and 5.8 MHz.

(1) The critical frequency of the F2 layer varies from 4.9 to 14.5 MHz (see Table I).

(2) The critical frequency of the F1 layer disappears in winter and varies from 3.2 to 6.5 MHz (Table II).

(3) $f_0$Es 50% changes from 1.8 to 4.6 MHz (Table III).

(4) $f_0$E 50% varies from 2.4 to 4.1 MHz (Table IV).

(5) The electron content of the D layer also changes.

Since we know that the critical frequency of the normal E layer $f_0$Es very well describes the solar angle, the flux density of the sun and is also responsible for selective absorption, we can take $f_0$Es 50% as a standard for our future statistic work. For each frequency and for suitable pairs of months we plot the absorption loss vs. $f_0$E 50%; through the points we then lay a smoothed curve and then we fit the curve by a straight line. An example is given in Fig. 20.

Figures 21–24 show the results for the eight frequencies and the months January and December; February and November; March and October; April to September.

By superposition we could estimate the deviative absorption as a parameter with a Gaussian distribution, having its maximum of 6 db at the critical frequencies of the E layer or F1 layer, and going down to 0 at about ± 0.6 MHz for the E layer and about 1.2 MHz for the F1 layer.

Fig. 25 gives nomograms for pairs of months (January and December, February and November, March and October, April to September) to estimate for a given $(f_0 E)_{noon}$ the mean noon-absorption loss for normal summer conditions, winter anomaly, spring and fall transitions including the deviative absorption by $f_0$E or $f_0$F1.

Freiburg 1.725 MHz

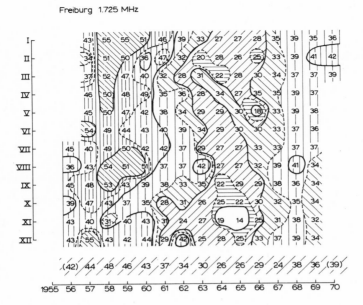

Fig. 12. Monthly mean values of *L* obtained at Freiburg "A1" at 1.725 MHz.

Freiburg 2.05 MHz

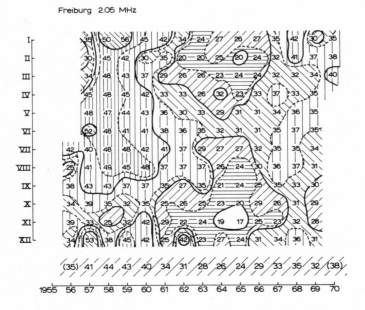

Fig. 13. Monthly mean values of *L* obtained at Freiburg "A1" at 2.05 MHz.

Freiburg 2.44 MHz

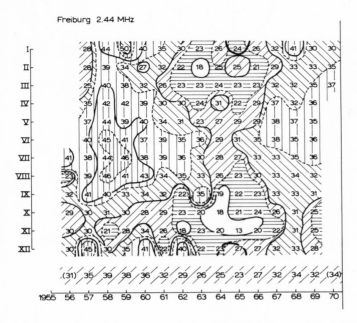

Fig. 14. Monthly mean values of $L$ obtained at Freiburg "A1" at 2.44 MHz.

Freiburg 2.90 MHz

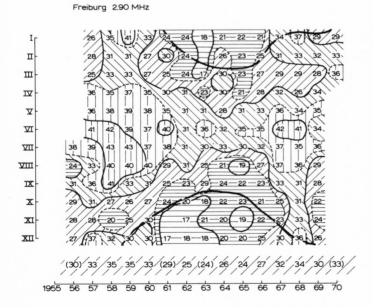

Fig. 15. Monthly mean values of $L$ obtained at Freiburg "A1" at 2.90 MHz.

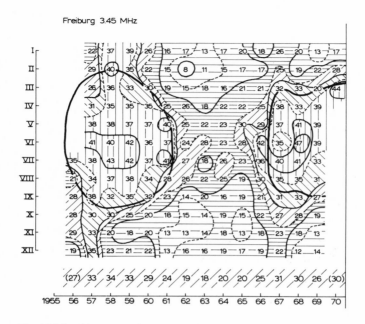

Fig. 16. Monthly mean values of $L$ obtained at Freiburg "A1" at 3.45 MHz.

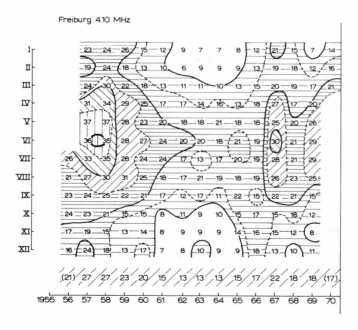

Fig. 17. Monthly mean values of $L$ obtained at Freiburg "A1" at 4.10 MHz.

Freiburg 4.9 MHz

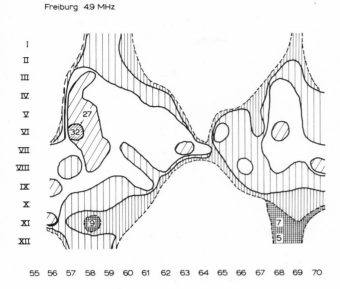

Fig. 18. Monthly mean values of *L* obtained at Freiburg "A1" at 4.90 MHz.

Freiburg 5.8 MHz

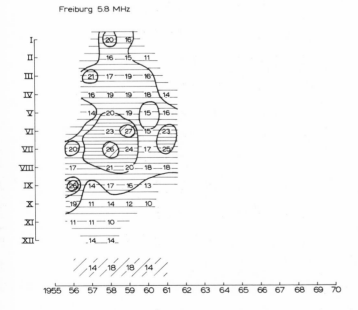

Fig. 19. Monthly mean values of *L* obtained at Freiburg "A1" at 5.80 MHz.

TABLE I

Freiburg, monthly mean of $f_oF2$ (1200 UT)

| Month | 1946 | 47 | 48 | 49 | 50 | 51 | 52 | 53 | 54 | 55 | 56 | 57 | 58 | 59 | 60 | 61 | 62 | 63 | 64 | 65 | 66 | 67 | 68 | 69 |
|---|---|---|---|---|---|---|---|---|---|---|---|---|---|---|---|---|---|---|---|---|---|---|---|---|
| I | 116* | 108 | 105 | 105 | 73 | 73 | 66 | 58 | 61 | 84 | 121 | 141 | 136 | 120 | 84 | 62 | 59 | 60 | 58 | 59 | 59 | 84 | 106 | 84 |
| II | | 130 | 114 | 124 | 107 | 76 | 69 | 59 | 57 | 60 | 96 | 128 | 135 | 136 | 124 | 84 | 68 | 59 | 62 | 62 | 63 | 91 | 111 | 106 |
| III | | 130 | 111 | 126 | 110 | 75 | 63 | 54 | 54 | 57 | 108 | 128 | 130 | 130 | 106 | 83 | 74 | 60 | 64 | 64 | 65 | 108 | 115 | 110 |
| IV | | 111 | 111 | 114 | 99 | 72 | 59 | 57 | 53 | 54 | 98 | 112 | 120 | 109 | 96 | 76 | 64 | 57 | 57 | 59 | 66 | 97 | 92 | 100 |
| V | | 94 | 87 | 93 | 82 | 68 | 57 | 52 | 52 | 57 | 87 | 96 | 94 | 94 | 85 | 68 | 64 | 58 | 53 | 56 | 66 | 80 | 81 | 82 |
| VI | | 86 | 89 | 77 | 71 | 66 | 56 | 51 | 49 | 57 | 78 | 86 | 81 | 81 | 81 | 65 | 61 | 56 | 52 | 54 | 64 | 69 | 71 | 80 |
| VII | 70 | 85 | 84 | 80 | 68 | 63 | 55 | 48 | 49 | 57 | 76 | 84 | 80 | 80 | 76 | 65 | 54 | 53 | 50 | 52 | 64 | 68 | 72 | 72 |
| VIII | 79 | 86 | 83 | 79 | 65 | 63 | 50 | 49 | 49 | 58 | 86 | 90 | 80 | 86 | 81 | 68 | 53 | 52 | 49 | 52 | 63 | 76 | 75 | 78 |
| IX | 86 | 98 | 96 | 104 | 66 | 67 | 58 | 55 | 49 | 66 | 104 | 108 | 113 | 93 | 98 | 69 | 59 | 56 | 56 | 56 | 68 | 84 | 82 | 84 |
| X | 112 | 127 | 114 | 120 | 82 | 84 | 63 | 64 | 64 | 87 | 132 | 142 | 118 | 108 | 82 | 78 | 72 | 64 | 68 | 68 | 84 | 110 | 108 | 103 |
| XI | 118 | 139 | 123 | 128 | 88 | 88 | 68 | 62 | 62 | 90 | 140 | 148 | 142 | 120 | 108 | 76 | 70 | 70 | 61 | 65 | 84 | 103 | 106 | 108 |
| XII | 112 | 116 | 117 | 117 | 76 | 64 | 60 | 56 | 56 | 86 | 139 | 145 | 132 | 112 | 98 | 68 | 62 | 58 | 60 | 57 | 78 | 104 | 104 | 88 |
| I–XII | 96 | 110 | 103 | 106 | 84 | 73 | 63 | 56 | 54 | 66 | 103 | 115 | 117 | 108 | 98 | 74 | 64 | 59 | 57 | 59 | 69 | 90 | 94 | 91 |

*116 indicates 11.6 MHz

TABLE II

Freiburg, monthly mean of $f_oF1$ (1200 UT)

| Month | 1948 | 49 | 50 | 51 | 52 | 53 | 54 | 55 | 56 | 57 | 58 | 59 | 60 | 61 | 62 | 63 | 64 | 65 | 66 | 67 | 68 | 69 |
|---|---|---|---|---|---|---|---|---|---|---|---|---|---|---|---|---|---|---|---|---|---|---|
| I |  |  |  | 39 | 35 | 36 |  |  |  |  |  |  |  |  |  | 37 | 38 | 39 | 40 |  |  |  |
| II |  |  | 40 | 41 | 39 | 38 | 40 |  |  |  |  | 49 |  | 40 |  | 40 | 40 | 40 | 40 |  |  |  |
| III | 47* | 42 | 48 | 44 | 44 | 41 | 41 | 42 | 44 | 49 |  | 52 | 51 | 45 | 44 | 43 | 43 | 42 | 42 |  | 50 | 50 |
| IV | 61 | 61 | 55 | 48 | 45 | 43 | 42 | 43 | 52 | 62 | 60 | 58 | 51 | 48 | 44 | 43 | 44 | 43 | 46 | 51 | 50 | 54 |
| V | 60 | 57 | 54 | 49 | 45 | 43 | 42 | 45 | 56 | 60 | 61 | 60 | 53 | 48 | 48 | 46 | 44 | 45 | 47 | 52 | 52 | 52 |
| VI | 58 | 54 | 52 | 50 | 45 | 44 | 43 | 46 | 54 | 60 | 59 | 57 | 54 | 48 | 47 | 45 | 43 | 45 | 48 | 50 | 53 | 51 |
| VII | 56 | 53 | 50 | 50 | 45 | 42 | 42 | 46 | 55 | 58 | 58 | 57 | 53 | 49 | 45 | 45 | 43 | 43 | 47 | 51 | 50 | 54 |
| VIII | 56 | 51 | 48 | 49 | 45 | 43 | 42 | 45 | 55 | 56 | 62 | 56 | 54 | 49 | 44 | 45 | 43 | 44 | 48 | 51 | 50 | 52 |
| IX | 49 | 53 | 47 | 47 | 44 | 43 | 41 | 45 | 53 | 53 | 65 | 54 | 49 | 47 | 44 | 43 | 42 | 42 | 45 | 47 | 50 | 47 |
| X |  |  | 43 | 44 | 40 | 40 | 39 | 43 |  |  |  | 46 | 43 | 43 | 42 | 41 | 41 | 41 | 44 |  |  |  |
| XI |  |  | 33 | 41 |  | 36 |  | 36 |  |  |  |  | 41 | 37 |  | 36 |  | 38 |  |  |  |  |
| XII |  |  | 33 | 36 | 34 |  |  |  |  |  |  |  |  | 36 | 32 | 34 |  |  |  |  |  |  |
| VI/VII | 57 | 54 | 51 | 50 | 45 | 43 | 43 | 46 | 55 | 59 | 59 | 57 | 54 | 49 | 46 | 45 | 43 | 44 | 48 | 51 | 52 | 53 |

*47 indicates 4.7 MHz.

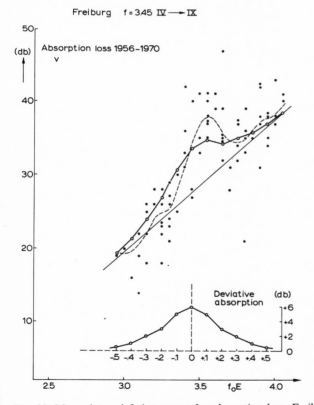

Fig. 20. Mass plot and fitting curves for absorption loss, Freiburg IV → IX, $f = 3.45$ MHz, mean deviative absorption included.

TABLE III

Freiburg, monthly mean of $f_o$Es 50% (1100,1200,1300 UT)

| Month | Year | | | | | | | | | | | | | | | |
|---|---|---|---|---|---|---|---|---|---|---|---|---|---|---|---|---|
| | 1955 | 56 | 57 | 58 | 59 | 60 | 61 | 62 | 63 | 64 | 65 | 66 | 67 | 68 | 69 | 70 |
| I | 18* | 22 | 35 | 34 | 32 | 32 | 27 | 29 | 28 | 27 | 27 | 26 | 28 | 21 | 22 | 32 |
| II | 23 | 32 | 33 | 34 | 34 | 30 | 29 | 29 | 18 | 29 | 28 | 28 | 32 | 35 | 26 | 25 |
| III | 27 | 29 | 36 | 37 | 36 | 29 | 31 | 31 | 29 | 27 | 30 | 30 | 34 | 35 | 34 | 33 |
| IV | 33 | 30 | 38 | 40 | 39 | 38 | 34 | 33 | 34 | 32 | 34 | 35 | 35 | 38 | 38 | 36 |
| V | 39 | 41 | 40 | 42 | 42 | 40 | 40 | 39 | 36 | 38 | 38 | 39 | 39 | 40 | 41 | 40 |
| VI | 44 | 39 | 46 | 46 | 46 | 44 | 44 | 42 | 40 | 38 | 38 | 40 | 40 | 42 | 45 | 43 |
| VII | 39 | 42 | 46 | 44 | 44 | 43 | 40 | 38 | 41 | 40 | 43 | 41 | 42 | 42 | 43 | 42 |
| VIII | 37 | 44 | 40 | 44 | 43 | 40 | 38 | 37 | 37 | 36 | 36 | 40 | 42 | 41 | 42 | 44 |
| IX | 33 | 42 | 40 | 40 | 38 | 36 | 35 | 35 | 33 | 34 | 34 | 34 | 37 | 39 | 36 | 38 |
| X | 28 | 35 | 36 | 35 | 35 | 34 | 32 | 34 | 30 | 31 | 33 | 33 | 32 | 35 | 35 | 36 |
| XI | 26 | 20 | 33 | 33 | 32 | 27 | 28 | 28 | 26 | 26 | 28 | 30 | 26 | 26 | 26 | 28 |
| XII | 26 | 26 | 35 | 32 | 31 | 28 | 28 | 29 | 28 | 27 | 26 | 29 | 27 | 21 | 28 | 30 |

*18 indicates 1.8 MHZ.

The correlation between $f_o$E 50% and $f_o$F1 50% for values from Freiburg is given by the following table:

| $f_o$E (MHz) | $f_o$F1 (MHz) |
|---|---|
| 2.5 | 3.0 |
| 3.0 | 4.0 |
| 3.5 | 5.0 |
| 4.0 | 6.0 |

We will not overlook the deviations of the loss curves from the straight lines (see Fig. 21–24). We hope that by introducing new parameters as the variations of the magnetic field, the sporadic E and the atmospheric warmings, we will then have a better adaptation.

## 1.4. Relation between ionospheric absorption and solar activity

A long series of A3-observations at an effective frequency of about 2 MHz has been made at Lindau. These observations have the advantage of being made continuously during daylight hours. From these measurements we have deduced representative noon values, avoiding SID-effects.

The dependence on solar activity was studied with the superposed epoch method, using relative minimum values of the Zurich sunspot number as reference (see Fig. 26).

Figure 26, obtained with the 1967 measurements, shows that the observed

TABLE IV

Freiburg, monthly mean of $f_0E$ 50% (1200 UT)

| Month | 1947 | 48 | 49 | 50 | 51 | 52 | 53 | 54 | 55 | 56 | 57 | 58 | 59 | 60 | 61 | 62 | 63 | 64 | 65 | 66 | 67 | 68 | 69 | 70 |
|---|---|---|---|---|---|---|---|---|---|---|---|---|---|---|---|---|---|---|---|---|---|---|---|---|
| I | 320* | 300 | 300 | 290 | 280 | 280 | 260 | 255 | 260 | 280 | 320 | 330 | 320 | 305 | 275 | 270 | 255 | 255 | 255 | 255 | 285 | 300 | 295 | 305 |
| II | 350 | 320 | 340 | 325 | 295 | 295 | 275 | 275 | 280 | 330 | 335 | 345 | 340 | 325 | 300 | 290 | 280 | 275 | 275 | 280 | 320 | 330 | 305 | 340 |
| III | 370 | 350 | 365 | 340 | 310 | 305 | 290 | 295 | 295 | 355 | 365 | 375 | 370 | 340 | 315 | 310 | 300 | 295 | 295 | 300 | 345 | 340 | 340 | 350 |
| IV | 390 | 380 | 375 | 355 | 350 | 320 | 315 | 310 | 320 | 370 | 390 | 395 | 385 | 365 | 335 | 330 | 315 | 315 | 310 | 325 | 355 | 350 | 360 | 360 |
| V | 400 | 400 | 375 | 365 | 355 | 325 | 310 | 320 | 330 | 370 | 400 | 400 | 390 | 375 | 345 | 335 | 325 | 320 | 315 | 330 | 355 | 360 | 365 | 370 |
| VI | 410 | 390 | 375 | 360 | 365 | 325 | 325 | 325 | 340 | 375 | 405 | 405 | 390 | 380 | 350 | 330 | 325 | 325 | 320 | 340 | 355 | 365 | 380 | 370 |
| VII | 400 | 380 | 375 | 350 | 360 | 330 | 320 | 320 | 335 | 380 | 405 | 400 | 390 | 375 | 350 | 330 | 325 | 320 | 325 | 340 | 355 | 355 | 380 | 370 |
| VIII | 390 | 370 | 370 | 330 | 360 | 325 | 320 | 315 | 335 | 380 | 395 | 400 | 390 | 375 | 345 | 330 | 325 | 315 | 310 | 335 | 365 | 355 | 370 | 380 |
| IX | 370 | 360 | 350 | 320 | 335 | 320 | 310 | 300 | 320 | 370 | 320 | 375 | 365 | 365 | 330 | 310 | 300 | 305 | 300 | 330 | 340 | 350 | 345 | 345 |
| X | 350 | 330 | 325 | 290 | 310 | 295 | 295 | 290 | 310 | 350 | 365 | 355 | 325 | 320 | 305 | 295 | 285 | 275 | 280 | 310 | 315 | 325 | 320 | 320 |
| XI | 310 | 310 | 315 | 280 | 285 | 275 | 275 | 265 | 295 | 325 | 340 | 320 | 310 | 290 | 270 | 265 | 265 | 255 | 260 | 280 | 300 | 300 | 300 | 310 |
| XII | 300 | 300 | 315 | 260 | 270 | 255 | 250 | 250 | 280 | 320 | 325 | 310 | 295 | 275 | 255 | 250 | 250 | 245 | 245 | 275 | 290 | 290 | 280 | 290 |
| I–XII | 365 | 350 | 345 | 320 | 320 | 305 | 295 | 295 | 310 | 350 | 370 | 365 | 355 | 340 | 315 | 300 | 295 | 290 | 290 | 310 | 330 | 335 | 335 | 345 |

*320 indicates 3.20 MHz.

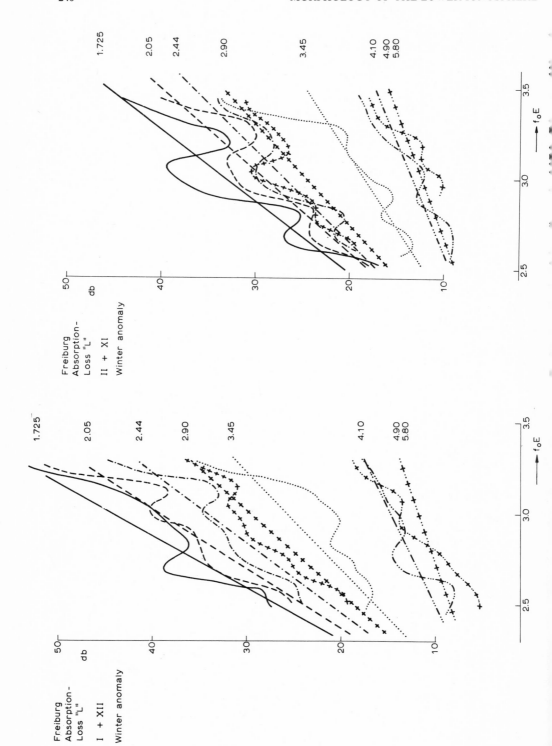

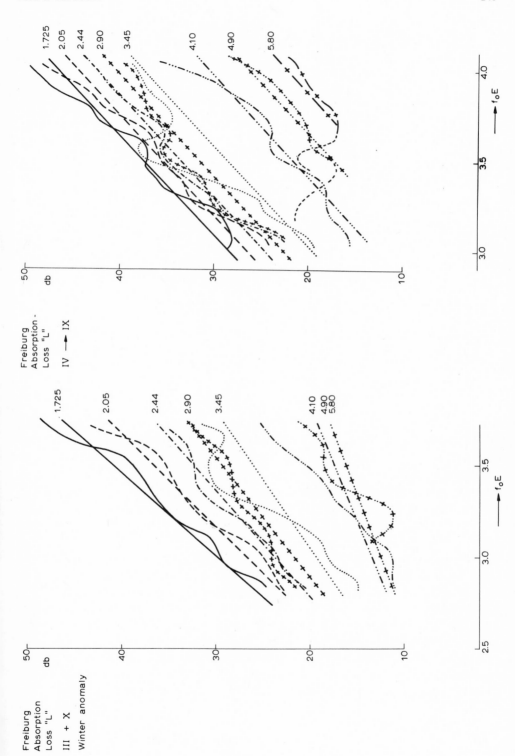

Fig. 23. Absorption loss $L$ depending on $f$ and $f_oE$ for March and October.

Fig. 24. Absorption loss $L$ depending on $f$ and $f_oE$ for April through September.

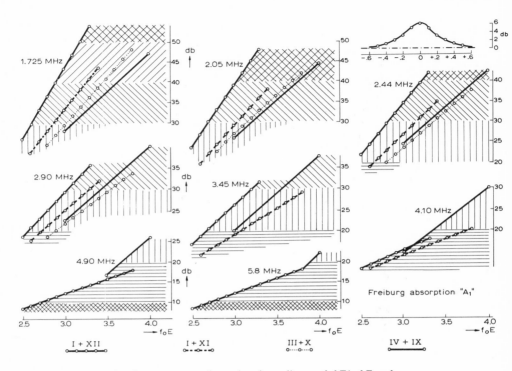

Fig. 25. Nomograms for the mean noon absorption depending on $f$, $f_oF1$, $f_oE$ and season.

absorption $A$ has a minimum at the same time as the sunspots. This is best seen by isolating into the $A$ curve a smoothed curve corresponding to the harmonic 27-day component $(\Delta A_1)$. For comparison, in Fig. 26 the corresponding curve of solar radio noise flux is given for six different frequencies. On frequencies higher than 1 GHz, the variation is noticeably similar to the one obtained with the sunspot numbers.

The first conclusion obtained from $\Delta A_1$ is that absorption follows the 27-day period without delay. This is in agreement with earlier findings by G. Langehesse (private communication).

Now, when looking for the remaining component, $\Delta A_2$, it appears that in 1967 there was a tendency towards rather regular variations within one week. Roughly similar variations also appear in the daily magnetic disturbance character, $\Sigma Kp$. There is, however, a delay time of a few days (two to four on the average). We have evidence that this consistent behaviour is related to the pattern of the solar wind existing during 1967.

The conclusion is that both the magnetic disturbance character and the fluctuations of absorption are related to this pattern; the effect on absorption follows, however, the magnetic disturbance by a few days.

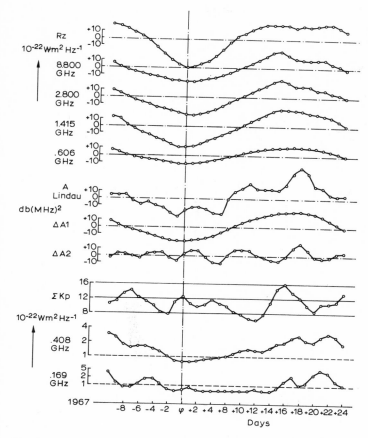

Fig. 26. 27-day periods in sunspots and radio noise flux for six different frequencies and absorption for Lindau. 7-day periodicity in absorption and in the magnetic disturbance character $\Sigma Kp$.

## 1.5. Seasonal effects in the relationship between absorption and magnetic disturbance character

In order to study these effects, superposed epoch diagrams have been established month by month for the years 1962 through 1968. Among these years, two classes of solar activity have been distinguished, namely: maximum activity (1966 through 1968); low activity (1964 and 1965). The diurnal sum of $Kp$ was used as an indicator, and maxima of this disturbance figure were taken as reference. The corresponding mean variation of absorption is shown in the upper part of Fig. 27.

There appears to be a net difference between summer and winter conditions. During the period from March through August, the magnetic effect upon absorption

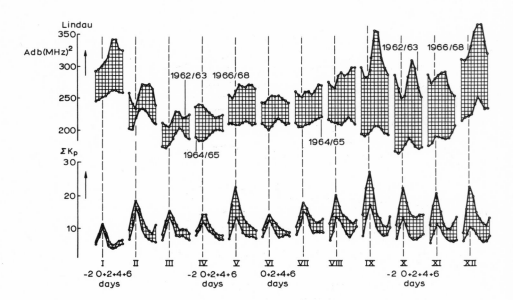

Fig. 27. Epochs for magnetic disturbances and variation of absorption at Lindau for every month and under different sunspot conditions.

is quite slight. There is, however, a tendency in most months towards absorption increase after a maximum of $\Sigma Kp$. In winter, a very large increase in absorption comes after a maximum of $\Sigma Kp$ with a delay of about two to four days, as shown in Fig. 27. Exceptionally strong effects are found in September and December. We should however, remember that magnetic disturbances also occur more often in September.

The conclusion is that a high absorption sensitivity in the magnetic disturbance character exists, preferentially in winter, with increased absorption a few days after the magnetic disturbance.

### 1.6. Solar radio noise flux effects

With maxima of 169 MHz solar flux as observed at Nancay, a superposed epoch study was made for winter months during 1966 and 1967, a period of great solar activity.

Figure 28 shows that each curve for an individual month has a clear minimum almost coincident with the maximum of 169 MHz solar radio noise. Thus, absorption and 169 MHz radio noise are *inversely* correlated. The lowest diagram shows the magnetic character to be only slightly related to 169 MHz noise, but in this case the correlation is positive.

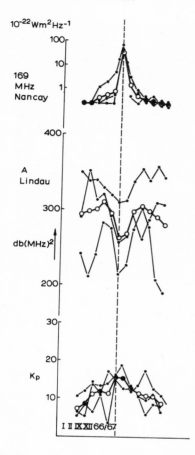

Fig. 28. Anti-correlation between radio noise flux at 169 MHz and absorption at Lindau.

## 2. THE NORMAL E LAYER

### 2.1. Methods of observation and layer parameters

The normal E layer is very interesting. It is the most regular of all the layers in the ionosphere. We find trace of it in the ionograms only in the day-time. The virtual height $h'$ is slightly above 100 km.

Though the normal E layer is very well controlled by the sun's height angle and solar activity, in the daily, seasonal and global behaviour of the E layer exist some anomalies, that have still to be explained theoretically.

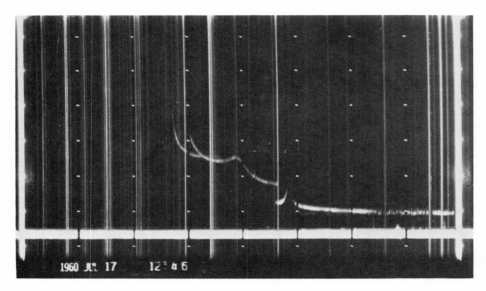

Fig. 29. Ionogram with normal E layer, low and high Es, F1 layer and F2 layer. The height scale (in ordinates) is linear. The frequency scale (in abscissa) is logarithmic. Height marks at distances of 150 km are given at frequencies 1.25, 1.75, 2.50, 3.55, 5.00, 7.05, 10.00 and 14.20 MHz (from right to left).

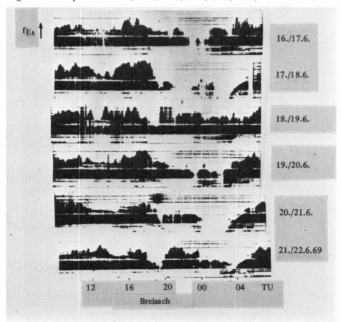

Fig. 30. Characteristic recording of $f_oE$ and $f$Es as functions of the time of day. These patterns are obtained from the normal ionograms by suppressing the height sweep and by gating the height range between 80 and 160 km. On the ordinates there is now the frequency scale. At least sixty ionograms per hour are taken and recorded on film that moves 5–10 mm/h. Time marks indicate the hours.

A characteristic ionogram is shown in Fig. 29, while Fig. 30 shows a characteristic recording of $f_o E$ and $f Es$ as functions of the time of day.

### 2.1.1. Thickness of the layer

The thickness of the lower part is about 15 km, as was determined from the shape of the virtual height curve, assuming a parabolic layer as typical for the electron density profile:

$$X_h = X_M \left[ 1 - \left( \frac{h - h_M}{2q} \right) \right]^2 \qquad (5)$$

where $X_h$ is the electron density at the height $h$, $h_M$ is the height of maximum and $2q$ is the half thickness of the parabolic layer.

The thickness of the upper part of the layer can sometimes be obtained from the retardation of echoes reflected from higher layers. Rocket observations have shown that the decrease of electron density on the top side of the E layer is much slower than the increase on the lower side.

The parabolic approximation can be used as an average description of the layer form on the lower side, but ledges and stratification in several layers appear rather often. $2q$ is slightly lower at low latitudes than at temperate latitudes.

### 2.1.2. Height of maximum electron density

The height of maximum electron density is about 120 km at medium latitudes and about 10 km lower at low latitudes.

#### 2.1.2.1. Diurnal and seasonal variation of $h'E$.
The increase of the minimum height $h'E$ at sunrise and sunset is essentially due to the decrease of the electron density and not to a true height variation. The diurnal variation of the true height has the same tendency but amounts only to a few kilometers.

The seasonal variation is mainly due to variations of the sun's height and is about 5–10 km, with maximum height in winter.

#### 2.1.2.2. Year-to-year variations of $h'E$.
There was some doubt whether a height variation with the solar cycle exists or not. In Freiburg, we have a long series of $h'E$ measurements. We tested the monthly mean of June and July and we found a clear variation with the sunspot cycle, shown in Fig. 31 and 32.

For each year, we calculated the mean of the height values from 0500 to 1800 UT, finding a clear variation within the sunspot cycle. Such mean is indicated by $h'E$ year. We did the same for the Lindau and Slough measurements as well. (See Fig. 33 and 34).

The variation of $h'E$ with the years and the time of day is similar for the three stations, as is shown in Fig. 35. The result is that around midday, for almost the

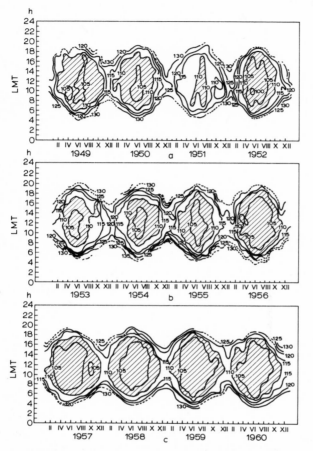

Fig. 31. Monthly medians of $h'$E: Freiburg, 48.1°N, 1949–1960. (Reproduced from Rawer and Suchy, 1967.)

whole solar cycle, we observe $100 < h'$E$ < 101$ km. Only during the years from 1961 to 1964 was there a clear enhancement in the virtual height of about 4 km.

To estimate the daily variation of $h'$E, shown in Table V, we calculated for Freiburg the quantity:

$$\Delta_1 h'\text{E} = h'\text{E} - h'\text{E year}$$

and also calculated the mean $\Delta h'$E day over all the years.

Finally, we have plotted in Table VI the quantity:

$$\Delta_2 h'\text{E} = \Delta_1 h'\text{E} - \Delta h'\text{E day}$$

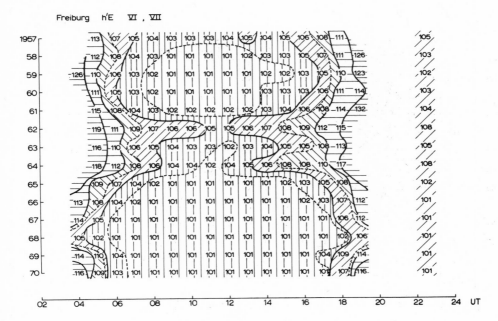

Fig. 32. Medians of $h'$E for the months of June and July from Freiburg.

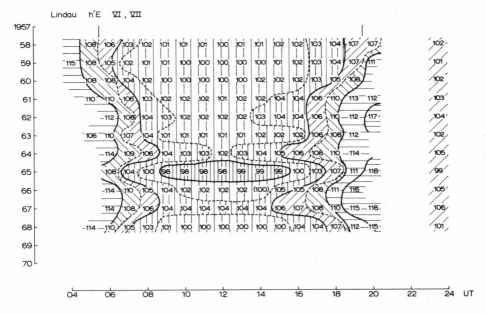

Fig. 33. Medians of $h'$E for June and July at Lindau.

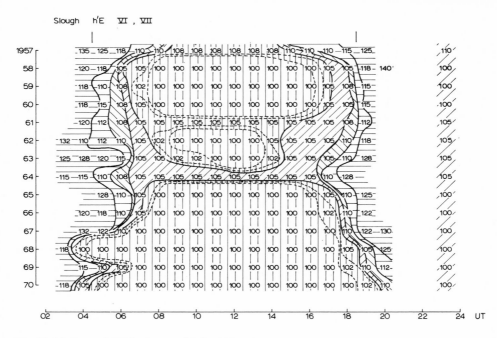

Fig. 34. Medians of $h'$E for June and July at Slough.

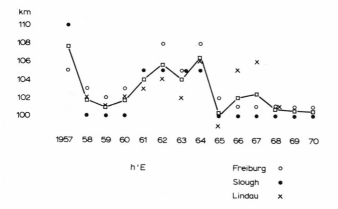

Fig. 35. Year-to-year variability for $h'$E.

TABLE V

Daily and mean daily variation for $h\,'E$ at Freiburg

| Year | UT | | | | | | | | | | | | | |
|---|---|---|---|---|---|---|---|---|---|---|---|---|---|---|
| | 05 | 06 | 07 | 08 | 09 | 10 | 11 | 12 | 13 | 14 | 15 | 16 | 17 | 18 |
| 1957 | +8 | +2 | 0 | -1 | -2 | -2 | -2 | -1 | 0 | -1 | 0 | +1 | +3 | +6 |
| 58 | +9 | +5 | +1 | 0 | -2 | -2 | -2 | -2 | -1 | 0 | 0 | +2 | +4 | +8 |
| 59 | +8 | +4 | +1 | 0 | -1 | -1 | -1 | -1 | -1 | 0 | 0 | +1 | +3 | +7 |
| 60 | +8 | +2 | 0 | -1 | -2 | -2 | -2 | -2 | -2 | 0 | 0 | 0 | +3 | +8 |
| 61 | +11 | +4 | 0 | -1 | -2 | -2 | -2 | -2 | -2 | -1 | 0 | +2 | +4 | +10 |
| 62 | +11 | +3 | +1 | -1 | -2 | -2 | -3 | -3 | -2 | -1 | 0 | +1 | +4 | +7 |
| 63 | +11 | +5 | +1 | 0 | -1 | -2 | -2 | -3 | -2 | -1 | 0 | 0 | +3 | +8 |
| 64 | +10 | +4 | 0 | -2 | -4 | -4 | -6 | -4 | -3 | -2 | 0 | 0 | +2 | +9 |
| 65 | +7 | +5 | +2 | 0 | -1 | -1 | -1 | -1 | -1 | -1 | 0 | +1 | +3 | +6 |
| 66 | +7 | +3 | +1 | 0 | 0 | 0 | 0 | 0 | 0 | 0 | 0 | +1 | +2 | +6 |
| 67 | +4 | 0 | 0 | 0 | 0 | 0 | 0 | 0 | 0 | 0 | 0 | 0 | 0 | +4 |
| 68 | +1 | 0 | 0 | 0 | 0 | 0 | 0 | 0 | 0 | 0 | 0 | 0 | 0 | +1 |
| 69 | +9 | +3 | 0 | 0 | 0 | 0 | 0 | 0 | 0 | 0 | 0 | 0 | +3 | +8 |
| 70 | +8 | +2 | 0 | 0 | 0 | 0 | 0 | 0 | 0 | 0 | 0 | 0 | 0 | +6 |
| Average values | +8 | +3 | 0 | 0 | -1 | -1 | -2 | -1 | -1 | 0 | 0 | +1 | +3 | +7 |

TABLE VI

The "noise" of the daily and the year-to-year variations for $h\,'E$ at Freiburg

| Year | UT | | | | | | | | | | | | | |
|---|---|---|---|---|---|---|---|---|---|---|---|---|---|---|
| | 05 | 06 | 07 | 08 | 09 | 10 | 11 | 12 | 13 | 14 | 15 | 16 | 17 | 18 |
| 1957 | 0 | -1 | 0 | -1 | -1 | -1 | 0 | 0 | +1 | -1 | 0 | 0 | 0 | -1 |
| 58 | +1 | +2 | +1 | 0 | -1 | -1 | 0 | -1 | 0 | 0 | 0 | +1 | +1 | +1 |
| 59 | 0 | +1 | +1 | 0 | 0 | 0 | +1 | 0 | 0 | 0 | 0 | 0 | 0 | 0 |
| 60 | 0 | -1 | 0 | -1 | -1 | -1 | 0 | -1 | -1 | 0 | 0 | -1 | 0 | +1 |
| 61 | +3 | +1 | 0 | -1 | -1 | -1 | 0 | -1 | -1 | -1 | 0 | +1 | +1 | +3 |
| 62 | +3 | 0 | +1 | -1 | -1 | -1 | -1 | -2 | -1 | -1 | 0 | 0 | +1 | 0 |
| 63 | +3 | +2 | +1 | 0 | 0 | -1 | 0 | -2 | -1 | -1 | 0 | -1 | 0 | +1 |
| 64 | +2 | +1 | 0 | -2 | -3 | -3 | -4 | -3 | -2 | -2 | 0 | -1 | -1 | +2 |
| 65 | -1 | +2 | +2 | 0 | 0 | 0 | +1 | 0 | 0 | -1 | 0 | 0 | 0 | -1 |
| 66 | -1 | 0 | +1 | 0 | +1 | +1 | +2 | +1 | +1 | 0 | 0 | 0 | -1 | -1 |
| 67 | -4 | -3 | 0 | 0 | +1 | +1 | +2 | +1 | +1 | 0 | 0 | -1 | -3 | -3 |
| 68 | -7 | -3 | 0 | 0 | +1 | +1 | +2 | +1 | +1 | 0 | 0 | -1 | -3 | -6 |
| 69 | +1 | 0 | 0 | 0 | +1 | +1 | +2 | +1 | +1 | 0 | 0 | -1 | 0 | +1 |
| 70 | 0 | -1 | 0 | 0 | +1 | +1 | +2 | +1 | +1 | 0 | 0 | -1 | -3 | -1 |

while Fig. 36 shows the results obtained at Freiburg for different years and Fig. 37 shows the observations of Slough and Lindau that confirm the results of Freiburg.

There is no doubt: for Freiburg, Slough and Lindau, the variation of $\overline{\Delta h'E}$ day is altered by a solar cycle. It is greatest at the beginning of the sunspot minimum and slight at the sunspot maximum.

Not only by sun and sunspots, but also by the phase (and perhaps by the declination) of the moon, the height of the E layer may be varied by ± 0.5 km; this lies, however, within the limit of our measuring accuracy.

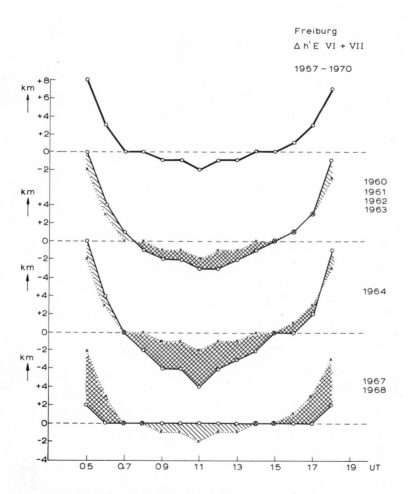

Fig. 36. Special daily variation for $h'E$ for different periods of years.

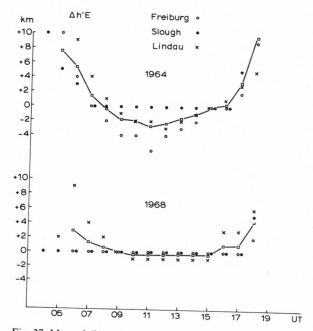

Fig. 37. Mean daily variation for $h'E$ for 1964 and 1968, computed from three stations.

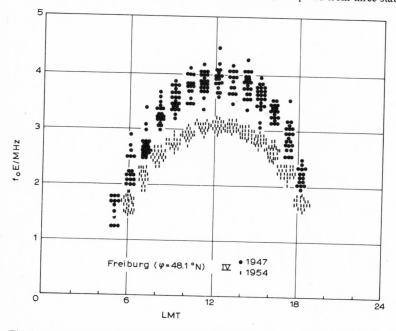

Fig. 38. Mass-plot of $f_oE$ as a function of the time of day for one year of high solar activity and one of low. (Reproduced from Rawer and Suchy, 1967.)

### 2.1.3. The critical frequency $f_oE$

*2.1.3.1. Solar control.* The critical frequency $f_oE$ is a function of the true height of the sun (Fig. 38). The day-to-day dispersion is small. Lunar tides are present. At sunset, $f_oE$ falls quickly according to a recombination law:

$$\frac{\partial N}{\partial t} = q(t) - \alpha N^2 \tag{6}$$

with $\alpha \simeq 10^{-14} \, \text{m}^3 \, \text{sec}^{-1}$.

To demonstrate how accurately the E layer follows the true height of the sun, we can make an interesting test. Plotting the monthly mean of $f_oE$ for some years as a function of the time of day, and searching for the axis of symmetry of the diurnal curve as a whole in superposing the before-noon and the afternoon values, we determine the time of the $f_oE$ maximum.

We determined the time of the $f_oE$ maximum for all months of the years 1968–1970 and compared these times with the true midday (Fig. 39–41). The difference is not more than $\pm$ 2 min.

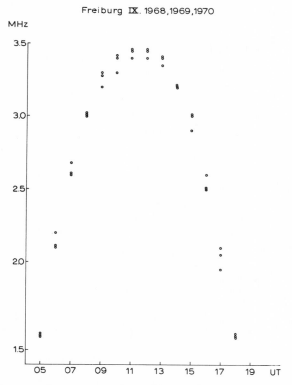

Fig. 39. Daily variation of $f_oE$ at Freiburg for September 1968, 1969 and 1970.

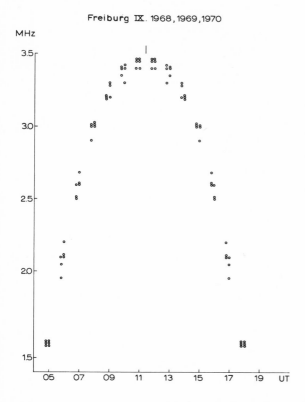

Fig. 40. Symmetrical daily variation of $f_o$E at Freiburg for September 1968, 1969 and 1970.

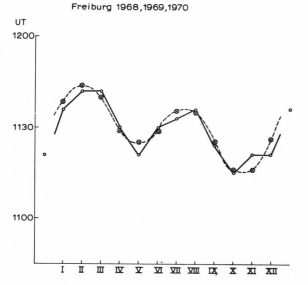

Fig. 41. Comparison between true noon and time of the $f_o$E maximum for three years, January to December, at Freiburg.

*2.1.3.2. Noon anomaly.*   Going further into detail, we will find that for a given month, especially for the values around noon (0900–1500 LMT) the $f_0E$ curve is not really symmetric with respect to the true noon, as is shown in Fig. 42.

At mid-latitudes, the values before noon are too low and the values after noon are too high. We have a time lag $\Delta t$ and since:

$$\Delta t = \frac{1}{2\alpha N_{(\text{noon})}} \tag{7}$$

$\Delta t$ should always be positive.

In Fig. 43, Appleton and Lyon (1961) show that negative values are regularly found near the magnetic equator. At mid-latitudes, negative values occur only in winter. Such negative delays can be explained with an important transport term in the balance equation:

$$\frac{\partial N}{\partial t} = q(t) - \alpha N^2 - \beta N - \frac{\partial}{\partial v}(Nv) \tag{8}$$

where $\beta$ is the attachment coefficient and $v$ is the effective electron velocity.

*2.1.3.3. Diurnal variation of $f_oE$.*   The diurnal variation of $f_0E$ can be determined according to a law of the form

$$f_{\text{cr}} = C \cos^P \chi \tag{9}$$

by a logarithmic plot, as shown in Fig. 44 and 45. $\chi$ is the zenith distance of the sun.

The exponent $p$ is given by the slope of the logarithmic curve. $p = 1/3$ is a good approximation at low and middle latitudes. Only in the auroral zone one finds $0.0 < p < 0.2$ for the auroral E layer, but this layer is not under solar control.

A variation of $p$ with the solar cycle has not yet been established.

*2.1.3.4. The seasonal variation.*   We may also describe the seasonal variation of $f_oE$ at a given hour by the cos $\chi$ power-law:

$$f_{\text{cr}} = C' \cos^{p'} \chi \tag{10}$$

Values for $p'$ are given in Fig. 46 and 47.

The total variability of $p'$ is great (about 0.1 to 0.3). Lower values are generally found near noon, but there is a tendency towards a pre-noon minimum of $p'$.

In Fig. 47 we observe an important latitude effect of the seasonal exponent $p'$. In the Far East, near Tokyo and Watheroo, we find two pronounced minima of $p'$. It seems that the position of the minima is rather near to the focus of the Sq current system which produces the diurnal variation of the geomagnetic field. It is highly

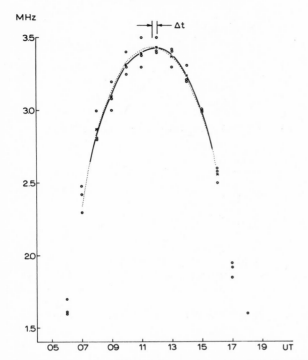

Fig. 42. "Noon anomaly" of $f_oE$ at Freiburg, March 1968, 1969 and 1970.

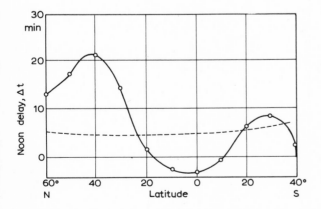

Fig. 43. Latitude variation of $\Delta t$. (Reproduced from Appleton and Lyon, 1961.)

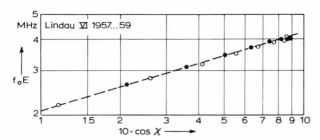

Fig. 44. Logarithmic plot of $f_oE$ vs. cos $\chi$, June 1957–1959. (Reproduced from Rawer and Suchy, 1967.)

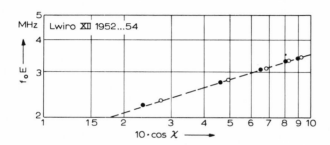

Fig. 45. Logarithmic plot of $f_oE$ vs. cos $\chi$, December 1952–1954. (Reproduced from Rawer and Suchy, 1967.)

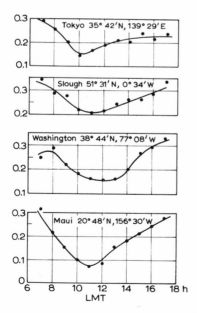

Fig. 46. Seasonal exponent $p'$ for different hours and stations located between 20° and 51°N. (Reproduced from Rawer and Suchy, 1967.)

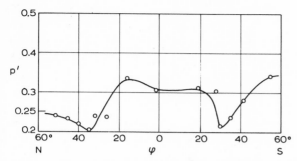

Fig. 47. Latitude variation of the seasonal exponent $p'$ from noon data; 1952–1954, Far East. (Reproduced from Rawer and Suchy, 1967.)

probable that the Sq current system is related to vertical drift movements which influence the ionization balance through the transport term. The minimum is present both on magnetically quiet and disturbed days, but is more accentuated on the latter.

*2.1.3.5. The latitude variation of $f_oE$.* When plotting the noon values of $f_oE$ as a function of latitude, the curve should be symmetrical with respect to the subsolar point. This seems to be true only in equinoctial months, but considerable asymmetries occur in summer and winter. In the logarithmic plot the effects appear as a shearing of northern and southern hemisphere values (Fig. 48).

The winter hemisphere gives relatively higher values. This, of course, is a consequence of winter enhancement and summer depression, which may be explained by vertical drift gradients, or by a decrease of the recombination coefficient $\alpha$ with height.

### 2.1.4. Solar activity

The influence of variable solar activity is well described by the correlation with the sunspot number or better, with the intensity $\Phi$ of solar radio noise (Fig. 49). Highest correlation is found at stations without summer or winter anomaly; these are at low latitudes. A good example is Lwiro. We find:

$$\frac{\overline{C}_E}{\text{MHz}} = 3.44 + \left\{ \frac{1}{200} \left( \frac{10^{22}\,\Phi}{\text{Wm}^{-2}\,\text{Hz}^{-1}} - 69 \right) \right\}^{0.6} \tag{11}$$

or, expressed in sunspots:

$$\frac{\overline{C}_E}{\text{MHz}} = 3.44[1 + 0.00687(\overline{R} - 8)^{0.7}] \tag{12}$$

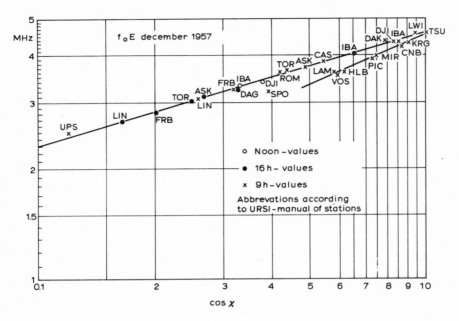

Fig. 48. Logarithmic plot of the latitude variation of $f_oE$ at noon, December 1957. (Reproduced from Rawer and Suchy, 1967.)

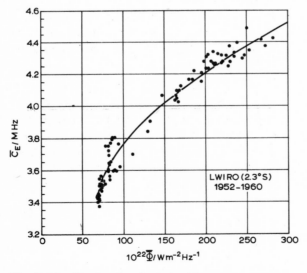

Fig. 49. Monthly $C_E$ values for Lwiro (Congo, $q = 2.3°S$; $\psi = -30°$), correlated with solar radio noise flux $\Phi$ (10-cm band), determined at Ottawa (Canada). (Reproduced from Rawer and Suchy, 1967.)

where $\overline{C}_E$, the mean value for the E layer of the constant $C$ of eq. 9, describes the absolute intensity of ionization. It is found by straight extrapolation to $\chi = 0$ (see Fig. 45).

At temperate latitudes the dispersion for individual months is much larger, but by introducing a latitude and a seasonal term, the dispersion can be reduced. Slightly different relations are also obtained for different sunspot cycles.

### 2.1.5. SID and $f_0E$

An instantaneous enhancement of $f_0E$ for a short time is found during and after a sudden ionospheric disturbance.

### 2.1.6. Magnetic disturbances and $f_0E$

The correlation between $f_0E$ and magnetic disturbances is shown in Fig. 50. There seems to be a decrease of $f_0E$ with magnetic disturbances at temperature latitudes, but an increase at low latitudes.

### 2.1.7. The origin of the normal E-layer ionization

The origin of the normal E-layer ionization is certainly due to solar wave radiation near 1000 Å, or in very soft X rays. Rocket and satellite observations support the X-ray hypothesis.

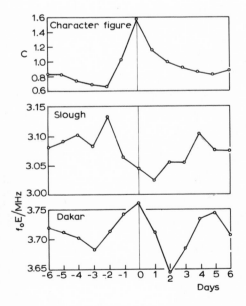

Fig. 50. Superposed epoch analysis of international magnetic character figure $C$ (upper curve), and $f_0E$ at Slough and Dakar. (Reproduced from Rawer and Suchy, 1967.)

## 2.2. *Statistical treatment for parameters with diurnal, seasonal and year-to-year variations*

In our ionospheric work we are mostly interested in having a quick way to find the *diurnal*, *seasonal* and *year-to-year* variations for different parameters. Since we can show only three dimensions in our figures (the third in line shading), we have to make three sets of charts. We can consider three cases:

(1) Seasonal variations depending on the year-to-year variation (the time of day or the sun distance $\chi$ is fixed);

(2) Diurnal variations depending on the year-to-year variation (the month or $\chi$ at noon is fixed);

(3) Diurnal variations depending on the seasonal variation (the year is fixed).

As an example, we treat normal E layer measurements at Freiburg for case 1. The parameter is $f_o$E 50% at noon, which depends on month and year (Fig. 51).

### 2.2.1. *The mean year-to-year variation*

At first we calculate the mean of $f_o$E 50% for each year and we get the mean year-to-year variability: $\overline{f_o E}$ 50% year. Fig. 52 shows the difference $\Delta_1 f_o$ E 50% = $f_o$ E 50% − $\overline{f_o E}$ 50% year .

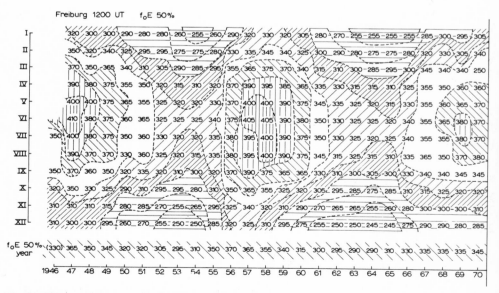

Fig. 51. $f_o$E at noon and mean year-to-year variability.

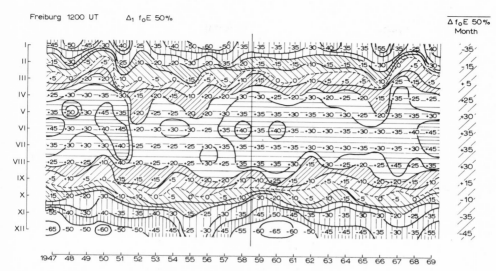

Fig. 52. $\Delta_1 f_0 E$ at noon and mean seasonal variability.

### 2.2.2. The mean seasonal variability

We now calculate the mean for each month over all the years and we get a mean seasonal description for $f_0 E$: $\overline{\Delta f_0 E \ 50\% \ month}$. The next step is to calculate the differences $\Delta_2 f_0 E \ 50\% = \Delta_1 f_0 E \ 50\% - \overline{\Delta f_0 E \ 50\% \ month}$ plotted in Table VII.

All these charts are valid for local noontime. We now make two tests, to prove whether our results confirm the findings by other scientists mentioned in the above chapter concerning the normal E layer.

Test (1): Plotting the mean year-to-year variability $\overline{f_0 E \ 50\% \ year}$, together with the noon values of $f_0 E$ for the months of June and July over the different years, we find a very good correlation both with the sunspot numbers and with the solar flux (Fig. 53).

Test (2): Plotting the mean seasonal variability $\overline{\Delta f_0 E \ 50\% \ month}$ and the special seasonal variabilities for 1950 and 1955 over the months, we observe a very great year-to-year dispersion (Fig. 54).

If we look for the symmetry of the mean curve as a whole, we find the June 22 ± 2 days. The normal E layer is not only a good clock, but a good calendar as well.

### 2.2.3. The diurnal variations

The diurnal variations correspond to case 2. Considering the diurnal variations, the parameter $f_0 E \ 50\%$ is now dependent on the hour and the years (Table VIII).

TABLE VII

The "noise chart": $\Delta_2 f_o E 50\%$ at noon (Freiburg, 1200 UT)

| Month\Year | 1947 | 48 | 49 | 50 | 51 | 52 | 53 | 54 | 55 | 56 | 57 | 58 | 59 | 60 | 61 | 62 | 63 | 64 | 65 | 66 | 67 | 68 | 69 |
|---|---|---|---|---|---|---|---|---|---|---|---|---|---|---|---|---|---|---|---|---|---|---|---|
| I    | -10 | -15 | -10 | +5  | -5  | +10 | 0   | -5  | -15 | -25 | -15 | 0   | 0   | 0   | 0   | +5  | -5  | 0   | 0   | -20 | 0   | +10 | -5  |
| II   | 0   | -15 | +10 | +20 | -10 | +5  | -5  | -5  | -15 | -5  | -20 | -5  | 0   | 0   | 0   | +5  | 0   | 0   | 0   | -15 | +15 | +20 | -15 |
| III  | 0   | -5  | +15 | +15 | -15 | -5  | -10 | -5  | -20 | 0   | -10 | +5  | +10 | -5  | -5  | +5  | 0   | 0   | 0   | -15 | +20 | +10 | 0   |
| IV   | 0   | +5  | +5  | +10 | +5  | -10 | -5  | -10 | -15 | -5  | -5  | +5  | +5  | 0   | -5  | +5  | -5  | 0   | -5  | -10 | +10 | 0   | 0   |
| V    | +5  | +20 | 0   | +15 | +5  | -10 | -5  | -5  | -10 | -10 | 0   | +5  | +5  | +5  | 0   | +5  | -5  | 0   | -5  | -10 | +15 | +15 | 0   |
| VI   | -10 | -5  | -5  | +5  | +15 | -15 | -5  | -5  | -5  | -10 | 0   | +5  | 0   | +5  | 0   | -5  | -5  | 0   | -5  | -5  | 0   | +5  | +10 |
| VII  | 0   | -5  | -5  | +5  | +5  | -10 | -10 | -5  | -10 | -5  | -10 | +5  | +5  | +5  | 0   | -5  | -5  | -5  | 0   | -5  | 0   | -5  | +10 |
| VIII | -5  | -10 | -5  | -20 | +5  | -10 | -5  | -10 | -5  | 0   | -5  | +5  | +5  | +5  | 0   | -15 | 0   | -5  | -10 | -5  | +15 | -5  | +5  |
| IX   | -10 | -5  | -10 | -15 | 0   | 0   | 0   | -10 | -5  | +5  | +5  | -5  | -5  | +5  | 0   | -5  | -10 | 0   | -5  | +5  | +5  | -5  | -5  |
| X    | -5  | -10 | -10 | -20 | 0   | 0   | +10 | +5  | +10 | +5  | +5  | 0   | -20 | -10 | 0   | +5  | 0   | -5  | +5  | +10 | +5  | +10 | -5  |
| XI   | -20 | -5  | +5  | -5  | 0   | 0   | -5  | +5  | +20 | +10 | +5  | 0   | -10 | -15 | -10 | 0   | +5  | 0   | +5  | +5  | +5  | -10 | 0   |
| XII  | -20 | -5  | -15 | -15 | -5  | -5  | 0   | 0   | +20 | +15 | 0   | -10 | -15 | -20 | -15 | -5  | 0   | 0   | 0   | +10 | +15 | +10 | -10 |

–10 indicates –0.10 MHz.

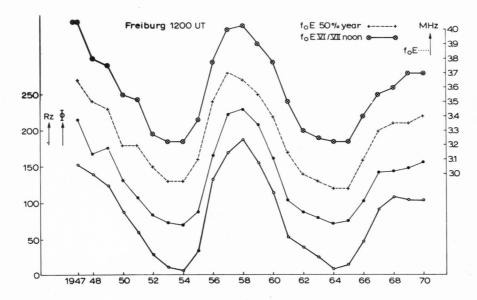

Fig. 53. Relations among $\overline{f_o E\,50\%\ \text{year}}$, $f_o E$ in June and July at noon, sunspots, and flux densities.

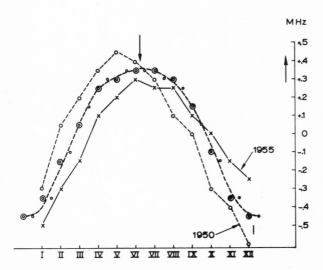

Fig. 54. Relative seasonal variation of mean values of $f_o E$ at noon and special variation for the years 1950 and 1955.

TABLE VIII

Diurnal variations of $f_o$ E $VI/VII$ (Freiburg, June/July)

| Year | UT | | | | | | | | | | | | | | | |
|------|-----|-----|-----|-----|-----|-----|-----|-----|-----|-----|-----|-----|-----|-----|-----|-----|
|      | 04  | 05  | 06  | 07  | 08  | 09  | 10  | 11  | 12  | 13  | 14  | 15  | 16  | 17  | 18  | 19  |
| 1957 | 149* | 250 | 302 | 335 | 366 | 385 | 395 | 398 | 400 | 400 | 389 | 370 | 344 | 308 | 255 | 168 |
| 58   | 168 | 250 | 302 | 339 | 360 | 379 | 390 | 400 | 402 | 398 | 390 | 370 | 345 | 310 | 258 | 162 |
| 59   | 178 | 250 | 298 | 332 | 352 | 372 | 380 | 390 | 391 | 389 | 378 | 361 | 338 | 305 | 258 | 172 |
| 60   | 172 | 235 | 278 | 314 | 338 | 355 | 368 | 374 | 376 | 375 | 360 | 348 | 320 | 288 | 240 | 174 |
| 61   | 155 | 220 | 260 | 290 | 312 | 329 | 340 | 345 | 350 | 346 | 340 | 325 | 300 | 268 | 222 | 165 |
| 62   | 146 | 205 | 248 | 280 | 300 | 315 | 328 | 331 | 330 | 328 | 320 | 308 | 288 | 255 | 212 | 148 |
| 63   | 144 | 212 | 252 | 274 | 295 | 310 | 318 | 322 | 328 | 322 | 318 | 302 | 282 | 252 | 215 | 144 |
| 64   | 146 | 202 | 238 | 262 | 288 | 300 | 312 | 320 | 322 | 320 | 312 | 292 | 272 | 239 | 198 | 150 |
| 65   | 148 | 200 | 242 | 270 | 290 | 308 | 318 | 322 | 324 | 319 | 310 | 300 | 280 | 242 | 200 | 160 |
| 66   | 154 | 210 | 250 | 290 | 310 | 326 | 340 | 340 | 340 | 335 | 325 | 310 | 290 | 252 | 208 | 154 |
| 67   | 164 | 226 | 275 | 302 | 325 | 340 | 345 | 348 | 353 | 348 | 338 | 330 | 310 | 275 | 222 | 164 |
| 68   | 165 | 232 | 275 | 310 | 331 | 348 | 355 | 358 | 355 | 350 | 345 | 338 | 316 | 285 | 230 | 162 |
| 69   | 164 | 235 | 285 | 311 | 335 | 252 | 366 | 372 | 375 | 370 | 355 | 342 | 315 | 280 | 236 | 175 |
| 70   | 162 | 232 | 284 | 315 | 337 | 352 | 361 | 370 | 372 | 368 | 358 | 342 | 325 | 285 | 238 | 162 |

*149 indicates 1.49 MHz.

To describe the year-to-year variability, we don't calculate the mean over the day, for it is much more interesting and simpler to take the noon value of $f_o$ E in June and July as a standard for the given month. We proceed in the same manner as above:

$$\Delta_1 f_o E \ VI/VII = (f_o E \ VI/VII - f_o E \ VI/VII)_{noon}$$

We calculate $\overline{\Delta f_o E \ VI/VII}$ hour, the mean diurnal behaviour (Table IX) and finally, $\Delta_2 f_o E \ VI/VII = \Delta_1 f_o E \ VI/VII - \overline{\Delta f_o E \ VI/VII}$ hour as shown in Table X, the "noise chart".

Again we test the results:

Test (3): We plot $\overline{\Delta f_o E \ VI/VII}$, the mean daily behaviour over the time of day (Fig. 55).

Fig. 55 confirms that: (1) the axis of symmetry for the curve as a whole corresponds exactly to midday; (2) the maximum is displaced towards the afternoon ("noon anomaly").

In Fig. 56, we plotted in logarithmic scales the mean seasonal behaviour $\overline{\Delta f_o E}$ 50% month over cos $\chi$ at noon and then calculated $p'$.

We very clearly see an s-curve fitting the points. This means: (1) winter

TABLE IX

$\Delta_1 f_0 E$ *VI/VII* (Freiburg, June/July) and mean diurnal variation

| Year | UT | | | | | | | | | | | | | | | |
|---|---|---|---|---|---|---|---|---|---|---|---|---|---|---|---|---|
| | 04 | 05 | 06 | 07 | 08 | 09 | 10 | 11 | 12 | 13 | 14 | 15 | 16 | 17 | 18 | 19 |
| 1957 | -25* | -15 | -10 | -6 | -3 | -1 | 0 | 0 | 0 | 0 | -1 | -3 | -6 | -9 | -14 | -23 |
| 58 | -23 | -15 | -10 | -6 | -4 | -2 | -1 | 0 | 0 | 0 | -1 | -3 | -5 | -9 | -14 | -24 |
| 59 | -21 | -14 | -9 | -6 | -4 | -2 | -1 | 0 | 0 | 0 | -1 | -3 | -5 | -9 | -13 | -22 |
| 60 | -21 | -14 | -10 | -6 | -4 | -2 | -1 | 0 | 0 | 0 | -2 | -3 | -6 | -9 | -14 | -20 |
| 61 | -18 | -13 | -9 | -6 | -4 | -2 | -1 | 0 | 0 | 0 | -1 | -2 | -5 | -8 | -13 | -18 |
| 62 | -18 | -12 | -8 | -5 | -3 | -1 | 0 | 0 | 0 | 0 | -1 | -2 | -4 | -7 | -12 | -18 |
| 63 | -19 | -12 | -8 | -5 | -3 | -2 | -1 | -1 | 0 | -1 | -1 | -3 | -5 | -8 | -11 | -19 |
| 64 | -17 | -12 | -8 | -6 | -3 | -2 | -1 | 0 | 0 | 0 | -1 | -3 | -5 | -8 | -12 | -17 |
| 65 | -17 | -12 | -8 | -5 | -3 | -2 | -1 | 0 | 0 | -1 | -1 | -2 | -4 | -8 | -12 | -17 |
| 66 | -19 | -13 | -9 | -5 | -3 | -1 | 0 | 0 | 0 | 0 | -2 | -3 | -5 | -9 | -13 | -19 |
| 67 | -19 | -13 | -8 | -5 | -3 | -1 | -1 | -1 | 0 | -1 | -2 | -2 | -4 | -8 | -13 | -19 |
| 68 | -19 | -13 | -8 | -4 | -3 | -1 | 0 | 0 | 0 | -1 | -1 | -2 | -4 | -8 | -13 | -20 |
| 69 | -22 | -14 | -9 | -6 | -4 | -3 | -1 | 0 | 0 | -1 | -2 | -3 | -6 | -9 | -14 | -20 |
| 70 | -21 | -14 | -9 | -6 | -4 | -2 | -1 | 0 | 0 | 0 | -1 | -3 | -5 | -9 | -13 | -21 |
| Average values | -19 | -13 | -9 | -6 | -3 | -2 | -1 | 0 | 0 | 0 | -1 | -3 | -5 | -8 | -13 | -20 |

* -25 indicates -2.5 MHz.

TABLE X

The "noise chart" for $f_0 E$ *VI/VII* (Freiburg, June/July)

| Year | UT | | | | | | | | | | | | | | | |
|---|---|---|---|---|---|---|---|---|---|---|---|---|---|---|---|---|
| | 04 | 05 | 06 | 07 | 08 | 09 | 10 | 11 | 12 | 13 | 14 | 15 | 16 | 17 | 18 | 19 |
| 1957 | -5* | -2 | -1 | 0 | 0 | +1 | +1 | 0 | 0 | 0 | 0 | 0 | -1 | -1 | -1 | -3 |
| 58 | -4 | -2 | -1 | 0 | -1 | 0 | 0 | 0 | 0 | 0 | 0 | 0 | 0 | -1 | -1 | -4 |
| 59 | -2 | -1 | 0 | 0 | -1 | 0 | 0 | 0 | 0 | 0 | 0 | 0 | 0 | -1 | 0 | -2 |
| 60 | -2 | -1 | -1 | 0 | -1 | 0 | 0 | 0 | 0 | 0 | -1 | 0 | -1 | -1 | -1 | 0 |
| 61 | +1 | 0 | 0 | 0 | -1 | 0 | 0 | 0 | 0 | 0 | 0 | +1 | 0 | 0 | 0 | +2 |
| 62 | +1 | +1 | +1 | +1 | 0 | +1 | +1 | 0 | 0 | 0 | 0 | +1 | +1 | +1 | +1 | +2 |
| 63 | 0 | +1 | +1 | +1 | 0 | 0 | 0 | -1 | 0 | -1 | 0 | 0 | 0 | 0 | +2 | +1 |
| 64 | +2 | +1 | +1 | 0 | 0 | 0 | 0 | 0 | 0 | 0 | 0 | 0 | 0 | 0 | +1 | +3 |
| 65 | +2 | +1 | +1 | +1 | 0 | 0 | 0 | 0 | 0 | -1 | 0 | +1 | +1 | 0 | +1 | +3 |
| 66 | 0 | 0 | 0 | +1 | 0 | +1 | +1 | 0 | 0 | 0 | -1 | 0 | 0 | -1 | 0 | +1 |
| 67 | 0 | 0 | +1 | +1 | 0 | +1 | 0 | -1 | 0 | -1 | -1 | +1 | +1 | 0 | 0 | +1 |
| 68 | 0 | 0 | +1 | +2 | 0 | +1 | +1 | 0 | 0 | -1 | 0 | +1 | +1 | 0 | 0 | 0 |
| 69 | -3 | -1 | 0 | 0 | -1 | 0 | 0 | 0 | 0 | -1 | -1 | 0 | -1 | -1 | -1 | 0 |
| 70 | -2 | -1 | 0 | 0 | -1 | 0 | 0 | 0 | 0 | 0 | 0 | 0 | 0 | -1 | 0 | -1 |

* -5 indicates -0.5 MHz.

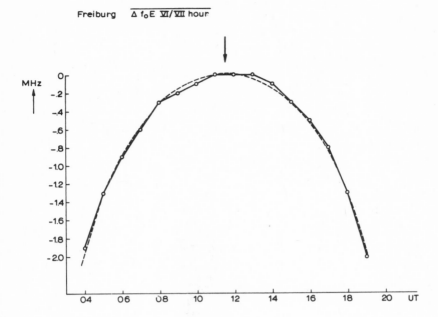

Fig. 55. Symmetry test and "noon anomaly" for $f_oE$.

| p' | Sunspots | | |
|---|---|---|---|
| | Min. | Max. | |
| | 0.20 → 0.30 | 0.18 → 0.27 | Without |
| | 0.25 | 0.27 | With |

"Noise" correction

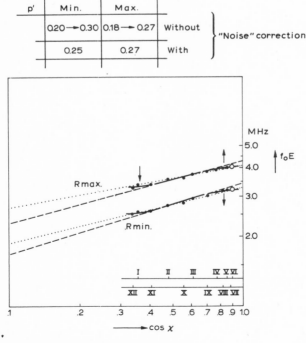

Fig. 56. Seasonal dependence on cos $\chi$.

enhancement and summer depression are easy to determine; (2) the value for $p'$ is 0.25. We may correct the value $p'$ for special years by the "noise chart". It is interesting to find from Fig. 30 that winter enhancement and summer depression seem to vanish in correspondence to very high sunspot numbers, and that summer depression is highest at the sunspot minimum.

In Fig. 57 we plotted, in logarithmic scales, the mean daily behaviour $f_oE$ $VI/VII$ day as a function of $\cos \chi$ and estimated $p$. We find $0.26 < p < 0.34$, depending on the sunspot number. Again we correct the $p$ values with the results of the "noise chart", and find for the whole sunspot cycle $p = 0.29$.

An interesting finding is that, for very low values of $\cos \chi$, the frequencies are too high. After correction with the noise chart we can state: at sunspot maximum, both the early morning and late afternoon values are correct, but at sunspot minimum, these values are much too high.

Finally, we will give the same treatment as we did above for the monthly mean of $f_oE$ values to the monthly mean of the sunspot numbers and the flux densities. We suppose that these values are free from variations with $\cos \chi$ (see Fig. 58 and 59).

We again calculate the mean for the different years, $\overline{R_z}$ year and $\overline{\Phi}$ year, and finally we determine $\Delta R_z$ and $\Delta \Phi$, as shown in Table XI and XII.

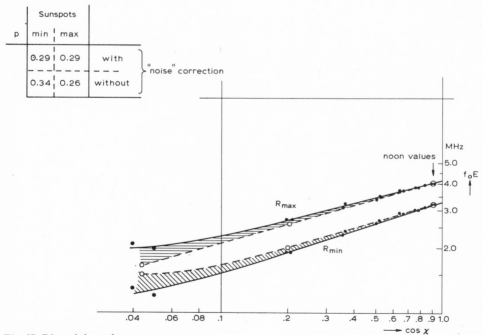

Fig. 57. Diurnal dependence on $\cos \chi$.

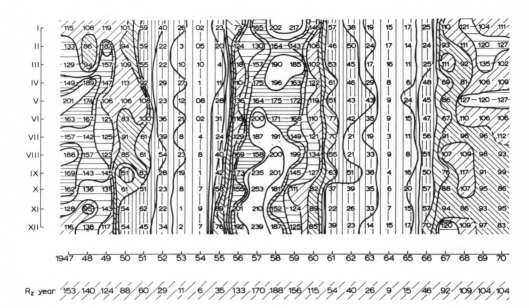

Fig. 58. The variation of $R_z$ (final relative sunspot number) with months and years.

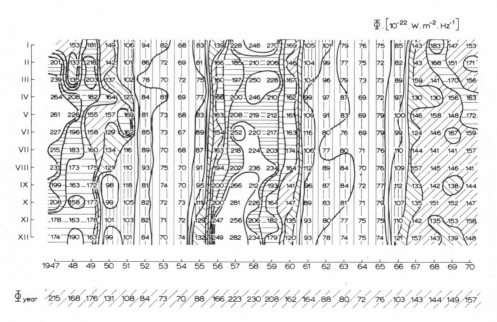

Fig. 59. The variation of $\Phi$ with months and years.

TABLE XI

The variation of $\Delta R_z$ with months and years

| Month Year | 1947 | 48 | 49 | 50 | 51 | 52 | 53 | 54 | 55 | 56 | 57 | 58 | 59 | 60 | 61 | 62 | 63 | 64 | 65 | 66 | 67 | 68 | 69 | 70 |
|---|---|---|---|---|---|---|---|---|---|---|---|---|---|---|---|---|---|---|---|---|---|---|---|---|
| I | -38 | -32 | -5 | +13 | -1 | +11 | +15 | -4 | -12 | -60 | -5 | +11 | +61 | +31 | +3 | -2 | -7 | +6 | +2 | -20 | +18 | +12 | 0 | +7 |
| II | -20 | -54 | +58 | +6 | -1 | -7 | -8 | -1 | -15 | -9 | -40 | -24 | -13 | -9 | -8 | +10 | -2 | +8 | -1 | -24 | +1 | +2 | +16 | +23 |
| III | -24 | -46 | +33 | +21 | -5 | -7 | -1 | +4 | -31 | -15 | -13 | +2 | +29 | -13 | -1 | +5 | -9 | +7 | -4 | -23 | +19 | -17 | +31 | -2 |
| IV | -4 | +49 | +23 | +25 | +32 | 0 | +16 | -5 | -24 | -23 | +5 | +8 | +7 | +7 | +7 | +6 | +3 | -1 | -9 | 0 | -23 | -28 | +2 | -5 |
| V | +48 | +34 | -18 | +18 | +48 | -6 | +1 | +2 | -7 | +3 | -6 | -13 | +16 | +4 | -3 | +3 | +17 | 0 | +9 | -3 | -6 | +18 | +16 | +23 |
| VI | +10 | +27 | -3 | -5 | +40 | +7 | +10 | -4 | -4 | -17 | +30 | -17 | +12 | -5 | +23 | +2 | +9 | 0 | 0 | -1 | -25 | +1 | +2 | +2 |
| VII | +4 | +2 | +1 | +3 | +1 | +10 | -3 | -2 | -9 | -4 | +17 | +3 | -7 | +6 | +16 | -9 | -7 | -6 | -4 | +8 | -1 | -13 | -8 | +8 |
| VIII | +35 | +17 | -1 | -3 | +1 | +25 | +12 | +2 | +5 | +36 | -12 | +12 | +43 | +19 | +1 | -19 | +7 | 0 | -7 | +3 | +15 | 0 | -6 | -11 |
| IX | +16 | +3 | +21 | -37 | +23 | -1 | +8 | -5 | +7 | +40 | +65 | +13 | -11 | +12 | +9 | +11 | +5 | -5 | +1 | +2 | -16 | +8 | -13 | -5 |
| X | +10 | -4 | +7 | -27 | -9 | -6 | -3 | +1 | +23 | +22 | +83 | -7 | -45 | -33 | -17 | -1 | +9 | -3 | +5 | +9 | -4 | -2 | -9 | -18 |
| XI | -25 | -45 | +19 | -34 | -8 | -7 | -10 | +3 | +54 | +68 | +40 | -36 | -32 | -26 | -22 | -14 | -3 | -2 | 0 | +9 | +2 | -23 | -11 | -9 |
| XII | -37 | -2 | -7 | -34 | -15 | +5 | -9 | +1 | +41 | +59 | +69 | -1 | -31 | -30 | -15 | -17 | -12 | +6 | +2 | +22 | +28 | +9 | -7 | -21 |

TABLE XII

The variation of $\Delta\Phi$ with months and years

| Month Year | 1947 | 48 | 49 | 50 | 51 | 52 | 53 | 54 | 55 | 56 | 57 | 58 | 59 | 60 | 61 | 62 | 63 | 64 | 65 | 66 | 67 | 68 | 69 | 70 |
|---|---|---|---|---|---|---|---|---|---|---|---|---|---|---|---|---|---|---|---|---|---|---|---|---|
| I |  | -15 | 5 | 18 | -2 | +10 | +9 | -2 | -5 | -27 | +5 | +18 | +62 | +7 | +1 | +13 | -1 | +4 | -1 | -18 | 0 | +39 | -2 | -4 |
| II | -14 | -35 | +42 | +11 | -7 | +2 | -1 | -1 | -7 | 0 | -38 | -20 | -2 | -16 | 0 | +11 | -3 | +3 | -4 | -21 | 0 | +24 | +2 | +14 |
| III | +20 | -32 | +27 | +6 | -6 | -6 | -3 | +2 | -13 | -6 | -6 | +20 | +20 | +5 | 0 | +8 | -1 | +1 | -3 | -14 | +16 | -3 | +21 | -1 |
| IV | +49 | +40 | +6 | +33 | +19 | 0 | +8 | -1 | -11 | 0 | -23 | +16 | +2 | 0 | -5 | +9 | +7 | -3 | -4 | -6 | -13 | -14 | +7 | +6 |
| V | +46 | +58 | -21 | +26 | +61 | -3 | 0 | -2 | -5 | -3 | -15 | -11 | +4 | -1 | +5 | +3 | +3 | -3 | +3 | -3 | +3 | +14 | -1 | +15 |
| VI | +12 | +28 | -18 | -2 | +54 | +1 | 0 | -3 | +1 | -1 | +29 | -10 | +2 | +1 | +12 | -8 | +3 | -3 | +3 | -4 | -19 | +2 | +18 | +2 |
| VII | 0 | +15 | -16 | +3 | +8 | +5 | -3 | -2 | -1 | -3 | -5 | -6 | -5 | +12 | +2 | 0 | -4 | -1 | 0 | +7 | +1 | -3 | -8 | 0 |
| VIII | +16 | +5 | -1 | -10 | +2 | +4 | +2 | 0 | -1 | +28 | -21 | +6 | +26 | +2 | +8 | +1 | +4 | -2 | +7 | +6 | +14 | +1 | -3 | -16 |
| IX | -16 | -5 | -4 | -33 | +10 | -3 | +1 | 0 | +7 | +34 | +43 | +12 | -15 | +8 | -8 | -1 | +4 | 0 | +6 | +9 | -10 | -2 | -11 | -13 |
| X | -7 | -10 | +1 | -32 | -3 | -2 | -1 | +3 | +23 | +34 | +58 | -4 | -44 | -21 | -15 | -5 | +1 | -1 | +1 | +4 | -8 | +7 | +3 | -10 |
| XI | -37 | -5 | +2 | -30 | -5 | -2 | -2 | +2 | +41 | +81 | +33 | -24 | -26 | -15 | -11 | -8 | -3 | +3 | +3 | +7 | -1 | -9 | +4 | +1 |
| XII | -41 | +22 | -13 | +32 | -7 | 0 | -3 | +4 | +44 | +83 | +59 | +4 | -29 | -27 | -92 | -11 | -10 | -6 | +3 | -2 | +18 | +14 | -1 | -10 |

The very peculiar and special form of these "noise charts" gives us a quick survey of whether similarly treated parameters are correlated, anticorrelated or non-correlated. By comparing Fig. 59 ($\Delta_2 f_o E$ 50%) with Table XI ($\Delta R_z$) or Table XII ($\Delta\Phi$), we see that good correlation between "sunspot noise" or "flux noise" and "$f_o E$ noise".

## 3. THE SPORADIC E LAYER

### 3.1. Methods of observation and principal parameters of the Es layer

The sporadic E layer, Es, is the most irregular and the thinnest layer of the ionosphere. At mid-latitudes, however, Es shows some regular features, i.e., a strong diurnal dependence on the time of day and a strong seasonal dependence. Moreover, there is an interesting variation with the sunspot cycle: both at sunspot minimum and sunspot maximum high values of $f_o$Es occur. Finally, day-to-day variations of $f_o$Es are correlated with the diurnal variations of the magnetic field.

#### 3.1.1. Thickness of Es
The thickness of Es varies from 400 m to a few kilometers. These values have been found from ionograms where good M-echoes were visible (Fig. 60). When the Es layer is transparent, the waves, depending on their frequencies, are partly reflected by the Es layer and partly penetrate it. One obtains the M-echo when the transmitted wave passes Es, is reflected from the bottom side of the F2 layer, is reflected from the upper side of the Es layer, again reflected from the bottom side of the F2 layer, passes the Es layer and arrives back to the receiver.

Recently, in-situ rocket measurements have given direct proof of the Es layer (see Fig. 61, 62 and 63).

#### 3.1.2. Height of Es
The minimum virtual height $h'$Es is close to the true reflection height, except in cases where retardation occurs in the E layer. The height range is $115 \pm 10$ km.

According to Rawer and Suchy (1967), the global behaviour of Es heights shows: (1) a small diurnal variation; (2) no seasonal variation; (3) no dependence on the sunspot number; (4) a lunar semidiurnal variation of about 1/2 km, with a rather constant phase over the globe.

#### 3.1.3. The characteristic frequencies of Es
$f_o$Es and $f_x$Es are the highest frequencies reflected from the Es layer for the ordinary or the extraordinary wave, respectively, $f_b$Es is the lowest frequency at which the Es layer becomes transparent (Fig. 64).

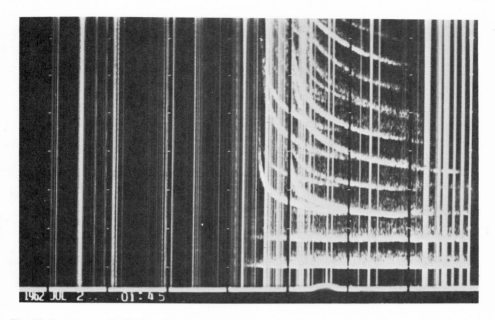

Fig. 60. Ionogram with "M" and "F+E" echoes. From below to the top we distinguish the following modes of propagation: Es, 2Es; F; F+Es; M; 2F and so on.

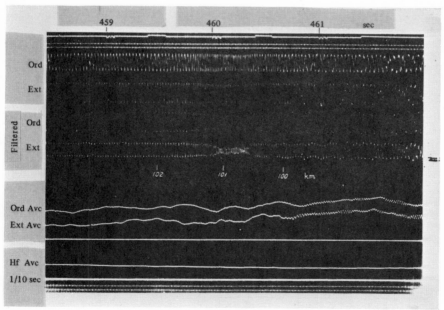

Fig. 61. Section of a dispersive (differential) Doppler record of rocket transmissions showing the crossing of a sporadic Es cloud. Beat curves for both characteristic polarizations i and e, here called "ord." and "ext."; lower curves: field strength. (Reproduced from Rawer and Suchy, 1967.)

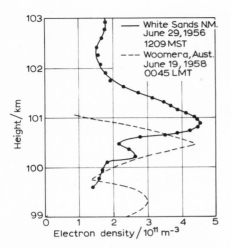

Fig. 62. Electron density profiles of Es layers, determined on two different occasions by the differential Doppler technique during a rocket ascent. (Reproduced from Rawer and Suchy, 1967.)

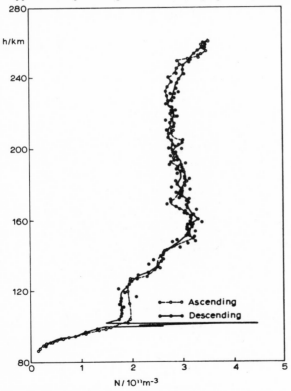

Fig. 63. Electron density profile obtained during rocket ascent and descent. (Reproduced from Rawer and Suchy, 1967.)

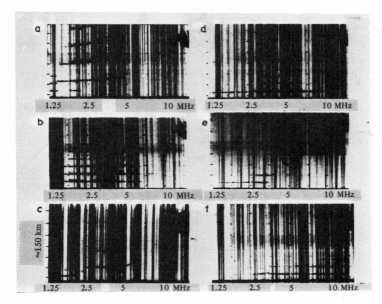

Fig. 64. Mid-latitude ionograms (Freiburg, 48°N); (a) and (b): summer night (June 1952); (c): winter day (January 1955); (d), (e) and (f): summer day (June 1952).

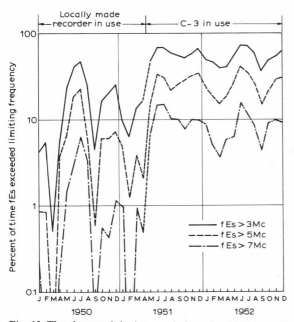

Fig. 65. The change of the ionosonde in spring 1951 brought about ten times more Es.

All these frequencies depend on the sensitivity of the ionosonde: the better the transmitter, the antennas and the receiver (phase detection), the higher $f_o$Es and $f_x$Es, and vice versa, the lower $f_b$Es.

An extreme example of Es measurements affected by equipment change is shown in Fig. 65.

Figure 66 gives a mass plot of all monthly measurements of the top frequency $f_o$Es for June 1958 at Freiburg. The dispersion of the individual points is very great.

Figure 67 gives a mass plot of all monthly measurements of the blanketing frequency $f_b$Es for June 1952 from Lwiro. The dispersion is still very great.

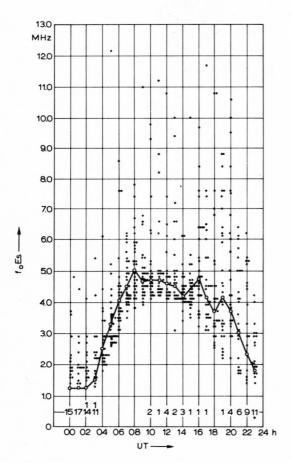

Fig. 66. Mass plot of all monthly measurements of the top frequency $f_o$Es in June 1958 at Freiburg. (Reproduced from Rawer and Suchy, 1967.)

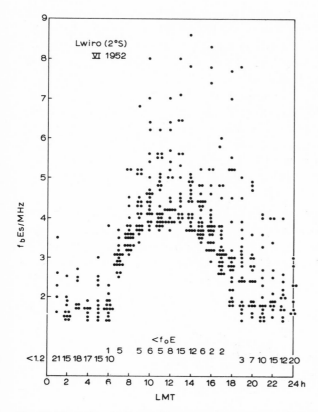

Fig. 67. Mass plot of all monthly measurements of the blanketing frequency $f_b$Es, Lwiro, June 1952. (Reproduced from Rawer and Suchy, 1967.)

## 3.2. The global behaviour of Es

The seasonal variation of $f_o$Es is important at temperate latitudes. Es is a summer phenomenon appearing in May through September on the northern hemisphere and in November through February on the southern one.

A lunar control was proposed in the past. As to the solar cycle, good correlations have not been found. The long-term cross correlation between different stations is bad. Considerable differences are found between different years. The dimensions of an Es-zone (which normally consists of many "clouds") are between 10 and 400 km. Drifts in different directions have been observed by the backscatter technique. The apparent speeds were between 50 and 200 m/sec.

The world-wide pattern of Es is rather well-known. World maps of the occurrence of high values of $f$ Es are shown in Fig. 68 for the day-time and in Fig. 69 for the night-time.

A belt of very high probability appears in the day-time; it follows the magnetic equator rather closely. At night, this belt practically disappears, but a zone of increased probability appears at high latitudes; the *auroral Es* and its appearance are closely related to aurorae. The layer profile is often rather thick, like day-time normal E. Auroral Es is probably caused by corpuscular radiation.

The *equatorial Es* gives the highest $f_0$Es values, but only the day-time and it is highly transparent. The echoes are mostly diffuse. The width of the equatorial Es belt is $\pm 7°$ magnetic dip, i.e., 400–500 km (Fig. 70).

A classification of different Es forms for temperate latitudes distinguishes three types in the day-time: (1) "h", high; (2) "c", cusp; and (3) "$l$", low. In the night-time, only the type "f", flat, is observed. Fig. 71 shows that "c", "h" and "f" types have maximum occurrence in the summer; "$l$" has its maximum in the early winter.

The 1968 ESSA Technical Report ERL 73-ITS 63 published maps of $f_0$Es for March, June, September and December and for high and low sunspot numbers (1958 and 1954), as well as for the lower decile, the median, and the upper decile,

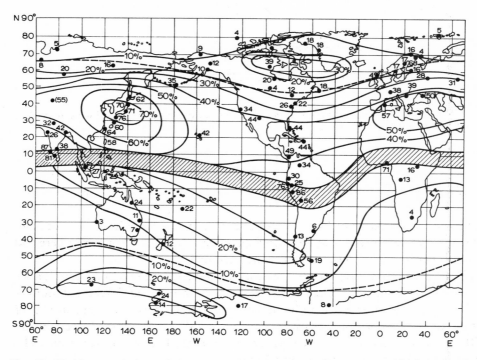

Fig. 68. World map of the probability of high top frequency ($f$Es $> 5$ MHz) during day-time. (Reproduced from Rawer and Suchy, 1967).

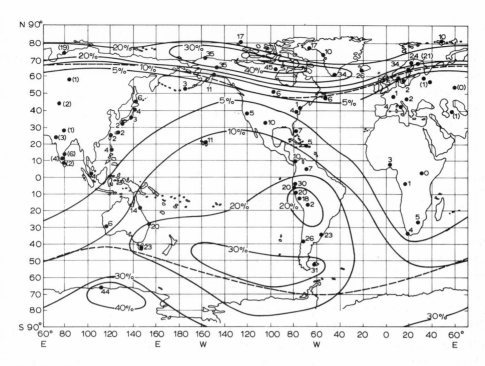

Fig. 69. World map of the probability of high top frequency ($f$Es $> 5$ MHz) during night-time. (Reproduced from Rawer and Suchy, 1967.)

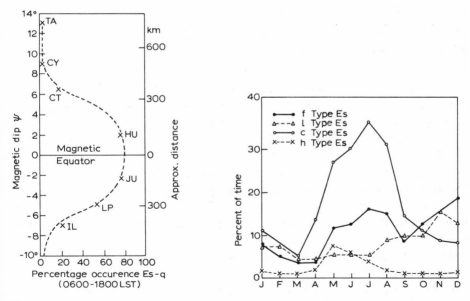

Fig. 70. Probability of occurrence of the q-type of Es in a latitude section crossing the magnetic (dip) equator in Southern America. (Reproduced from Rawer and Suchy, 1967.)

Fig. 71. Occurrence of the different Es-types at a temperate latitude as a function of the season. Washington, 1958. (Reproduced from Rawer and Suchy, 1967.)

as shown in Fig. 72, 73 and 74. Only the Es ionization over Europe and Africa around noon in June 1954, and June 1958, is given.

It was found that

(1) The equatorial Es is changing with the solar cycle, as is shown in the following table:

|            | Lower decile | Median | Upper decile |
|------------|--------------|--------|--------------|
| Year 1954  | lower        | higher | higher       |
| Year 1958  | higher       | lower  | lower        |

1954 was a year of very low sunspot numbers, while 1958 was at the sunspot maximum. The lower decile of the equatorial Es, which is near to the normal E, shows the normal behaviour: low values in the sunspot minimum and high values in the sunspot maximum. But the median values and the upper decile of the equatorial Es show just the contrary: high values for the sunspot minimum and low values for the sunspot maximum.

(2) $f_o$Es at the temperate latitudes also varies with the solar cycle.

We can notice a big difference between the northern summer and the southern winter. Moreover, for the northern summer $f_o$Es is higher at sunspot minimun 1954. But if we search for the fine structure of the Es results we obtained from ionospheric stations, it is almost lost in these charts.

### 3.3. The fine structure of Es over Europe

Smith (1968) stated that the principal features of intense sporadic E at temperate latitudes are: (1) seasonal distribution; (2) diurnal distribution; (3) geographic variation; (4) magnetic correlation; (5) year-to-year variability; and he also stated that it appears probable that a series of boundary conditions has to be fulfilled to produce intense sporadic E.

I feel that, since we have in Europe such well-working ionospheric stations that have been running now for more than 10 or 20 years, we should at least be able to describe, as well as possible, the behaviour of Es, correlating it to the above-mentioned parameters in order to have an idea of what happens in the sporadic E layer over Europe.

The chain of European stations from which our statistical data were taken are shown in Table XIII.

Searching for the boundary conditions that produce a high sporadic E, we know that for the temperate latitude of the northern hemisphere most Es is found in June and July. To make the statistics more meaningful, the following work is based on the months June and July mixed together.

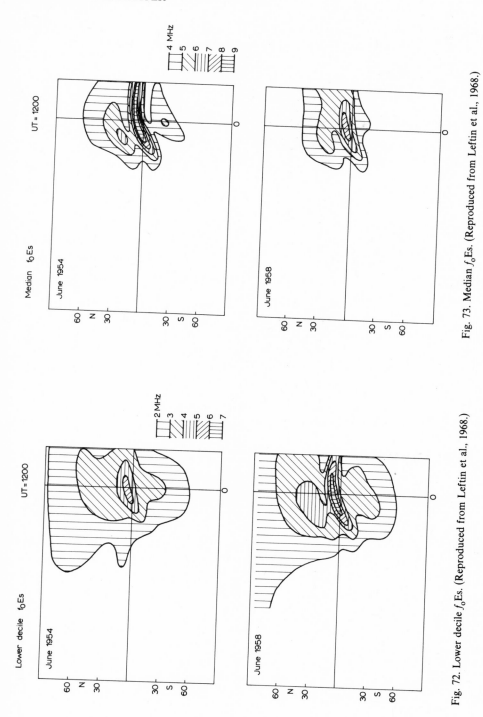

Fig. 73. Median $f_oEs$. (Reproduced from Leftin et al., 1968.)

Fig. 72. Lower decile $f_oEs$. (Reproduced from Leftin et al., 1968.)

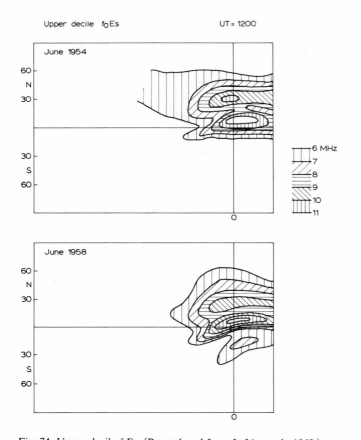

Fig. 74. Upper decile $f_o$Es. (Reproduced from Leftin et al., 1968.)

TABLE XIII

European ionospheric stations (north to south)

| Slough | 51° 31′N | 00° 34′W |
|--------|----------|----------|
| Lindau | 51° 39′N | 10° 7.5′E |
| Dourbes | 50° 06′N | 04° 36′E |
| Prague | 49° 59′N | 14° 33′E |
| Freiburg | 48° 03′18″N | 07° 35′24″E |
| Genoa | 44° 17′45″N | 07° 53′18″E |
| Rome | 41° 50′N | 12° 30′E |
| (Dakar | 14° 46′N | 17° 25′W) |

### 3.3.1. The raw data

The rough material for these statistics are the published data for all hours and all days between 1957 and 1970. When not published, we calculated the monthly means: $f_oE$ 50%, $f_oEs$ 50%, $f_oEs$ 25%, $f_oEs$ 10% and $f_oEs$ 3%. As an example, the results for Genoa (June 1959) are plotted in Fig. 75.

After smoothing using the weights 1/4, 1/2 and 1/4, we obtained charts for the different probabilities of $f_oEs$ as a function of the year and the time of day (Fig. 76

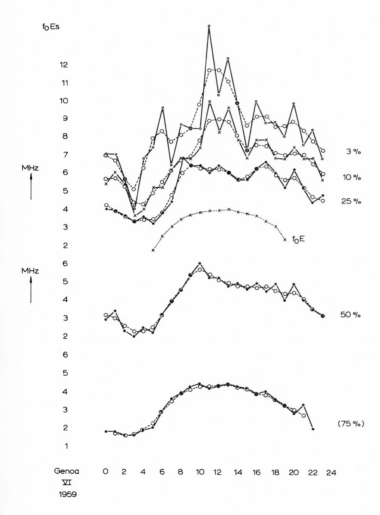

Fig. 75. Es parameters and smoothing process for Genoa.

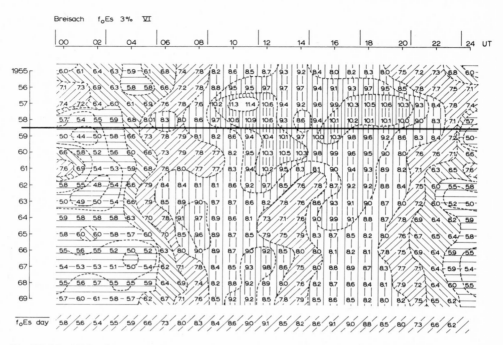

Fig. 76. $f_o$Es 3% for Breisach in June, noon values, depending on the year and the time of day.

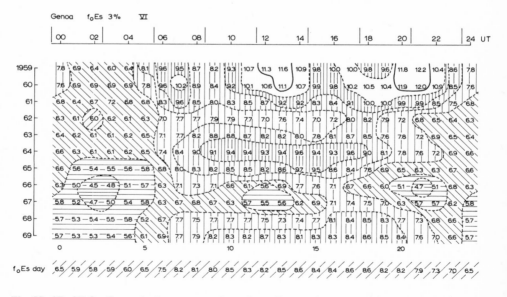

Fig. 77. $f_o$Es 3% for Genoa in June, noon values, depending on the year and time of day.

and Fig. 77). The smoothing is done over the time of the day. To have the smoothed value for $\overline{1100}$ we take one quarter of the values for 1000 and 1200 and add one half of the value of 1100, i.e.:

$$\overline{X}_{1100} = \frac{X_{1000}}{4} + \frac{X_{1100}}{2} + \frac{X_{1200}}{4} \tag{13}$$

### 3.3.2. The daily variation

As described above, we calculated the mean daily variation $\overline{f_o \mathrm{Es}\,\%\ \mathrm{day}}$. In Fig. 78 and Fig. 79, there are two sets of curves concerning the different probabilities for Breisach and Prague. The great similarity between different stations is astonishing.

Fig. 80 and Fig. 81 show the daily variation of $\overline{f_o\ \mathrm{Es}\ \mathrm{day}}$ in June and July for all

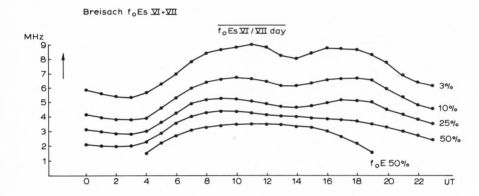

Fig. 78. Sets of curves concerning the different probabilities for Breisach.

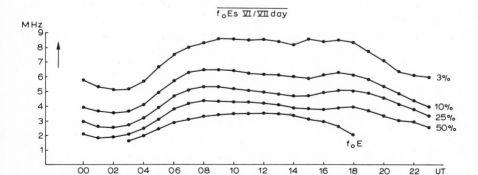

Fig. 79. Sets of curves concerning the different probabilities for Prague.

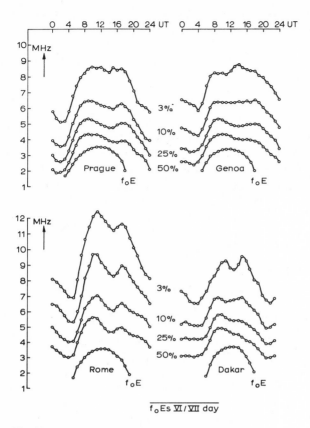

Fig. 80. Daily variation of $f_o$Es day.

probabilities and for all stations. We can observe a morning maximum between 10 and 12 UT, a daily minimum around 14 UT and an afternoon maximum between 16 and 20 UT. The sensitivity of the ionosonde plays a certain role (see Lindau, with higher absolute values for $f_o$Es), but the relative changes in the diurnal behaviour and the different probabilities of $f_o$Es are very similar for all stations.

### 3.3.3. The year-to-year variation.

To describe the year-to-year variation, the differences $f_o\mathrm{Es} - \overline{f_o\mathrm{Es\ day}} = \Delta_1 f_o\mathrm{Es}$ are calculated. An example is given in Fig. 82 for Genoa in June and 3% probability. Such difference charts are calculated for all stations and all percentages for the months of June and July. From the charts, we derive the mean over all hours $\overline{\Delta f_o\mathrm{Es\ year}}$ for the months of June and July also and for all stations and percentages. The results are shown in Fig. 83 and 84. These results are very

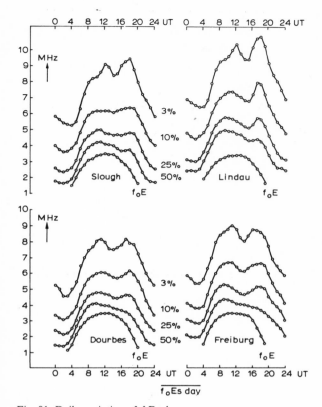

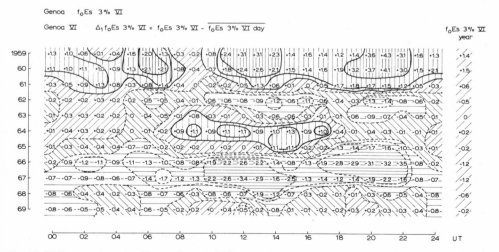

Fig. 81. Daily variation of $f_o$Es day.

Fig. 82. Difference chart for Genoa, June and 3%.

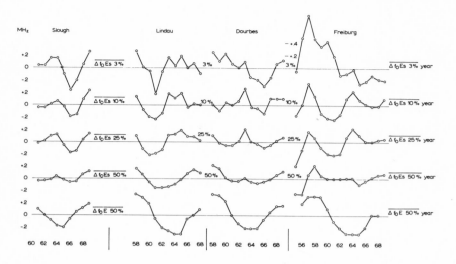

Fig. 83. Year-to-year variation for different ionospheric stations and different probabilities of $f_o$Es, compared with the $f_o$E variation.

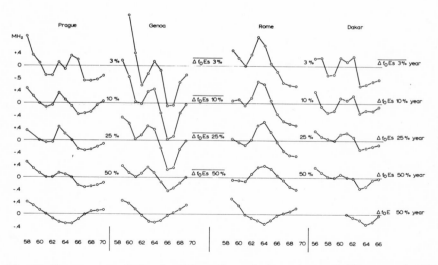

Fig. 84. Year-to-year variation for different ionospheric stations and different probabilities of $f_o$Es, compared with the $f_o$E variation.

interesting. $\Delta f_o E$ year shows only—and for all stations—a clear solar-cycle dependence; maxima and minima are correlated.

$\overline{\Delta f_o Es}$ has two maxima for all percentages from Slough to Rome, one in the sunspot maximum and the other in the sunspot minimum. There is good correlation among all these stations. Dakar, a test station outside the temperate zone, has quite a different behaviour.

### 3.3.4. Long-term influences on the daily variation

The last step is to calculate the "noise charts":

$$\Delta_2 f_o \, Es = \Delta_1 f_o \, Es - \overline{\Delta f_o \, Es \; year}$$

for all stations, percentages and so on (Fig. 85).

There is an effect controlled by long-term influences which, for example, causes the morning peak to be higher in 1963–1965, while the 1968–1969 values around midday were higher. In the year after the sunspot maximum, the afternoon peak is relatively higher and so on. Fig. 86 shows values for the years 1963–1965 and 1968–1969. The stations from Slough to Genoa are very similar, so that we are allowed to give them the same treatment. The Rome results are different but there the influence for the periods 1963 to 1965 and 1968 to 1969 is just the opposite.

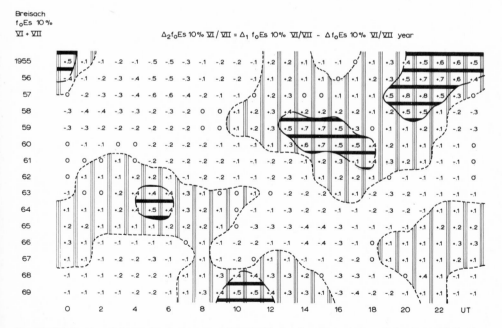

Fig. 85. "Noise chart" for Breisach (June and July, 10%) showing the variation of the diurnal behaviour in the years (solar cycle).

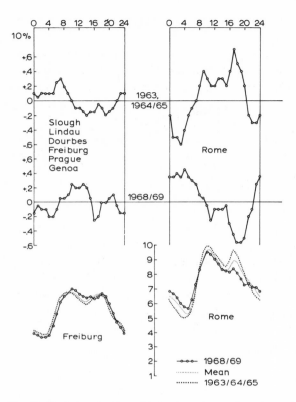

Fig. 86. Variations in diurnal behaviour during the solar cycle.

Freiburg Ⅵ/Ⅶ   f_bEs 50%

| Year | 00 | 01 | 02 | 03 | 04 | 05 | 06 | 07 | 08 | 09 | 10 | 11 | 12 | 13 | 14 | 15 | 16 | 17 | 18 | 19 | 20 | 21 | 22 | 23 | f_oE 50% noon |
|---|---|---|---|---|---|---|---|---|---|---|---|---|---|---|---|---|---|---|---|---|---|---|---|---|---|
| 1957 | 1.6 | 1.4 | 1.2 | 1.3 | 2.0 | 2.8 | 3.5 | 4.0 | 4.5 | 4.6 | 4.7 | 4.4 | 4.4 | 4.4 | 4.4 | 4.0 | 3.8 | 3.5 | 3.6 | 2.6 | 2.2 | 2.2 | 1.8 | <1.6 | 4.0 |
| 58 | 1.2 | 1.2 | 1.2 | 1.2 | 2.0 | 2.9 | 3.6 | 4.0 | 4.4 | 4.4 | 4.4 | 4.4 | 4.3 | 4.3 | 4.2 | 4.0 | 4.0 | 3.7 | 3.1 | 2.8 | 2.4 | 1.8 | <1.4 | 1.2 | 4.0 |
| 59 | 1.3 | 1.2 | 1.2 | 1.3 | 1.9 | 2.5 | 3.6 | 4.0 | 4.3 | 4.3 | 4.4 | 4.4 | 4.4 | 4.4 | 4.0 | 4.0 | 3.7 | 3.6 | 3.2 | 2.2 | 1.8 | 1.6 | 1.6 | 1.4 | 3.9 |
| 60 | 1.4 | 1.2 | 1.2 | 1.2 | 1.9 | 2.6 | 3.4 | 3.7 | 4.1 | 4.0 | 4.3 | 4.2 | 4.3 | 4.2 | 4.0 | 3.8 | 3.5 | 3.4 | 2.8 | 2.4 | 1.8 | 1.6 | 1.6 | <1.4 | 3.8 |
| 61 | 1.2 | 1.2 | 1.2 | 1.2 | 1.8 | 2.5 | 3.0 | 3.5 | 3.8 | 4.1 | 4.0 | 4.0 | 3.9 | 3.8 | 3.7 | 3.7 | 3.6 | 3.6 | 3.4 | 3.0 | 2.2 | 1.7 | 1.6 | <1.4 | 3.5 |
| 62 | 1.2 | 1.2 | 1.2 | 1.2 | 1.9 | 2.5 | 3.1 | 3.5 | 3.8 | 3.7 | 3.8 | 3.8 | 3.6 | 3.6 | 3.6 | 3.4 | 3.4 | 3.4 | 2.8 | 2.4 | 1.6 | 1.4 | 1.3 | | 3.3 |
| 63 | 1.2 | 1.2 | 1.4 | 1.3 | 2.0 | 2.6 | 3.4 | 3.6 | 3.8 | 4.0 | 4.0 | 3.8 | 3.8 | 3.6 | 3.6 | 3.5 | 3.2 | 3.2 | 3.3 | 2.5 | 2.2 | 1.8 | 1.3 | | 3.3 |
| 64 | 1.4 | <1.4 | 1.3 | 1.2 | 1.8 | 2.6 | 3.9 | 3.6 | 4.0 | 3.7 | 3.7 | 3.6 | 3.6 | 3.5 | 3.6 | 3.6 | 3.4 | 3.3 | 3.1 | 3.4 | 2.6 | 1.9 | <1.6 | <1.6 | 3.2 |
| 65 | 1.4 | 1.4 | 1.4 | 1.2 | 1.7 | 2.6 | 3.1 | 3.6 | 3.8 | 3.8 | 3.9 | 3.7 | 3.7 | 3.6 | 3.4 | 3.4 | 3.3 | 3.2 | 3.2 | 2.4 | 2.0 | 1.8 | <1.7 | <1.6 | 3.2 |
| 66 | 1.6 | 1.3 | 1.4 | 1.2 | 1.9 | 2.6 | 3.4 | 3.5 | 4.1 | 4.0 | 4.0 | 4.0 | 4.0 | 3.9 | 3.8 | 3.6 | 3.4 | 3.4 | 3.2 | 2.7 | 2.2 | 1.8 | <1.8 | <1.6 | 3.4 |
| 67 | 1.5 | 1.3 | 1.2 | 1.4 | 1.9 | 2.6 | 3.4 | 3.7 | 3.9 | 4.0 | 4.1 | 4.0 | 4.0 | 3.8 | 4.0 | 3.7 | 3.5 | 3.4 | 3.1 | 3.1 | 2.8 | 2.5 | 2.2 | 2.0 | 3.5 |
| 68 | 1.2 | 1.2 | 1.2 | 1.4 | 2.0 | 2.7 | 3.4 | 3.8 | 4.0 | 4.0 | 4.0 | 4.1 | 4.1 | 3.9 | 3.8 | 3.8 | 3.6 | 3.4 | 3.2 | 2.7 | 2.6 | 2.0 | 2.0 | <1.6 | 3.6 |
| 69 | 1.8 | 1.6 | 1.8 | 1.8 | 2.0 | 2.9 | 3.4 | 4.1 | 4.1 | 4.2 | 4.3 | 4.2 | 4.0 | 4.1 | 3.9 | 3.9 | 3.4 | 3.4 | 3.4 | 2.5 | 2.4 | 2.1 | 1.8 | 1.8 | 3.8 |
| 1970 | 1.2 | 1.2 | 1.2 | 1.2 | 2.0 | 2.8 | 3.4 | 3.8 | 4.1 | 4.2 | 4.2 | 4.2 | 4.1 | 4.0 | 3.9 | 3.9 | 3.6 | 3.4 | 3.3 | 2.8 | 2.4 | 2.6 | 2.4 | <1.6 | 3.7 |

00   02   04   06   08   10   12   14   16   18   20   22   UT

Fig. 87. Freiburg $f_b$Es 50%, June and July (34 = 3.4 MHz).

### 3.4. Es parameters over Freiburg, compared with the $f_oE$ 50% as a standard

This section concerns the relations existing among $f_o E$, $f_b Es$, $f_o Es$ 50% and $f_oE$ 25% for Freiburg. The year-to-year variation and the diurnal behaviour are considered.

As we showed in section 2, $f_oE$ 50% is very well correlated to the position of the sun and the sunspot number. Since the variation of the position of the sun is almost the same in the months of June and July and the variation of the sunspot cycle is expressed by the noon value of $f_oE$ 50%, this value seems to be ideal for being compared with the daily and year-to-year variations of $f_b Es$ 50%, $f_o Es$ 50%, $f_oEs$ 25%. Therefore, in the following statistical analysis, we used the noon values for $f_oE$ 50% for the months of June and July.

The rough material is shown in Fig. 87, 88 and 89.

The next set of charts (Fig. 90, 91 and 92) shows the differences from the standard. For each time of day we calculate $\overline{\Delta f_o Es\ day}$ or $\overline{\Delta f_b Es\ day}$. In this manner, we obtain the mean diurnal variation for the different percentages.

The third set of charts (Fig. 93, 94 and 95) shows the differences from the mean diurnal variation and from these charts we can compute the mean year-to-year variation $\overline{\Delta \Delta f_o Es\ 50\%\ year}$ or $\overline{\Delta \Delta f_b Es\ 50\%\ year}$. The last step leads to the "noise", i.e., for example, the differences from the mean year-to-year variation:

$$\Delta_3 f_b\ Es\ 50\%\ VI/VII = \overline{\Delta_2 f_b\ Es\ 50\%\ VI/VII} - \overline{\Delta \Delta f_b\ Es\ 50\%\ year} \qquad (14)$$

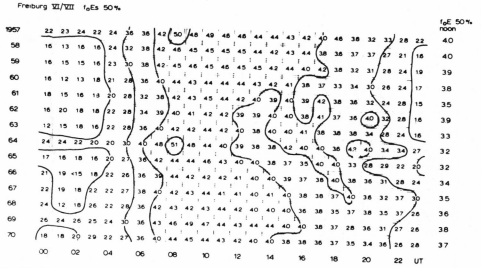

Fig. 88. Freiburg $f_oEs$ 50%, June and July (32 = 3.2 MHz).

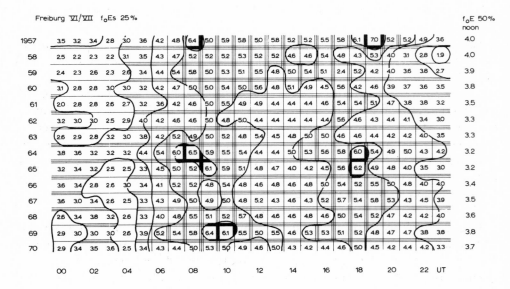

Fig. 89. Freiburg $f_o$Es 25%, June and July (58 = 5.8 MHz).

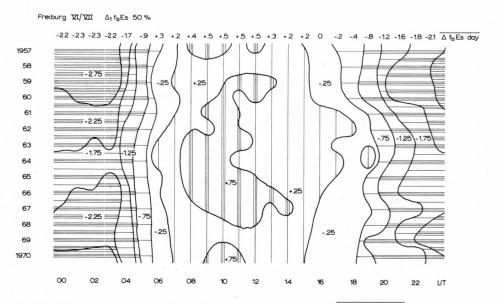

Fig. 90. Differences from the standard $f_o$E, and mean daily variation $\overline{\Delta f_b\,\text{Es}}$ 50% day.

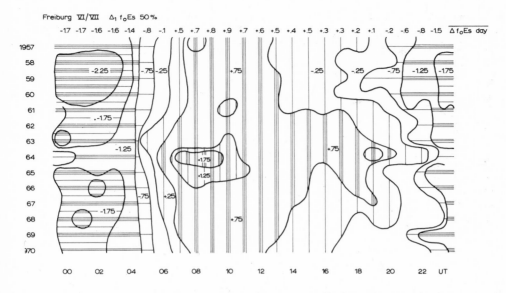

Fig. 91. Differences from the standard $f_o$E, and mean daily variation $\overline{\Delta f_o \text{ Es}}$ 50% day.

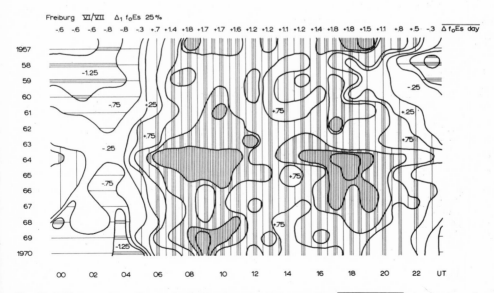

Fig. 92. Differences from the standard $f_o$E, and mean daily variation $\overline{\Delta f_o \text{ Es}}$ 25% day.

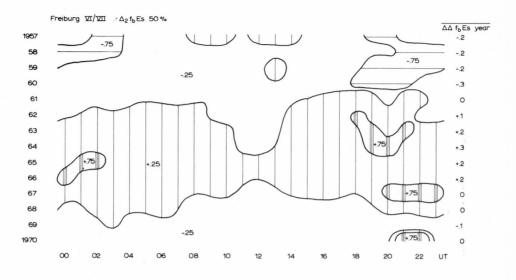

**Fig. 93. Difference** chart and mean year-to-year variation $\overline{\Delta f_b \, \text{Es} \, 50\% \, \text{year}}$.

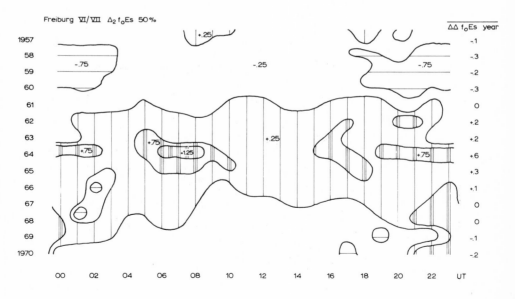

**Fig. 94. Difference** chart and mean year-to-year variation $\overline{\Delta f_o \, \text{Es} \, 50\% \, \text{year}}$.

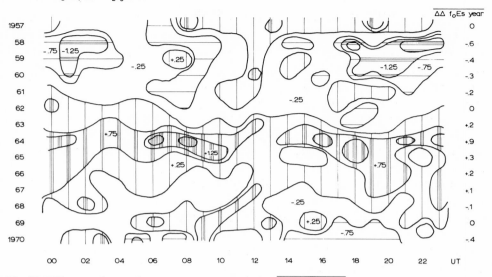

Freiburg Ⅵ/Ⅶ    $\Delta_2 f_o Es$  25%

Fig. 95. Difference chart and mean year-to-year variation $\overline{\Delta f_o\,Es}\,25\%$ year.

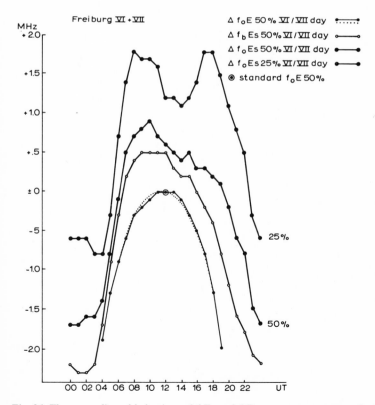

Fig. 96. The mean diurnal behaviour of $f_b Es$ and $f_o Es$ parameters, compared with the noon value of $f_o E$.

In Fig. 96 we have plotted the results showing the mean diurnal behaviour. The results are interesting: we can see, in fact, that just at the time when $f_o$E is minimum, we have the morning peak for the Es parameters, while when the Es parameters have their minimum, $f_o$E is the highest.

Figure 97 shows that such a mean diurnal behaviour is a global feature: in fact similar variations were observed in Japan.

Figure 98 shows the results for the mean year-to-year variability. A dependence on the solar cycle is justified only for the normal E layer. For $f_b$ Es 50%, $f_o$ Es 50% and $f_o$Es 25% we find two maxima, one correlating with high sunspot values and the other with low ones.

Figure 99 is another representation of the differences between $f_o$E and, respectively, $f_b$Es and $f_o$Es. For the diurnal variation we now take the symmetrical $f_o$E curve as a standard. We can see very clearly the correlation among the Es types but also the anticorrelation with $f_o$E.

Finally, we can examine the "noise charts" and find that once again there is a strong deviation from the diurnal behaviour in the years 1963–1965 (Fig. 100).

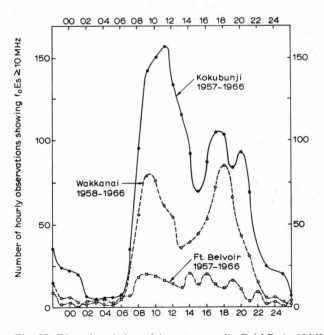

Fig. 97. Diurnal variation of intense sporadic E ($f_o$ Es $\geqslant$ 10 MHz), as observed at three ionosonde stations. (Reproduced from Smith, 1968.)

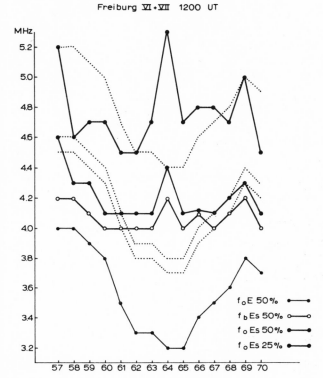

Fig. 98. Year-to-year variation for the E-layer parameters $f_o$E 50%, $f_b$Es 50%, $f_o$Es 50% and $f_o$Es 25%.

### 3.5. Statistical analysis concerning the Es classification evaluated at Freiburg

The statistical analysis about these classifications show the following results for Es over Freiburg, for the months June and July:

(1) Type "h" (see Fig. 101): the classification "h" describes an Es trace (usually a day-time type) showing a discontinuity in height, with the normal E layer at or above $f_o$ E. The low frequency end of the Es trace lies clearly above the high frequency end of the normal E trace.

There is an interesting variation with the solar cycle. The minimum of "h" cases took place in 1963 to 1965, while the maximum (40% probability) was around 0800 and 1600 in the years 1959 to 1961.

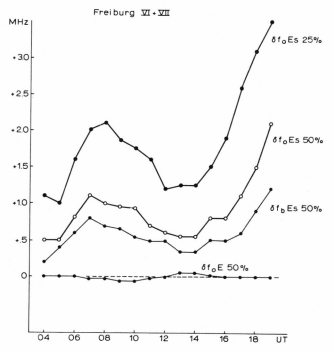

Fig. 99. Differences from the symmetrical $f_oE$ curve for $f_bEs$ and $f_oEs$ parameters.

(2) Type "c" (Fig. 102): the classification "c" corresponds to an Es trace showing a relatively symmetrical cusp at or below $f_oE$. The classification "cusp" has a predominant large maximum from 0600–1000 UT in the sunspot minimum, followed by a smaller maximum in the late afternoon at 1800 UT. The maxima have probabilities greater than 90%. In times when the type "h" has the maximum probability of 40%, type "c" goes down to 60%.

(3) Type "$l$" (Fig. 103): the classification "$l$" means a flat Es trace at or below the normal E layer minimum virtual height. Type "$l$" was found only in years with very high sunspot numbers, with a low probability of about 30% around noon.

Around sunrise and sunset, type "$l$" changes to night type "f" (see Fig. 104) with higher probability and a maximum of 80% between 2000 and 2100 UT.

Around midnight, the probability for Es is 70% for the sunspot minimum 1964–1965. At sunspot maxima the probability is 60%.

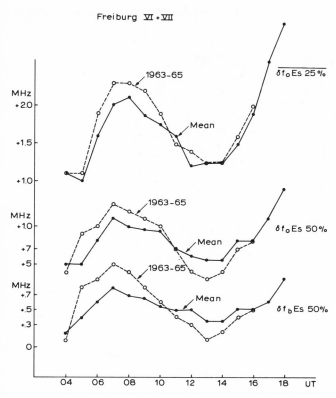

Fig. 100. Deviation from the mean diurnal behaviour in the years 1963–1965 for the $f_b$Es and $f_o$Es parameters.

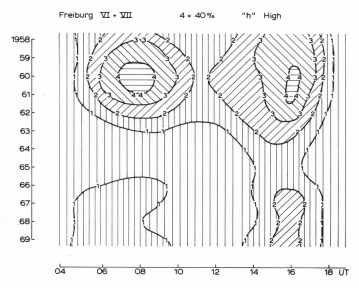

Fig. 101. Distribution of Es "h" for the months of June and July, as a function of the year and time of day.

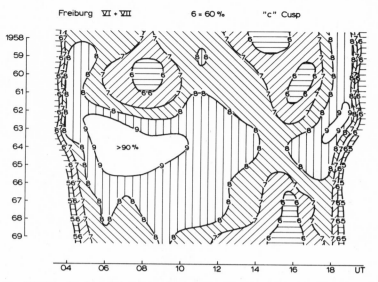

Fig. 102. Distribution of Es "c" for the months of June and July as a function of the year and time of day.

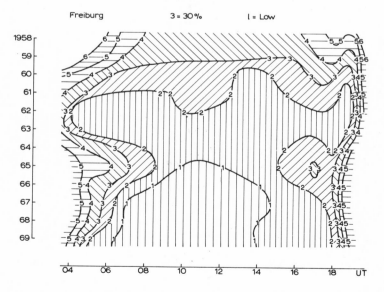

Fig. 103. Distribution of Es "*l*" for the months of June and July, as a function of the year and time of day.

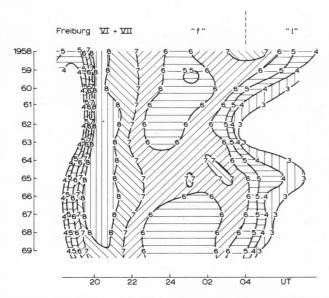

Fig. 104. Distribution of Es "f" for the months of June and July, as a function of the year and time of day.

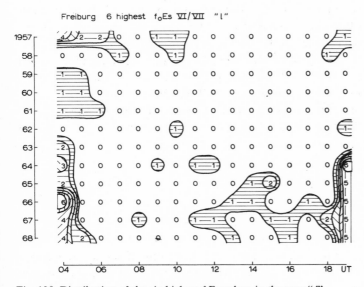

Fig. 105. Distribution of the six highest $f_0$Es values in the case "$l$".

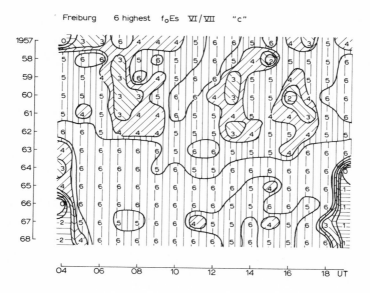

Fig. 106. Distribution of the six highest $f_o$Es values in the case "c".

We wondered whether these classifications had any relation to the highest $f_o$Es frequencies measured at a given time. Accordingly, for the months of June and July we searched out the three highest values for every hour and for all the years, and then noted the classifications. We found that in day-time, the highest values are normally *not* in the cases "*l*" (Fig. 105).

The classification "c" is mostly correlated with the six highest values of June and July. But from 1959 to 1961 and 0800 and 1600 UT, when type "h" has its maxima, only 3 out of the 6 highest values are marked "c" (Fig. 106).

Fig. 107 shows that up to 50% of the highest $f_o$E values are to be found in case "h" between 1957 and 1962. Such result appears to be interesting and should be confirmed by other European stations.

## 3.6. The day-to-day variation of $f_o$Es over Europe

### 3.6.1. The raw data

Figure 108 shows an example of our observations, in order to explain what we observe on a given day. In Fig. 108, (*a*) shows us a recording of the top frequency $f_o$Es for the entire day. We can see at glance that there are fast and major changes in the critical frequencies. Periods of some minutes up to three hours are visible. Changes from over 2 up to 10 MHz can occur in a short time. How can we condense

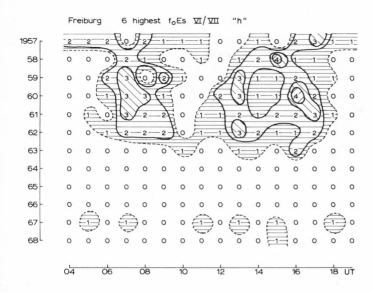

Fig. 107. Distribution of the six highest $f_oEs$ values in the case "h".

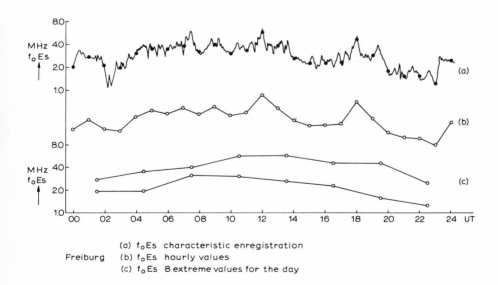

(a) $f_oEs$ characteristic enregistration
Freiburg (b) $f_oEs$ hourly values
(c) $f_oEs$ 8 extreme values for the day

Fig. 108. $f$ Es recording: hourly measurements and three-hour value.

this complex information into a simple description of the behaviour in a given day? In Fig. 108, (b) shows us the hourly values of $f_o$Es for the same day as they are normally published by ionospheric stations. We still have great variations and we know that the hourly control of Es gives only statistical information. In Fig. 108, (c) proposes a new daily description for $f_o$Es. From the hourly values we take the extreme values for the upper and lower limit of $f_o$Es for 3-h periods. This is taken as information for the given day.

We know that a correlation exists between the production of sporadic E and the variation of the magnetic field, as was stated above. But the statistics with the $\Sigma K$, which describes the quiet and disturbed days very well, is not good. To get a better description of the changes of the magnetic field over Europe, we took the 3-h $K$ values for three stations: Wingst, Dourbes and Genoa. We plotted them versus time (Fig. 109), and calculated the mean. After smoothing with the factors 1/4, 1/2, 1/4, we have the lower curve $(\overline{\overline{K}})$, which we can take as indicative for Europe.

Figure 110 is an example of the raw data covering 1962–1968 for the months of June and July. Such a figure shows that: (1) the condensation to eight values per day still yields sufficient information for our task of comparing the Es variation with the magnetic variation; (2) to describe Es by the extreme values is worthwhile, since the difference between maximum and minimum also seems to be a good parameter for investigating the sporadic E; (3) sporadic E is correlated for large distances, like Lindau–Rome; and (4) the correlation between Es and the magnetic field is not described by $\Sigma K$.

But our rough material is still too complex and abundant. We condense it once

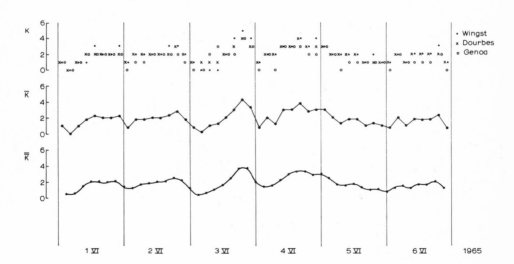

Fig. 109. The magnetic variation $\overline{\overline{K}}$ representative of European conditions.

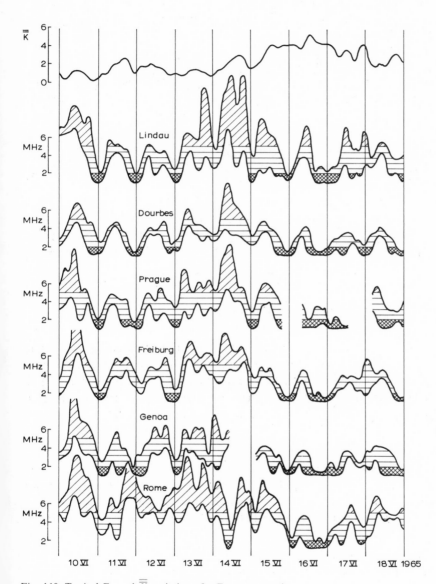

Fig. 110. Typical Es and $\overline{\overline{K}}$ variations for European stations.

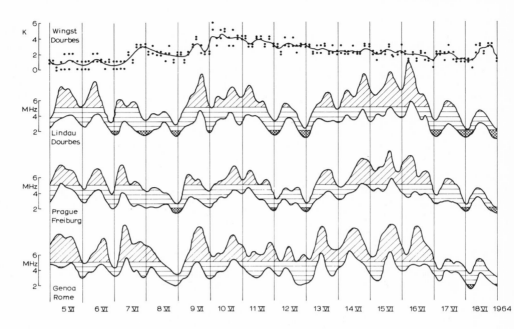

Fig. 111. Condensed Es variations for three pairs of stations.

more and take the stations Lindau–Dourbes, Prague–Freiburg, and Genoa–Rome together (Fig. 111). Here we see again the correlation with $\Sigma K$ is sometimes bad, since in the period 9–11 June 1964 all European stations have high Es while $\Sigma K$ is also high. The correlation among the stations is very good. So, it turns out that the law "Magnetically quiet days are correlated with maximum of $f_0$Es" is not always valid.

### 3.6.2. The 27-day period

Nevertheless, as a first task, we will look for a 27-day period in $f_0$Es. In 1963, we had three minima for $K$ in the months of June and July. We superposed the minima as shown in Fig. 112. The anti-correlation between $K$ and $f_0$Es works well in 1963.

### 3.6.3. The anti-correlation between $f_0$Es and $\Sigma K$

The next task was to superpose epochs showing somewhat similar variations of $K$, for different days and years. In 1962 the anti-correlation did not work (Fig. 113). In 1968 (Fig. 114) we again observe an anti-correlation. For the other years, the results were dubious.

Somewhat confused after these first results, we looked again into our "charts" and found an interesting law: when the daily variation of $\overline{\overline{K}}$ is similar for a group of several days, high values of $f_0$Es prevail. But, if the following days are dissimilar,

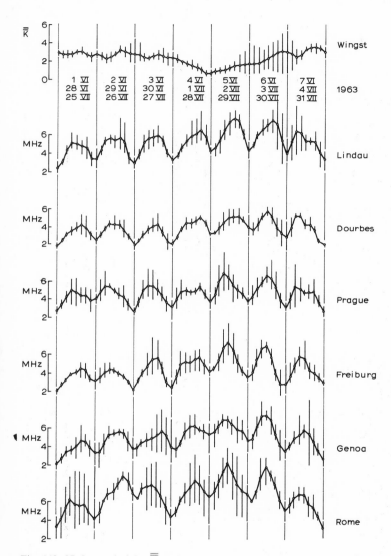

Fig. 112. 27-day period for $\overline{\overline{K}}$ and Es.

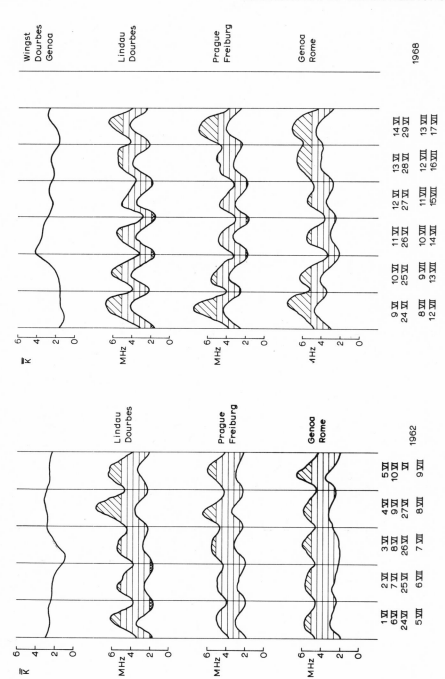

Fig. 114. Superposed epochs for $\overline{K}$ and Es variations.

Fig. 113. Superposed epochs for $\overline{K}$ and Es variations.

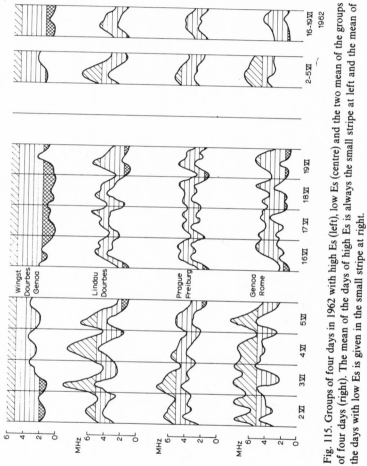

Fig. 115. Groups of four days in 1962 with high Es (left), low Es (centre) and the two mean of the groups of four days (right). The mean of the days of high Es is always the small stripe at left and the mean of the days with low Es is given in the small stripe at right.

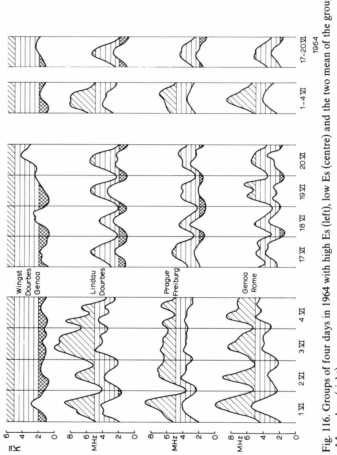

Fig. 116. Groups of four days in 1964 with high Es (left), low Es (centre) and the two mean of the groups of four days (right).

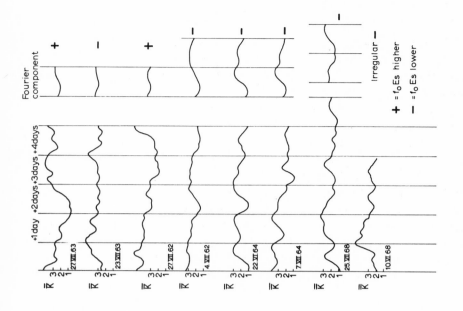

Fig. 118. Daily magnetic variations and occurrence of Es.

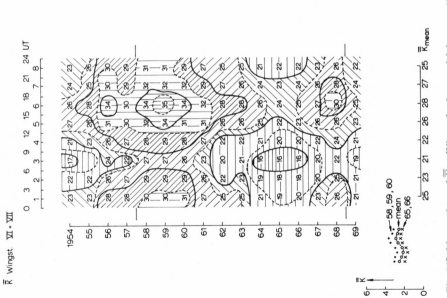

Fig. 117. Mean values of $\overline{\overline{K}}$ at Wingst for June and July, years 1954 to 1969, and mean daily variation over all years (23 means 2.3).

then the critical frequencies $f_o$Ės are relatively low.

### 3.6.4. The normal daily magnetic variation and the production of Es

Our next task was to search out with the above criteria a group of four days for every year. We can see in Fig. 115 and 116 that this law holds well for all the years we investigated.

We also calculated the average $\overline{\overline{K}}$ and $f_o$Es variation for all years, and the result was very clear: $f_o$Es is controlled by the normal daily variation of $\overline{\overline{K}}$ (bay disturbance). We then had to estimate what the normal daily magnetic variation is. For this purpose, we computed monthly mean values for $\overline{\overline{K}}$ at Wingst (Fig. 117). Fig. 117 is a chart for the $\overline{\overline{K}}$ values. The mean over all years fits very well with the $\overline{\overline{K}}$ variation we obtained for the days with high $f_o$Es.

Finally, we found that the mechanism between the magnetic field and $f_o$Es must have some sort of memory for the last day. If, for example, the first and third day are similar, and if the second and fourth day are also similar, there is less occurrence of Es. A qualitative correlation between groups of days with special daily magnetic variations and occurrence of Es is shown in Fig. 118.

### 3.7. The variation of h'Es

From the preceding discussions, we may notice that there is not much information on the Es layer height. It is a pity that up to now many stations have been evaluating the Es height in steps of 10 km. At Freiburg, we try to have these heights as exact as possible.

Fig. 119 shows the h'Es values in June and July for 1957 to 1970. Interesting variations are visible. As usual we calculate: the year-to-year variability $\overline{h'}$ year, the daily variation $\overline{h'}$ day and the "noise chart" (Fig. 120 and Fig. 121). Since the "noise chart" showed a very high level of noise, after smoothing we obtained Fig. 122. The results are encouraging:

(1) The year-to-year variation follows more or less the findings of the normal E layer (Fig. 123). Only h'Es is from 5–12 km higher than h'E. (The values from Slough and Dourbes confirm our results.) Again, the virtual heights in the year-to-year variation show a maximum from 1961 to 1964 (Fig. 124).

(2) The daily variation follows the height of the normal E layer in the sunlit hours very well. Its minimum occurs during the night. This night-time minimum has a shallow maximum (110 km) around 1961–1963 and falls down to 102 km from 1965 to 1968. Thereafter we find high values again, as is shown in Fig. 119.

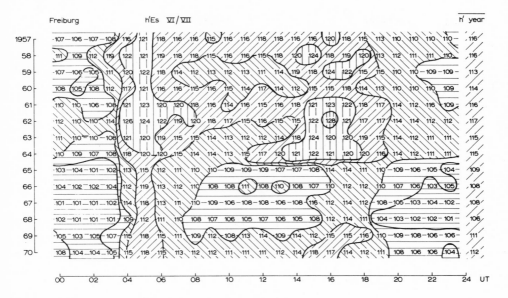

Fig. 119. *h'* Es values in June and July for 1957 and 1970.

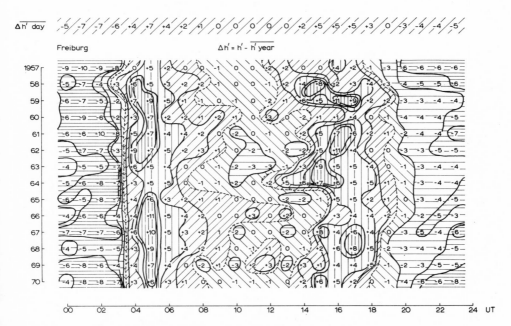

Fig. 120. The daily variation in different years and the mean daily variation.

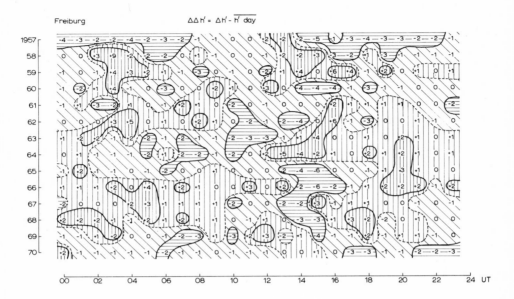

Fig. 121. The "noise chart".

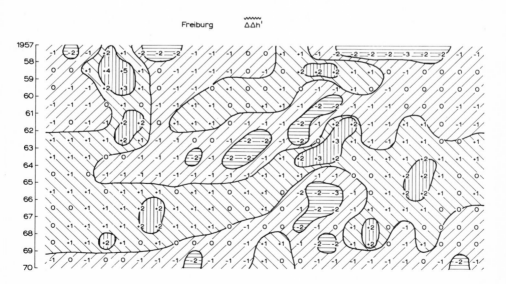

Fig. 122. The smoothed "noise chart".

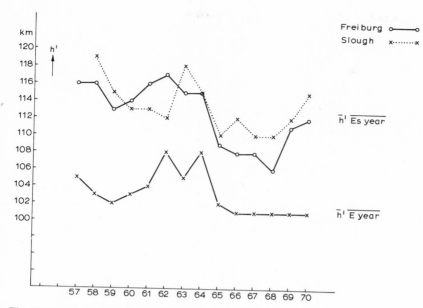

Fig. 123. The mean year-to-year variation for $h'$Es compared with $h'$E.

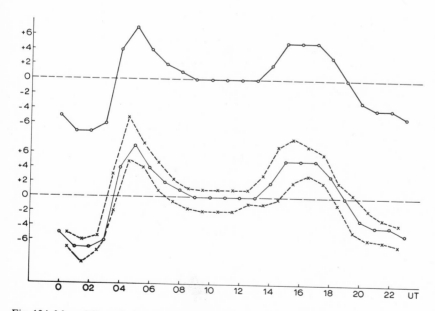

Fig. 124. Mean $h'$Es variation during the day, at Freiburg, for the years 1957–1970. Below: with "noise".

REFERENCES

Appleton, E. V. and Lyon, A. J., Studies of the E-layer of the ionosphere, II: Electromagnetic perturbations and other anomalies. *J.Atmos.Terr.Phys.*, **21**:73–99, 1961.

Landmark, B. and Lied, F., Electron density profiles in the lower E layer deduced from a study of ionospheric cross modulation. In: *Some Ionospheric Results Obtained During the International Geophysical Year* (W. J. G. Benyon, Editor). Elsevier, Amsterdam, pp. 221–230, 1960.

Leftin, M., Ostrow, S. M. and Preston, C., Numerical maps of $f_oEs$ for solar cycle. *Proc. 2nd Sem., Cause and Structure of Temperature Latitude Sporadic E, 19–22 June 1968, Veil, Colo., Vol. 1.* ESSA Nat. Cent. Atmos. Res., Boulder, Colo., pp. 1–21, 1968.

Rawer, K. and Suchy, K., Radio-observations of the ionosphere. In: *Handbuch der Physik, XLIX/2* (S. Flugge, Editor). Springer, Berlin, 630 pp., 1967.

Schwentek, H., The variation of ionospheric absorption from 1956 till 1963. *J.Atmos.Terr.Phys.*, **25**:733–735, 1963.

Smith, E. K., Some unexplained features in the statistics for intense sporadic E. *Proc. 2nd Sem., Cause and Structure of Temperature Latitude Sporadic E., 19–22 June 1968, Veil, Colo., Vol. 1.* ESSA Nat. Cent. Atmos. Res., Boulder, Colo., 1968.

BIBLIOGRAPHY

Here below there is a selected list of papers published on the subject.

*The D Region*

Appleton, E. V. and Piggott, W. R., Ionospheric absorption measurements during a sunspot cycle. *J.Atmos.Terr.Phys.*, **5**:141–172, 1954.

Bain, W. C., Observations on the propagation of very long radio waves reflected obliquely from the ionosphere during a solar flare. *J.Atmos.Terr.Phys.*, **3**:141–152, 1953.

Belrose, J. S., Electron density measurements in the D-region by the method of partial reflection. In: *Electron Density Distribution in Ionosphere and Exosphere* (E. Thrane, Editor). North-Holland, Amsterdam, pp. 41–49, 1964.

Belrose, J. S. and Burke, M. J., Study of the lower ionosphere using partial reflection. *J.Geophys.Res.*, **69**:2799–2818, 1964.

Benkova, N. P., Principles results of U.S.S.R. ionospheric absorption measurements in the I.G.Y.–I.G.C. using the pulse method. *J.Atmos.Terr.Phys.*, **27**:47–52, 1965.

Beynon, W. J. G. and Jones, E. S. O., Meteorological influences in ionospheric absorption measurements. *Proc. R. Soc.,Ser.* A, **288**:558–563, 1965.

Beynon, W. J. G. and Jones, E. S. O., Some medium latitude radio wave absorption studies. *J. Atmos. Terr.Phys.*, **27**:761–773, 1965.

Bibl, K., Paul, A. K. and Rawer, K., Die Frequenzabhängigkeit der ionosphärischen Absorption. *J.Atmos.Terr.Phys.*, **16**:324–339, 1959.

Bibl, K., Paul, A. K. and Rawer, K., Some results of absorption measurements at Freiburg (Germany). *J.Atmos. Terr.Phys.*, **27**:145–154, 1965.

Bossolasco, M. and Elena, A., Absorption de la couche D et température de la mésosphère. *C.R. Acad.Sci.Paris*, **256**:4491–4493, 1963.

Bracewell, R. N. and Bain, W. C., An explanation of radio propagation of 16 Kc/sec in terms of two layers below E layer. *J.Atmos.Terr.Phys.*, **2**:216–225, 1952.

Bracewell, R. N., Budden, K. G., Ratcliffe, J. A., Straker, T. W. and Weekes, K., The ionospheric propagation of low and very-low frequency radio waves over distances less than 1000 km. *Proc.Inst.Electr.Eng.,Lond.,Part* III, **98**:221–236, 1951.

Brown, J. N. and Watts, J. M., Ionosphere observations at 50 Kc. *J.Geophys.Res.*, **55**:179–181, 1950.

Davies, K. A., A study of 2 mc/s ionospheric absorption measurements at high latitudes. *J.Atmos.Terr. Phys.*, **23**:155–169, 1962.

Deeks, D. G., D-region electron distributions in middle latitudes deduced from the reflection of long radio waves. *Proc.R.Soc., Lond. Ser.* A **291**:413–437, 1966.

Delobeau, F. and Suchy, K., L'absorption ionosphéric à Dakar. *J.Atmos.Terr.Phys.*, **9**:45–50, 1956.

Dieminger, W., On the causes of excessive absorption in the ionosphere of winter days. *J.Atmos.Terr. Phys.*, **2**:340–349, 1952.

Dieminger, W., Structure of the D-region from partial radio reflection observations. In: *Wind and Turbulence in Stratosphere, Mesosphere and Ionosphere* (K. Rawer, Editor). North-Holland, Amsterdam, 178–200, 1967.

Findlay, J. W., Moving clouds of ionization in region E of the ionosphere. *J.Atmos.Terr.Phys.*, **3**:73–78, 1953.

Gardner, F. F. and Pawsey, J. L., Study of the ionospheric D-region using partial reflection. *J.Atmos.Terr.Phys.*, **3**:321–344, 1953.

Gnalalingam, S. and Weekes, K., D-region echoes observed with a radio wave of frequency 1.4 Mc/s. In: *The Physics of the Ionosphere*. The Physical Society, London, pp. 63–70, 1955.

Gregory, J. B., The relation of forward scattering of very high frequency radio waves to partial reflection of medium frequency waves at vertical incidence. *J.Geophys. Res.*, **62**:383–388, 1957.

Helliwell, R. A., Ionospheric virtual height measurement at 100 Kilocylces. *Proc.Inst.Radio Eng. N.Y.*, **37**:887–889, 1949.

Jackson, J. E. and Kane, J. A., Measurement of ionospheric electron densities using an RF probe technique. *J.Geophys.Res.*, **64**:1074–1075, 1959.

Landmark, B. and Lied, F., Observations of the D-region from a study of ionospheric cross modulation. *J.Atmos.Terr.Phys.*, **23**:92–100, 1962.

Lange-Hesse, G., Radio-Polarlicht (Radio-Aurora). *Z.Geophys.*, **34**:323–352, 1968.

Lauter, E. A. and Bossy, L., A survey of A3 - absorption measurements at low and medium frequencies. *Ann.Géophys.*, **22**:289–299, 1966.

Naismith, R. and Bramley, E. N., Time delay measurements on radio transmissions. *Wireless Eng.*, **28**:271–277, 1951.

Piggott, W. R. and Thrane, E. V., The effect of irregularities in collision frequency on the amplitude of weak partial reflections. *J.Atmos.Terr.Phys.*, **28**:311–314, 1966.

Piggott, W. R. and Thrane, E. V., The electron densities in the E and D-regions above Kjeller. *J.Atmos.Terr.Phys.*, **28**:467–479, 1966.

Rao, M. K., Mazumdar, S. C. and Mitra, S. N., Investigation of ionospheric absorption at Delhi. *J.Atmos.Terr.Phys.*, **24**:245–256, 1962.

Schwentek, H., Bestimmung eines Kennwertes für die Absorption der Ionosphäre aus einer automatisch-statistischen Analyse von Feldstärkeregistrierungen, *Archiv.Elektr.Übertragung*, **12**:301–308, 1958.

Schwentek, H., Über die Eigenschaften regulärer Tagesgänge der Absorption in der Ionosphäre. *Nachrichtentechn.Z.*, **18**:94–98, 1965.

Seddon, J. C., Differential absorption in the D and lower-E regions. *J.Geophys.Res.*, **63**:209–216, 1958.

Thomas, L., The winter anomaly in ionospheric absorption. *J.Atmos.Terr.Phys.*, **23**:301–317, 1962.

Thrane, E. V. and Piggott, W. R., The collision frequency in the E and D-regions of the ionosphere. *J.Atmos.Terr.Phys.*, **28**:721–737, 1966.

Volland, H., Diurnal phase variation of VLF waves at medium distances, *Radio Sci.*, **68D**:225–238, 1964.

Watts, J. M. and Brown, J. N., Some results of sweep-frequency investigation in the low frequency band. *J.Geophys.Res.*, **59**:71–86, 1954.

*The normal E layer*

Beynon, W. J. G. and Brown, G. M., Geomagnetic distortion of region E. *J.Atmos.Terr.Phys.*, **14**:138–166, 1959.

Bibl, K., L'ionization de la couche E, sa mesure et sa relation avec les éruptions solaires. *Ann.Géophys.*, **7**:208–214, 1951.

Eyfrig, R., Kritische Bemerkungen zur Beschreinbung der Ionization der ionosphärischen E-Schicht. *Geofis. Pura e Appl.*, **45**:179–184, 1960.

Eyfrig, R., Non seasonal variation in the E-layer ionization. *Nature*, **196**:758–760, 1962.

Friedman, H., The sun's ionizing radiations. In: *Physics of the Upper Atmosphere* (J. A. Ratcliffe, Editor). Academic Press, London, pp. 133–218, 1960.

Harnischmacher, E., L'influence solaire sur la couche E normale de l'ionosphère. *C.R.Acad. Sci. Paris*, **230**:1302, 1950.

Hasegawa, M., On the position of the focus of the geomagnetic $S_q$ current system. *J.Geophys. Res.*, **65**:1437–1447, 1960.

Kundu, M. R., Solar radio emission on centimeter waves and ionization of the E-layer of the ionosphere. *J.Geophys.Res.*, **65**:3903–3907, 1960.

Minnis, C. M. and Bazzard, G. H., Some indices of solar activity based on ionospheric and radio noise measurements. *J.Atmos.Terr.Phys.*, **14**:213–228, 1959.

Naismith, R., Bevan, H. C. and Smith, P. A., A long term variation in the relationship of sunspot numbers to E-region character figures. *J.Atmos.Terr.Phys.*, **21**:167–173, 1961.

Pichler, H., Studie zur Ionization der E-Schicht. *Geofis. Pura e Appl.*, **33**:146–152, 1956.

Rawer, K. and Argence, E., Considerations critiques relatives à la formation de la région E de l'ionosphère. *Ann.Géophys.*, **9**:1–25, 1953.

Rawer, K. and Argence, E., Origin of the ionospheric E-layer. *Phys. Rev.*, **94**:253–256, 1954.

Robinson, B. J., Experimental investigations of the ionospheric E-layer. *Rep.Progr.Phys.*, **22**:241–279, 1959.

Scott, J. C. W., The solar control of the E and F1 layers at high latitudes. *J.Geophys.Res.*, **57**:369–386, 1952.

Shimazaki, T., World-wide measurements of horizontal ionospheric drifts. *Ionosph.Space Res. in Japan*, **7**:95–109, 1960.

## The sporadic E layer

Axford, W. I., Note on a mechanism for the vertical transport of ionization of the ionosphere. *Can.J.Phys.*, **39**:1393–1396, 1961.

Axford, W. I., The formation and vertical movement of dense ionized layers in the ionosphere due to neutral wind shears. *J.Geophys.Res.*, **68**:769–779, 1963.

Bibl, K., Sur le mécanisme d'ionization de la couche sporadic Es de l'ionosphère. *Ann.Géophys.*, **16**:148–151, 1959.

Bibl, K., Dynamic characteristics of the ionosphere and their coherency with the local and planetary magnetic index. *J.Geophys.Res.*, **65**:2333–2342, 1960.

Bowles, K. L. and Cohen, R., A study of radio wave scattering from sporadic E near the magnetic equator. In: *Ionospheric Sporadic E* (E. Smith and S. Matsushita, Editors). Pergamon, London, pp. 51–77, 1962.

Chadwick, W. B., Variations in frequency of occurrence of sporadic E, 1949–1959. In: *Ionospheric Sporadic E* (E. Smith and S. Matsushita, Editors). Pergamon, London, pp. 182–193, 1962.

Dueno, B., Sporadic E as observed from Mayaguez, P. R. by backscatter sounders. In: *Ionospheric Sporadic E* (E. Smith and S. Matsushita, Editors). Pergamon, London, pp. 110–112, 1962.

Dungey, J. W., The influence of the geomagnetic field on turbulence in the ionosphere. *J.Atmos.Terr. Phys.*, **8**:39–42, 1956.

Dungey, J. W., Effect of the magnetic field on turbulence in an ionized gas. *J.Geophys.Res.*, **64**:2188–2191, 1959.

Egan, R. D. and Peterson, A. M., Backscatter observations of sporadic E. In: *Ionospheric Sporadic E* (E. Smith and S. Matsushita, Editors). Pergamon, London, pp. 89–109, 1962.

Hines, C. O., The formation of midlatitude sporadic E layers. *J.Geophys.Res.*, **69**:1018–1019, 1964.

Heissler, L. H. and Whitehead, J. D., Rapid variations in the sporadic E-region. *J.Atmos.Terr.Phys.*, **24**:753–764, 1962.

Istomin, V. C., Ions of extra-terrestrial origin in the earth ionosphere. In: *Space Research III* (W. Priester, Editor). North-Holland, Amsterdam, pp. 209–220, 1963.

Knecht, R. W. and McDuffie, R. E., On the width of the equatorial Es belt. In: *Ionospheric Sporadic E* (E. Smith and S. Matsushita, Editors). Pergamon, London, 214–218, 1962.

Leighton, H. I., Shapley, A. H. and Smith, E. K., The occurrence of sporadic E during the IGY. In: *Ionospheric Sporadic E* (E. Smith and S. Matsushita, Editors). Pergamon, London, 166–177, 1962.

McNicol, R. W. and Gipps, G. de V., Characteristics of the Es-region at Brisbane, *J.Geophys.Res.*, 56:17–31, 1951.

Matsushita, S., Ionospheric variations associated with geomagnetic disturbances, I: Variations at moderate latitudes and the equatorial zone, and the current system for the $S_D$ field. *J.Geomagn.Geolectr.*, 5:109–135, 1953.

Matsushita, S., Lunar tidal variations in the sporadic E-region. *Ionosph.Res. Japan*, 7:45–52, 1953.

Matsushita, S., Sequential Es and lunar effects on the equatorial Es. *J.Geomagn.Geoelect.*, 7:91–95, 1955.

Narcisi, R. S. and Bailey, A. D., Mass spectrometric measurements of positive ions at altitudes from 64 to 112 km. In: *Space Research V* (D. G. King-Hele, P. Muller and G. Righini, Editors). North-Holland, Amsterdam, pp. 753–754, 1965.

Oksman, J., Über den s-Typ der sporadischen E-Schicht in Sodankylä. *Klienheubacher Ber.*, 10:131–134, 1965.

Penndorf, R. and Coroniti, S. C., Polar Es. *J.Geophys. Res.*, 63:709–802, 1958.

Rawer, K., Unregelmässigkeit und Regelmässigkeit der sporadischen E-Schicht. *Geofis. Pura e Appl.*, 32:170–244, 1955.

Rawer, K., Correlation of Es-characteristics in time and space. *AGARDograph 34*:67–80, 1958.

Rawer, K., Structure of Es at temperate latitudes. In: *Ionospheric Sporadic E* (E. Smith and S. Matsushita, Editors). Pergamon, London, pp. 292–343, 1962.

Seddon, J. C., Propagation measurements in the ionosphere with the aid of rockets. *J.Geophys.Res.*, 58:323–335, 1953.

Seddon, J. C., Electron densities in the ionosphere. *J.Geophys.Res.*, 59:463–466, 1954.

Seddon, J. C., Pickar, A. D. and Jackson, J. E., Continuous electron density measurements up to 200 km. *J.Geophys.Res.*, 59:513–524, 1954.

Skinner, N. J. and Wright, R. W., The reflection coefficient and fading characteristics of signals returned from the Es layer at Ibadan. In: *Ionospheric Sporadic E* (E. Smith and S. Matsushita, Editors). Pergamon, London, pp. 37–49, 1962.

Thomas, J. A., Report on recent Es work in Brisbane. In: *Ionospheric Sporadic E* (E. Smith and S. Matsushita, Editors). Pergamon, London, 123–130, 1962.

Tsuda, T., Sato, T. and Maeda, K., Formation of sporadic E layers at temperate latitudes due to vertical gradients of charge density. *Radio Sci.*, 1:212–225, 1966.

Whitehead, J. D., The formation of the sporadic E-layer in the temperate zones. *J.Atmos.Terr.Phys.*, 20:49–58, 1961.

Wright, J. W., Murphy, C. H. and Bull, G. V., Sporadic E and the wind structure of the E-region. *J.Geophys.Res.*, 72:1443–1460, 1967.

# TECHNIQUES FOR THE MEASUREMENTS OF THE ELECTRON DENSITY AND TEMPERATURE IN THE D REGION

GIAN CARLO RUMI

*Istituto Elettrotecnico Nazionale Galileo Ferraris, Turin (Italy)*

## SUMMARY

Two techniques for the measurements of the electron density profile and the electron temperature profile in the D region of the ionosphere, one based on cross modulation and the other on a peculiar anomaly of LF propagation, are presented and discussed. Their theoretical background is given and some pertinent experimental results are illustrated.

In this context, specific reference is made to three experimental operations: Experiment Luxembourg that was developed at the University of Alaska in the late fifties; Experiment LIMBO that was carried on at Cornell University, in Ithaca, New York during the late sixties; and some experimental findings in the LF link between Rugby and Turin that were obtained at the Istituto Elettrotecnico Nazionale Galileo Ferraris during the first months of 1970.

The conclusion derived from these experimental operations is that the cross-modulation technique is much more complex than the LF technique, but the latter yields either the electron density profile or the electron temperature profile and is operative only between 60 and 70 km of altitude, while the former can operate fully between 40 and 70 km.

## 1. FOREWORD

The determination of $N(h)$ (i.e., the electron density profile) in the D region of the ionosphere is a difficult task. While $N(h)$ in the E region can be obtained without much difficulty through the use of a standard ionospheric sounder, much more complex techniques are required for the D region. There, the plasma (angular) frequency is of the same order of magnitude as the electron collision frequency. It follows that the refractive index $M$ assumes a complex form, namely (in mks units):

$$M^2 = (\mu - j\chi)^2 = 1 - \frac{X}{1 - jZ} \tag{1}$$

where $\mu$ and $\chi$ are its real and imaginary parts; $X^{1/2} = f_p/f$; $Z = \nu/2\pi f$; $f_p$ is the plasma frequency, $f$ the operating frequency, $\nu$ the electron collision frequency, and the geomagnetic field has been neglected.

Equation 1 reduces to:

$$M^2 = 1 - X \tag{2}$$

in the region E where indeed $Z \ll 1$. The operating frequencies taken into consideration are not much higher than the plasma frequencies; otherwise, it would not make sense to speak of an ionosphere. A standard sounder makes use of eq. 2 and detects the (virtual) height where $M = 0$, that is $X = 1$, and in turn $f = f_p$. Since $f$ is a known quantity and $f_p = 9\sqrt{N}$, $N$ can be determined at that particular (virtual) height. On the other hand, there are no sounders that—making use of eq. 1—can probe the D region of the ionosphere.

Nevertheless, the D region can still be successfully probed by radio wave techniques; namely, partial reflection in the upper D region and cross modulation in the lower D region.

The partial reflection technique operates on the eddies of ionization that are produced by turbulence. The cross-modulation approach is based upon the interaction between two waves of different frequencies that interact in the D region on account of its non-linearity. Unfortunately, such measurements are not very efficient, in spite of all the care with which they may be carried on.

Other methods of investigation have been used in the D region: i.e., balloons and rockets. But balloons are efficient up to 30 km only, and rockets are a source of great perturbations in the region that they explore. Furthermore, operations with rockets and balloons are not a daily affair.

There is also a new and rather simple technique that leads to the determination of $N(h)$ in the D region at least during a disturbance: it is based on LF propagation and will be illustrated in a following section.

For the measurement of the electron temperature in the lowest part of the ionosphere, considerations of the same kind can be made.

As this is not a review, I will not present a detailed analysis of all the techniques that can be used for the measurements of the electron density and temperature in the D region. Instead I will concentrate on the cross-modulation and on the LF approach, for the simple reason that I have some familiarity with such techniques. A well-known (in Italy) Italian writer would say that everybody, both old and young, likes to speak of the things he is familiar with, and not only of those. Section 2 of this presentation will deal with Experiment Luxembourg and Experiment LIMBO, which exemplify the application of the cross-modulation technique. Section 3 will interpret some records of the LF link between Rugby and Turin in terms of $N(h)$. A few conclusive remarks will be given in section 4.

## 2. THE CROSS-MODULATION TECHNIQUE

### 2.1. Introduction

The technique of cross modulation is an application of the Luxembourg Effect that was first detected in 1933. It was then noticed that the Beromünster radio program was received at Eindhoven together with the Luxembourg program which was broadcasted on a completely different frequency. A kind of a radio smuggling, one might say.

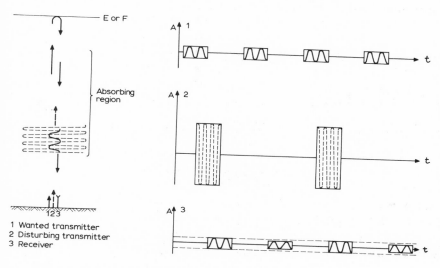

Fig. 1. Scheme illustrating the technique of the cross-modulation experiment.

The Luxembourg Effect was interpreted as being the result of the non-linearity of the medium through which both programs were propagating. The technique that was derived from this effect can be described in a few words with the help of Fig. 1. Two pulses of different radio frequencies propagate vertically in the D region, one on its way up to the ionosphere and the other on its way back from the ionosphere. If one is a short pulse from a high-power transmitter (the disturbing transmitter), it affects the collision frequency in a thin moving slab of the D region. This changes the attenuation experienced by the other pulse, which is from a low-power transmitter (the wanted transmitter). If the low-power transmitter is pulsed at twice the rate of the high-power transmitter, alternate pulses from the low-power transmitter will be recorded with slightly different intensities because of the change in attenuation produced by the high-power transmitter. The difference in amplitude for successive pulses at the receiver therefore depends on both the electron density $N$ and the electron collision frequency $\nu$ (i.e., on the temperature) at the level where the pulses from the weak and strong transmitter pass each other. This level can be determined by the relevant delay time. From the amount of the detected cross modulation, $N$ and $\nu$ can be deduced as a function of the height.

This classical Luxembourg interaction applies independently to the ordinary and to the extraordinary component of the wanted wave, if they are perfectly separated by the use of circularly polarized antennas. If perfect separation is not achieved or if linearly polarized antennas are used in reception, then in addition to the Luxembourg interaction, one observes a cross modulation of the Faraday rotation of the wanted echo. In this context the geomagnetic field cannot be neglected, but only the quasi-longitudinal propagation is considered.

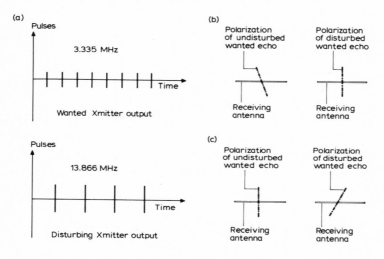

Fig. 2. Illustration of the mechanism of cross modulation of Faraday rotation.

Fig. 2 illustrates two specific cases which give a simplified explanation of the mechanism of the cross modulation of the Faraday rotation. The plane of polarization of the wanted wave is assumed to make an angle of almost 90° with the polarization plane of the receiving antenna. Let us first consider the case illustrated in Fig. 2b. When the wanted wave is not affected by the disturbing wave, the receiving antenna collects a signal. When the wanted wave is affected by the disturbing wave on every second pulse (as shown in Fig. 2a), no signal is collected. The disturbing transmitter changes the collision frequency of the electrons at the height of interaction; therefore it changes the propagation velocity of both components of the wanted wave. The change in velocity for the ordinary component is different from that of the extraordinary component, therefore the polarization plane of the wanted wave rotates with respect to what it would be in the undisturbed conditions. In time, the receiver will receive a succession of pulses having the same repetition rate as the disturbing transmitter. Since the synchronous detection system that analyzes the output of the receiver is tuned just on the repetition frequency of the disturbing transmitter, the signal modulation will be recorded. The same reasoning applies to the case of Fig. 2c, and there is an equal likelihood of positive or negative Faraday cross modulation at any given height.

In connection with this technique, two specific experiments will be considered: Experiment Luxembourg and Experiment LIMBO. Experiment Luxembourg was put into operation at College, Alaska during 1960. The parameters that characterize Experiment Luxembourg are rather different from the ones that have characterized other cross-modulation experiments. The specifications for the transmitters and for the receiver are listed in Table I.

TABLE I

Specifications for transmitter and receiver

| | |
|---|---|
| **Disturbing transmitter** | |
| Peak power | 200 kW (at time of operation 100 kW) |
| Frequency | 17.5 MHz |
| Pulse length | 50 $\mu$sec |
| Repetition rate | 37.5 cps |
| Antenna | 16 (three-element) Yagi array |
| **Wanted transmitter** | |
| Peak power | 10 kW |
| Frequency | 4.86 MHz |
| Pulse length | 50 $\mu$sec |
| Repetition rate | 75 cps |
| Antenna | 4 horizontal half-wave dipoles forming the sides of a square and producing circular polarization. |
| **Receiver** | |
| Bandwidth | 50 kHz |
| Antenna | 4 horizontal half-wave dipoles forming the sides of a square and producing circular polarization. |

Experiment LIMBO was put into operation at Ithaca, New York in the late sixties. The major items of the equipment used had the specifications shown in Table II.

*2.2. Theoretical compendium*

The following analysis should take into account the geomagnetic field; but only the quasi-longitudinal propagation will be considered. Therefore, any linearly polarized wave will be resolved into a right hand circularly polarized wave (+) and

TABLE II

Specifications of equipment used in Experiment LIMBO

| | | | |
|---|---|---|---|
| **Disturbing Transmitter** | | Wanted antenna | linear polarization in |
| Peak power | 500 kW | Inverted V | the NE–SW direction |
| Frequency | 13.866 MHz | | (Note that this |
| Pulse length | 50 μsec | | was also used as a |
| Repetition rate | 37.5 cps | | receiving antenna) |
| | | **Receivers** | |
| **Disturbing antenna** | | Bandwidth | 30 kHz |
| 16 (three-element) | linear polarization in | | |
| Yagi array | the geomagnetic | | |
| | east-west direction | | |
| Gain | 20 db | Receiving antenna | |
| | | Two inverted V's | cross polarized in the |
| **Wanted Transmitter** | | | SE–NW and SW–NE |
| Peak power | 30 kW | | directions one of which |
| Frequency | 3.335 MHz | | also served as the |
| Pulse length | 50 μsec | | wanted antenna in |
| Repetition rate | 75 cps | | transmission |
| | | **Recorder** | |
| | | Mod.4784, seven channels, | |
| | | Sangamo magnetic tape recorder | |

into a left hand circularly polarized wave $(-)$. Specifically, reference is made to a thin slab. Its output is:

$$E_+ = \frac{E_1}{2}(\vec{a}_x - j\vec{a}_y)\exp\left(jpt - \frac{jp\mu_+\Delta z}{c} - \frac{p\chi_+\Delta z}{c}\right) \qquad (3)$$

and:

$$E_- = \frac{E_1}{2}(\vec{a}_x + j\vec{a}_y)\exp\left(jpt - \frac{jp\mu_-\Delta z}{c} - \frac{p\chi_-\Delta z}{c}\right) \qquad (4)$$

where: $E_1$ = amplitude of linearly polarized wave at the input; $\vec{a}_x$, $\vec{a}_y$ = unit vectors respectively in the x and y directions; $\Delta z$ = small step in the z direction; $p$ = operating angular frequency; $c$ = velocity of light; $\mu_+$, $\mu_-$ = real part of the refractive index; and $\chi_+$, $\chi_-$ = imaginary part of the refractive index.

Let us first consider $E_+$ alone, and let $p\chi_+/c = K_+ \equiv$ absorption coefficient. This quantity is a known function of $\nu$ and $N$. Any perturbation of the medium in the slab $\Delta z$, that is characterized by a specific value of $\nu$ and $N$, will be accompanied by a perturbation of $E_+$. A rhythmic perturbation is indeed produced by the disturbing transmitter and is then detected by the receiver of the wanted wave.

According to Crompton et al. (1953), $\nu \propto T$, the electron temperature. The energy of the disturbing transmitter that is dissipated in the layer $\Delta z$ increases $T$, hence $\nu$.

There is also a secondary effect of the disturbing wave that becomes significant under special conditions at very low heights and consists in a change of $N$. This change is related to the dependence of $N$ on the temperature via the attachment coefficient (Chanin et al., 1959).

The point to be emphasized is that, given a disturbing power and the characteristics of the medium in the slab $\Delta z$ in terms of $\nu$ and $N$, it is possible to compute the change in $E_+$ that is produced by the disturbing power. Conversely, a measurement of the change in $E_+$ that is produced by a given disturbing power can lead to the determination of $\nu$ and $N$. This reasoning concerns the Luxembourg interaction and applies individually to the right hand and left hand component of a linearly polarized wave.

For an analysis of the Faraday interaction, $E_+$ and $E_-$ must be considered together. From this point of view the disturbing power produces an increase in the electron temperature in the layer of thickness $\Delta z$. This in turn changes $\nu$ (and eventually $N$), and ultimately $\mu_+$ and $\mu_-$. But the perturbation in $\mu_+$ is different from the perturbation in $\mu_-$. The net result is a rotation $\delta_\nu \theta$ of the polarization plane of the wanted wave at the receiving antenna; hence, a change in the intensity of the received wanted signal.

Even in this case it is emphasized that, given a disturbing power and the characteristics of the medium in the slab $\Delta z$ in terms of $\nu$ and $N$, it is possible to compute the change in the polarization angle. Conversely, a measurement of $\delta_\nu \theta$ can lead to the determination of $\nu$ and $N$ in the slab where the disturbing power has been dissipated.

In place of the cumbersome formulas that relate the Luxembourg cross modulation ($\delta_\nu E_+/E_+$ or $\delta_\nu E_-/E_-$) and the Faraday cross modulation ($\delta_\nu \theta$) to $\nu$ and $N$, it is better to present the corresponding diagrams of Fig. 3. They give $\delta_\nu E_+/E_+$, $\delta_\nu E_-/E_-$ and $\delta_\nu \theta$ at different heights for $\nu_0(h)$, as is Fig. 4, $\nu_1(h) = 1.5\nu_0(h)$, and $\nu_2(h) = 0.5\nu_0(h)$, when $N = 10^9$ el m$^{-3}$, the disturbing power equals 440 kW, the

TABLE III

Temperature $T$ versus height $h$

| $h$ (km) | $T(^\circ\mathrm{K})$ | $h$ (km) | $T(^\circ\mathrm{K})$ | $h$ (km) | $T(^\circ\mathrm{K})$ |
|---|---|---|---|---|---|
| 40 | 261 | 54 | 280 | 68 | 218 |
| 42 | 267 | 56 | 271 | 70 | 210 |
| 44 | 273 | 58 | 263 | 72 | 201 |
| 46 | 279 | 60 | 254 | 74 | 192 |
| 48 | 283 | 62 | 245 | 76 | 183 |
| 50 | 283 | 64 | 236 | 78 | 174 |
| 52 | 283 | 66 | 227 | 80 | 166 |

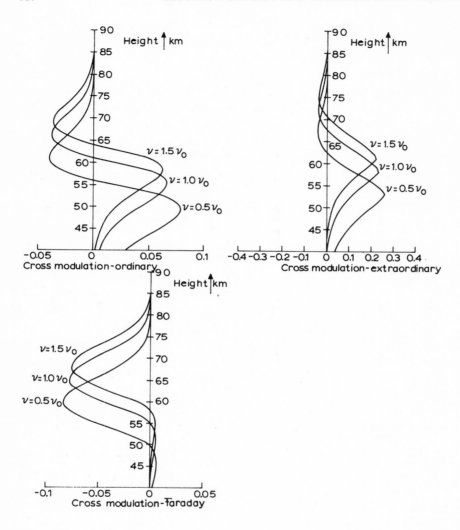

Fig. 3. Computed profiles of cross modulation versus height.

disturbing antenna gain equals 100, $\Delta z = 7.5$ km, the integration time equals 50 $\mu$ sec or $(10^3/\nu)$—whichever is shorter—and the neutral gas temperature is taken as in Table III. It is important to note that all the coefficients of cross modulation are proportional to $N$, to the disturbing power, to the gain of the disturbing antenna, to $\Delta z$ and to the integration time, so that the diagrams of Fig. 3 can be applied to widely different experimental arrangements. The coefficients of cross modulation presented in Fig. 3 were computed without taking into account possible changes of $N$ that may be produced by the heating of the medium.

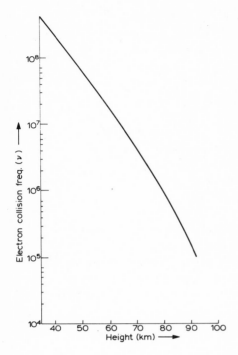

Fig. 4. Standard profile of $\nu$ (electron collision frequency) versus height.

## 2.3. Experimental results

For the reliability of the experimentation it is necessary to discriminate between the two circularly polarized components of the wanted wave when the Luxembourg cross modulation is recorded, and to single out the linearly polarized wave that is the sum of the two, when the Faraday cross modulation is observed. The task is not too easy from an experimental point of view, because there are so many causes of contamination among the various quantities.

With Experiment Luxembourg only the circularly polarized waves were considered. In Experiment LIMBO the schedule of operations consisted of 4 successive intervals of 4 min each; precisely:

Interval 1.          Reception on linearly polarized antenna
                     Wanted transmitter = On
                     Disturbing Transmitter = On

Interval 2.                    Reception on linearly polarized antenna
                               Wanted transmitter = Off
                               Disturbing Transmitter = On

Interval 3.                    Reception on circularly polarized antenna
                               Wanted transmitter = On
                               Disturbing Transmitter = On

Interval 4.                    Reception on circularly polarized antenna
                               Wanted transmitter = On
                               Disturbing Transmitter = Off

The results of Experiment Luxembourg indicated the existence of about 4 · $10^7$ el/m$^3$ at an altitude of 48 km under quiet conditions in the summer of 1960 over central Alaska. The electron collision frequency there was found to be equal to $1.8 \cdot 10^7$ sec$^{-1}$ near the height of 60 km (Rumi, 1961). Some of the records of this experiment suggested that above Alaska the electrons at 40 km may have reached a temperature of almost 1200°K. Such a conclusion sounds quite surprising and probably not convincing. Maybe intercloud discharges were the source of the mysterious heating process (Rumi, 1962a,b).

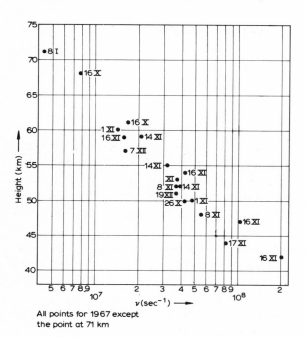

All points for 1967 except
the point at 71 km

Fig. 5. Measured values of $\nu$ (electron collision frequency) versus height obtained on different dates.

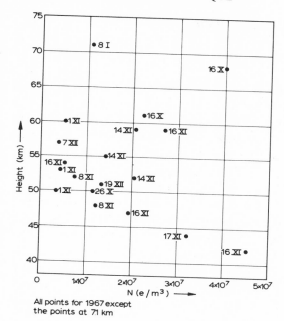

Fig. 6. Measured values of $N$ (electron density) versus height obtained on different dates.

The best results of Experiment LIMBO concerning cross modulation are presented in Fig. 5 and 6. Fig. 5 gives measured values of $\nu$ versus height obtained on different dates, and Fig. 6 gives measured values of $N$ versus height obtained on different dates also. Each point in these figures was derived from the analysis of 16 independent experimental data. Their reliability seems to be limited only by the validity of the theoretical formulas used for the determination of $\nu$ and $N$.

The data, besides supplying a measurement of $\nu$ and $N$, raise the question of why cross modulation was detected almost every day in October and November 1967, but was practically absent in December 1967 and January 1968. LF radio measurements by Doherty (1968) showed the importance of associative detachment and dissociative attachment in the subionosphere. According to Doherty, these processes directly control the neutral ozone and atomic oxygen concentration and indirectly, the electron density, $N$ tends to decrease as the concentration of $O_3$ increases and the concentration of O decreases. Moreover, Doherty (1968) states: "Below 65 km, the mesosphere is cooler in winter than in summer, suggesting that the atomic oxygen concentration is reduced. The rate of associative detachment would be correspondingly reduced, and lower electron concentration would be expected." The trend shown by the data of Experiment LIMBO at the end of 1967 and the beginning of 1968 appears to agree with Doherty's findings, since both Luxembourg and Faraday cross modulation are proportional to the electron density.

## 3. THE LF TECHNIQUE

### 3.1. Introduction

The LF technique for the measurement of $N$ and $\nu$ is based on the peculiarity of LF propagation through the D region. The phenomenon can be described in terms of the refractive index $M$. A complete analysis of the behavior of $M$ in the D region would require consideration of the geomagnetic field, together with the electron collision frequency and the electron density. For the sake of simplicity the geomagnetic field will be neglected in the following discussion. Therefore, the final conclusions will be valid only for the type of propagation that is independent of that field. It is the case of ordinary, quasi-transverse propagation that is characterized by having the electric field of the wave quasi-parallel to the geomagnetic field. Then the Lorentz force on the free electrons that are compelled to vibrate along the geomagnetic field is negligible, and the latter is quite ineffective.

The analytical expression of the refractive index for ordinary, quasi-transverse propagation is given by eq. 1. The behavior of $\mu$ and $\chi$ shall be studied as a function of $N$ and $\nu$ for frequencies in the LF band that are strongly affected by the D region, where $N = 10^6 - 10^{10}$ el m$^{-3}$ and $\nu = 2 \cdot 10^5 - 2 \cdot 10^8$ sec$^{-1}$. Fig. 7 shows $\mu(N)|_\nu$ and Fig. 8 shows $\chi(N)|_\nu$ for $f = 60$ kHz. They also can be seen to represent respectively $\mu(N)|_h$ and $\chi(N)|_h$ being $\nu = \nu(h)$, as shown in Fig. 4.

The trend of $\mu$ for $N \simeq 5 \cdot 10^8$ el m$^{-3}$ is worth considering. It is characterized by a value that is very close to unity as in free space, and by a very small gradient with respect to the height. At the same time $\chi \simeq 0$. Therefore, when $N(h)$ has a very steep profile and $N \simeq 5 \cdot 10^8$ el m$^{-3}$, a 60 kHz wave impinging on the D region from the bottom side will hardly be reflected in that region. More generally, if $N(h)$ changes from $N(h) < 5 \cdot 10^8$ el m$^{-3}$ to $N(h) > 5 \cdot 10^8$ el m$^{-3}$, a 60 kHz signal that impinges on the D region from the bottom side will be reflected rather strongly at first, then very weakly, and finally strongly again.

The upshot of these considerations is the following. When an LF signal intensity is recorded at a distance of about 1000 km from the transmitting station (the link being oriented in the magnetic north–magnetic south direction, and located at relatively high latitudes, so that the mode of propagation is ordinary, quasi-transverse), every time the records indicate a quiet behavior that is interrupted by a deep crevass, it can be argued that the electron density profile has gone through the point $(N,h)$ defined by that crossing of the curves $\mu(N)|_\nu$ which was called "chiasmic spot" (Rumi, 1970 a). If more than one frequency is used, more than one point in the electron density profile can be determined. For the validity of this interpretation, such a profile must be quite steep, since this hypothesis is implicit in the development above. According to the available information, this condition often seems to be met in the lower part of the D region (Deeks, 1966).

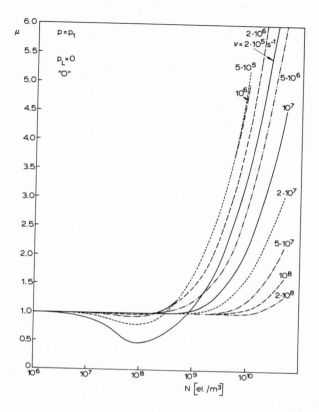

Fig. 7. The real part $\mu$ of the refractive index versus the electron density $N$ for several values of the collision frequency $v$. Case of ordinary, transverse propagation for $p = 2\pi \cdot 60 \cdot 10^3$ rad/sec.

## 3.2. Theoretical developments

Figures 7 and 8 apply to 60 kHz. Similar diagrams can be derived for different frequencies. It is desirable to find in analytical form the value of $N$ that is related to the conditions $\mu = 1$, $\chi \simeq 0$, and $d\mu/dh|_N = 0$. Equation 1 yields:

$$2\mu^2 = 1 + \frac{\alpha}{\alpha^2 + \beta^2} + \left(1 + \frac{1 + 2\alpha}{\alpha^2 + \beta^2}\right)^{1/2} \qquad (5)$$

and:

$$2\chi^2 = -1 - \frac{\alpha}{\alpha^2 + \beta^2} + \left(1 + \frac{1 + 2\alpha}{\alpha^2 + \beta^2}\right)^{1/2} \qquad (6)$$

where: $\alpha = -mp^2\varepsilon_0/Ne^2$; $\beta = mpv\varepsilon_0/Ne^2$; $m, e$ = mass, charge of an electron; and $\varepsilon_0 = 8.85 \cdot 10^{-12}$ F m$^{-1}$.

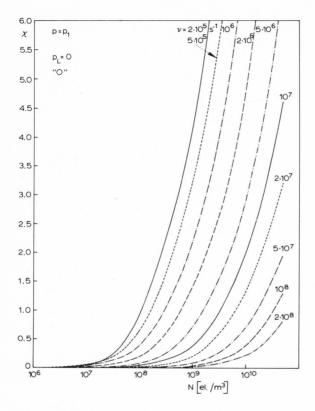

Fig. 8. The imaginary part $\chi$ of the refractive index versus the electron density $N$ for several values of the collision frequency $v$. Case of ordinary, transverse propagation for $p = 2\pi \cdot 60 \cdot 10^3$ rad/sec.

According to eq. 5 the condition $d\mu/dh|_N = 0$ can be written as:

$$\beta^2 = \frac{4\alpha^3 + 3\alpha^2}{4\alpha + 1} \tag{7}$$

Furthermore, according to eq. 5, $\mu = 1$ when:

$$\beta^2 = \frac{-4\alpha^3}{4\alpha + 1} \tag{8}$$

Besides $\alpha = \beta = 0$, a solution of the system of eq. 7 and 8 is: $\alpha = -3/8$ and $\beta^2 = -27/128$. This solution indicates that there is no real ionospheric height where the two conditions $d\mu/dh|_N = 0$ and $\mu = 1$ are simultaneously verified. In place of $\mu = 1$, the condition $\mu = (1 + \varepsilon)$ shall be considered, where $\varepsilon$ is a small quantity with respect to 1. Then eq. 8 is substituted by:

$$\beta^2 \xi = -\xi\alpha^2 + 2\alpha + 4\varepsilon\alpha + 0.5$$

$$\pm (4\alpha^2 + 16\varepsilon^2\alpha^2 + 0.25 - \xi\alpha^2 + 16\varepsilon\alpha^2 + 2\alpha + 4\varepsilon\alpha)^{1/2} \qquad (9)$$

where $\xi \equiv (16\varepsilon^2 + 8\varepsilon)$. In deriving eq. 9, $\varepsilon^2$ has been neglected in comparison to $2\varepsilon$. By combining eq. 7 and 9 the following equation is obtained:

$$64\xi^2\alpha^4 + 16\xi^2\alpha^2 + 64\xi^2\alpha^3 - 128\xi\alpha^3 - 256\varepsilon\xi\alpha^3 - 64\varepsilon\xi\alpha^2 - 112\xi\alpha^2 - 128\xi\varepsilon\alpha^2$$

$$- 32\xi\varepsilon\alpha - 32\xi\alpha - 3\xi = 0 \qquad (10)$$

If $\varepsilon = 0.2$ and, consequently, $\xi = 2.24$, eq. 10 yields:

$$+0.110 < -\alpha < +0.120 \qquad (11)$$

Substituting the value of $\alpha = -0.115$ into eq. 7 yields:

$$\beta = 0.249 \qquad (12)$$

Finally, $\alpha = -0.115$ corresponds to:

$$N = 0.107 f^2 \qquad (13)$$

and $\beta = 0.249$ corresponds to:

$$\nu = 13.60 f \qquad (14)$$

The height $h$ is associated to $\nu$ according to the curve of Fig. 4. It is interesting to check that, for $f = 60$ kHz, $N = 3.85 \cdot 10^8$ el m$^{-3}$ and $\nu = 8.16 \cdot 10^5$ sec$^{-1}$, in excellent agreement with the values of $N$ and $\nu$ at the chiasmic spot.

### 3.3. Experimental evidence

In the early part of 1970, fifteen phase anomalies that were interpreted as due to a swelling of the $N(h)$ profile in the D region were recorded at the Istituto Elettrotecnico Nazionale Galileo Ferraris (IENGF) in Turin, Italy (Rumi, 1970 a). The transmitting station, MSF, was located at Rugby, England and was operating on 60 kHz. Signal strength records simultaneous with phase records are available. Two of them are shown in Fig. 9. With the exception of the data obtained on February 28, at 1400 CEMT and on March 7, at 1705 CEMT, the expected crevasses in the behavior of the signal strength were indeed recorded. On the other hand, the two exceptions are not against the theoretical expectations since, according to the phase analysis, the events of February 28 and of March 7 were not associated with a crossing of the chiasmic spot, and therefore should not produce a crevass in the signal strength record.

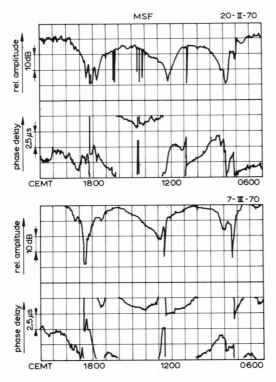

Fig. 9. Examples of amplitude and phase records for MSF.

There is also some other supporting experimental evidence, possibly related to some frequencies lower than 60 kHz, that is worth considering. Already in the early thirties, Namba (1933) made reference to the "sunset phenomenon", that is "a pronounced crevass at the sunset period" in the "recorded signal intensity curve". A sunrise phenomenon of this type was also noticed in the morning hours. According to the preceding discussion, this sunset phenomenon might be associated with a crossing of the chiasmic spot. This explanation is advanced tentatively. A careful analysis of some sets of experimental data would be required before a more definite statement could be made.

In any case, it appears that the explanation of the phenomenon advanced by Namba—namely that "when the angle of incidence of the wave just coincides with the Brewster's angle of the reflecting layer, the reflection ratio and the phase change at reflection vary markedly"—is quite untenable. Indeed, this mechanism of the Brewster angle would work quite well for a uniform layer, but the ionosphere cannot be interpreted as such even if $N$ does not increase significantly with height, since in any case $\nu$ decreases rapidly. In the D region the Brewster angle or better, the pseudo-Brewster angle changes gradually but significantly with height, so that if a wave escapes reflection at a given level, it is definitely reflected at a small fraction of the wavelength above it. On the contrary, the chiasmic spot represents a deep hole in the ionosphere.

Equations 13 and 14 together with Fig. 4 can be used for the determination of $N(h)$ in the D region of the ionosphere at some specific time. It is just enough to observe or record a sharp crevass in the signal strength that is followed at an interval of 40–60 min by a broad crevass. Then it can be stated that the electron density profile at the onset of the disturbance, midway between the transmitter and the receiver, went through the point of $N$ and $v = v(h)$ given by eq. 13, 14, and Fig. 4. One point of the profile for each frequency used can be determined.

But this information is not relevant per se. Rather, it must be used to deduce $N(h)$ before and after the onset of the disturbance. In this context it is important to carefully examine the behavior of the signal strength during the recovery time after the sharp crevass, which can be interpreted as a playback at reduced speed of what happened at the onset of the disturbance. Since the effective recombination coefficient in the region of interest is relatively well-known, and (see Rumi, 1970b):

$$N^- = \left(\frac{q}{\alpha_i + \frac{\alpha_D}{\lambda}}\right)^{1/2} - \frac{2\exp\{-2[q/(\alpha_i + \alpha_D/\lambda)]^{1/2} \cdot (\alpha_i + \alpha_D/\lambda)t\}}{\left(\frac{q}{\alpha_i + \frac{\alpha_D}{\lambda}}\right)^{-1/2} \cdot \exp\{-2[q/(\alpha_i + \alpha_D/\lambda)]^{1/2} \cdot (\alpha_i + \alpha_D/\lambda)t\} + C_1}$$

$$\equiv N\lambda$$

(15)

where: $N^-$ = negative-ion density; $t$ = time; $q$ = ionization rate; $\alpha_i$ = ion-ion recombination coefficient (known quantity); $\alpha_D$ = ion-electron recombination coefficient (known quantity); $\lambda$ = negative-ion density/electron density (known quantity); and $C_1$ = constant for initial conditions, any information that leads to the determination of $q$ and $C_1$ will result in a full knowledge of $N$ before, during and after the disturbance.

When the signal strength during the recovery process reaches its lowest value, the electron density is equal to $0.107f^2$ according to eq. 13. Then the time elapsed from the beginning of the recovery—which develops just after the peak of the disturbance—is almost equal to the interval between the two consecutive minima at the bottom of the two consecutive crevasses in the signal strength records. When the trace of the signal strength, during the recovery process, has just gone through the characteristic broad crevass, the electron density is equal to:

$$N = 0.072f^2$$

(16)

This value has been deduced by putting $\mu = 1$ and $\beta = 0.249$ into eq. 5 and is taken as representative of a boundary of the chiasmic spot in Fig. 7. The corresponding time elapse from the beginning of the recovery is almost equal to the interval between the minimum at the bottom of the sharp crevass that is simultaneous to the very rapid onset of the disturbance, and the end of the following broad crevass in the signal-strength records.

Such two values of $N$ and $t$, introduced into eq. 15, bring about a determination of $q$ and $C_1$. Once these two unknown quantities are determined, the complete history of $N$ from disturbed to quiet conditions is given by eq. 15. By putting $t = 0$, the value of $N_{max}$ at the peak of the disturbance is obtained. By putting, say, $t = 7.200$ sec, the value of $N_q$ under quiet conditions is derived.

This type of analysis applied to some data of the LF link between Rugby and Turin, leads to the conclusion that the electron density at an altitude of about 70 km almost above Paris, France, during the day hours of the first months of 1970, was of the order of $10^8$ el/m$^3$ under quiet conditions, and of the order of $10^9$ el/m$^3$ just after a sudden disturbance.

Up to this point the determination of $N$ has been considered. In the process, the electron collision frequency (i.e., the electron temperature) was taken as known. It is worth pointing out that, if $N$ is actually evaluated with some other technique, the method of measurement outlined above could be used for the determination of $v(h)$.

Finally, the validity and the precision of the determination of $N$ through the discussed procedure must be examined. For the validity it is important to recall the underlying assumptions that were implied in deriving eq. 13 and 14. First of all, the profile $N(h)$ had to be quite steep in the region of interest. But, if on the contrary this profile varies gradually with the height, the predicted crevasses in the signal strength will not appear and the technique will be automatically discarded. Secondly, the geometry of the LF link with respect to the geomagnetic field must be such that it operates on a quasi-transverse ordinary mode. An often quoted analysis by Booker (1935) indicates when a specific case can be considered quasi-transverse.

The precision of the method is limited by the inaccuracy of the function $v(h)$ in Fig. 4, and by the width of the chiasmic spot. The available information about $v$ as a function of $h$ between 85 and 60 km is presented by Whitten and Poppoff (1965). For any given value of $v$, $h_{max} \leqslant h_{min} + 5 \cdot 10^3$ m. The relative width of the chiasmic spot can be obtained by combining eq. 13 and 16: that is, $\Delta N/N = (0.107 - 0.072)/0.107 = 34\%$. Comparison with published data on the ionization of the lower ionosphere indicates that the precision of the proposed method of measurement is equal if not better than the precision of all other techniques presently in use.

## 4. CONCLUSION

Experiment Luxembourg, Experiment LIMBO and the experiment based on LF signal strength yielded some information on $N(h)$ and $v(h)$. Which technique is preferable, cross modulation or LF? From the point of view of simplicity and immediacy the LF approach is definitely superior. When the kind of information that can be obtained is considered, it appears that in the height range between 60 and 70 km the LF technique can compete with the cross-modulation technique in order to determine $N(h)$ or $\nu(h)$. But when both these profiles are simultaneously desired, cross modulation must be chosen even in the region between 60 and 70 km. Below 60 km, the technique of cross modulation must be used.

## ACKNOWLEDGEMENTS

Experiment Luxembourg was developed at the Geophysical Institute of the University of Alaska under the sponsorship of the USAF Geophys. Res. Directorate (Contract No.AF 19(604)-3880). Experiment LIMBO was developed at Cornell University under the sponsorship of the National Science Foundation (Grant GP-5452). The LF research was carried on at the Istituto Elettrotecnico Nazionale Galileo Ferraris in Turin; the LF data were supplied by Prof. S. Leschiutta. Prof. C. Egidi encouraged this research.

## REFERENCES

Booker, H. G., The application of the magneto-ionic theory to the ionosphere. *Proc.R.Soc.*, *Ser.* A, **150**:267–286, 1935.
Chanin, L. M., Phelps, A. V. and Biondi, M. A., Measurement of the attachment of slow electrons in oxygen. *Phys.Rev. Lett.*, **2**:344–346, 1959.
Crompton, R. W., Huxley, L. G. H. and Sutton, D. J., Experimental studies of the motions of slow electrons in air with application to the ionosphere. *Proc.R.Soc.*, *Ser.* A, **218**:507, 1953.
Deeks, D. G., D-region electron distributions in middle latitudes deduced from the reflection of long radio waves. *Proc.R.Soc.*, *Ser.* A, **291**:413–437, 1966.
Doherty, R. H., Importance of associative detachment and dissociative attachment in the lower ionosphere as shown by LF radio measurements. *J.Geophys.Res.*, **73**:2429–2440, 1968.
Namba, S., General theory of the propagation in the upper atmosphere. *Proc.I.R.E.*, **21**:238, 1933.
Rumi, G. C., Preliminary results of Experiment Luxembourg. *J.Atmos.Terr.Phys.*, **23**:101–105, 1961.*
Rumi, G. C., Experiment Luxembourg: cross modulation at high latitude, low height, 1: Theoretical aspects. *IRE Transactions on Antennas and Propagation*, **AP-10**:594–600, 1962a.
Rumi, G. C., Experiment Luxembourg: cross modulation at high latitude, low height, 2: Experimental aspects. *IRE Transactions on Antennas and Propagation*, **AP-10**:601–607, 1962b.
Rumi, G. C., LF and VLF phase antinomies, 1: An interpretation. *Alta Frequenza*, **39**:721–726, 1970a.
Rumi, G. C., LF and VLF phase antinomies, 2: Some deductions. *Alta Frequenza*, **39**:997–999, 1970b.
Whitten, R. C. and Poppoff, I. G., *Physics of the Lower Ionosphere*. Prentice-Hall, Englewood Cliffs, N.J., p. 169, 1965.

*Note: the raw data reported in that paper must be reinterpreted according to the theoretical findings given in the author's paper "Experiment Luxembourg" (Rumi, 1962a).

# MID-LATITUDE SPORADIC E

MARIO BOSSOLASCO AND ANTONIO ELENA

*Istituto Geofisico e Geodetico, Università di Genova, Genoa (Italy)*

## SUMMARY

We find a significant correlation among Es-layer ionization over four stations in Japan. Such a correlation shows that Es patches can have horizontal dimensions greater than 1000 km. Peaks of Es ionization over such large areas appear to be independent of geomagnetic activity. An attempt to explain this spatial correlation is presented.

## 1. INTRODUCTION

At middle latitudes, $f$Es, the maximum plasma frequency of the Es layer, is highly variable in time and space. The formation of Es appears as a highly localized phenomenon; cross-correlation between distant stations gives a correlation number of 0.5 for a distance of about 200–300 km (Rawer, 1962; Koizumi, 1969; Goodwin and Summers, 1970; Raghava Reddi and Ramachandra Rao, 1970).

Conversely, from a statistical standpoint, the sporadic E layer shows considerable regularities, as far as the daily, annual, quasi-biennial and solar cycle variations are concerned (Bossolasco and Elena, 1963, 1964, 1966; Samardjiev, 1966).

## 2. DATA AND METHOD

This paper deals with a comparison among sporadic E-layer ionization at the Japanese stations of Wakkanai, Akita, Kokubunji and Yamagawa, for the summer months from 1965 to 1970 included. Information on the station geography is given in Table I.

The daily, annual, quasi-biennial and solar cycle variation of $f$Es is greatly enhanced over the Far East, probably because of the great contrast between the Asiatic continent and the near Pacific ocean (Bossolasco and Elena, 1963, 1964, 1966); therefore a significant spatial correlation of Es ionization, if true, will be more easily detected here than elsewhere.

TABLE I

Geographical coordinates and relative distances between the Japanese stations

| Station | Coordinates | | Stations | Separation |
|---|---|---|---|---|
| Wakkanai | 45° 24′N | 141° 41′E | Wakkanai–Akita | 640 km |
| Akita | 39° 44′N | 140° 08′E | Akita–Kokubunji | 430 km |
| | | | Kokubunji–Yamagawa | 1090 km |
| Kokubunji | 35° 42′N | 139° 29′E | Wakkanai–Kokubunji | 1080 km |
| Yamagawa | 31° 12′N | 130° 37′E | Wakkanai–Yamagawa | 1830 km |
| | | | Akita–Yamagawa | 1270 km |

Since individual hourly values of $f$Es show great scatter, for our purpose we have first determined the median value of hourly values of $f$Es for each day; such a median will be indicated by $m(f$Es$)$.

Continuous recordings show that large time variations of $f$Es can occur, the duration of a flash often being only a few minutes long; whereas in other cases high Es ionization can be present for many hours (Rawer, 1962; Bossolasco and Elena, 1964). The changes in Es characteristics during day-time are almost wholly due to movements of small-scale frontal irregularities; at night, changes are also linked to the formation and dissipation of Es layers having horizontal dimensions of the order of a few hundred km. The small-scale frontal irregularities apparently consist of ridges and troughs of enhanced and diminished Es-layer electron density (Goodwin and Summers, 1970).

The daily median $m(f$Es$)$, utilized in this study, mainly gives the persistent Es ionization.

## 3. THE SPATIAL CORRELATION OF THE ES LAYER

A significant spatial correlation of $m(f$Es$)$ at the Japanese stations of Wakkanai, Akita, Kokubunji, Yamagawa during some groups of days in 1968 and 1970 appears from Fig. 1. The results are also numerically presented in Tables II and III.

Table II shows some cases in which maxima of $m(f$Es$)$ occur on the same day at least at three stations. Table III refers to cases where maxima of $m(f$Es$)$ occur at least at three stations with a delay of no more than one day. In Tables II and III the increases of $m(f$Es$)$, related to the simultaneous maxima, are given both as difference $\Delta$ and as ratio $r$ of individual values of $m(f$Es$)$ to the corresponding 15-day-running median values of $m(f$Es$)$.

Figure 1 shows that sometimes the Es ionization over the Far East can increase considerably and reach values of $m(f$Es$)$ up to 8 MHz and over, i.e., more than twice the normal values. Such maxima, at least on some days, are highly correlated geographically, although the absolute maxima of $m(f$Es$)$ are not always strictly simultaneous. The delay of about one day is likely to be due to drift of patches of Es ionization.

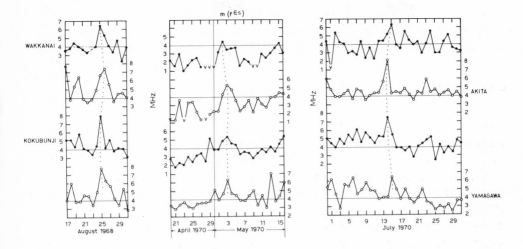

Fig. 1. The behaviour of $m(f\text{Es})$ at the Japanese stations during some days of summer 1968 and summer 1970.

TABLE II

Some cases of simultaneous maxima of $m(f\text{Es})$ at least at three Japanese stations: $\Delta$ and $r$ are respectively differences and ratios of individual $m(f\text{Es})$ to the corresponding 15-day running medians

|  | $Ap$ | Wakkanai | | Akita | | Kokubunji | | Yamagawa | |
|---|---|---|---|---|---|---|---|---|---|
|  |  | $\Delta$ (MHz) | $r$ | $\Delta$ (MHz) | $r$ | $\Delta$ (MHz) | $r$ | $\Delta$ (MHz) | $r$ |
| 1965 August 6 | 3 | — | — | 2.3 | 1.53 | 3.5 | 1.81 | 2.2 | 1.58 |
| 1970 May 3 | 10 | — | — | 3.0 | 2.25 | 2.0 | 1.58 | 2.3 | 1.59 |

TABLE III

Some cases where maxima of $m(f\text{Es})$ occur at least at three stations, with a possible delay of no more than one day: $\Delta$ and $r$ are respectively differences and ratios of individual $m(f\text{Es})$ to the corresponding 15-day running medians

|  | $Ap$ | Wakkanai | | Akita | | Kokubunji | | Yamagawa | |
|---|---|---|---|---|---|---|---|---|---|
|  |  | $\Delta$ (MHz) | $r$ | $\Delta$ (MHz) | $r$ | $\Delta$ (MHz) | $r$ | $\Delta$ (MHz) | $r$ |
| 1965 May 14 | 3 | — | — | 2.6 | 1.68 | 5.1 | 2.19 | — | — |
| May 15 | 4 | — | — | — | — | — | — | 2.7 | 1.63 |
| 1967 July 22 | 3 | 4.4 | 2.10 | 2.6 | 1.52 | — | — | — | — |
| July 23 | 14 | — | — | — | — | 4.5 | 1.92 | — | — |
| 1968 August 25 | 4 | 2.4 | 1.60 | — | — | 3.8 | 1.90 | 3.2 | 1.71 |
| August 26 | 4 | — | — | 2.9 | 1.64 | — | — | — | — |
| 1970 July 14 | 9 | — | — | 3.8 | 1.86 | 3.5 | 1.88 | — | — |
| July 15 | 5 | 2.3 | 1.77 | — | — | — | — | — | — |

Moreover, from Tables II and III, where the geomagnetic index *Ap* for the days of maximum *m*(*f*Es) was also reported, it appears that peaks of Es ionization on large areas are independent of geomagnetic activity.

## 4. DISCUSSION

Despite the uncertainties due to the lack of data on some days, a parallelism in the behaviour of *m*(*f*Es) at different Japanese stations appears to be ascertained, at least for some groups of days. The agreement is sometimes better among more distant stations. Such a parallelism extends to distances much greater than the values found for the horizontal dimensions of Es layers by Goodwin and Summers (1970). Therefore, in some cases the dimensions of Es layers are exceptionally large. Moreover, rocket results tend to show a larger horizontal structure of Es than radio measurements do. On the other hand, from a statistical standpoint, the correlation range of 200–300 km (Rawer, 1962; Koizumi, 1969), as a mean value, remains true.

Gravity waves in the neutral atmosphere are the source of wind shears responsible for Es. Such waves can be generated by an upward propagation of energy from the tropospheric and stratospheric wind systems. Some evidence of an enhancement of Es activity, due to atmospheric waves at the ionospheric heights and originating at the frontal regions during the passage of cold fronts at ground level, was recently found by Shrestha (1971). Following the wind-shear theory, the strong Es ionization over Japan could be partially due to the corresponding maximum of the geomagnetic horizontal component *H* (Whitehead, 1971).

The authors (Bossolasco and Elena, 1963, 1964), already stated that the anomalous behaviour of the Es layer over the Far East is most probably due to the enormous contrast existing there between the atmospheric conditions over land (Asia) and sea (Pacific). In summer, the air masses lying over Asia are warmer than those over the sea; thus, there is a large-scale, strong upwards movement of air over the continent, which, at its eastern border, reaches the troposphere, giving rise to wavelike perturbations upon contact with the colder oceanic air masses. Over and near Japan, because of the upwards propagation of these disturbances, gravity waves can arise as a source of very intense wind shears which are responsible for the enhancement of the Es ionization.

In order to check this explanation, upper atmospheric data are needed. Since only low troposphere data are available to us at the present, from a preliminary analysis of same we have found that during the summers of 1967, 1968 and 1969, about 60% of the cases of enhanced Es ionization over Japan occurred in the presence of low level cold fronts there.

Similar to the F2 region, some evidence of a coupling between ionospheric and tropospheric irregularities is also found for the Es layer.

## REFERENCES

Bossolasco, M. and Elena, A., Sporadic E-layer ionization and sunspot cycle. *Geofis.Pura Appl.*, **56:** 142–149, 1963.

Bossolasco, M. and Elena, A., Ricerche di Aeronomia II. *Geofis.Meteorol.*, **XIII:**115–126, 1964.

Bossolasco, M. and Elena, A., Une variation presque biennale dans l'ionosphère. *C.R.Acad.Sci. Paris*, **263 B:**95–97, 1966.

Goodwin, G.L. and Summers, R.N., Es-layer characteristics determined from spaced ionosondes. *Planet.Space Sci.*, **18:**1417–1432, 1970.

Koizumi, T., Some characteristics of the Es-layer in Japan. *J.Radio Res.Lab.*, **16:**17–49, 1969.

Raghava Reddi, C. and Ramachandra Rao, B., Characteristics of the sporadic E-layer at Waltair, using the phase path technique. *Radio Sci.*, **5:**679–683, 1970.

Rawer, K., Structure of Es at temperate latitudes. In: *Ionospheric Sporadic E* (E.K.Smith, Jr. and S.Matsushita, Editors), Pergamon, Oxford, pp. 292–343, 1962.

Samardjiev, D., The effect of solar activity on the boundary frequencies of the Es-layer. *Pure Appl. Geophys.*, **64:**126–132, 1966.

Shrestha, K.L., Sporadic E and atmospheric pressure waves. *J.Atmos.Terr.Phys.*, **33:**205–211, 1971.

Whitehead, J.D., Production and prediction of sporadic E. *Rev.Geophys.Space Phys.*, **8:**65–144, 1970.

# WINDS AND TURBULENCE IN THE METEOR ZONE

HEINZ G. MULLER

*Department of Physics, University of Sheffield, Sheffield (Great Britain)*

## SUMMARY

Phase coherent radar observations of the drift of meteor trails provide one of the most direct means of monitoring upper atmospheric winds continuously and over long periods of time. Both C.W. and pulsed techniques have been applied successfully in the past, and for a given mean transmitter power the same theoretical sensitivity may be achieved with the two methods. In single station experiments, wind components are usually resolved by observing meteor drifts in different directions. The inherent disadvantages of this technique can be overcome by the use of two independent and geographically separated radars with overlapping aerial beams. Wind shears may also be studied with single or multistation meteor equipment.

The general structure of the upper atmospheric wind in the meteor region is described, dividing the wind into two main components: (a) the regular component comprising prevailing wind and well-defined oscillations like those caused by atmospheric tides; and (b) the irregular one which consists of a spectrum of oscillations that may be ascribed to comparatively ordered motion, such as caused by the presence of short-period internal gravity waves; in addition to the latter, small-scale and short-period motions exist which may be interpreted as the results of atmospheric turbulence. Both components are described in some detail, and attention is drawn to the coupling between different atmospheric levels. Such a coupling is indicated by some recent observations of long-period oscillations, possibly the effects of planetary waves on a global scale.

The nature of the irregular fluctuations in the wind is illustrated by the results obtained with various techniques, and reference is made to the implications of turbulence theory. The diurnal and seasonal variations in shear- and turbulent parameters at meteor heights are described and related to features of the regular component in certain cases.

## 1. RADIO METEOR TECHNIQUES FOR THE STUDY OF UPPER ATMO-SPHERIC WINDS

### 1.1. Introduction

A number of techniques are used at present to investigate movements of neutral atmosphere at heights between some 80 and 110 km, but of these only the radio meteor method permits reliable continuous studies of upper atmospheric winds over long periods.

This technique has the advantage of a steady influx of meteors into an atmospheric layer of appreciable thickness, and although the number of meteors formed in this layer varies diurnally and seasonally, the available radio echo rates never fall below the minimum level required for an accurate determination of the basic features of a wind system.

Limitations exist in the uneven distribution of meteors in the vertical, as is illustrated by the two diagrams in Fig. 1, showing that most meteors are observed between some 90 and 105 km altitude. Winds can thus be resolved more accurately at some heights than at others, and the comparatively sharp peak in the distribution indicates that it is even permissible to use measurements irrespective of height in order to obtain average wind parameters applying to an average height, usually between 95 and 100 km.

All radio techniques applied in this field of study depend on the scattering of incident electromagnetic waves from the ionised matter in a meteor trail, and we realise that the free electrons in the plasma are the important elements in the scattering process. The radio echo from a meteor trail at the wavelengths used in drift studies is essentially of the diffraction type, where destructive interference reduces the returns from most parts of the trail so that only the first Fresnel zone contributes effectively to the echo. In meteor wind studies the size of the first Fresnel zone is usually of the order of 1 km, and this region is commonly identified as the specular reflection point on a meteor trail.

It had always been assumed in the past that below some 110 km the ionised matter in a meteor trail shares the motion of the atmosphere in which it is formed, so that the apparent drift of a meteor relative to a radar station may be interpreted as the sole effect of the upper atmospheric wind. It is questionable whether this assumption is justified in all cases since we must obviously consider the effect of the geomagnetic field on the movement of a plasma at these altitudes. The situation has recently been reviewed by Kaiser et al. (1969) in a theoretical paper, and their results essentially confirm that below 95 km the effects of the geomagnetic field may be ignored because the collision frequency of the electrons (and ions) is much greater than their gyrofrequency, and the ionised medium will therefore behave in the upper atmospheric wind as part of the neutral gas. Above 95 km, the mobility of the electrons and ions changes appreciably, but in the case of meteors unless an

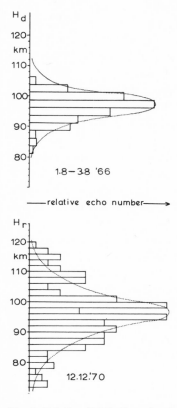

Fig. 1. Relative distribution of meteor echoes as a function of height. In the top diagram, heights are based on the ambipolar diffusion coefficient calculated from the decay of underdense echoes; in the bottom one, the heights are real heights obtained with the height and direction finding equipment described in the text.

extended plasma cloud is very closely aligned with the direction of the magnetic field (within a degree or two), the overall motion of the electrons and ions still follows the circulation of the neutral air. We may, therefore, assume with confidence that the general drifts studied by the radio meteor method are representative of neutral air motions.

*1.2. The basic principle of observation*

Most of the measurements of upper atmospheric winds in the meteor region have been carried out in terms of radio backscatter studies (radar). The average duration of meteors is not sufficiently long to permit range changes to be determined by the time delay of the echo; instead, the Doppler shift of the radio echo wave is utilised where the rate of change of the relative echo phase is representative of the magnitude of the radial velocity component with respect to the observing station.

In Fig.2 we see the principle of such radial drift studies illustrated. Two positions of a steadily drifting meteor trail are shown, and it can be seen that characteristic changes in relative echo phase are associated with the radial displacement of the trail, and these can be measured with suitable electronic techniques. A radial velocity of magnitude $V_R$ will cause a phase gradient $d\phi/dt$ of the relative echo phase $\phi$, and the two quantities are related by the equation $V_R = (\lambda/4\pi)(d\phi/dt)$. The gradient $d\phi/dt$ may be determined by comparing the phase of the echo wave with that of the transmitted wave during the lifetime of a meteor. Clearly, this would render a conventional radar, where the phase of the transmitted wave is not coherent, unsuitable for such studies. Instead, coherent phase radar systems are used, basically in two forms: (1) the bistatic C.W.; and (2) the phase-coherent pulsed radar. Whilst the first pioneering wind measurements appear to have been made at Stanford (U.S.A.) by Manning et al. (1950), many subsequent observations have been carried out systematically since, notably in the 1950's at Adelaide (Australia) by Elford (1959) using a C.W. system for the Doppler comparison, and at Jodrell Bank (U.K.) by Greenhow and Neufeld (1961) with the aid of a coherent pulse radar. In more recent years more stations have come into operation in the U.S.S.R., the U.S.A., France and the U.K., and numerous data concerning upper atmospheric winds are now accumulating.

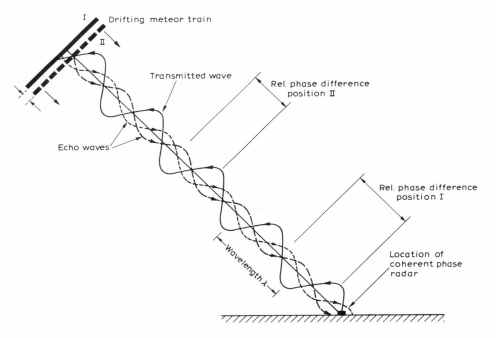

Fig. 2. Changes in relative echo phase, associated with the drift of a meteor trail relative to a coherent phase radar station. The diagram is not to scale.

### 1.3. C.W. radar techniques

Both C.W. and pulsed techniques have been favoured by individual research groups in the past, each stressing the particular merits of their chosen technique. In order to provide some kind of comparison it appears appropriate to describe briefly the basic pattern and mode of operation of the respective apparatus and, beginning with the C.W. technique, we will refer to the French system developed at the Centre National d'Etudes des Télécommunications situated at Garchy, France (Spizzichino, 1971). The principle of the system is illustrated in Fig. 3; the equipment consists of a transmitting and a receiving station which are geographically separated, since it is essential that the direct (ground) wave from the transmitter $T_1$ located at Garchy be sufficiently attenuated when it reaches the receiver at Sens-Beaujeu in order to prevent saturation of the receivers connected to the aerials $R_1$ to $R_3$, otherwise the comparatively small signals from meteor trails may become completely obliterated.

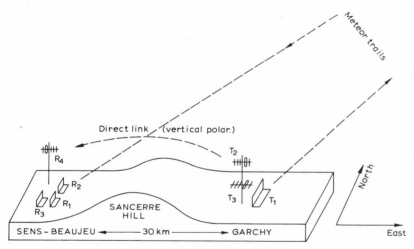

Fig. 3. Basic layout of the French meteor wind radar at Garchy (after a diagram by Spizzichino, 1971). $T_1$ = main transmitting antenna; $R_1$, $R_2$, $R_3$ = receiving antennae; $T_2$, $R_4$ = vertically polarized links for the transmission of the direct field; $T_3$ = transmitting antenna sending a wave with the same amplitude as $T_1$, but with opposite phase, in order to cancel the direct field received by $R_1$, $R_2$, $R_3$.

In the French exercise, the spacing between the two sites is 30 km and the Sancerre Hill is a convenient geographical feature which aids in the suppression of the ground wave. As a further means of attenuating the direct signal at the receivers, a directional aerial $T_3$ transmits a signal of the same polarisation as the main radar wave, but of opposite phase; this signal is then added to the receiver input to eliminate the residual effect of the ground wave from $T_1$. However, in order to facilitate a Doppler comparison, it is essential that the echo wave from a meteor trail be combined with a reference signal locked in phase to the main transmitted wave, and such a reference signal is communicated by radio link from $T_2$ to $R_4$, using vertical polarisation in order to avoid coupling effects with the remaining signals which are all horizontally polarised. The main transmitting aerial $T_1$, using a corner reflector, has a narrow beam, 26° wide in azimuth and 20° in elevation, with the beam axis inclined 45° and directed due east. A continuous wave is transmitted at a frequency of 29.8 MHz, at a power level of 5 kW.

The aerials $R_1$, $R_2$, $R_3$ are of similar construction but smaller in size and serve to receive the echo wave from a meteor trail. Since there are three receiving aerials on the corners of a triangle, the location of a meteor reflection point can be determined by comparing the relative echo phase angles at $R_1$, $R_2$ and $R_3$, while the phase difference between the echo wave at $R_1$ and the reference wave at $R_4$ is monitored continuously in a phase meter to provide a basis for drift velocity determinations.

An intriguing feature of the French system is found in the way the echo range is measured. This method involves the use of two additional waves transmitted at frequencies close to the main transmitter frequency. By measuring the phase difference between the corresponding echo signals at $R_1$, and utilising the reference signal at $R_4$, the distance of the specular reflection point on a meteor trail may be determined. It was found that by using a third frequency the range accuracy could be further improved, leading to an ultimate resolution of about 0.3 km. There is no point in pushing the resolution any further, because of the size of the first Fresnel zone which contributes essentially to the echo wave from a meteor trail and the extent of which ranges from a few hundred meters to about 1 km for most radar systems employed in such studies.

Up to recently, measurements made with the French system were confined to the direction E–W so that only the zonal wind component could be resolved, but at present efforts are being made to duplicate the equipment in order to obtain the meridional (N–S) component of the wind so that the total magnitude and direction of the wind vector may be determined.

### 1.4. Pulsed techniques

An alternative method for meteor drift measurements depends on the transmitting of phase-coherent pulses of short duration at comparatively high power levels. Such a system has been developed by Muller (1970) at Sheffield, and we shall describe its main features as a typical example of coherent pulse equipment. The apparatus consists essentially of two stable oscillators (Fig.4), one for the transmitter and one for the receiver, both contributing to a reference signal against which the meteor echo wave is compared in order to establish the rate of change $d\phi/dt$, of relative echo phase. The transmitter oscillator signal is pulse modulated and amplified, in order to maintain phase coherence from pulse to pulse, and the amplified signal is passed through an automatic transmit-receive (T.-R.) switch to one of two aerials pointing in directions 90° apart, and which are used alternately to resolve two orthogonal wind components. The same aerial is used for transmitting and receiving; the received signal passing through the T.-R. switch and various amplifiers into two phase-sensitive detectors controlled by two reference signals in quadrature in order to establish the magnitude and sign of $d\phi/dt$, the latter indicating whether the meteor is approaching or receding.

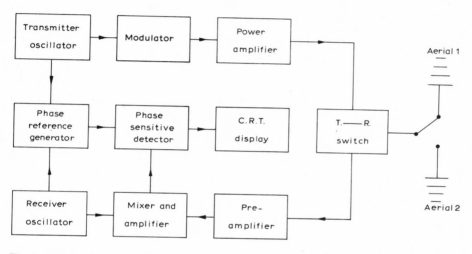

Fig. 4. Simplified block diagram of coherent phase radar equipment at Sheffield.

The Sheffield system incorporates two independent radar installations operating at 25 and 36 MHz respectively, using 25 μsec pulses at a rate of 450 Hz, the maximum power being of the order of 200 kW at 36 MHz. Fixed Yagi-type aerials are directed NW, SW and NE, respectively, their beamwidth being about 25° in azimuth and elevation at 36 MHz and their beam tilt being about 30° with respect to the horizontal plane. The principal meteor collecting areas resulting from the shape of the main aerial beam are shown in Fig. 5.

In order to establish azimuth and elevation of the specular reflection point on a meteor trail, three aerials are placed on the corners of a right-angled triangle and spaced at three wavelengths from each other, facilitating a phase comparison similar to that incorporated in the French C.W. system at Garchy, described in section 1.3.

Fig. 6 shows the general arrangement of the meteor wind equipment at Sheffield. The twin Yagi aerials used in this exercise are quite conspicuous and the picture is dominated by the tall mast which carries aerials used in the remote station experiments described in section 1.8.

Information concerning drift velocity, range and direction of the echo point is displayed on a matrix of 8 cathode ray tubes, incorporating 14 channels altogether.

Fig. 5. Principal meteor collection regions used in the Sheffield meteor wind experiment, relative to the British Isles.

Fig. 7 shows a typical meteor echo: at the top left, the output from the two phase-sensitive detectors (referred to above) is displayed—the signals being proportional to the sine and cosine respectively, of the relative echo phase angle—together with 20 millisec time markers, so that the phase gradient $d\phi/dt$ can be extracted for the computation of radial velocity components. The adjacent tube on the right displays (negative going) the echo amplitude as a function of time, and at the bottom right two traces provide range information; the top scale covering 0–667 km, and the bottom one serving as a vernier, spread over 0–33.3 km, and permitting a maximum range resolution of 0.4 km. The remaining displays are employed in conjunction with the height and direction-finding equipment. In order to describe the method we refer to Fig. 8 showing on the ground plane the position of three individual aerials $R_1$, $R_2$ and $R_3$ spaced at the distances $d_x$ and $d_y$. The shaded area represents the

Fig. 6. General view of the Sheffield meteor wind radar site at Edgemount near High Bradfield.

plane which contains the line of sight to the meteor echo point P whose azimuth relative to the direction $R_1R_2$ is $\theta$, and $\phi$ representing the elevation of the line of sight. The signal received at $R_1$ is used as a reference and the phase difference with respect to that signal, of the waves at $R_2$ and $R_3$ is a function of the angles $\alpha$ and $\beta$ shown in the diagram. The phase difference $T$, involving the aerials $R_1$ and $R_3$, is read off the top right hand polar display, and the phase angle $B$ related to $R_1$ and $R_2$ is read off the polar display immediately below. As seen in Fig. 8, $T$ and $B$ are functions of $\phi$ and $\theta$, and both these quantities may be computed from the displayed phase angles. The remaining tubes of the display show individual sin- and cos-components contributing to $T$ and $B$, as well as the amplitude of the signals at $R_2$ and $R_3$ in order to resolve any changes in the quantities with time, which a polar-type display fails to show. It is easy to see that the height of the echo point P follows as $h = r \sin \phi$, which can be determined accurately within about 1 km.

*1.5. Comparison between C.W. and pulsed techniques*

It can be shown that the optimum theoretical sensitivity of a backscatter radar is independent of its mode—whether C.W. or pulsed—and only dependent on the mean power which is the actual output of a C.W. system, or the peak pulse power

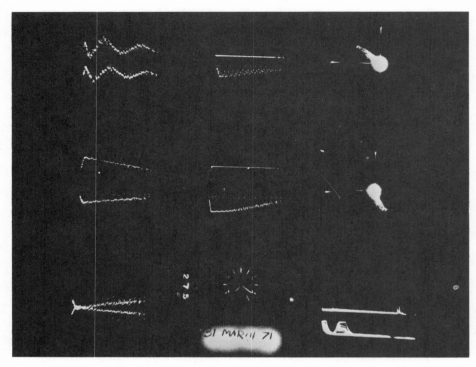

Fig. 7. Typical meteor echo record obtained at Sheffield. Details of the various signals displayed are given in the text.

multiplied by the duty cycle. The situation will be appreciated if we consider that although a pulsed radar can develop much greater power levels than a C.W. radar, a finite bandwidth has to be employed in order to resolve echo pulses without excessive distortion, and that bandwidth must be at least equal to the reciprocal of the pulse length. It is well known that the system noise power is proportional to its bandwidth and the signal-to-noise ratio, the figure of merit of a system, cannot therefore be improved by increasing the peak transmitter power for a given mean power. In a C.W. system, on the other hand, the maximum bandwidth can be much smaller than in a pulsed system, with a resulting reduction in noise power so that a good signal-to-noise ratio may be achieved with comparatively low power levels.

In practice, we find that the performance of a C.W. system is likely to be reduced owing to the presence of man-made radio interference which is easily picked up by a low-noise receiver. There is increasing congestion of stations in the band of radio frequencies used for meteor wind studies. In a pulsed system the meteor echo amplitudes are usually much greater than the interference signal which itself rarely exceeds the noise level, so that the performance of the equipment remains unaffected. A further disadvantage of C.W. systems lies in the presence of short-

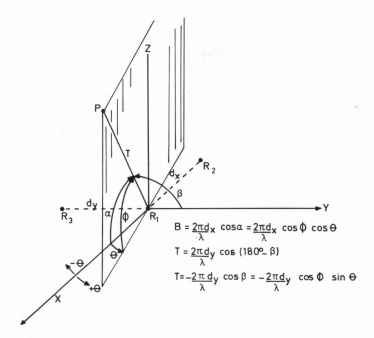

$$B = \frac{2\pi d_x}{\lambda} \cos\alpha = \frac{2\pi d_x}{\lambda} \cos\phi \, \cos\theta$$

$$T = \frac{2\pi d_y}{\lambda} \cos(180° - \beta)$$

$$T = -\frac{2\pi d_y}{\lambda} \cos\beta = -\frac{2\pi d_y}{\lambda} \cos\phi \, \sin\theta$$

Fig. 8. Geometry applying to the Sheffield height and direction finding equipment, illustrating the dependence of the echo azimuth $\theta$ and echo elevation $\phi$ on the measurable quantities $T$ and $B$, explained in the text.

range echo signals due to reflection from aircraft, and these may at times obliterate echoes from meteor trails. In a pulsed system such unwanted echoes can be most effectively suppressed by blanking the receiver range, for example between 0 and 100 km.

Perhaps the best way to compare the efficiency of one system with that of another is to look at the published figures for the hourly rates of meteor echoes. Such a comparison was carried out in the past by Greenhow (1954), who quotes a rate of about 50 usable echoes per day for the early Adelaide C.W. system with an average power of a few hundred watts, whereas he himself observed a rate of 3,000 usable echoes for his pulsed system with an average power of only 150 W! Data recently published by Spizzichino (1971) show that the peak hourly value in the diurnal rate curve is about 120, dropping to 26 at the time of minimum meteor activity for their 5 kW C.W. system. Greenhow (1954) observed a peak rate of 284/h at the same time of the year, decreasing to a minimum of 54 for a mean power of 150 W, and our own counts at Sheffield show a peak of 486/h, dropping to 60/h for a peak pulse power of 100 kW and a mean of 1350 W. It thus appears that the comparison of two C.W. systems with two pulsed systems, all operating, incidentally, near frequencies of 30 MHz, comes out in favour of pulsed radar as far as the available meteor echo rate is concerned, on which the reliability of a wind measurement ultimately depends. A further advantage of a pulsed system is that the transmitter and the receiver may be located at the same place, and the same aerial may be used for transmission and reception, whereas a C.W. system has to be essentially bistable, i.e., transmitter and receiver must be well separated geographically, preferably utilising topographical features such as hills or valleys in order to reduce the effect of the direct wave at the receiving end.

*1.6. Problems associated with the determination of individual wind components*

As described in section 1.4, meteor echoes are usually obtained in two directions involving aerials whose azimuths differ by 90°. In order to resolve individual wind components, records are taken alternately on such aerials where, depending on the available echo number, each aerial is in operation for a period ranging between some 5 and 15 min. The records thus obtained are grouped according to heights and then averaged. It is clear that fluctuations in the wind whose time constant is less than about 10 min will be smoothed out in the records. Smoothing will also take place if the wind changes rapidly with height, since height resolution is limited by instrumental factors and the physical size of the first Fresnel zone on the meteor trail to about 1 km. Although these limitations are not too serious, we are concerned about the degree of smoothing which results from a mode of operation that depends on collecting areas of meteors which are geographically separated by several hundred km (see Fig. 5). It is clear that if components are obtained in this fashion in order to establish the total magnitude and direction of the wind, only the largest features with horizontal scales in excess of 100 km will be resolved. Such scales apply, for example, to tidal components in the spectrum of the variable upper atmospheric wind, but there are numerous short period modes whose scales are comparatively small, and records of these will be seriously affected if the primary data are obtained in such widely separated regions.

In order to overcome such difficulties, a new system has now started operating in the U.K. where the Sheffield radar and a coherent pulse radar, similar to the Sheffield system recently completed by the British Meteorological Office, have their aerial beams directed toward a common collecting area. Fig. 9 shows the geometry applying to this system, avoiding all the disadvantages of separate meteor collecting regions and thus permitting the resolution of small-scale variations in the upper atmospheric wind. We understand that a similar approach is planned in France; it involves a mobile radar unit that can be moved into various locations, relative to the main installation described in section 1.3.

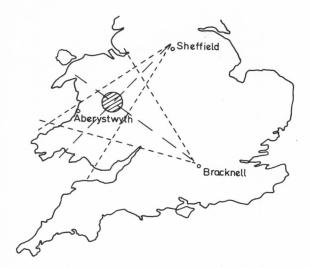

Fig. 9. Overlapping aerial beams from two similar radar systems situated at Sheffield and Bracknell for the study of the small-scale wind structure over a confirmed area.

## 1.7. Measurements of wind shears in single station experiments

In recent years a special investigation has been carried out at Sheffield involving the analysis of wind shears in the meteor zone. Interest in this matter was stimulated when, during routine observations of meteor winds, significant gradients in the ratio $d\phi/dt$ of relative echo phase were observed, clearly showing that changes in radial velocity occurred whilst an echo was being recorded. Such an acceleration can be readily explained as the effect of reflection point motion along a trail which is being distorted by a wind shear. Details of the analysis of records resolving the magnitude and distribution together with the theory of reflection point motion are given in a paper by Muller (1968).

While it is comparatively simple to extend meteor wind studies in utilising the second derivative in the echo phase variation resolved with phase coherent radar equipment, regrettably the number of stations possessing such equipment is quite limited at present. There are, however, numerous non-phase coherent radars in operation, particularly for the study of the radio aurora, and it is not generally known that these can be adapted to the study of the amplitude and scale of a variable wind system in the meteor region. The above-mentioned trail distortion due to wind shears not only causes a movement of the specular reflection point from which the initial echo is obtained, but may also result in the production of a second, or more, reflection points as the distortion of the trail progresses. It is easy to imagine, for example, a sinusoidal trail distortion where reflection points exist wherever the tangent to the trail is perpendicular to the line of sight from the radar.

A certain time elapses before a second reflection point is formed, and this time depends on the structure of the wind which is responsible for the distortion. As soon as two reflection points exist jointly, we have two echo waves which interfere with one another at the receiver, thus giving rise to regular amplitude fading whose frequency depends on the differential motion of the two reflection points, since each will produce a characteristic Doppler shift in echo frequency that is proportional to the radial drift velocity. Kaiser (1968) has shown that in a statistical analysis of the delay in the onset of fading and the fading frequency, both the scale and the amplitude of a quasi-periodic wind profile may be deduced, and this method has been applied successfully at Sheffield by Phillips (1969) who also considered an increase of trail distortion with increasing height.

*1.8. The utilisation of multiple reflections from a meteor trail*

In section 1.7 we were concerned with the motion of a reflection point on a meteor trail, as well as the presence of several reflection points on a distorted trail, giving rise to characteristic interference patterns. Multiple reflection points may also be obtained on undistorted trails by suitably spaced receivers, using the main transmitter to illuminate the whole of a meteor trail which may exceed a length of 10 km. Such a system, incorporating two receivers, had been used at Jodrell Bank by Greenhow and Neufeld (1959a) for the study of wind variations over short distances, and a similar system, involving three spaced receivers, is now part of the Sheffield wind experiment. Fig. 10 shows the geometry applying to the Sheffield system where two remote stations are used, located 5 and 20 km, respectively, SW of the main site. For special studies, a third station located 50 km SW of Sheffield is being operated by members of the Newchapel Observatory near Stoke-on-Trent. As it can be seen in Fig. 10, the technique depends on the knowledge of the speed at which the meteor trail is initially formed in the atmosphere, and this can be determined from the echo amplitude fluctuations due to Fresnel-type diffraction. The height of the first reflection point is accurately determined by the technique described in section 1.4 and the difference in height between successive reflection points follows from the relations given in Fig. 10. Doppler information is recorded from all reflection points, and radial velocities are thus known at three points spaced along the same meteor trail. Using data for many individual trails, a quasi-instantaneous wind profile may be obtained within a very short time. The method is particularly suitable for the study of the irregular component in the upper atmospheric wind.

*1.9. On the possibility of a network of meteor wind stations on a global scale*

Because of the relatively low cost of equipping and maintaining meteor wind stations compared with some other techniques, such as rocket contaminant release

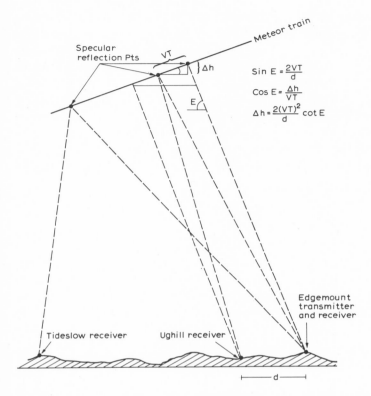

$$\sin E = \frac{2VT}{d}$$

$$\cos E = \frac{\Delta h}{VT}$$

$$\Delta h = \frac{2(VT)^2}{d} \cot E$$

Fig. 10. Geometry applying to the Sheffield multi-station experiment for the determination of fine structure in the upper atmospheric wind. The equations illustrate how height differences may be calculated from the time delay $T$ in the occurrence of echoes and the meteorite velocity $V$, if the elevation angle $E$ is known.

experiments, it appears feasible to establish, in the course of time, a global network of stations for the synoptic observation of winds in the meteor zone. It is clear from the material presented in this paper that a considerable technical standard has been obtained in the engineering of radar systems, and meteorological authorities should realise the considerable potential lying in the installation of such systems for continuous observations which are impracticable, for example, at university departments where all the early observations of meteor winds took place. It is gratifying to note that steps in this direction are already being taken, for example, by the British Meteorological Office who are now using meteor wind equipment regularly in the U.K., and intend to operate this radar in the equatorial zone in the near future. A number of systems exist elsewhere and they could be utilised in a global co-operative scheme over selected periods of time. Most stations are in middle northern latitudes, but equipment is now being completed in Jamaica (19°N) in conjunction with the Sheffield meteor wind experiment. For some time now, regional co-operation has existed between various stations in Europe, and a number of simultaneous meteor wind runs have been conducted, supplemented by other techniques.

Whilst the number of meteor wind stations is gradually increasing, special efforts are being made at various institutions concerned with such measurements to rationalise observations and introduce digital techniques in the analysis of data, relying on the use of electronic computers for the production of final wind parameters. The process of obtaining winds in the meteor zone is thus being speeded up appreciably and perhaps the meteorologist's dream of an anemometer in the upper atmosphere will soon become reality in the form of a meteor radar producing wind data in real time.

## 2. NEUTRAL WINDS IN THE METEOR ZONE

### 2.1. Introduction

Section 2 is concerned with the region of the atmosphere where air motions can be conveniently studied in terms of the drift of meteor trails subjected to the effect of the neutral atmospheric gases. The height range in which meteors occur extends between some 80 and 110 km, and most meteors are observed between 90 and 100 km. According to present-day classification of the atmosphere, the meteor zone partly joins with the mesosphere and its upper part merges with the thermosphere; the latter is characterised by a positive temperature gradient and adiabatic lapse rate, and it extends well above the ceiling of the meteor region. The earliest studies of air motion in this region were made on long-enduring visual meteor trains which last long enough to permit significant displacements to be observed from the surface of the earth. Likewise, the movement of noctilucent clouds provided a means of assessing atmospheric circulation at these heights; but due to the intermittent nature of these phenomena, regular observations were quite out of the question. Since meteors may also be studied by radio techniques, using the ionised matter of the trail as a scattering agent for electromagnetic waves, attractive possibilities exist in the study method of upper atmospheric circulation by the drift of radio meteors. The first experiments concerning movements of meteor trails utilising the basic ability of a radar system to measure range, were made around 1949, but the resolution was generally too poor then to obtain significant wind patterns. About the same time, special equipment was specifically developed to measure small displacements of meteor trails accurately as a function of time: a number of remarkable features were observed in the upper atmospheric wind, and particularly, the presence of regular oscillations whose period is tied to the rotation period of the earth. Although the radio meteor technique is a very versatile one, it still has certain disadvantages because information obtained from a single observation is essentially confined to a restricted region in space, and many individual measurements are therefore necessary in order to resolve the three-dimensional structure of the wind. In recent years, fortunately, meteor wind measurements have been supplemented by other techniques which enable us to obtain vertical profiles of the wind in a single experiment. These techniques include the contaminant release method involving artificial chemical trails injected into the upper atmosphere by rockets, or sensors ejected from rockets. Moreover, in recent years, the technique of ejecting chaff from rockets has been perfected and may now be applied to heights up to 100 km. In all these cases, tracking from a ground based station is necessary, involving photographic techniques in the case of chemical trails, and radar in the case of ejected sensors. Trail release experiments cover a considerable height range terminating well above the meteor zone, and they thus provide an abundance of valuable data. Unfortunately, the technique is limited to observations at night or under twilight conditions, and by its very nature it is discontinuous. A large part of the data concerning this region of the atmosphere is still recorded by radio methods, and most of the material presented in this paper is based on such studies.

*2.2 The general structure of the upper atmospheric wind*

It has been known for some time that the flow of neutral gas on the rotating earth is predominantly zonal (either W to E or E to W) in the lower parts of the atmosphere. Meridional (S → N or N → S) components are observed occasionally, but their magnitude is usually small compared with that of the zonal component. The latitudinal and seasonal variation in the zonal flow are well illustrated by Fig. 11. The structure of the zonal wind can be explained as the result of meridional thermal gradients, and several workers have applied the so-called thermal wind equation to compute winds from a given temperature distribution in the atmosphere. The top section of Fig. 11 which covers the meteor zone, displays data that are based on contaminant release experiments, notably by Rosenberg (see Kent and Wright (1967) for detailed references), and there is still some uncertainty in the data due to the presence of tidal components that cannot be completely eliminated in such artificial trail studies.

An interesting feature of the wind in the meteor zone is its horizontal stratification. This is essentially a consequence of the thermal structure of the region which has a controlling effect on the atmospheric gases because any tendency of vertical motion is counteracted by the effect of buoyancy forces. The relatively simple system of a steady, predominantly zonal, circulation observed in the lower and middle atmosphere breaks down as we enter the meteor zone. Considerable wind oscillations are observed above 80 km, and meridional components reach amplitudes comparable to those of the zonal wind. It is useful to distinguish between several individual components of the wind between 80 and 110 km, and the following scheme may be used:

| Regular component | Irregular component |
|---|---|
| prevailing winds | short period ($T \ll 24$ h)–non tidal winds |
| tidal winds (period $T \leqslant 24$ h) | turbulent wind |
| long period winds ($T > 24$ h) | |

Little is known at present about the global distribution of the regular components, mainly because of the limited number and uneven distribution of observing stations across the globe. The description of the individual components given below will, therefore, contain a certain amount of bias towards areas where frequent observations are being made, and this applies particularly to middle latitudes in the northern hemisphere.

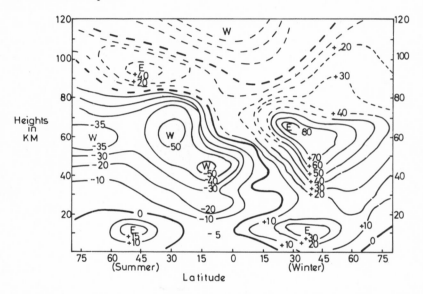

Fig. 11. Zonal wind pattern versus heights and latitude after Kent and Wright (1967), using data by Kantor and Cole (1964). (See Kent and Wright (1967) for further details.)

### 2.3. The resolution of individual components in the upper atmospheric wind

The radio meteor method has been quite successfully used in the past to resolve winds over long periods in a quasi-continuous manner. As a typical example, in Fig. 12 we see the results of a 3-day meteor wind run at Sheffield, in August 1966. The two components shown were obtained by operating the Sheffield coherent pulse radar alternately on two aerials, directed NW and SW respectively. A smooth curve based on a polynomial has been fitted to each data set, involving the method of least squares. Even without detailed analysis, the main features of the wind are quite evident. We notice a pronounced oscillation of a period of 12 h whose phase is systematically shifted between the two wind profiles, and suggests a component which may be represented by a clockwise rotating vector of constant magnitude. We also notice an oscillation whose period is much longer, apparently close to about two days. The mean of the fluctuations is significantly shifted, with respect to the zero velocity axis, indicating the presence of a prevailing component, or at least one whose period is very large compared with the duration of the observations. Because of the distinct periodic character of the data, it is appropriate to subject these to harmonic analysis. This can now be done conveniently with the aid of electronic computers using various built-in subroutines. The results of such an analysis are shown in Fig. 13 and 14, representing the amplitude spectra corresponding to the two profiles shown in Fig. 12. The peak near 12 h is quite pronounced on both diagrams; there is some response near 24 h, and peaks exist near a period of 2 days. There are also various responses in the spectrum for periods of less than 12 h, particularly for the SE component.

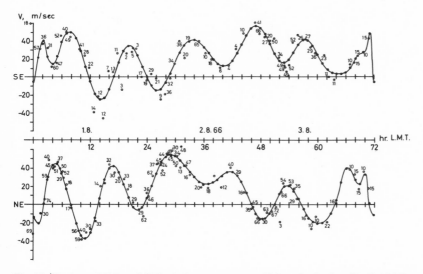

Fig. 12. Horizontal SE- and NE-wind components at 97 km obtained at Sheffield. The numbers refer to the number of echoes contributing to each data point, and the smooth curves represent the best fitting polynomials to the data.

Referring to the scheme introduced in section 2.1, we may now identify individual components in these particular sets of data. We ascribe the deviation of the mean from zero to the effect of the prevailing wind. We note that the mean is positive using the ionospheric convention where the wind direction is the same into which the wind blows, and of comparable magnitude for both profiles, amounting to a predominantly zonal flow directed towards the east. The 12 h component may be interpreted as the effect of the solar semidiurnal tide, which may be represented by

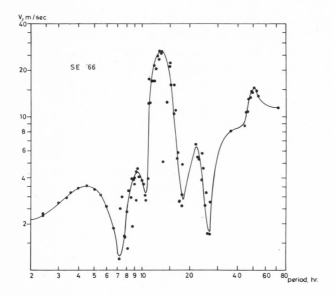

Fig. 13. Amplitude spectrum of the SE-wind component at Sheffield, based on the SE-wind profile in Fig. 12.

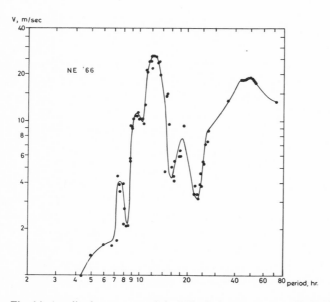

Fig. 14. Amplitude spectrum of the NE-wind component at Sheffield, based on the NE-wind profile in Fig.12.

a clockwise rotating wind vector. There is not much evidence of a diurnal component on these records, but the peaks near 20 h may well be representative of such a mode since rapid phase shifts have been reported for this component in the past (Spizzichino, 1971); they are equivalent to an apparent shift in the period of oscillation. The peaks in the two amplitude spectra near a period of 2 days indicate the presence of a long period oscillation in the upper atmospheric wind, and a detailed study of the relationship between the phases observed on each of the profiles in Fig. 12 shows that the oscillation is associated with a propagating wave which appears to exist on a global scale, and which may be compared with the so-called planetary waves established in the lowest levels of the atmosphere.

The short-period spectra may well be representative of the dominant modes of internal atmospheric gravity waves, such as described by Hines (1960).

## 2.4. The prevailing wind

Investigation of the prevailing component of the wind, using radio meteor data obtained at various stations and over long periods, shows that the general behaviour of this component conforms to the pattern shown in Fig. 11 although, as already mentioned before, a significant meridional wind component exists at meteor heights. A comprehensive study of the prevailing wind has recently been made by Lysenko et al. (1972) including meteor wind data obtained in the U.S.S.R., Australia and Antarctica. The results of this analysis are shown in Fig. 15. The authors have also succeeded in establishing a regular circulation pattern over Eurasia and the Arctic, for which numerous data are now available from many individual stations. These patterns are shown in Fig. 16, and include data taken at 7 stations in the U.S.S.R., supplemented by records taken at Sheffield. Some of the vectors shown are not identified by letters, and these have been obtained by extrapolating from known wind patterns in the lower parts of the atmosphere, based on the assumption of a regular vertical structure, such as the one shown in Fig. 11. It is seen that the prevailing wind shows very regular features in the northern hemisphere. Unfortunately there is no corresponding coverage yet at southern latitudes due to the lack of observing stations.

## 2.5. The periodic tidal wind

### 2.5.1. The diurnal component

A regular 24-h component has been observed for many years at Adelaide, using the radio meteor method (Elford, 1959), but similar observations in the northern hemisphere have rarely yielded consistent patterns concerning this component. This is partly due to the comparatively short vertical wavelength of the propagating mode of the diurnal tide, which is described in the dynamical theory of the

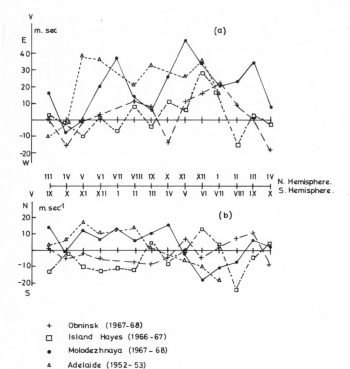

Fig. 15. Seasonal variations of the prevailing winds for the northern and southern hemisphere: (a) the zonal components; (b) the meridional components. (After Lysenko et al., 1972.)

atmosphere; see, for example, Kato (1966a,b) and Lindzen (1966). Since most of the meteor wind data in the past represent averages over the 80–105 km region, the diurnal tidal component becomes greatly reduced in the records due to excessive smoothing in the averaging process. Furthermore, past data were usually obtained by harmonic analysis of 24-h data sets resolving a 24-h component as the fundamental mode of oscillation, but, because of the presence of long-period oscillations, the wind is never strictly periodic in 24 h, and components based on such time-limited data series are of limited significance. Shifts in the actual period of oscillation of this component have been observed at Sheffield (see Fig. 13 and 14), and have been confirmed by the French meteor wind group (Spizzichino, 1971), which, incidently, made valuable contributions to the resolution of the height variation of the diurnal tide and was able to demonstrate the presence of propagating modes with vertical wavelengths around 25 km, and evanescent modes where the phase does not change with altitude and the energy decreases rapidly with increasing height. This result is in agreement with the structure of the diurnal tide described in the dynamical theory of the atmosphere. It is evident, though, that our

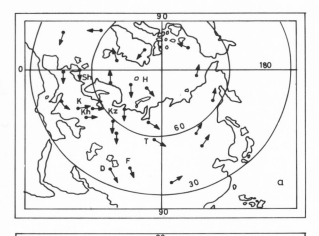

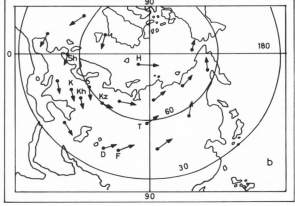

Fig. 16. Tentative models of prevailing circulation in the meteor zone over Eurasia and the Arctic. a: autumn; b: winter. *Sh* = Sheffield, *Kz* = Kazan; the remaining stations are identified by the same letters as in Fig. 15. (After Lysenko et al., 1972.)

knowledge of this tidal component is still very limited, and there is a strong case for extended observations of upper atmospheric winds with the radio meteor method, particularly designed for the resolution of this component. Efforts in this direction have been made in France (Spizzichino, 1971) and the U.K. (Muller, 1970), but more observations are needed in different parts of the globe in order to analyse the latitudinal structure of the 24-h tide.

### 2.5.2. The semidiurnal component

Oscillations with a period close to 12 h usually dominate the spectrum of wind oscillations, particularly when continuous records are taken, such as in meteor wind studies. Fig. 13 and 14 clearly show the existence of this mode in the records, and because of its large amplitude, details of this oscillation are readily obtained. It is now understood that this component results from the effect of the solar semidiurnal tide, which is thermally forced by the absorption of solar radiation in the ozone layers of the atmosphere. A detailed account of this tidal component is given by Chapman and Lindzen (1970) who explain the dominance of the semidiurnal tide over the diurnal tide at certain atmospheric heights, an unsolved problem for a considerable length of time. But although the main features of this tide are satisfactorily explained by the theory a number of features still need accounting for, particularly the pronounced seasonal changes in amplitude and changes evident from the radio meteor data. In Fig. 17 we see such seasonal variation displayed in a polar harmonic diagram, using data recorded at various locations over 12 years. It is seen that a distinct structure exists in the phase of this oscillation throughout the years and, as yet, only the most superficial explanation of these features is available from the theory.

In recent years increasing efforts have been made in studying the height variation of wind features by the radio meteor method. This has led to the resolution of the phase and amplitude dependence of the 12-h oscillation on atmospheric height. Some of the most recent results have been published by Massebeuf et al. (1969) and are based on meteor wind studies at Garchy, France. Fig. 18 illustrates the distinct increase of the amplitude of the 12-h tide over a considerable height range, as it is predicted in tidal theory. The phase increases with height, in agreement with earlier observations by Greenhow and Neufeld (1956), although Massebeuf et al. (1969) point out that on other occasions the phase does not exhibit significant variations with height. A gradual increase in phase, corresponding to a wave of downward phase propagation, is compatible with the theoretical $S_2^2$ tidal mode dealt with in tidal theory.

### 2.6. Short-period oscillations in the wind

Regular oscillations with periods ranging from about 30 min to several hours in the wind at meteor heights have been reported in the past (Greenhow and Neufeld, 1960; Muller, 1966; Spizzichino, 1971), and it appears that these can be satisfactorily explained as the effects of internal atmospheric gravity waves (Hines, 1960), rather than higher harmonics of the tidal oscillations. A recent analysis by Spizzichino (1971) reveals that the phase of these oscillations increases linearly with height (see, for example, Fig. 19), corresponding to waves of constant downward phase propagation, consistent with gravity waves which propagate their energy upwards. These waves have vertical wavelengths between some 10 and 30 km, which is evident in the vertical wind profiles such as shown in Fig. 19. A general increase of amplitude with height is also observed, and this is expected for a net upward flux

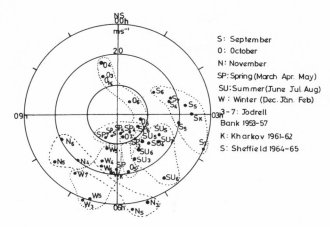

S: September
O: October
N: November
SP: Spring (March Apr. May)
SU: Summer (June Jul. Aug)
W : Winter (Dec. Jan. Feb)
3-7: Jodrell
Bank 1953-57
K: Kharkov 1961-62
S: Sheffield 1964-65

Seasonal averages of the semidiurnal wind over 12 years

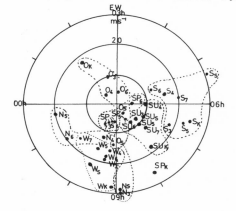

Fig. 17. Seasonal changes in the structure of the semidiurnal tidal wind, using data from various stations over a period of 12 years.

of energy in an atmosphere of steadily decreasing density. Statistical analyses of meteor wind data indicate great horizontal wavelengths, in excess of 100 km, in accordance with the theory of these waves given by Hines (1960). The actual wind profile may be rendered complex due to the presence of several modes of gravity waves interfering with one another, and these can be separated by harmonic analysis, provided that sufficient data are available for a given period of time.

*2.7. Long-period oscillations in the wind and the connection with other atmospheric parameters*

It is evident from Fig. 12, 13 and 14 that the upper atmospheric wind possesses a strong oscillating component whose period exceeds one day and does not fall into

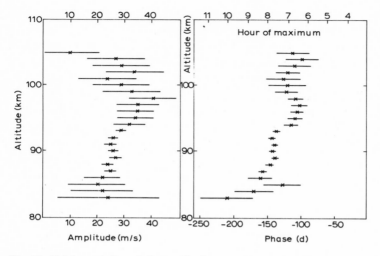

Fig. 18. Height variation of the amplitude and phase of the semidiurnal wind component at Garchy, France, 13–15 September, 1966. (After Massebeuf et al., 1969.)

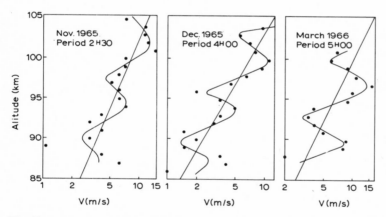

Fig. 19. An example of propagating gravity waves, evident from radio meteor data at Garchy, France. (After Spizzichino, 1971.)

the category of tidal oscillations. Several additional series of meteor wind observations at Sheffield have clearly shown that such oscillations exist for several years during the same season, and that they have a broad spectrum ranging from some 35 to 72 h. Fig. 20 shows the result of harmonic analysis of a 130-h data set recorded at Sheffield in July/August 1969, where it is seen that these long-period modes have amplitudes comparable to those observed in the 12-h and 24-h tidal oscillations. Although peaks are occasionally found near a period of 2 days (see Fig. 13 and 14), it does not appear appropriate to see an explanation of these oscillations in terms of tides whose period is naturally linked to the earth's period of rotation. Instead, we may consider that these oscillations are the effects of planetary-type waves, existing on a large scale. The characteristic phase relationship between orthogonal components observed at Sheffield, would clearly support the assumption of a propagating wave of large horizontal extent (see Muller, 1972). It is of considerable interest to look for the source of energy of such long-period waves, and it appears reasonable to assume that it lies in the lower tropospheric levels, since it has been known for some time that similar waves exist in the lowest atmospheric levels. A comparison between the variation of tropospheric parameters, like atmospheric pressure and wind, reveals certain corresponding features, which are evident by comparing Fig. 20 and 21 where the amplitude spectra of both wind and pressure oscillations recorded near the main meteor collecting areas show considerable similarity, as far as the longer periods are concerned. However, we are finding it difficult to explain the presence of a long-period wave system in the meteor zone as a result of an upward propagation of planetary wave energy, since such propagation is, in general, severely restricted due to the structure of the intervening atmosphere where the prevailing winds represent barriers at which most of the planetary wave energy is reflected. The situation has been studied theoretically in considerable

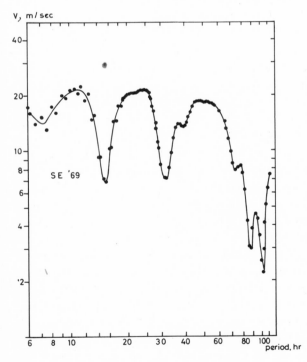

Fig. 20. Amplitude spectrum of the SE-wind component at 97 km, at Sheffield during the period 29.7 to 4.8, 1969.

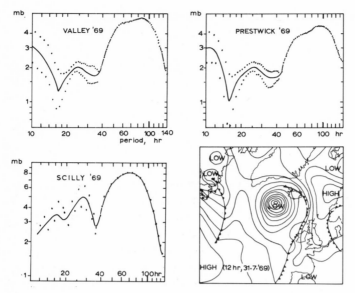

Fig. 21. Amplitude spectrum of sea-level atmospheric pressure oscillations at Prestwick, Valley and Scilly, based on data recorded by the British Meteorological Office during the period 27.7 to 4.8, 1969, together with an isobar chart showing frontal weather systems.

detail by Charney and Drazin (1961) who predict that at certain times of the year appreciable amounts of planetary wave energy may reach meteor heights, but unfortunately we cannot rely on this result safely as far as the Sheffield data are concerned, since these were taken during July–August when the conditions for upward propagation are most unfavourable.

Alternatively, we may consider that internal atmospheric gravity waves, referred to in a previous section, play an important part in the coupling between the two atmospheric levels. These waves may be generated by the passing of frontal systems (such as seen on the isobar chart in Fig. 21) and travel obliquely up to the meteor zone, acting as a carrier for a modulation pattern superimposed by tropospheric features and undergoing a demodulation, due to non-linear processes like viscous dissipation in the meteor region, which results in the production of local thermal gradients which, in turn, are responsible for a wind system bearing the same pattern as the modulating agent. Although it will be difficult to trace the source of gravity-wave energy in detail, the atmospheric pressure profiles at sea level should provide a useful reference. The coupling between the upper and the lower atmosphere is evident from phenomena observed with other techniques; for example, ionospheric soundings. Observers in this field of study ought not to be discouraged by the difficulties which still exist in theoretically founding their results.

## 3. WIND SHEARS AND TURBULENCE IN THE METEOR ZONE

### 3.1. Introduction

It has been known for some time that the atmospheric motions between some 80 and 110 km altitude show a distinct horizontal stratification, and that they exhibit very regular features which are usually described in terms of the prevailing- and regular-periodic wind components. However, we also notice the presence of less regular motions, and at one time they were thought to be solely the effects of turbulence existing at these heights.

Some of the earliest indications of irregular motions were found in the shape of distorted enduring visual meteor trains, of which the one shown in Fig. 22 is a classical example. Close inspection of Fig. 22 will, however, reveal a regular pattern underlying the apparently irregular shape of the train, and Hines (1960) argued quite convincingly that this regular pattern is the result of the presence of internal atmospheric gravity waves propagating with a vertical component through the meteor zone. There still remains the problem of explaining the small-scale distortion observed on such trains, since they fall outside the range of motions, due to the dominant modes of gravity waves at meteor heights. It appears reasonable to ascribe these variations to the effects of small-scale eddies, such as those discussed in turbulence theory.

It is obviously of considerable interest to study such distorted meteor trains in detail in order to extract information concerning ordered wave motion as well as the less regular turbulent motion in the upper atmosphere, but there is a severe restriction on this kind of observation, due to the scarcity of such enduring trains. Experimenters have, therefore, introduced techniques involving the release of contaminants into the upper atmosphere to produce artificial clouds which behave similarly to long-enduring meteor trains. These contaminants are usually ejected from rockets, occasionally from shells fired by large guns. At an early stage of these experiments barium and sodium were the accepted contaminants; however, their use was restricted to the times of twilight, and in more recent years T.M.A. has become increasingly popular as a contaminant in such studies because of its chemiluminescent properties which permit its use throughout the hours of darkness.

While the number of individual observations has been greatly increased in this manner, it is still difficult to monitor the irregular component of the upper atmospheric wind continuously by this technique, mainly due to the lack of facilities during the day time, but also because of the considerable cost of rocket-born experiments. Furthermore, the technique depends on a clear sky in order to be able to take photographic records of trails from several locations on the ground. We must stress, of course, the considerable value of contaminant release observations in the upper atmosphere since this method is the only one where, in a planned experiment, irregularities down to the smallest scales can be examined.

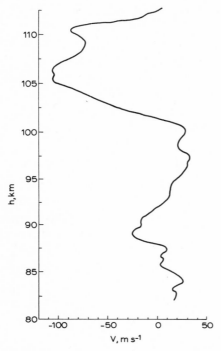

Fig. 22. The distortion of an enduring visual meteor train in the upper atmospheric wind (After Liller and Whipple, 1954.)

A completely different approach to the problem involves the study of irregularities in the wind by the radio meteor method. Whilst it is impossible with this technique to obtain an extended instantaneous wind profile covering the total height range of the meteor zone, detailed information may be obtained from individual meteor trails over smaller distances, and by treating numerous results statistically, it is possible to construct a quasi-instantaneous wind profile, at least at times when meteor activity is high. Such a profile, obtained from records taken during less than 30 min is shown in Fig. 23. Individual wind components were measured with the Sheffield meteor wind radar described elsewhere by Muller (1970) involving special height-finding equipment with an ultimate resolution of about 1 km. We realise, however, that due to the finite time required for the observation, a certain amount of smoothing occurs, resulting in the loss of detail concerning fluctuations whose time constant is small compared with about 30 min. We also note that the meteor echo does not come from a well-defined point on a meteor trail. In the case of meteor wind observations the echo is essentially of the diffraction type and originates from the Fresnel zone whose extent is of the order of 1 km on the trail, and therefore no fine detail can be studied within this limit.

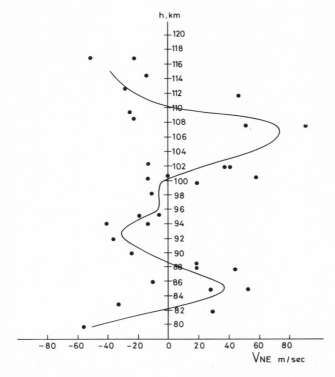

Fig. 23. Vertical wind profile obtained at Sheffield with the aid of coherent pulse radar equipment, during the period 0330-0400 on 12 December, 1970.

Apart from this semi-direct method involving radio meteor data containing wind amplitude and height, variations over smaller distances may, alternatively, be obtained by carefully analysing the Doppler shift of the meteor echo wave relative to the transmitted wave as a function of time. Characteristic phase gradients indicating changes in the recorded radial velocity of a meteor trail relative to the radar can be observed, and are found to be associated with progressive distortion of the trail, due to wind shears resulting from the irregular wind component. Simultaneously observations of the drift of several reflection points on the same meteor trail may also be made with the aid of spaced receivers, and data thus obtained are particularly suitable for statistical analysis, utilising the basic results of the turbulence theory, in order to determine turbulent parameters of the irregular wind component in the meteor zone.

### 3.2. The nature of the irregular component of the upper atmospheric wind

Radio meteor data obtained in the past (Greenhow and Neufeld, 1961; Muller, 1966) show that the amplitude of the irregular component of the upper atmospheric wind is generally comparable with that of the more regular component, including oscillations caused by the diurnal and semidiurnal tides. A value between 10 and 30 m/sec may be taken as typical. The irregular component shows some distinct features: short-period fluctuations ranging from about 30 min to several hours, and horizontal scales varying between about 1 and over 100 km, where we are conscious that the more direct release experiments described above resolve even shorter periods and smaller scales down to about 50 m. Quasi-periodic variations of the wind velocity exist in the vertical, and superimposed we find variations so irregular that they appear almost random.

The observable regular features within this component can be almost completely described as the effects of internal gravity waves. The apparently random nature of these waves, where seen as a whole, would indicate a comparatively broad spectrum of waves, but a detailed analysis shows that the modes propagating in the meteor zone are comparatively restricted, as far as their horizontal and vertical wavelength is concerned. We speak in terms of dominant modes and recognise, as shown by Hines (1960), that modes of certain scales are severely attenuated due to viscous effects in the upper parts of the region, and that temperature gradients are responsible for severe reflections affecting certain modes at lower levels. The dominant modes which survive in the meteor zone are well defined, and show a characteristic pattern of phase propagation from which the predominant direction of energy transport may be inferred.

Subtracting the known effects of internal atmospheric gravity waves, the residual features of the so-called irregular wind component may be visualised as the net effects of atmospheric turbulence. We consider a turbulent system as one where irregular motion results in a continuous mixing of atmospheric properties to an extent which is far greater than could be accounted for by the natural movement of the molecules in the atmospheric gas. A characteristic feature of such a system is the presence of eddies of a variety of sizes, and which may be either isotropic or anisotropic in structure. We generally assume that a turbulence system is homogeneous, but it need not necessarily have a three-dimensional isotropy. Important contributions to the theory of turbulence have been made in the more recent past by Batchelor (1953) and Kolmogoroff (1941a and b) and a valuable account of the features of the theory which are relevant to upper atmospheric wind data is given by Elford and Roper (1967). They draw attention to the behaviour of a parcel of air in the atmosphere: the latter restricts the vertical movement of such a parcel through the action of buoyancy forces in an atmosphere possessing a specific thermal structure which applies in particular to the meteor region. After a characteristic time $t_g = \{(g/t)[(dT/dz) + \Gamma]\}^{-1/2}$ ($T=$ absolute gas temperature, $\Gamma$ = adiabatic lapse rate = $(\gamma - 1)T/\gamma H$ with $\gamma$ as the adiabatic index and $H$ the scale height, $g$ = acceleration due to gravity, $z$ = height), the parcel of air will be brought to rest. The corresponding vertical displacement $L$ of an eddy depends on the time

$t_g$, its scale $l$, and the rate $\varepsilon$ at which turbulent energy is dissipated in the atmosphere per unit mass. It can be shown that $L = (\varepsilon l)^{1/3} t_g$. For eddies in the meteor zone $t_g \simeq 40$ sec and is virtually independent of height and the size of the eddy. The rate of dissipation $\varepsilon$ does not appear to vary much with height, and a value of the order of $10^{-1}$ W/kg may be taken as representative in the 80–100 km region. It follows then that $L$ will remain below about 1 km for eddy sizes ranging from a few hundred metres to as much as 100 km! This means that the larger eddies are all strongly anisotropic. This fact should be born in mind when applying the results of turbulence theory in the interpretation of observations of wind fluctuations in the meteor zone, since the theory often only considers isotropic turbulence and is therefore strictly applicable only to the smallest observed scales of less than about 1 km in the horizontal and the vertical.

An interesting feature of upper atmospheric turbulence is its sudden cessation at levels close to, or just above, 100 km altitude. This phenomenon has been investigated in some detail using contaminant release data; see, for example Blamont and De Jager (1961). They refer to the importance of the Reynolds number $R_e \equiv Va/\eta$ where $V$ is a characteristic velocity, $\eta$ the molecular kinematic viscosity, and $a$ a characteristic length in a turbulent system. Turbulence is only generated if $R_e$ exceeds a certain critical magnitude, and the absence of turbulence above a certain height is ascribed to an undercritical value of Reynolds number: for example, if $R_e < 2000$. We must remember, though, that the Reynolds number is important in cases where viscous effects are responsible for the generation of turbulence, usually where the general flow is confined to channels bounded by solid walls so that viscous stresses are set up in the system. Hines (1963) pointed out that such viscous stresses provide an unlikely source of turbulent energy in the upper atmosphere, and he refers to the relevant Richardson number $R_i \equiv (g/T) [(dT/dz) + \Gamma](dv/dz)^{-2}$ with the same meaning for the symbols as above. The equation takes into account the stabilising effect due to buoyancy forces in the presence of temperature gradients, and also the opposing effect of a wind shear which tends to cause instabilities.

### 3.3. *The examination of the characteristics of turbulence at meteor heights*

Turbulence is usually described in terms of statistical parameters, such as averages, because of the extreme variability of individual phenomena. In a statistical analysis of data applying to a turbulence system, the mean square velocity difference between points of given separation $\xi$ plays an important part, and is expressed in terms of the so called velocity structure function:

$$D(\xi) = \overline{[V(x + \xi) - V(x)]^2}$$

where $V(x)$ and $V(x+\xi)$ are the turbulent velocity at the position $x$, and $(x+\xi)$, respectively. $D(\xi)$ is a time average. Care must be taken that the turbulent velocity data are not contaminated by tidal- or prevailing wind components possessing significant gradients vertically or spatially.

For a homogeneous and isotropic turbulence system, the structure fuction $D(\xi)$ may be related to the rate $\varepsilon$ of dissipation of turbulent energy:

$$D(\xi) = 4.82\chi(\varepsilon\xi)^{2/3}$$

where $\chi$ is a constant near unity; see Batchelor (1953), and Zimmerman (1969). We can see that a characteristic power law exists between $D$ and $\xi$, if $\varepsilon$ is a constant quantity. The latter assumption appears to be justified between 80 and 100 km altitude. We may now enquire to what extent the actual data obtained with the contaminant-release and the radio-meteor methods are compatible with the theory.

### 3.3.1. Contaminant release data

Figure 24 shows the velocity structure function for a sodium trail experiment carried out by Blamont and De Jager (1961), together with a best fitting straight line to the linear section of the data series in a log–log co-ordinate system. Such a co-ordinate system is very convenient for the graphical examination of the power law $D(\xi)$, since the exponent is equivalent to the slope of the best-fitting straight line to the data point. Surprisingly, in this case, the slope is observed to be 1.5, in considerable contrast to the expected slope of 0.67 for a turbulent region. There are various other trail release data, obtained independently, which also show considerable departure of the exponent in the power law from 0.67; see, for example, Zimmerman (1969). It has been argued that the observed slopes are all close to the value 4/3, predicted by Tchen (1954) for a turbulence system where a general shear is superimposed on the turbulent motion. However, difficulties arise if we consider the maximum vertical mixing distances of less than 1 km as derived above. Elford and Roper (1967) and Muller (1968) have drawn attention to the fact that the steeper slope of the graph may be attributed to the dominant effect of internal atmospheric gravity waves, which represent a much more ordered motion than that applying to a purely turbulent system.

A special analysis has been carried out by Justus (1966) who took cross sections through contaminant release data in the horizontal plane so that a vertical shear would not affect the structure function $D(\xi)$. Fig. 25 shows that in this case the slope of the best fitting straight line to the data points is that expected for turbulence and that the relationship is constant over a wide range of distances extending between about 100 m and some 7 km and possibly even at larger separations.

### 3.3.2. Radio meteor data

Figure 26 shows data obtained at Adelaide (Elford and Roper, 1967) by the radio meteor method, presented in a similar fashion as the contaminant-release data

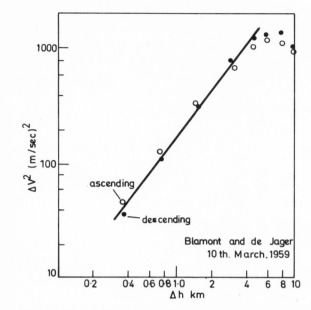

Fig. 24. The velocity structure function obtained by Blamont and De Jager (1961) in an analysis of contaminant release data.

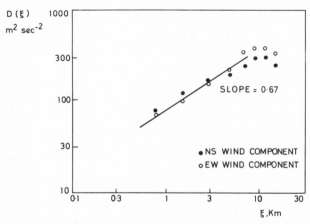

Fig. 25. The velocity structure function obtained by Justus (1966) in an analysis of horizontal variations of the turbulent wind, using contaminant release data.

described in the preceding paragraph. We can see that the exponent in the power law relating the structure function to differences in height is close to 1.4, ruling out a purely turbulent regime. Fig. 27 shows some recent results obtained at Sheffield by Muller (1968), based on meteor trail distortion due to the presence of wind shears. The slope of the best-fitting line to the data points is 1.57, much greater than the slope of 0.67 for a turbulent system, but the slope is also appreciably greater than Tchen's slope of 4/3 for a turbulent model including a uniform wind shear. The Sheffield data obviously lie in the sector bounded by a slope of 4/3 and 2.0, where the latter value applies to the case of a linear or a long-wave sinusoidal shear profile. The Sheffield results thus confirm the dominating effect of ordered wave motions in the upper atmosphere, such as internal gravity waves of longer wavelengths, like the dominant modes described by Hines (1960), the wavelengths of which are of the order of 10 to 20 km.

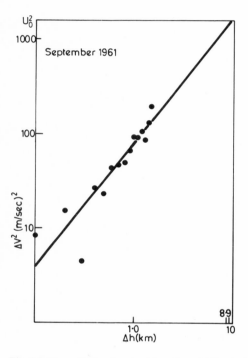

Fig. 26. The velocity structure function obtained at Adelaide in an analysis of radio meteor data.

Greenhow and Neufeld (1960) approached the problem of resolving the turbulent wind in a different fashion. They subtracted the known prevailing and tidal wind components from the observed wind fluctuations, and subjected the residue to a time correlation analysis. Fig. 28 shows their correlogram from which the time constant of the wind fluctuations can be readily obtained. We can see that the zero of the correlation coefficient is reached for a time-constant of 6000 sec. The characteristic scale of the corresponding turbulent system follows, using one of the basic concepts of turbulence theory, as the product of this time and the r.m.s. velocity fluctuations, which Greenhow and Neufeld (1960) found to be about 25 m sec$^{-1}$. The resultant value of 150 km would indicate a highly anisotropic turbulent

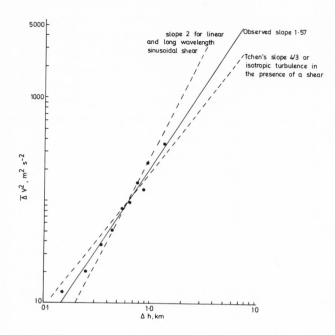

Fig. 27. The velocity stucture function obtained at Sheffield in an analysis of radio meteor data.

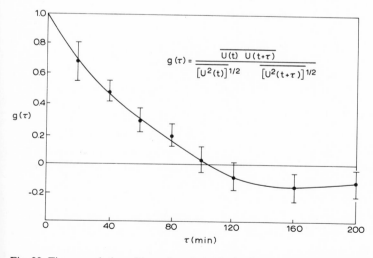

Fig. 28. Time correlation of irregular wind velocity fluctuations, obtained at Jodrell Bank by Greenhow and Neufeld (1960).

system; in fact, we feel more inclined now to reinterpret their results in the light of the gravity-wave hypothesis, and the value of 150 km would then correspond to the scale of the dominant horizontal wave modes in the meteor region.

In order to investigate the vertical structure of the irregular wind Greenhow and Neufeld (1959b) used a two-station technique utilising Doppler-drift information from two reflection points on the same meteor trail. In their analysis they obtained vertical scales of the order of 10 km, which we may again ascribe to the structure of internal gravity waves rather than turbulent features.

### 3.4. The diurnal and seasonal variation of shear and turbulent parameters

It has been known for some time that the diurnal and seasonal variation in the influx of solar energy into the atmosphere causes corresponding effects on a variety of atmospheric parameters, including wind velocity. It is not surprising, then, to find appreciable variations in the irregular component of the upper atmospheric wind, particularly in the seasonal patterns. Diurnal changes are more difficult to observe since it is necessary to monitor the irregular wind continuously over 24 h, and the only successful technique in such investigations is the radio-meteor method.

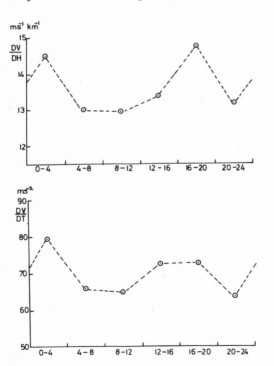

Fig. 29. Diurnal variations of the vertical shear (top) and radial velocity gradients (bottom) at Sheffield, based on averages taken over 12 months.

Using the distortion of meteor trails, the magnitude of the vertical shear has been calculated, and Fig. 29 shows the results of an analysis by Muller (1968), involving data recorded at Sheffield and averaged over a period of 12 months. In the top section of Fig. 29, we notice a regular diurnal variation in the wind shear magnitude with peaks in the early hours of the morning and evening; and in the diagram below, changes in radial velocity which also show a distinct semidiurnal structure are displayed. Both parameters are subject to the effects of gravity waves as well as turbulence, and it is not possible to separate the effects in an elementary fashion. The presence of two peaks in 24 h would perhaps suggest that gravity waves are generated by partial dissipation of the 12-h tidal energy, and turbulence may be the result of a break-down of gravity wave evergy.

Seasonal changes in shear and turbulent parameters are well-marked and appear to be correlated with individual components of the wind. In Fig. 30 we see the rate $\varepsilon$ of dissipation of turbulent energy per unit mass at Adelaide as it varies seasonally, and a comparison may be made with various components of the regular wind. At Adelaide, the dissipation rate is obviously well-correlated with the 24-h tidal wind. A similar analysis carried out at Sheffield does not confirm this result. Fig. 31 shows the seasonal variation of $\varepsilon$ at Sheffield, displaying features similar to those found at Adelaide, but the 24-h tidal wind energy shows little correlation with the turbulent dissipation rate for a 12-month period, as seen in Fig. 32. It would however appear

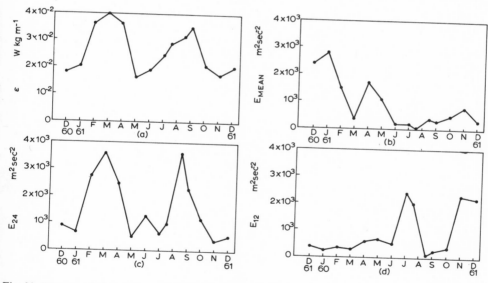

Fig. 30. The rate of dissipation of the turbulent energy per unit mass (a), calculated from the Adelaide radio meter data together with various quantities (b = prevailing wind energy; c = diurnal wind energy; d = semidiurnal wind energies) derived from the steady component of the meteor wind at Adelaide. The mean flow energies are the sums of the squares of the appropriate zonal and meridional wind amplitudes at 91 km. (After Elford and Roper, 1967.)

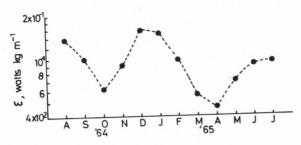

Fig. 31. The rate of dissipation of turbulent energy per unit mass, calculated from the Sheffield radio meteor data.

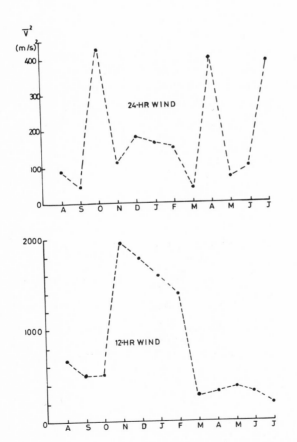

Fig. 32. Energy of the diurnal wind (top) and the semidiurnal wind (bottom), obtained at Sheffield during a period of 12 months.

from the Sheffield data that a connection exists between $\varepsilon$ and the 12-h tidal wind energy (compare Fig. 31 and 32.) Although a correlation between tidal wind energy and the turbulent dissipation rate could indicate a generation of turbulence by a degenerative process involving tidal winds and gravity waves, it is by no means proved that gravity waves are, in fact, a result of a break-down of tidal oscillations in the upper atmosphere. It is more likely that the source of gravity waves lies in the lower levels of the atmosphere where they may be generated by the passing of frontal weather systems or where prevailing winds blow over mountains ridges and they reach meteor heights by oblique propagation, the atmosphere acting as a filter and passing those modes which we recognise as the dominant modes in the meteor region.

## REFERENCES

Batchelor, G. K., *The Theory of Homogeneous Turbulence.* Cambridge University Press, 123 pp., 1953.

Blamont, J. E. and De Jager, C., Upper atmospheric turbulence near the 100 km level. *Ann. Geophys.*, **17**:134–144, 1961.

Chapman, S. and Lindzen, R. S., *Atmospheric Tides.* Reidel, Dordrecht, 200 pp., 1970.

Charney, J. G. and Drazin, P. G., Propagation of planetary-scale disturbances from the lower to the upper atmosphere. *J.Geophys.Res.*, **66**:83–109, 1961.

Elford, W. G., A study of winds between 80 and 100 km at medium latitudes. *Planet. Space Sci.*, **1**:94–101, 1959.

Elford, W. G. and Roper, R. G., Turbulence in the lower thermosphere. In: *Space Research VII* (R.L.Smith-Rose, Editor). North Holland, Amsterdam, pp. 42–54, 1967.

Greenhow, J. S., Systematic wind measurements at altitudes of 80–100 km using radio echoes from meteor trails. *Philos. Mag.*, **45**:471–490, 1954.

Greenhow, J. S. and Neufeld, E. L., The height variation of upper atmospheric winds, *Philos. Mag.*, **1**:1157–1171, 1956.

Greenhow, J. S. and Neufeld, E. L., Measurements of turbulence in the 80–100 km region from radio echo observations of meteors. *J.Geophys.Res.*, **64**:2129–2133, 1959a.

Greenhow, J. S. and Neufeld, E. L., Measurements of turbulence in the upper atmosphere. *Proc.Phys.Soc.*, **74**:1–12, 1959b.

Greenhow, J. S. and Neufeld, E. L., Large scale irregularities in high altitude winds. *Proc.Phys.Soc.*, **75**:228–234, 1960.

Greenhow, J. S. and Neufeld, E. L., Winds in the upper atmosphere. *Q.J.R. Meteorol.Soc.*, **87**:472–489, 1961.

Hines, C. O., Internal atmospheric gravity waves at ionospheric heights. *Can.J.Phys.*, **38**:1441–1481, 1960.

Hines, C. O., The upper atmosphere in motion. *Q.J.R. Meteorol. Soc.*, **89**:1–41, 1963.

Justus, C. G., *Ph.D. Thesis.* Georgia Institute of Technology, Atlanta, Georgia, U.S.A., 1966.

Kaiser, T. R., Upper atmospheric wind parameters from radio meteors. In: *Proceedings of the Workshop on Methods of Obtaining Winds and Densities from Radar Meteor Trail Returns* (A. A. Barnes, Jr. and J. J. Pazniokas, Editors), Office of Aerospace Research, U.S. Air Force, 177–193, 1968.

Kaiser, T. R., Pickering, W. M. and Watkins, C. D., Ambipolar diffusion and motion of ion clouds in the Earth's magnetic field. *Planet.Space Sci.*, **17**:519–552, 1969.

Kantor, A. J. and Cole, A. E., Zonal and meridional winds up to 120 km. *J.Geophys.Res.*, **69**:5131, 1964.

Kato, S., Diurnal atmospheric oscillations. *J.Geophys. Res.*, **71**:3201–3209, 1966a.

Kato, S., Thermal excitation in the upper atmosphere. *J.Geophys.Res.*, **71**:3210–3214, 1966b.

Kent, G. S. and Wright, R. W. H., Movements of ionospheric irregularities and atmospheric winds. Univ. West Indies, Kingston, Jamaica, *Sci. Rep.*, 12, 48 pp., 1967.

Kolmogoroff, A. N., The local structure of turbulence in incompressible viscous fluid for very large Reynolds numbers. *C.R.Acad.Sci.U.S.S.R.*, **30**:301, 1941a.

Kolmogoroff, A. N., Dissipation of energy in locally isotropic turbulence. *C.R.Acad.Sci.U.S.S.R.*, **32**:16 pp., 1941b.

Liller, W. and Whipple, F. L., High-altitude winds by meteor-train photography. *Sp.Supp.J.Atmos.Terr. Phys.*, **1**:112–130, 1954.

Lindzen, R. S., On the theory of the diurnal tide. *Month.Weather Rev.*, **94**:295–301, 1966.

Lysenko, I. A., Orlyansky, A. D. and Portnyagin,Yu. I., A study of wind regime by the meteor radar method at the altitude of 100 km by the meteor-radar method. *Philos.Trans.R.Soc.*, A**271**:601–610, 1972.

Manning, L. A., Villard, O. G. and Peterson, A. M., Meteoric echo study of upper atmospheric winds. *Proc.Inst.Radio Engin.*, **38**:877–883, 1950.

Massebeuf, M., Revah, I. and Spizzichino, A., Zonal semi-diurnal tide and prevailing wind measured at Garchy-sens-Beaujeu (France) from June to September 1966. *Note Tech.* P/38. Groupe de Recherche Ionosphérique, Issy-les-Moulineaux, France, 25 pp., 1969.

Muller, H. G., Atmospheric tides in the meteor zone. *Planet.Space Sci.*, **14**:1253–1272, 1966.

Muller, H. G., Wind shears in the meteor zone. *Planet. Space Sci.*, **16**:61–90, 1968.

Muller, H. G., The Sheffield meteor wind experiment. *Q.J.R. Meteorol. Soc.*, **96**:195–213, 1970.

Muller, H. G., Long-period meteor wind oscillations. *Philos.Trans.R.Soc.*, A**271**:585–598, 1972.

Phillips, E., Wind structure from the amplitude fluctuations in persistent radio meteor echoes. *Planet.Space Sci.*, **17**:553–559, 1969.

Spizzichino, A., Wind profiles in the upper atmosphere from meteor observations. *Note Tech., Etud. Spat.Transm./Rech. Spat.Radioelectr.*, 55. Groupe de Recherche Ionosphérique, Issy-les-Moulineaux, France, 1971.

Tchen, C. M., Transport processes as foundations of the Heisenberg and Obukhoff theories of turbulence. *Phys.Rev.*, **94**:4–14, 1954.

Zimmerman, S. P., Turbulent atmospheric parameters determined from radio meteor trails. *Air Force Cambridge Res.Rep.* 69-0162. U.S. Air Force, 1969.

# COMPOSITION STUDIES IN THE THERMOSPHERE BY MEANS OF MASS SPECTROMETERS

ULF VON ZAHN

*Institute of Physics, University of Bonn, Bonn (Germany)*

## SUMMARY

In the first chapter an introduction to upper atmospheric processes is given which can be studied by means of in situ mass spectrometric measurements. Particular attention is paid to vertical transport and photodissociation processes. The second chapter deals with technological aspects and instruments. Recent developments in ion source technology are discussed in detail as well as modern calibration procedures for space mass spectrometers. The results of upper atmosphere measurements are discussed in the third chapter. Main subjects here are the problem of atomic oxygen density measurements and the strong seasonal variation in helium concentration in the thermosphere as an indicator for transport phenomena.

## 1. UPPER ATMOSPHERIC PROCESSES INVESTIGATED IN SITU BY MEANS OF MASS SPECTROMETERS

### 1.1. Introduction

A full understanding of the properties of the upper atmosphere and of the processes acting in this region can only be obtained by taking into account the fact that the upper atmosphere is actually a weakly ionized plasma. Thus it consists of a neutral part and an ionized part, the latter commonly called the ionosphere. There are numerous ways in which these parts interact with each other. However, at altitudes below 200 km, the degree of ionization of this plasma is rather low ($< 10^{-8}$ at 100 km, $< 10^{-4}$ at 200 km). Therefore, it is not surprising that there are many processes taking place in the neutral atmosphere which can be dealt with, while altogether neglecting the presence of charged particles in the atmosphere.

Mass spectrometers have been used extensively for in situ measurements of aeronomic parameters of both the neutral and ionized atmosphere, predominantly at altitudes between 70 and 250 km. We will concentrate here on neutral atmosphere measurements performed by mass spectrometers: What do we expect to learn from them and what have we already learnt?

Mass spectrometers are basically instruments to determine the absolute number densities $n_i$ (number of particles per unit volume; dimension $cm^{-3}$) of various gases contained in a gas mixture. From these values we can easily derive the total number density, relative composition and molecular weight of the gas sample. We will show later that atmospheric temperatures can also be determined from mass spectrometer measurements. The composition of dry air at ground level is listed in Table I.

TABLE I

Ground-level composition of dry air

| Gas | Fraction by volume |
| --- | --- |
| Nitrogen ($N_2$) | 0.78084 |
| Oxygen ($O_2$) | 0.20946 |
| Argon (Ar) | 0.00934 |
| Carbon dioxide ($CO_2$) | 0.00033 |
| | 0.99997 |
| Neon (Ne) | $1.82 \cdot 10^{-5}$ |
| Helium (He) | $5.24 \cdot 10^{-6}$ |
| Krypton (Kr) | $1.14 \cdot 10^{-6}$ |
| Xenon (Xe) | $8.7 \cdot 10^{-8}$ |

The resulting mean molecular weight is $\overline{M} = 28.966$ g $mole^{-1}$. In the free atmosphere we have to add to the constituents listed in Table I trace gases like $H_2O$, $O_3$, $CH_4$, $N_2O$, $H_2$, $SO_2$, etc. which taken all together do not contribute more than $10^{-5}$ to the total balance (except for the comparatively high water content of the troposphere).

*1.2. Altitude dependence of pressure and number densities below 85 km*

In order to derive the altitude dependence of the pressure, we start out with the hydrostatic equation:

$$dp = -g\rho dz \qquad (1)$$

where $p$ is the (hydrostatic) pressure (dyn $cm^{-2}$), $g$ the gravity acceleration (cm $sec^{-2}$), $\rho$ the mass density ($g \cdot cm^{-3}$) and $z$ the vertical coordinate (cm).

Equation 1 can be combined with the equation of state for an ideal gas:

$$p = nkT \qquad (2)$$

where $n$ is the number density ($cm^{-3}$), $k$ the Boltzmann constant $= 1.38 \cdot 10^{-16}$ erg $°K^{-1}$ and $T$ the absolute temperature ($°K$), to yield:

$$\frac{dp}{p} = -\frac{dz}{H} \qquad (3)$$

where $H = (nkT)/\rho g$.

The density $\rho$ is a function of the molecular masses of the various constituents $m_i$ and their number densities $n_i$. Thus, with:

$$\rho = \sum_i n_i m_i$$

$$n = \sum_i n_i$$

and the mean molecular mass:

$$\overline{m} = \frac{\rho}{n}$$

we obtain:

$$H(z) = \frac{kT(z)}{\overline{m}(z) \cdot g(z)} = \frac{RT(z)}{\overline{M}(z) \cdot g(z)} \tag{4}$$

where $R$ = molar gas constant = $8.31 \cdot 10^7$ erg mole$^{-1}$ °K$^{-1}$ and $\overline{M}$ = mean molecular weight of air, which for dry air is 28.966 g mole$^{-1}$.

The quantity $H$ has the dimension of a length and is called the scale height. Using eq. 2 and 3, one can easily derive the following expressions for the pressure and number density

$$p(z) = p(z_0) \cdot \exp\left\{ - \int_{z_0}^{z} \frac{dz}{H(z)} \right\} \tag{5}$$

$$n(z) = n(z_0) \cdot \frac{T(z_0)}{T(z)} \cdot \exp\left\{ - \int_{z_0}^{z} \frac{dz}{H(z)} \right\} \tag{6}$$

where $z_0$ is the lower boundary of the altitude range.

For a limited altitude interval, we may assume isothermal conditions and $g$, $\overline{M}$= constant and obtain:

$$p(z)/p(z_0) = n(z)/n(z_0) = \exp\left( \frac{z_0 - z}{H} \right)$$

Thus, $H$ determines the altitude interval over which the pressure decreases to $1/e$ of its lower boundary value.

The local scale height at the surface is easily calculated from eq. 4:

$$H(0) = \frac{8.31 \cdot 10^7 \cdot 290}{29.0 \cdot 981} \text{ cm} = 8.5 \text{ km}$$

Bearing in mind that up to 80 km the mean temperature of the atmosphere is about 240°K, we can use as a rule of thumb that the pressure drops by one order of magnitude for each 16-km altitude increase in the lower and middle atmosphere, up to 80 km. Hence at 80 km we expect a pressure of 10 dyn cm$^{-2}$ to which a mean free path of about 0.5 cm corresponds. Furthermore, convection and turbulence are strong enough throughout this region to keep the composition of dry air completely constant (see Table I).

### 1.3. Altitude dependence of number densities and composition above 85 km

When reaching altitudes above 85 km, we enter what is called the upper atmosphere. It is a region of our atmosphere that is distinctively different from the regions below. The differences are all due to the rather low gas density in the upper atmosphere, which leads to: (a) the dominance of diffusion processes over convection and turbulence; (b) notable changes in composition due to photochemical processes; (c) the dominance of eddy conduction and heat conduction as energy transport processes; (d) creation of a permanent ionization and subsequent interactions between the neutral and the ionized part of the atmosphere.

For composition studies in the upper atmosphere, (a) and (b) are of greatest importance. Therefore we shall look at these processes in greater detail.

#### 1.3.1. Transport processes

Vertical gradients in composition are very large compared to horizontal gradients, and transport processes in vertical direction affect the air composition very strongly. In the following, we will deal therefore mainly with this type of process.

We start our investigation of vertical transport processes by first writing down the continuity equation for an atmospheric constituent, assuming that horizontal flux divergence may be neglected:

$$\frac{\partial n_i(z)}{\partial t} = -\frac{\partial \phi_i(z)}{\partial z} + Q_i(z) - L_i(z) \tag{7}$$

where $n_i$ is the number density of the $i$-th constituent, $\phi_i$ is the vertical flux induced by transport processes, and $Q_i$ and $L_i$ are the photochemical production and loss rates, respectively. Here we will only consider two transport mechanisms: molecular diffusion and turbulent mixing. Vertical transport of minor constituents by non-linear processes of gravity waves (Hodges, Jr., 1970), for example, will be neglected. Thus, we can write:

$$\phi_i(z) = n_i(z) \cdot u_i(z) + n_i(z) \cdot v_i(z) \tag{8}$$

where $u_i$ and $v_i$ are the vertical velocities of the $i$-th constituent, due to molecular diffusion and turbulent mixing, respectively.

#### 1.3.1.1. Molecular diffusion.
Molecular diffusion of the $i$-th constituent in a stratified multiple gas mixture may be described by the equation (Curtiss and Hirschfelder, 1949; Chapman and Cowling, 1952):

$$\sum_j p_i p_j (u_i - u_j)/D_{ij} = p \left\{ \rho_i F_i - \frac{\partial p_i}{\partial z} - \frac{\rho_i}{\rho} \sum_j \left( \rho_j F_j - \frac{\partial p_j}{\partial z} \right) \right\} \tag{9}$$

where $p_i$ = partial pressure = $n_i \cdot k \cdot T$; $D_{ij}$ = *mutual* diffusion coefficient for the $i$-th and $j$-th gases (cm$^2$ sec$^{-1}$); $p$ = total pressure; and $F_i$ = external force per unit mass (dyn g$^{-1}$).

According to Colegrove et al. (1966), for a steady-state atmosphere the last term in eq. 9 produces only corrections of a few percent and thus can be neglected. Substituting $F_i = -g$ and $\sum_i n_i = n$, yields:

$$\sum_j n_i n_j (u_i - u_j)/n D_{ij} = -n_i \left\{ \frac{1}{H_i} + \frac{1}{p_i} \cdot \frac{\partial p_i}{\partial z} \right\}$$

where: $H_i = (kT)/(m_i g)$. Since: $\partial p/p = \partial n/n + \partial T/T$ we find:

$$\sum_j n_i n_j (u_i - u_j)/n D_{ij} = - \left\{ \frac{n_i}{H_i} + \frac{n_i}{T} \cdot \frac{\partial T}{\partial z} + \frac{\partial n_i}{\partial z} \right\} \tag{10}$$

Equation 10 does not account for thermal diffusion, and the introduction of thermal diffusion into eq. 10 in a rigorous manner is rather complicated. This process, on the other hand, adds only relatively small effects to the overall behaviour of the atmosphere and can be neglected except for the lightest gases, helium and hydrogen. We therefore will not pursue this point further (for details see Colgrove et al., 1966; Kockarts, 1971). We can now define an *average* molecular diffusion coefficient $D_i$ by $D_i^{-1} = \sum_{j \neq i} n_j/n D_{ij}$ and solve eq. 10 for the density profile:

$$\frac{\partial n_i}{\partial z} = - \left\{ \frac{n_i}{H_i} + \frac{n_i}{T} \cdot \frac{\partial T}{\partial z} + \frac{\phi_i}{D_i} - \frac{n_i}{n} \sum_{j \neq i} \frac{\phi_j}{D_{ij}} \right\} \tag{11}$$

The mutual diffusion coefficients $D_{ij}$ can be computed from (Chapman and Cowling, 1952):

$$D_{ij} = \frac{3}{2(\sigma_i + \sigma_j)^2 (n_i + n_j)} \left\{ \frac{kT(m_i + m_j)}{2\pi m_i m_j} \right\}^{1/2}$$

where $\sigma_i$ is the collision diameter of the $i$-th constituent. Measured values of $D_{ij}$ have been tabulated by Colegrove et al. (1966) and Chapman and Cowling (1970) for example. We say that a gas mixture is in a state of diffusive equilibrium if the divergence of each diffusive flux $\phi_i$ vanishes. For many applications we can use the more stringent assumption that the diffusive fluxes $\phi_i$ are zero. Under this condition eq. 11 reduces to:

$$\frac{1}{n_i} \cdot \frac{\partial n_i}{\partial z} = - \left\{ \frac{1}{H_i} + \frac{1}{T} \cdot \frac{\partial T}{\partial z} \right\} \tag{12}$$

or:

$$-\frac{\partial z}{H_i} = \frac{\partial n_i}{n_i} + \frac{\partial T}{T} = \frac{\partial p_i}{p_i}$$

which yields the partial pressure and number density profile under diffusive equilibrium conditions:

$$p_i(z) = p_i(z_0) \cdot \exp\left\{ - \int_{z_0}^{z} \frac{dz}{H_i(z)} \right\} \qquad (13)$$

$$n_i(z) = n_i(z_0) \cdot \frac{T(z_0)}{T(z)} \cdot \exp\left\{ - \int_{z_0}^{z} \frac{dz}{H_i(z)} \right\} \qquad (14)$$

The important difference to the mixed atmosphere described by eq. 5 and 6 is that, under diffusive equilibrium conditions, each constituent follows an altitude profile characterized by its own molecular mass through $H_i = kT/m_i g$. The partial pressure and number density of light gases like helium and atomic oxygen will therefore decrease slowly with altitude, whereas the pressure of argon, for example, will decrease rather rapidly.

This process of diffusive (or gravitational) separation is, together with the photodissociation of $O_2$, the most important cause for the drastic differences in composition of the upper atmosphere with respect to the lower atmospheric layers. Fig. 1, taken from a recently published model atmosphere, exemplifies the influence of this process on the composition of upper atmosphere air. In Fig. 1, $T_\infty$ is the

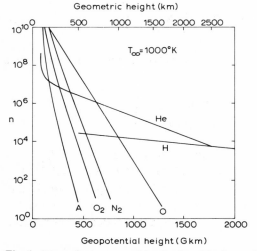

Fig. 1. Atmospheric composition in the altitude range 100 to 2500 km (upper scale) for an exosphere temperature $T$ of 1000°K. Note the diffusive separation of the different gases and the extended altitude range in which atomic oxygen becomes the dominant constituent. (From Jacchia, 1971.)

temperature which the atmosphere attains at altitudes above 400 km. This so-called exospheric temperature is (almost) constant with altitude because of the high thermal conductivity of the gas.

J.C.G. Walker (private communication, 1966) pointed out that for an atmosphere in diffusive equilibrium the ratio of number densities of three constituents can be written as:

$$\frac{n_1^2(z)}{n_2(z) \cdot n_3(z)} = \frac{n_1^2(z_0)}{n_2(z_0) \cdot n_3(z_0)} \cdot \exp\left\{ (m_2 + m_3 - 2m_1) \int_{z_0}^{z} \frac{g}{kT} dz \right\}$$

If the three constituents fulfill the condition $2m_i = m_2 + m_3$ the ratio $n_1^2/n_2 n_3$ becomes independent of altitude, temperature, and time, as long as the lower boundary conditions $n_1(z_0)$, $n_2(z_0)$, $n_3(z_0)$ remain constant. The most important example is:

$$\frac{n^2(N_2)}{n(O) \cdot n(Ar)} \tag{15}$$

Over the short times of sounding rocket flights, the lower boundary conditions can be assumed to be invariant. Hence, the altitude dependence of the ratio (15) provides for a quantitative check on how well the measured composition represents an atmosphere in diffusive equilibrium.

It is worth mentioning that (15) is also a constant for a fully mixed atmosphere. Therefore the constancy of (15) is a necessary, but not sufficient condition for diffusive equilibrium.

For another sensitive test of diffusive equilibrium conditions we write eq. 12 in the form:

$$k \cdot \frac{\partial(n_i T)}{\partial z} = -n_i m_i g$$

which yields after integration:

$$T(z) = \frac{1}{n_i(z)} \left\{ n_i(z_0) \cdot T(z_0) - \frac{m_i}{k} \int_{z_0}^{z} n_i(z) \cdot g(z) dz \right\}$$

Therefore from each measured constituent profile $n_i(z)$ one can calculate a temperature profile $T(z)$ by (downward) integration of the number density profile. The upper boundary value $T(z_0)$ has to be taken from an appropriate atmosphere model. The integration converges rapidly and the calculated $T(z)$ becomes reasonably independent of $T(z_0)$ at about $z_0 - 20$ km.

Under ideal equilibrium conditions the temperature profiles derived from the various constituents must be alike. No test for equilibrium conditions on this basis has been published so far.

A very important question is how long it takes for the atmosphere to adjust its composition from a mixing distribution to a diffusive equilibrium distribution. Mange (1957) has shown that this time constant $\tau_i$ is given by: $\tau_i = A_i \overline{H}^2/D_i$ where the constant $A_i$ depends strongly on the gradient of scale height $d\overline{H}/dz$ and may be calculated at each level for a specific change $n_i(t)/n_i(t = 0)$. Because $D_i$ is inversely proportional to the number densities, the diffusion times $\tau_i$ decrease rapidly with increasing altitudes. Nicolet (1960) calculated that at an altitude of 115 km more than one day but less than one week is required for the various constituents to establish diffusive equilibrium. Hence, at altitudes above 120 km the atmosphere should come rather close to a diffusive equilibrium distribution.

The lowest altitude at which diffusive separation sets in is determined by the strength of turbulence and in the next paragraph we will investigate this latter effect.

*1.3.1.2. Turbulence (eddy diffusion).*   Turbulent mixing is responsible for the fact that the atmosphere remains well mixed, i.e., $\overline{M} = $ constant, up to altitudes of about 90 km. In addition, turbulence will produce a downward heat transport in a statically stable atmosphere, and dissipation of turbulent energy will noticeably heat the atmosphere.

We have little firm knowledge about the forces which cause this turbulent exchange, and in the same manner the spatial and temporal scales of these motions are rather uncertain. Theories describing turbulent processes in the upper atmosphere are often complicated and difficult to compare with experimental results.

An approach to the evaluation of the effects of turbulence was proposed by Lettau (1951) and has been further developed and used by Colegrove et al. (1965, 1966), Shimazaki (1967), Shimazaki and Laird (1970), and Keneshea and Zimmerman (1970). In their work, turbulence is considered as a kind of macroscopic diffusion process to which the mathematical formalism developed for molecular diffusion is applied. One can think of turbulence as causing an exchange of many parcels of air, that may have any size and shape, which move as coherent units, but conserve a certain property $r$ (for example ratio of $Ar/N_2$) during a displacement of typical length $l$, the mixing length. After this displacement, exchange with the atmosphere dissolves the air parcel, and its property $r$ is mixed with the respective property $r'$ of the ambient atmosphere.

If seen in the framework of ordinary diffusion theory, we may call "eddy diffusion" the process of turbulent mixing. One has, however, to emphasize that in this context the word "eddy" does not imply anything like a rotary motion, but instead has the more general meaning of macroscopic exchange.

Following this line of thought, the vertical flux $\phi_i$ due to molecular and eddy diffusion (eq. 8) is expressed as:

$$\phi_i = -D_i\left\{\frac{n_i}{H_i} + \frac{n_i}{T}\frac{\partial T}{\partial z} + \frac{\partial n_i}{\partial z} - \frac{n_i}{n}\sum_j \frac{\phi_j}{D_{ij}}\right\} - K\left\{\frac{n_i}{\overline{H}} + \frac{n_i}{T}\frac{\partial T}{\partial z} + \frac{\partial n_i}{\partial z}\right\} \quad (16)$$

Here the second term on the right-hand side describes mass transport by eddy diffusion, and $K$ is the so-called eddy diffusion coefficient ($cm^2$ $sec^{-1}$). If eddy diffusion by far exceeds molecular diffusion ($K \gg D_i$), the atmosphere attains a mixing distribution with an altitude dependence characterized by $\bar{H} = kT/\bar{m}g$. The upper boundary of the mixing region is called the turbopause and is by definition where $K = D_i$. A consequence of this definition is that we have to think of the turbopause as being located at slightly different altitudes for the various atmospheric constituents.

Equation 16 can easily be solved for the number density profile $\partial n_i/\partial z$. But this offers no new insights here and hence will not be performed.

Two important questions can be asked directly. First, at what altitude is the turbopause, i.e., does $K$ equal $D$; and second, what is the altitude dependence of $K$ in this important region? Today we assume that the turbopause is somewhere between 100 and 105 km, but we know almost nothing about its "fine structure". In addition, because of the lack of experimental data, Colegrove et al. (1966) and Shimazaki and Laird (1970) use in their atmosphere models a single value for $K$ over the entire altitude range of interest (Shimazaki and Laird also have one model with $K$ decreasing below 100 km). This choice, however, was disputed by Keneshea and Zimmerman (1970) who argue that $K$ ought to drop rapidly at and above the turbopause level (Fig. 2).

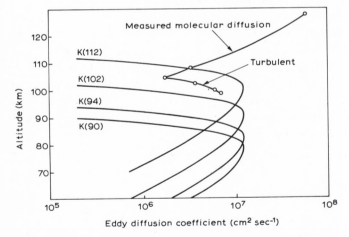

Fig. 2. Measured molecular and turbulent diffusion coefficients, and theoretical eddy diffusion coefficients used for atmospheric model calculations. The numbers in parentheses indicate the altitude at which turbulence ceases. (From Keneshea and Zimmerman, 1970.)

Mass spectrometric determinations of the ratio $n(Ar)/n(N_2)$ at and about turbopause altitudes offer the most direct means to determine the absolute value of $K$ and, for a limited range, its altitude dependence. We will discuss respective results later on.

Measurements of the ratio $n(O)/n(O_2)$ have also been used by atmospheric model makers to derive eddy diffusion coefficients. This ratio is *not* well-suited to this purpose because it is strongly affected by photochemical processes which contain some unknowns, and furthermore, a reliable determination of atomic oxygen number densities by means of mass spectrometers is much, much more difficult than in the case of argon and nitrogen.

*1.3.1.3. Global wind system.* At the solstice, the heat input to the upper atmosphere by absorption of solar radiation exhibits a strong global asymmetry. Johnson and Gottlieb (1970) suggested that this radiation unbalance sets up a large-scale circulation pattern, which moves air in the thermosphere and mesosphere from the summer to winter polar regions with a return circulation at lower altitudes. Horizontal winds towards the winter pole with velocities of 1 m sec$^{-1}$ at 60 km, increasing to 30 m sec$^{-1}$ at 120 km above the equator are balanced by vertical movements above the polar regions, with velocities ranging from less than 1 cm sec$^{-1}$ in the mesosphere up to 10 cm sec$^{-1}$ at an altitude of 120 km. The latter value implies a vertical flux of helium of the order of $2 \cdot 10^8$ cm$^{-2}$ sec$^{-1}$ which is a factor 100 larger than the estimated average escape flux for helium (see section 1.3.1.4.).

Vertical movements above the turbopause can induce notable changes in atmospheric compositions because under diffusive equilibrium conditions, the composition is strongly altitude dependent. For example, a downward motion will bring helium-rich air to lower altitudes, and we expect to observe an anomalously high ratio $n(He)/n(N_2)$ above the turbopause when the downward transport occurs faster than diffusive equilibrium can be re-established for these constituents.

Johnson and Gottlieb (1970) did some approximate calculations on the effectiveness of this circulation system for the transport of helium. Their results indicate that such a circulation pattern may explain the observation of a helium-enriched zone over the winter poles, the so-called helium-bulge (Jacchia and Slowey, 1968; Keating and Prior, 1968). Mass spectrometer measurements again are the most direct way, though not the only means, to investigate these seasonal variations in upper air composition.

*1.3.1.4. Miscellaneous transport processes.* It is well known that during geomagnetic storms energy is deposited in the lower thermosphere, and this leads to an expansion of the heated atmospheric region. The vertical motion caused by this expansion may result in drastic changes in the composition of the higher atmospheric layers, in particular when the expansion occurs in a time period that is short when compared to the time required for re-establishing diffusive equilibrium. Again mass spectrometers, in particular satellite-borne instruments, can be used to trace out the temporal and spatial behaviour of these atmospheric disturbances.

Another mechanism for the vertical transport of minor constituents was proposed by Hodges, Jr. (1970). It is based on the fact that owing to the propagation of internal gravity waves, large amplitude fluctuations of number densities and temperature occur in the lower thermosphere. Hodges pointed out that these density fluctuations do not cancel in the time average, producing a diffusion-type transport: lighter gases are carried upward and heavier gases downward. Owing to this transport phenomenon, the altitude profile of minor constituents, in particular that of helium, may show noticeable deviations from a diffusive equilibrium distribution in the lower thermosphere.

Finally, one more vertical transport process shall be mentioned. It is the upward flow through the thermosphere which sustains the escape of the light gases helium and atomic hydrogen from the earth's gravitational field. The Maxwellian velocity distribution for these atoms extends with a small, but noticeable component beyond the escape velocity $v_c = 10.8$ km sec$^{-1}$ (at 500 km altitude). Therefore, atoms contained in the high energy tail of the Maxwellian distribution are continually evaporating into space. Another important escape process works through ionizing the atoms in the upper ionosphere and accelerating these ions outwards along the open magnetic field lines, extending from the polar regions to outer space (Axford, 1968; Banks and Holzer, 1968).

For the case of helium, Nicolet (1957) from the terrestrial production of He estimated the vertical flux to be $2 \cdot 10^6$ cm$^2$ sec$^{-1}$. As we have shown in section 1.3.1.3 this flux is much smaller than those induced by vertical air motions. Therefore, it no longer seems likely that the escape flux can be measured in an easy and direct manner. In the case of hydrogen, we simply lack any mass spectrometric measurement to compare with theoretical predictions. Therefore, we will not discuss further details of this transport process, and refer the reader to the excellent review of this subject by Kockarts (1971).

### 1.3.2. Photochemical processes

The source of energy for photochemical processes is short wave solar radiation. Fig. 3 shows that above 85 km, wavelengths with $\lambda < 1750$ Å are absorbed. Note that the graph is valid for overhead sun. The altitude of absorption will get higher as the sun comes closer to the horizon.

In Table II, ionization thresholds are listed for the most important atmospheric gases. We see that most gases have ionization thresholds below 1000 Å, except for NO and $O_2$. The latter has a threshold very close to 1000 Å. We therefore make a somewhat arbitrary, but convenient division between ultraviolet radiation (UV) with wavelengths longer than 1000 Å and extreme ultraviolet radiation (EUV) with wavelengths smaller than 1000 Å. Absorption of UV leads mostly to *dissociation* processes, whereas absorption of EUV leads mostly to *photoionization*.

Absorption of EUV is one of the important heating processes for the atmospheric layers above 120 km. It has a strong and direct influence on the temperature

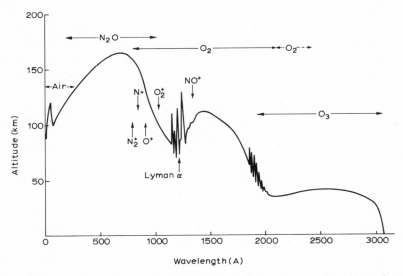

Fig. 3. Altitude at which the intensity of solar radiation drops to $1/e$ of its value outside the earth's atmosphere, for vertical incidence (From Hines et al., 1965.)

TABLE II

Photoionization thresholds for atmospheric gases

|        | Wavelength (Å) | Ionization potential (eV) |
|--------|----------------|---------------------------|
| NO     | 1340           | 9.25                      |
| $O_2$  | 1027           | 12.07                     |
| H      | 912            | 13.59                     |
| O      | 910            | 13.61                     |
| $CO_2$ | 900            | 13.77                     |
| N      | 852            | 14.54                     |
| $N_2$  | 796            | 15.58                     |
| Ar     | 787            | 15.76                     |
| He     | 504            | 24.56                     |

structure of the thermosphere. The composition of the neutral upper atmosphere, however, is affected by these processes only in a rather complex manner. Composition studies of the neutral atmosphere are therefore not an efficient way to learn more about details of EUV absorption, e.g., how much of this energy is transformed to heat, and in which way this is done. Notable examples are number density measurements of trace constituents like NO and N which are produced through secondary reactions following the photoionization of $N_2$, O and $O_2$.

Photodissociation, on the other hand, modifies the composition of the atmosphere in two very direct ways: firstly, through the formation of atomic oxygen, and secondly through destruction in the altitude range 70 to 100 km of almost all trace gases like $H_2O$, $O_3$, $H_2$ which make the physics of the atmosphere so rich and interesting.

Atomic oxygen is produced in the thermosphere by the reaction:

$$O_2(^3\Sigma_g^-) + h\nu \rightarrow O(^3P) + O(^1D)$$

where absorption takes place in the Schumann—Runge continuum ($1300 < \lambda < 1750$ Å). Recombination occurs through three-body collision:

$$O + O + M \rightarrow O_2 + M$$

where the third body M carries away the excess energy and momentum. This process becomes effective only below 90 km altitude (Johnson and Gottlieb, 1970), so that the lifetime of O atoms at 100 km are of the order of one month, and at 120 km, more than a year (Nicolet, 1960).

On the other hand, the lifetime of $O_2$ molecules in unattenuated sunlight will be inversely proportional to the absorption cross section in the Schumann—Runge continuum (which is well known), and the solar flux (which is not well known). If the sun radiated at 1450 Å like a black body of 6000°K, the lifetime of $O_2$ would be less than one hour, for 5000°K less than one day, for 4500°K of the order of 10 days, and at 4000°K about four months. In Fig. 4 a summary is presented of recent solar UV flux measurements. A black body temperature of about 4600°K is indicated at 1450 Å which implies that an $O_2$ molecule will survive many days in unattenuated sunlight before it dissociates. Therefore $O_2$ is fairly stable in the thermosphere too.

Atomic oxygen is of paramount importance for both the neutral and ionized part of the upper atmosphere for the following reasons:

(1) Owing to its small atomic weight and to diffusive separation, O becomes the dominant constituent of the neutral atmosphere from 160 km to at least 500 km. For example, generally 90% of all particles are O atoms at 300 km altitude.

(2) O, like $N_2$ and $O_2$, has little possibility of radiating in the infra-red. Thus, most of the heat deposited in the thermosphere by EUV absorption must be carried downward by heat conduction. But since this transport mechanism becomes ineffective in the lower thermosphere, huge temperature gradients are building up before the downward energy transport through heat conduction can balance the energy input into the thermosphere. These steep gradients, in turn, result in rather high exospheric temperatures of $(1000\pm500)$ °K. Contrary to terrestrial conditions the upper atmospheres of Venus and Mars contain sizable amounts of $CO_2$ which radiate away "in-situ" excess energy. This leads to rather lower exospheric temperatures: Venus 650°K, Mars 400°K.

(3) The atoms act as carriers for chemical energy, because the dissociation energy of 5.1 eV for $O(^3P) + O(^3P)$ is released only when the atoms

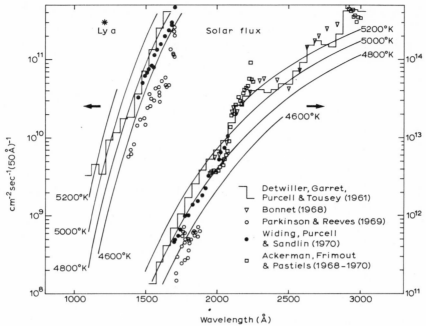

Fig. 4. Flux of solar photons at one astronomical unit distance from the sun. Note the marked flux discontinuity at 2000 Å. Recent measurements indicate considerably smaller flux in the important region around 1500 Å compared to earlier data obtained by Detwiler et al. (1961). (From Ackerman, 1971.)

recombine into molecules. Recombination, however, takes place at quite different locations than dissociation.

(4) O is very reactive and at lower altitudes destroys trace constituents like $O_3$ , $O_2^-$ , $O_4^+$ , etc.

(5) For the physics of the ionosphere, the transition from a molecular atmosphere to an atomic atmosphere above 160 km or so leads to drastic changes in ion density and composition because atomic ions recombine much slower than molecular ions.

Because of the great importance of atomic oxygen for aeronomy, considerable effort has been expanded to calculate its altitude profile. Examples of this effort are Nicolet (1960), Colegrove et al. (1966), Hesstvedt (1968), Shimazaki and Laird (1970), Keneshea and Zimmerman (1970). Here I will only mention the two basic difficulties arising in these model calculations: the densities found in the thermosphere are mainly governed by the production rates and transport fluxes in the vicinity of the turbopause. The production rates, however, are uncertain because we do not know the solar flux in the Schumann—Runge continuum well enough (see Fig. 4), and the transport processes are still difficult to evaluate because the effective eddy diffusion coefficient $K$ and its altitude dependence is another poorly-known parameter.

In Fig. 5 comparison is made between the predictions of a number of model atmospheres with respect to the ratios $n(O)/n(O_2)$ and $n(O_2)/n(N_2)$. We realize that theoreticians still provide us with a wide choice of possible solutions and it ought to be one of the main tasks for mass spectrometer measurements to discriminate between these various models.

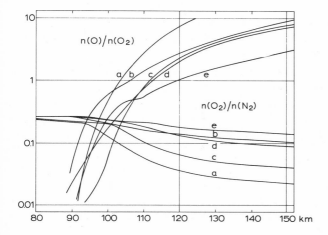

Fig. 5. Calculated ratios $n(O)/n(O_2)$ and $n(O_2)/n(N_2)$ in lower thermosphere; (a) Zimmerman et al. (1971); cessation of turbulence at 94 km; (b) Jacchia (1971); exospheric temperature = 900°K which represents average conditions; (c) Zimmerman et al. (1971); cessation of turbulence at 102 km; (d) Colegrove et al. (1966); ratio $n(O)/n(O_2)$ fixed to a value of 2 at 120 km; (e) CIRA (1965) Mean Atmosphere.

As regards the photodissociation of minor constituents, later we will see that mass spectrometers have so far contributed no data to this rather important point. Therefore I will omit here a discussion of this subject.

### 1.3.3. The six variations in upper atmosphere density

The data on atmospheric density derived from satellite drag measurements have quite clearly shown six different types of density variations in the upper atmosphere. Roemer (1971) has reviewed this aspect of aeronomy during last year's course. The different types of variations are: (1) the diurnal variation; (2) the variation with the 27-day solar rotation period; (3) the semi-annual variation; (4) the seasonal/latitudinal variation; (5) the variation with the 11-year cycle of solar activity; (6) the variation connected with geomagnetic activity.

Naturally, all these variations are superimposed on each other and one needs a tremendous amount of observational data before one can separate their individual amplitudes, periods and phases. For the study of *total density* variations one can use all the numerous observations on satellite drag. For the study of the respective *composition* changes, however, the situation is rather bad. In order to single out one of the six variations, one would need an extensive series of rocket flights. This, however, is rather expensive and does not necessarily assure success because an unexpected severe geomagnetic storm, for example, may strongly, but unwantedly, influence part of the results.

Some of these effects can be investigated by satellite-borne mass spectrometers; in particular, variations with geomagnetic activity and seasonal/latitudinal variations. On the other hand, because of the peculiarities of satellite orbits, it is rather difficult to investigate the diurnal variation in composition with a single satellite. Another limitation of satellite techniques is their altitude range which does not allow measurements in the denser part of the thermosphere (below 150 km).

So, it appears that the study of the six well-known variations by means of rocket-borne mass spectrometers better concentrates on some characteristic features within these effects, like the seasonal variation of the ratio $n(He)/n(N_2)$. At the same time we should give up the hope that the diurnal variation can be determined by a single pair of two sounding rocket flights, as has been tried in the past by many workers, including myself. To obtain a significant result, a minimum of six rockets, but preferably a dozen, would have to be launched within 24 hours.

## 2. TECHNOLOGICAL ASPECTS AND INSTRUMENTS

### 2.1. General problem areas

Since the very first flight of a mass spectrometer into the upper atmosphere in 1949 (O'Day, 1954), vacuum technology, electronics design, and calibration methods have advanced considerably so that today it is quite easy to build better instruments than in the early days of space mass spectrometry. There remain, however, two principal problems for which an ideal solution is still to be found: first, measurements under conditions of thermal non-equilibrium between the object of measurement (air sample) and sensor (mass spectrometer); and second, measurements in the presence of reactive gases (like O). We will elaborate on these points in turn.

#### 2.1.1. Thermal non-equilibrium

The temperature of a mass spectrometer ion source $T_s$ will be somewhat higher than room temperature: e.g., in the 300—350°K range due to heating by the filament. Gas temperatures $T_a$ in the lower thermosphere under average geophysical conditions are 325°K at 120 km, 790°K at 200 km and 900°K at 500 km (Jacchia, 1971). Mass spectrometric measurements have been limited so far to altitudes above 110 km because conventional instruments can operate only at pressures below $10^{-4}$ Torr, and because of the assumption of free molecular flow conditions around the payload which is contained in the following mathematical models. It is immediately evident then, that there is a temperature difference between the air and the instrument which results in a difference between the gas density at the ion source and the ambient atmosphere. Therefore, the densities measured in the actual ion source have to be corrected in order to derive the ambient number densities.

The second and much more serious reason for thermal non-equilibrium stems from the fact that the mass spectrometer travels through the air with a velocity $v_r$ which is high when compared to the average thermal velocity of the gas particles $\bar{v}_a = (8RT_a/\pi M)^{1/2}$. For a typical lower thermosphere temperature of $T_a = 500°$K and nitrogen ($M = 28$), this amounts to $\bar{v}_a = 0.61$ km/sec. On the other hand, this velocity is reached by a rocket after a free fall of 20 km. After 120 km of free fall, the velocity is 1.5 km/sec and the nitrogen molecules hit the ion source with an energy equivalent to a temperature of 3000°K. At satellite speeds of 8 km/sec, this equivalent temperature is 85,000°K (for $N_2$)!

In order to derive a relationship between the particle densities in the ambient atmosphere and inside the ion source, we will first consider the model of the so-called *closed* ion source. This source consists of a spherical enclosure (= "accommodation sphere") which is connected with the atmosphere by a small orifice (see Fig. 6). The number density inside the sphere is sensed by the ion source proper which is connected to the accommodation sphere at its lower end. The following assumptions are made: (1) the mean free path is large compared to the chamber dimensions; (2) the area of the orifice is much smaller than the surface of the accommodation sphere; (3) no absorptions or reactions of gas particles at or with the walls take place.

If the instrument is moving with constant velocity through the atmosphere, the particle flux into the source $\phi_{in}$ must equal the flux leaving the source $\phi_{out}$. Under assumption 2, particles make many collisions with the walls inside the accommodation sphere before they leave the sphere again. They will have reached complete thermal equilibrium with the surface temperature $T_s$ and have a velocity distribution according to the usual Maxwell—Boltzmann distribution.

The flux $\phi_{out}$ leaving the enclosure through the orifice is thus given by kinetic theory:

$$\phi_{out} = \frac{1}{4}n_s\bar{v}_s$$

where $n_s$ is the number density inside the ion source and $\bar{v}_s$ is the average thermal velocity at source temperature $= (8RT_s/\pi M)^{1/2}$.

The flux $\phi_{in}$ into the enclosure is similarly given by:

$$\phi_{in} = \frac{1}{4}n_a\bar{v}_a F(S)$$

where $n_a$ is the ambient number density, $\bar{v}_a$ the average thermal velocity at ambient gas temperature $= (8RT_a/\pi M)^{1/2}$, $S$ the speed ratio $v_r/v_{pa}$, $v_r$ the component of rocket velocity *normal* to entrance orifice and positive towards the inside of the

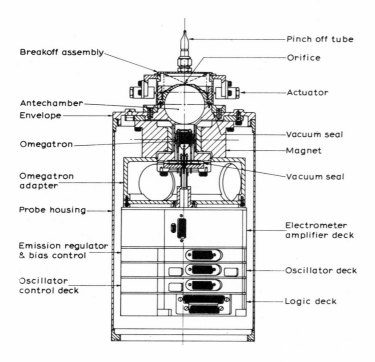

Fig. 6. Closed ion source with accommodation sphere (here called antechamber) attached to omegatron mass spectrometer. Ion-source cap is ejected at altitudes of about 140 km for this instrument. (From Simmons et al., 1968.)

enclosure and $v_{pa}$ the most probable thermal velocity of ambient particles $= (2RT_a/M)^{1/2}$.

The velocity correction function $F(s)$ can be calculated by integrating the Maxwell—Boltzmann distribution in a moving coordinate system, multiplied by the component of molecular velocity normal to the orifice of the enclosure (Schultz et al., 1948). Expressed in terms of the speed ratio $S$, it is:

$$F(S) = \exp(-S^2) + \sqrt{\pi} \cdot S[1 + \mathrm{erf}(S)] \tag{17}$$

where:

$$\mathrm{erf}(S) = 2(\sqrt{\pi})^{-1} \int_0^S \exp(-x^2)\,dx$$

Equating $\phi_{in}$ and $\phi_{out}$ results in the basic relationship between the ion source and the ambient density:

$$n_s = n_a (T_a/T_s)^{1/2} F(S) \tag{18}$$

Equation 18 makes allowances for both causes of thermal non-equilibrium; i.e., the difference between gas and source temperature as well as high instrument speed. In general, the second effect is the greater because of the behaviour of the $F(S)$ function (see Table III).

TABLE III

Functional dependence of $F(S)$ on $S$

| $S$ | $F$ |
|-----|-----|
| $-3$ | $10^{-5}$ |
| $-1$ | $0.09$ |
| $0$ | $1$ |
| $+1$ | $3.63$ |
| $+3$ | $10.6$ |

$F(S) = 3.54S$ is a good approximation for $S > 2$. Note that by definition $S \equiv v_r/v_{pa}$, which equals $1.13v_r/v_a$

It is evident from eq.18 that the number density inside a forward-looking ion source is greatly enhanced over the ambient density ("ram effect"), whereas a source looking into the wake of the payload senses a much decreased density. An ion source mounted at the side of a spinning rocket or tumbling payload therefore experiences drastic density modulations. Even if the rocket travelled with only $v_r = v_{pa}$, the density modulation inside the source could reach a ratio of $3.63/0.09 = 40/1$ during one-half revolution of the rocket. Fig. 7 shows a well-resolved example of such density modulation.

For the data analysis of an actual flight experiment one has to derive $n_a$ from the measured $n_s$ and $T_s$, as well as the parameters $T_a$ and $S$. $T_a$ is usually taken in first approximation from an appropriate model atmosphere. To obtain $S$, we have to calculate first $v_r$, for which we need to know the angle between the normal to the source orifice and the instantaneous velocity vector, the so-called angle of attack for the instrument. This information is derived from an attitude-sensing system in the payload, typically consisting of a combination of magnetometers and solar or moon sensors. The higher the vehicle speed and the larger the angle of attack, the more important the accuracy of such attitude data becomes.

The shape of the density modulation curve shown if Fig.7 is given by the velocity distribution of the ambient particles, the depth of modulation is a direct measure for the local ambient temperature $T_a$. Mass spectrometers therefore offer the interesting opportunity to determine the local gas temperature by evaluating the observed density modulation inside the ion source.

2.1.2. Measurement of reactive gases and trace constituents

Closed ion sources as described in the foregoing paragraph offer the great

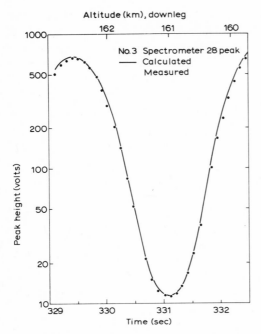

Fig. 7. Example of density modulation inside closed ion source due to spinning mass spectrometer. The calculated modulation profile, derived from ion source model and attitude information, closely agrees with the measured profile. Data were taken with spectrometer fixed-tuned to molecular nitrogen. (From Hedin and Nier, 1966.)

advantage of having the well-known relationship (eq.18) between $n_s$ and $n_a$. Thus they solve in principle the problem of thermal non-equilibrium conditions. Equation 18, however, holds only under the assumption of negligible absorption processes and chemical reactions of the particles at the source walls. This requirement is certainly not fulfilled for reactive gases and radicals like O, H, OH, etc. Therefore, we have to face two difficulties: firstly, a certain percentage of these particles will be lost inside the source and we have somehow to estimate the loss coefficient; secondly, these loss processes lead to many reaction products like $CO_2$, CO and $H_2O$ which grow so abundant inside the ion source that it becomes absolutely impossible to measure the true content of these trace gases in the free atmosphere.

   In order to minimize the number of wall collisions, it is tempting to use a wide *open*, nude ion source instead of the closed source. We will schematize its geometry by a flat plate moving through the atmosphere with a velocity component $v_r$ normal to the plate. We again write the particle density $n_s$ in the close vicinity of the plate (as sensed by the electron beam) as the sum of those particles $n_{in}$ moving toward the plate and those leaving the plate $n_{out}$ after one reflection: $n_s = n_{in} + n_{out}$.

Assuming full accommodation during the collision with the wall and diffuse reflection with Maxwellian energy distribution afterwards, we get:

$$n_s = \frac{n_a}{2}\left\{1 + \text{erf}(S) + (T_a/T_s)^{1/2} F(S)\right\}$$ (19)

But what is probably even more important here is the ratio of $n_{out}/n_{in}$, which is:

$$\frac{n_{out}}{n_{in}} = \frac{F(S)}{1 + \text{erf}(S)}\sqrt{\frac{T_a}{T_s}}$$ (20)

For sounding rockets in the lowest part of the thermosphere, $T_a \simeq T_s$, but $S$ can be about 3. It follows that:

$$\frac{n_{out}}{n_{in}} = \frac{10.6}{1 + 1} \cdot 1 \simeq 5$$

Due to the low velocity of the reflected particles after accommodation, the total number density in front of the plate is mainly determined by the reflected particles. The same is true at higher altitudes where $T_a \gg T_s$, even so, $S$ may become small close to apogee. We can summarize by stating that the wide open ion source still dominantly determines the density of *reflected* particles. We can use eq.19 for the calculation of the ambient density $n_a$ only with the assumption of negligible loss processes at the wall. In particular, for reactive gases this assumption is highly questionable. This is the crux of many mass spectrometer measurements, and we will now discuss the attempts to overcome this difficulty.

## 2.2. Instruments

### 2.2.1. Ion sources
Because of the general problems outlined above, the most important part of a neutral gas mass spectrometer for space applications is its ion source. Early designs (Townsend, 1952) were semi-closed, which caused great losses of O inside the source and made a quantitative deduction of $n_a$ from $n_s$ rather difficult. Schaefer (1963) was the first to fly a wide open ion source for the purpose of minimizing the probability of surface recombination of atomic oxygen. A schematic view of this source is shown in Fig. 8. Similar sources were employed with mass spectrometers onboard the satellites OGO-2 and -4 by Hinton et al. (1969).

Schaefer was the first to show that the ratio $n(O)/n(O_2)$ is larger than 1 at an altitude of 120 km. Because of the lack of attitude sensors in his payload, he was not able to derive accurate absolute number density values. One disadvantage of Schaefer's wide open ion source was that it does not provide a good approximation to the idealized flat plate, nude source model which forms the basis for eq.19. Therefore, the research groups of the Universities of Minnesota and Bonn have used semi-open sources which resemble a small recess in a flat plate and for which

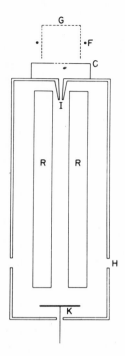

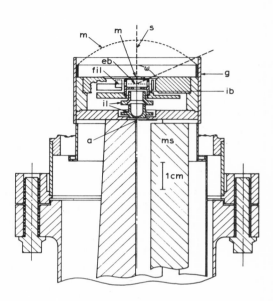

Fig. 8. Wide open ion source which essentially consists only of high transparency grids attached to massenfilter; $G$ = accelerating grid; $F$ = filament; $C$ = ground plane; $I$ = ion inlet port; $R$ = rods of quadrupole field; $H$ = breather holes; $K$ = ion collector. (From Schaefer and Nichols, 1964.)

Fig. 9. Semi-open ion source where ionization volume lies inside a small cylindrical recess. $fil$ = filament; $eb$ = electron beam; $ib$ = ionization box; $il$ = ion lenses; $m$ = high transparency grids; $a$ = entrance aperture to analyzing field; $ms$ = monopole spectrometer; $g$ = ground shield; $s$ = rocket spin axis.

accurate correction formulas have been given by Hedin et al. (1964). A schematic view of one of our sources is shown in Fig. 9.

For the measurement of inert gases like $N_2$, Ar, and He, the closed source is definitely the optimal design because the assumptions for its correction function (eq.18) are best fulfilled. An advanced design has been already shown in Fig. 6.

The standard material used for these sources was stainless steel. Attempts have been made to suppress the background gases by gold-plating the ion source parts, but they have met with little real success. Interesting experiments have been performed to enhance or suppress recombination of O to $O_2$ at the ion source walls. Reber and Harpold (1967) have coated the interior surfaces of closed ion sources with silver oxide in an effort to obtain a high coefficient of recombination for atomic oxygen. The flight results, however, indicated considerably less recombination than expected. Philbrick et al. (1970) employed ion sources made of pure titanium in order to reduce the amount of recombination of atomic oxygen. This attempt was successful insofar as the recombination coefficients indeed turned out to be very small. Even in a closed source with many wall collisions of gas particles, large mass 16 peaks were observed. However, pronounced adsorption and slow desorption processes of atomic oxygen strongly affected the measurements and made the data analysis rather complex.

The hot filament used inside the ion sources for emitting the electrons is a potential source of background gases. This may be true in particular in an oxygen-rich gas mixture as found in the upper atmosphere. In order to learn more about the relative importance of surface reaction at the filament versus ion source walls, Trinks (1971) built an electron impact ion source without filament. As shown in Fig.10, an electron current (of the order of 40 $\mu$A) is drawn from the last dynode of an electron multiplier and directed across the flat front plate of the instrument. The multiplier is excited by a small UV lamp so that there is no part of the spectrometer with an elevated temperature. This type of instrument has been flown four times so far; we experienced vehicle failures three times and at the fourth try the ion source cap did not eject for unknown reasons. Really bad luck!

The latest approach to this problem was taken by my group at Bonn. The starting point for this development was the evidence that almost all our problems are caused by those particles which become reflected, desorbed or otherwise emitted from the ion source walls. In order to completely suppress those unwanted particles, Offermann and Trinks (1971) built a source whose outside and inside surfaces are cooled by helium down to temperatures of 8°K and lower (except for the filament). The atmospheric gas enters this source (see Fig. 11) through a small orifice in the front plate. All incoming particles that are not ionized in their passage through the electron beam immediately behind the entrance orifice are removed from the gas phase by the efficient cryo-pumping of the ion-source walls. Thus the influence of the walls in the measured density becomes very small. The very encouraging results of the first flight of such an instrument will be reported later. Because of its inherently high pumping speed the Offermann source also holds great promise for future mass spectrometric measurements at low altitudes, i.e., say, above 70 km.

### 2.2.2. Analyzing fields

For measurements in the upper atmosphere, mass spectrometers need only give medium resolving power because the neutral air spectrum is rather simple. A resolution of 40 (i.e., a 10% valley between 40 and 41) seems sufficient. In addition, the analyzing field should be small and suited to a rugged design. These require-ments are so undemanding that almost any type of analyzing field arrangement can be adapted for space applications.

If we consider the use of mass spectrometers for the investigation of both the neutral and ionized atmosphere, the most frequently used analyzing fields are the

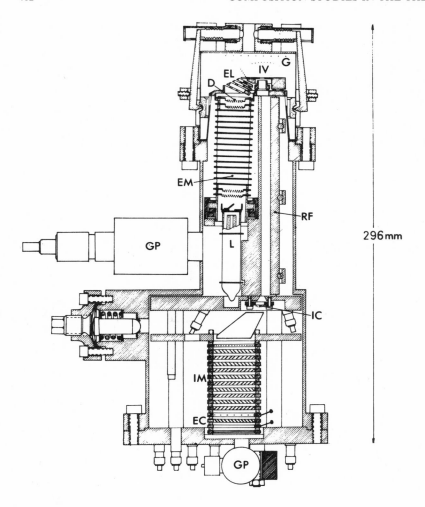

Fig. 10. Monopole spectrometer employing filament-less electron impact ionization source. UV lamp *L* excites electron multiplier *EM*. Current extracted from last dynode *D* is guided by lenses *EL* and front grid *G* through ionization volume *IV*. *RF* = electrode of monopole spectrometer. Intermediate ion collector *IC* collects 50% of analyzed ion current. Rest goes to ion multiplier *IM* and collector *EC*. *GP* = ion getter pumps. (From Trinks, 1971.)

quadrupole massenfilter (Paul et al., 1958) and monopole spectrometer (Von Zahn, 1963) types. They are used by the following research groups: University of Michigan, U.S. Air Force Cambridge Research Laboratories (AFCRL), University of Bonn, Max Planck Institut Heidelberg, Goddard Space Flight Center (GSFC). Single-focusing magnetic spectrometers have been used extensively by the University of Minnesota, and a few times by the University of Bern; double focusing magnetic spectrometers by the GSFC, and recently also by the University of

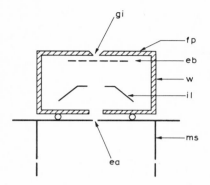

Fig. 11. Helium-cooled ion source for neutral atmosphere measurements. $gi$ = gas inlet; $fp$ and $w$ = helium cooled frontplate and walls of ion source (at $\sim 8°K$); $il$ = ion lens; $ea$ = entrance aperture; $ms$ = mass spectrometer case at room temperature. (From Offermann and Von Zahn, 1971.)

Minnesota. Many successful flights using omegatrons have been made by one group at the University of Michigan and the GSFC. Russian scientists still prefer the Bennet RF spectrometer for neutral atmosphere measurements, but in the western world it is now only used for ion composition measurements. Detailed description of space mass spectrometers have been published by Townsend (1952), Schaefer and Nichols (1961), Taylor et al. (1962), Spencer and Reber (1963), Bailey and Narcisi (1966), Niemann and Kennedy (1966), and Mauersberger et al (1968b). Actual instrumental layouts are shown in Fig. 6, 10, 12, 13 and 14.

Again, it is the opinion of the author that almost any analyzing field can be adapted to space use. Each one has its small advantages and disadvantages. It would be too time consuming and would put too much importance on the details if we discussed all of them. So we had better turn to the necessary ion current detection system.

### 2.2.3. Detector systems

During the past decade it was standard practice to collect the mass analyzed ion currents in Faraday cups which were connected to electrometer circuits for amplification of the collected ion currents. Assuming a sensitivity of $10^{-4}$ A Torr$^{-1}$ for the spectrometers, the major constituents $N_2$, O, and $O_2$, can be detected up to about 200 to 250 km without a multiplier. Because of its lower abundance, Ar will be lost at altitudes between 140 and 180 km. In the case of satellite instruments, even He can be measured without a multiplier because the longer time available for measurements allows significantly smaller ion currents to be detected, as compared with the faster scanning sounding rocket instruments.

The demand for higher sensitivity and shorter scan time has led to the introduction of multipliers as a current detection system. The original motivation was the desire for helium density measurements, and Fig. 12 and 13 show examples of

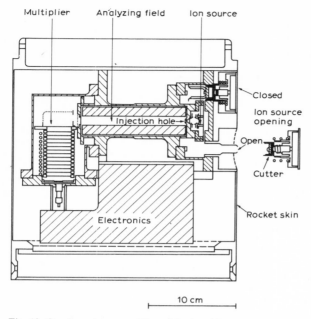

Fig. 12. Quadrupole massenfilter with closed ion source and multiplier, particularly designed for helium measurement. (From Müller and Hartmann, 1969.)

instruments where multipliers serve in particular for this purpose.

Since our first flight of a multiplier equipped spectrometer in 1966 (Hartmann et al., 1968), we have developed this technique further so that all masses are measured by means of the multiplier. This allows us to determine Ar profiles up to 200 km (Von Zahn and Schneppe, 1971), and provides a highly interesting check on ion currents at mass positions $14(N^+)$ and $30(NO^+)$, as well as the behaviour of all the background gases.

The gain of such a multiplier does not need to be large because these instruments are basically limited in their sensitivity by current statistics. Let us assume that one mass peak is scanned in 0.1 sec and that it needs a total number of 100 ions to produce a reasonably well defined peak. This indicates that the useful minimum current is about $10^{-16}$ A. Any good modern electrometer can measure $10^{-12}$ A with acceptable time constants. Thus, a gain of $10^4$ for the multiplier fully satisfies our needs because it cannot help improve the current statistics. As already mentioned, spectrometers without multipliers may have a sensitivity of $10^{-4}$, A Torr$^{-1}$, so that by adding a multiplier this sensitivity increases to an useful limit of the order of 1 A Torr$^{-1}$.

The long-time stability of these low gain multipliers is fairly good. But even so, we have provided in our instruments for a double collector system which allows us to determine the instantaneous gain of the multiplier also during flight. As shown in

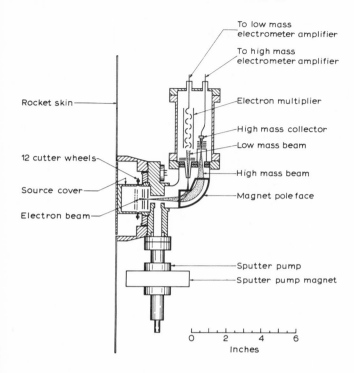

Fig. 13. Single focusing magnetic mass spectrometer with semi-open ion source. An electron multiplier is used for detection of masses smaller than 8. The 1 liter sec$^{-1}$ ion getter pump maintains a pressure differential between source and analyzer during flight. (From Nier, 1967.)

Fig. 10, the ion current passes through a grid system behind the analyzing field which collects about 50% of the ion current and feeds a linear electrometer. The remaining 50% enters the multiplier, becomes converted into electrons, and is amplified until it reaches the electron collector which feeds a logarithmic electrometer. At intermediate pressures of $10^{-7}$ to $10^{-6}$ Torr, the signals can be read off both electrometers and the ratio of collected electron to ion currents determines the multiplier gain (see Fig.15).

This double collector system allows us to measure the mass dependence of the multiplier gain. This is of paramount importance for species like atomic oxygen ions for which we have no laboratory calibrations facilities. Experience has shown us that the multiplier gain for $O^+_2$ ions and $O^+$ ions can be quite different for one multiplier, but almost alike for another. Fig. 15 gives a rather interesting example of two rocket flights performed within 4 hours with mass spectrometers of identical design. The ion current ratios $i^+$ $(16^+)/i(32^+)$ are unity at almost the same altitude for both flights. Yet the current ratios $e^-(16^+)/e^-(32^+)$ measured behind the multipliers differ by a factor 4 at the same altitudes for both flights. For a discussion

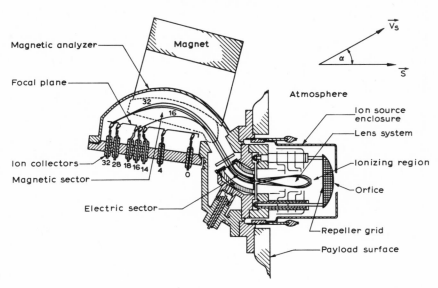

Fig. 14. Double focusing magnetic mass spectrometer which is permanently tuned to six mass numbers. (From Cooley and Reber, 1969.)

of miscellaneous techniques and accessories like programmed scanning, use of ion-getter pumps, ejectable ion source caps, mounting requirements for the instruments etc., the reader is referred to Von Zahn (1968).

### 2.3. Calibration procedures

The ion current measured at the collectors behind the analyzing field is given by:

$$I_i = S_{ni} n_i \tag{21}$$

where $I_i$ is the ion current measured when the mass number $M_i$ is focussed to the collector (A), $n_i$ the number density of $M_i$ inside the ion source (cm$^{-3}$) and $S_{ni}$ the so-called sensitivity factor (A cm$^3$).

For laboratory calibrations eq.21 is commonly written in terms of partial pressure $p_i$:

$$I_i = S_{pi} p_i \tag{22}$$

where $p_i$ is the partial pressure $i$-th gas (Torr) and $S_{pi}$ the sensitivity factor (ATorr$^{-1}$)

The sensitivity $S_{pi}$ is given by:

$$S_{pi} = A W_i p_i f_{Ti} G_i$$

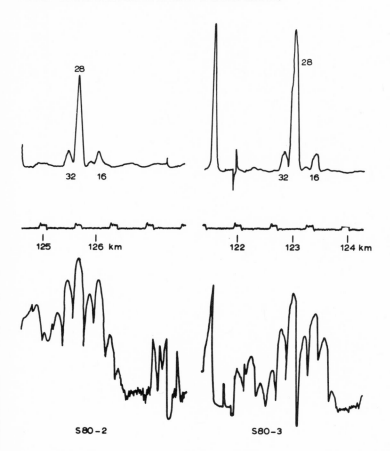

Fig. 15. Flight spectra taken with monopole spectrometers and semi-open ion sources. Programmed scan was used which allowed to scan only across the important constituents. Upper traces show $i^+$ - electrometer output (linear characteristic). Lower traces show $e^-$ -electrometer output (with logarithmic characteristic). Zeroline of $i^+$ -electrometer showed small regular modulation with frequency of rocket spin during flight. Note the apparent increase of $16^+$ at $e^-$ -electrometer of S80-3.

Further information on experiments:

| Payload | ESRO S80-2 | ESRO S80-3 |
|---|---|---|
| Date of launch | Feb. 7, 1971 | Feb. 7, 1971 |
| Time of launch | 00.22.12 CET | 04.45.00 CET |
| Apogee | 226 km | 214 km |
| Current ratios | | |
| 16/ 32 at $i^+$ -collector | 0.89 | 1.0 |
| 16/ 32 at $e^-$ -collector | 1.03 | 4.8 |

where $A$ is the dimension factor (independent of mass), $W_i$ the electron impact ionization cross section, $f_{Ti}$ the transmission factor describing the probability for any ion formed in the ion source to reach the ion collector and $G_i$ the gain of the multiplier system, including transmission through deflection system in front of multiplier. $G_i = 1$, if no multiplier is employed.

One has to emphasize that $W_i$ is not the total ionization cross section for a certain molecule like $O_2$; rather, it is the cross section to form a specific molecular ion like $O_2^+$, or $O_2^{2+}$. Thus, in general, we need literature values for both the total as well as dissociative ionization cross sections in order to calculate $W_i$.

Because neither $f_T$ nor $G$ can be calculated from straight theory, one has to measure $S_p$ and $G$ in the laboratory for all the gases of interest, and this is called a spectrometer calibration. During such calibration, one also has to check at what pressure the transmission factor starts depending on pressure.

In the past many calibration systems used Bayard–Albert gauges for pressure reference which in turn were calibrated against McLeod manometers. This method has the disadvantages of chemical reactions at the hot gauge filament upon introduction of oxygen, plus strongly mass-dependent gauge sensitivity and considerable power dissipation into the calibration chamber. Therefore, we at Bonn calibrate our spectrometers by means of a gas flow method (see Fig. 16). A membrane manometer is used as a reference pressure instrument, and its absolute

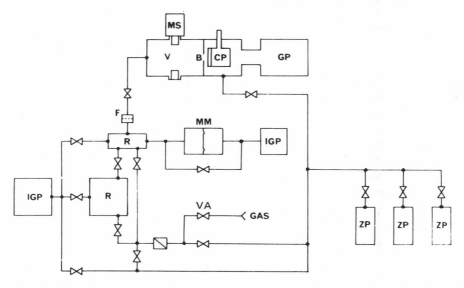

Fig. 16. Calibration system for mass spectrometer using gas flow method. Gas is introduced through porous glass plate $F$ into calibration volume $V$ to which mass spectrometer $MS$ is attached. Gas is pumped through orifice $B$ into main pump $CP$ which is helium-cooled. 220 liter $sec^{-1}$ ion getter pump $GP$ is used as standby pump. Pressure readings are derived from membrane manometer $MM$. Gas from mixing system (not shown) is let into reservoir $R$ through valve $VA$.

calibration is obtained by volume expansion techniques (Müller, 1967; Schneppe, 1969). The accuracy of the absolute pressure determination inside the calibration volume is about 2% in the range $10^{-4}$ Torr to $10^{-7}$ Torr, and 3% at $10^{-8}$ Torr.

Gross (1970) showed that the hot filaments of mass spectrometers "pump" molecular oxygen with pumping speeds of the order of some decilitres/sec. Therefore, the pumping speed of the orifice B should be of the order of 50 liter sec$^{-1}$ to make the influence of the filament negligible. In order to make also back streaming through orifice B negligible, the pump behind B must have a pumping speed much greater than 50 liter sec$^{-1}$, preferably some 2000 liter sec$^{-1}$. Because of this requirement for high pumping speeds, and also because of its zero background, we use helium-cooled cryopumps as the main pump CP. The helium reservoir of the pump shown in Fig. 17 is under atmospheric pressure and at 4.2°K. This temperature, however, is still too high to give a sufficient pumping speed for good $H_2$ and He calibrations. Therefore, we have now installed a cryopump working with low pressure helium which reaches 2.8°K and which should allow easy $H_2$ and He calibrations.

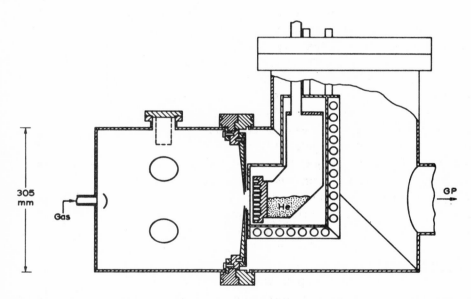

Fig. 17. Helium-cooled main pump *CP* of Fig. 16 in greater detail. Helium here is under atmospheric pressure (4.2°K), but lower pressures and temperatures are required for $H_2$ and He calibrations.

A very special problem is the determination of the mass spectrometer sensitivity for atomic oxygen $S_p(O)$. It has been impossible to carry out an absolute calibration of this kind because no source for atomic oxygen was available which provided a well-known pressure or flux of O atoms at pressures of $10^{-6}$ Torr. Therefore, the value of $S_p$ (O) only was calculated using:

$$S_p(O) = S_p(O_2) \frac{W(O)}{W(O_2)} \cdot \frac{f_T(O) \cdot G(O)}{f_T(O_2) \cdot G(O_2)}$$

where $S_p(O_2)$ is the measured sensitivity for $O_2$, $W(O)$ and $W(O_2)$ are the literature values for respective ionization cross sections, $f_T(O_2) \cdot G(O_2)$ is the measured value of transmission times the multiplier gain for $O_2$ and $f_T(O) \cdot G(O)$ is the value of transmission times the multiplier gain for O interpolated on the transmission curve.

So we can see that it becomes a rather long way before one can quote a sensitivity for atomic oxygen which is a prerequisite for the calculation of absolute atmospheric O densities.

The difficulty in obtaining a valid calibration of the spectrometer sensitivity for atomic oxygen led Niemann (1969) to design a very promising atomic oxygen beam generator for the laboratory evaluation of mass spectrometers used in upper atmosphere research. As outlined in Fig. 18, the atomic oxygen is generated by thermal dissociation of molecular oxygen on the surface of a tungsten filament heated to 2800°K. A symmetrical bidirectional beam is produced to permit continuous monitoring of the particle flux in the beam while spectrometer calibrations or other experiments are being conducted. Fig. 19 shows details of the actual atomic oxygen source. At flux levels below $10^{13}$ particles/cm$^2$ sec, relative O concentrations of more than 90% were obtained. At higher fluxes, the degree of dissociation was somewhat lower. Fig. 20 presents the direct electrometer output of a mass spectrometer attached to this beam apparatus. The ratio of ion currents $16^+/32^+$ still has to be divided by the ratio of ionization cross sections $W(O)/W(O_2) = 0.77$ (at 75 eV electron energy) in order to obtain the density ratio $n(O)/n(O_2)$. It is also of great interest to see here the same large amounts of $CO_2$ (44$^+$) and CO (28$^+$) present under laboratory conditions, and which we always find during mass spectrometer experiments in space (Von Zahn, 1967). So it appears that Niemann's technique provides us for the first time with a powerful source of atomic oxygen, the flux of which is reasonably well known. It remains to be verified, however, whether the surfaces of an ion source calibrated at such beam apparatus are stable enough to maintain the sensitivity $S_p(O)$ unchanged from lab calibration throughout the flight.

I wish to conclude this chapter with the general remark that the task of performing a good spectrometer calibration today is at least as involved as building the instrument. Careful calibrations require bakable UHV systems preferably without oil diffusion pumps, expensive pressure monitors or flow meters, high speed pumps and atomic beam apparatus. The amount of spectra collected during one complete instrument calibration (about 10 gases or gas mixtures at many pressure settings and read-out of 2 or 3 electrometers having more than one sensitivity range) is so great that digitized data collection and automatic data evaluation becomes

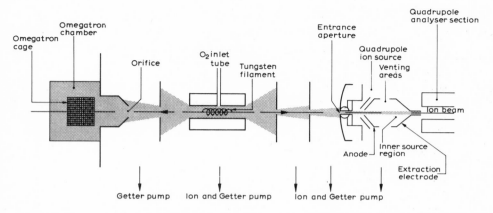

Fig. 18. Schematic layout of atomic oxygen beam apparatus. Two mass spectrometers are used to study the behaviour of atomic oxygen beam, an omegatron on the left side and a quadrupole filter on the right side. (From Niemann, 1969.)

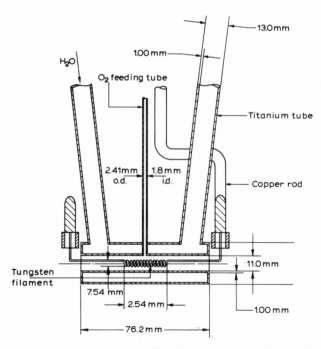

Fig. 19. Cross-sectional view of atomic oxygen source. Tungsten filament is heated to 2800°K and walls are water-cooled. (From Niemann, 1969.)

## TABLE IV

Number densities at 150 km (from Von Zahn, 1970)

| Payload | Date | Time | Latitude | $n(N_2)$, $10^9$ cm$^{-3}$ | $n(O_2)$, $10^9$ cm$^{-3}$ | $n(O)$, $10^9$ cm$^{-3}$ | Method used[*1] | Reference |
|---|---|---|---|---|---|---|---|---|
| | Aug. 23, 1961 | 1004 | 33° N | | 4.2 | | EUV | Hinteregger, 1962 |
| | March 6, 1962 | (0600) | 33° N | | 3.1 | | EUV | Jursa et al., 1963 |
| NASA 6.06 | Nov. 20, 1962 | 1700 | 38° N | 19.6[*2] | | | MS | Spencer et al., 1970[*3] |
| NC3 115 F | June 6, 1963 | 0730 | 33° N | 19.8 | 2.05 | 6.60 | MS | Hedin et al., 1964 |
| | July 10, 1963 | 1004 | 33° N | 21. | 3.3 | 21. | EUV | Hall et al., 1965 |
| NASA 6.08 | July 20, 1963 | 1647 | 38° N | 35.5 | | | MS | Spencer et al., 1970 |
| NASA 6.09 | Jan. 28, 1964 | 2154 | 38° N | 11.3 | | | MS | Spencer et al., 1970 |
| | March 2, 1965 | 1713 | 32° N | | 2.1 | | EUV | Hall et al., 1967 |
| | March 3, 1965 | 0719 | 32° N | | 2.7 | | EUV | Hall et al., 1967 |
| | March 4, 1965 | 1715 | 32° N | | 2.4 | | EUV | Hall et al., 1967 |
| NASA 4.127 | April 15, 1965 | 0345 | 33° N | 21.9 | 2.56 | 2.81 | MS | Hedin and Nier, 1966 |
| NA 14.244 | Sept. 15, 1965 | ? | 38° N | | 1.9 | | EUV | Weeks and Smith, 1968 |
| NASA 18.03 | Nov. 9, 1965 | 1316 | 59° N | 28.5[*2] | | | MS | Spencer et al., 1970 |
| NASA 18.02 | Nov. 10, 1965 | 0100 | 59° N | 23.7[*2] | | | MS | Spencer et al., 1970 |
| ESRO CO7/2 | Dec. 11, 1965 | 0443 | 40° N | 11.6 | 1.49 | 11.0 | MS | Mauersberger et al., 1968a |
| | March 2, 1966 | 1259 | 38° N | 33 | 2.1 | 21. | EUV | Hall et al., 1967 |
| NASA 8.25 | March 2, 1966 | 1300 | 38° N | 24.0[*4] | | | MS | Reber and Cooley, 1967; Pelz and Newton, 1969 |
| NASA 18.05 | Aug. 26, 1966 | 1327 | 38° N | 35.4 | | | MS | Spencer et al., 1970 |
| NASA 18.06 | Aug. 26, 1966 | 1347 | 38° N | 37.3 | | | MS | Taeusch and Carignan, 1968* |
| NASA 18.22 | Aug. 27, 1966 | 2300 | 38° N | 27.4 | | | MS | Spencer et al., 1970 |
| NASA 4.181 | Nov. 30, 1966 | 0445 | 32° N | 26.3 | 2.87 | 9.74 | MS | Kasprzak et al., 1968 |
| NASA 4.180 | Dec. 2, 1966 | 1409 | 32° N | 26.7 | 2.47 | 6.45 | MS | Kasprzak et al., 1968 |
| BB3/ 1 | Dec. 12, 1966 | 1305 | 59° N | 32.4[*5] | | | MS | Müller and Hartmann, 1969; Von Zahn and Gross, 1969 |
| ETR-1474 | Jan. 24, 1967 | 0326 | 28° N | 21.4 | | | MS | Spencer et al., 1970 |
| ETR-1828 | Jan. 24, 1967 | 0618 | 28° N | 20.2 | | | MS | Spencer et al., 1970 |
| ETR-1165 | Jan. 24, 1967 | 0935 | 28° N | 24.2 | | | MS | Spencer et al., 1970 |
| ETR-0381 | Jan. 24, 1967 | 1400 | 28° N | 23.3 | | | MS | Spencer et al., 1970 |
| ETR-0611 | Jan. 24, 1967 | 1712 | 28° N | 25.2 | | | MS | Spencer et al., 1970 |
| ETR-0851 | Jan. 24, 1967 | 2126 | 28° N | 23.4 | | | MS | Spencer et al., 1970 |
| ETR-1942 | April 25, 1967 | 0155 | 28° N | 32.9 | | | MS | Spencer et al., 1970 |
| ETR-4803 | April 25, 1967 | 1425 | 28° N | 31.8 | | | MS | Spencer et al., 1970 |
| NASA 4.179 | June 21, 1967 | 1249 | 32° N | 34.5 | 2.28 | 7.56 | MS | Krankowsky et al., 1968 |
| NASA 4.212 | July 20, 1967 | 0200 | 32° N | 18.6 | 1.40 | 2.10 | MS | Krankowsky et al., 1968 |
| NASA 4.211 | July 20, 1967 | 1224 | 32° N | 27.1 | 2.46 | 2.70 | MS | Krankowsky et al., 1968 |
| NASA 18.50 | Sept. 18, 1967 | 1408 | 38° N | 28.1 | | | MS | Spencer et al., 1970 |
| ESRO S19/1 | Oct. 4, 1967 | 1401 | 40° N | 28.3 | 2.00 | 16.1 | MS | Bitterberg et al., 1970 |
| ESRO S19/2 | Oct. 10, 1967 | 1511 | 40° N | 40.5 | | | MS | Bitterberg et al., 1970 |
| SL 407 | Nov. 1, 1967 | 1430 | 31° S | | 1.7[*6] | | EUV | Wildman et al., 1969 |
| NASA 18.53 | March 17, 1968 | 0253 | 18° N | 25.8 | | | MS | Spencer et al., 1970 |
| NASA 18.49 | March 17, 1968 | 0452 | 18° N | 27.5 | | | MS | Spencer et al., 1970 |
| NASA 18.51 | Aug. 8, 1968 | 1403 | 38° N | 21.7 | | | MS | Spencer et al., 1970 |
| NASA 18.56 | Aug. 9, 1968 | 0155 | 38° N | 27.5 | | | MS | Spencer et al., 1970 |
| BB3/2 | Dec. 4, 1968 | 1251 | 59° N | 19.2 | 2.56 | 11.8 | MS | Gross and Von Zahn, 1971 |
| NASA 4.272 | Feb. 4, 1969 | 0835 | 59° N | 31.1 | 3.49 | 8.96 | MS | Hickman and Nier, 1970 |

[*1] EUV, solar EUV absorption measurement; MS, mass spectrometric measurement.
[*2] Adjusted for upleg-downleg difference.
[*3] $n(N_2)$ values from this reference have been increased by a factor of 1.2 (see text).
[*4] Geoprobe mass spectrometer.
[*5] Average of both mass spectrometers.
[*6] Extrapolated from 148-km altitude upward.

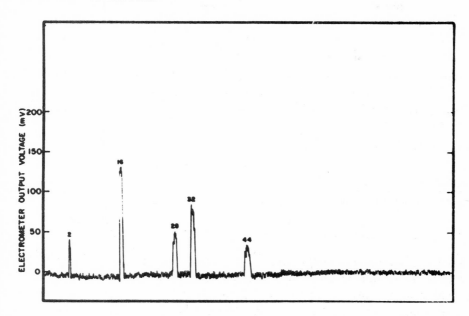

Fig. 20. Typical spectra obtained with quadrupole massen-filter attached to the atomic oxygen beam apparatus of Niemann. Flux of $O_2$ into apparatus was $10^{14}$ molecules $cm^{-2}$ $sec^{-1}$. (From Niemann, 1969.)

almost mandatory. This part of the overall experiment will become increasingly more important as we continue to leave behind us the "pioneer decade" of space mass spectrometry which, I would say, ended in the late 1960's.

## 3. DISCUSSION OF RESULTS OBTAINED SO FAR

### 3.1. Mean value of number densities at 150 km altitude

Most experiments with neutral gas mass spectrometers in the upper atmosphere have been performed in the altitude range 150–200 km. Von Zahn (1970) reviewed the results obtained at 150 km. His conclusions will be summarized briefly in the following paragraph.

Table IV (from Von Zahn, 1970) presents a summary of absolute number density determinations of $n(N_2)$, $n(O_2)$, and $n(O)$. It is reproduced here to give an impression about the scatter in the original data. But what I consider to be more important is a comparison of the results from the various research groups, which is presented in Table V.

We note a rather impressive agreement of all groups with respect to the mean number densities of $N_2$, $O_2$, and Ar. It is also quite evident that the COSPAR International Reference Atmosphere (see CIRA, 1965) requires considerable revision as regards its composition.

TABLE V

Number density averages at 150 km (from Von Zahn, 1970)

| Group | Method | $n(N_2)$, $10^9$ cm$^{-3}$ | $n(O_2)$, $10^9$ cm$^{-3}$ | $n(O)$, $10^9$ cm$^{-3}$ | $n(Ar)$, $10^7$ cm$^{-3}$ |
|---|---|---|---|---|---|
| Univ. Michigan and GSFC | MS | 26.2 (22*) | | | |
| Univ. Minnesota | MS | 25.7 (8) | 2.45 (8) | 5.87 (8) | 5.4 (8) |
| Univ. Bonn | MS | 26.4 (5) | 2.02 (3) | 13.0 (3) | 4.4 (4) |
| AFCRL and other groups | EUV | 27 (2) | 2.61 (9) | 21  (2) | |
| Averages | | 26.2 (37) | 2.46 (20) | ? (13) | 5.1 (12) |
| CIRA 1965 Mean Atmosphere | | 33.4 | 4.78 | 14.2 | 18.0 |

* Numbers in parentheses are the number of measurements.

## 3.2. The problem of atomic oxygen

Table V clearly points out the great discrepancies which still exist among the results from the various groups concerning the mean atomic oxygen density. EUV absorption measurements indicate more than three times higher O densities than the mass spectrometer measurements of the Minnesota group, for example. This variance exists in spite of the excellent agreement on the densities of the other constituents. The O densities in the thermosphere depend mostly on the solar flux in the Schumann–Runge continuum and on the eddy diffusion coefficient at 100 km altitude. The explanation that solar flux variations at 1450 Å cause the different $n(O)$ results can be ruled out. If the eddy diffusion coefficient shows strong temporal and spatial variations, then this must show up also in corresponding changes of the argon densities. The group averages on Ar, however, agree nicely.

There are other oddities too. Let us take an average of only the mass spectrometer determinations of $n(O)$. It comes out to be $n(O) = 8 \cdot 10^9$ cm$^{-3}$. Now we calculate the total density from the measured mean densities of Table V using $n(N_2) = 2.6 \cdot 10^{10}$ cm$^{-3}$, $n(O_2) = 2.5 \cdot 10^9$ cm$^{-3}$, $n(Ar) = 5 \cdot 10^7$ cm$^{-3}$, and $n(O) = 8 \cdot 10^9$ cm$^{-3}$, and obtain $\rho = 1.55 \cdot 10^{-12}$ g cm$^{-3}$. This density, however, is considerably below the value of about $\rho = 2.0 \cdot 10^{-12}$ g cm$^{-3}$ derived from a number of satellite drag measurements at the 150 km altitude. So we have to explain a considerable density deficit between these two sets of data. Another serious argument was raised by Von Zahn (1967) after Colegrove et al. (1966) had published their model atmospheres which took into account eddy diffusion in the lower thermosphere. It became apparent that none of the models was able to fit in an acceptable way both the ratios $n(O)/n(O_2)$ and $n(Ar)/n(N_2)$, as measured by mass spectrometers. In order to fit the measured $n(O)/n(O_2)$ values, one needs considerably higher eddy diffusion coefficients than those required for fitting the $n(Ar)/n(N_2)$ data. The Ar and $N_2$ measurements between 120 and 150 km are, on the other hand, the easiest measurements a space mass spectrometer can make. Therefore, I consider them to be much more reliable than the $n(O)$ measurements.

For these reasons and for some other less serious ones I proposed some time ago (Von Zahn, 1970) to reject for the time being all absolute values of $n(O)$ determined by means of rocket-borne mass spectrometers. The new reference set of average number densities and total densities which I proposed is compared in Table VI with CIRA (1965) and a recent model atmosphere of Jacchia (1971). Note that the ratio $n(O)/n(O_2)$ in this new set is a factor of 3 larger than in the CIRA (1965) model.

TABLE VI

Air composition at 150 km

| Parameter | Von Zahn (1970) | CIRA 1965 (average) | Jacchia (1971) $(T_\infty = 900°)$* | |
|---|---|---|---|---|
| $n(N_2)$ | $26 \cdot 10^9$ | $33.4 \cdot 10^9$ | $25.4 \cdot 10^9$ | $cm^{-3}$ |
| $n(O)$ | $23 \cdot 10^9$ | $14.2 \cdot 10^9$ | $23.6 \cdot 10^9$ | $cm^{-3}$ |
| $n(O_2)$ | $2.5 \cdot 10^9$ | $4.78 \cdot 10^9$ | $2.68 \cdot 10^9$ | $cm^{-3}$ |
| $n(Ar)$ | $0.05 \cdot 10^9$ | $0.18 \cdot 10^9$ | $0.04 \cdot 10^9$ | $cm^{-3}$ |
| $\rho$ | $1.96 \cdot 10^{-12}$ | $2.18 \cdot 10^{-12}$ | $1.95 \cdot 10^{-12}$ | $g\ cm^{-3}$ |
| $M$ | 22.85 | 25.17 | 22.73 | $g\ mole^{-1}$ |

*Approximate average temperature for the interval 1961 to 1969 in the system of this model.

Half a year after I had proposed this new reference set, we had the first successful flight of a mass spectrometer equipped with an Offermann source. Preliminary data analysis (Offermann and Von Zahn, 1971) indicates that the measured ratio of $n(O)/n(O_2)$ was 3.5 times higher than the CIRA (1965) value at 120 km! So we now hope that final results of this flight will substantiate this high ratio and that further flights of this type of instrument will yield similar results.

The fact that rocket-borne spectrometers record O densities that are much too low is caused by chemical reactions at the ion source walls and probably also by direct adsorption. The flux of O atoms into the source integrated over the time of a sounding rocket flight is too small to stabilize the surface conditions inside the source. This situation is different for the case of satellite-borne spectometers. A very notable example is a mass spectrometer onboard the satellite OGO-6 which was launched June 5, 1969 into a nominal 400 km perigee/1100 km apogee orbit with 82° inclination. The spectrometer employed a gold-plated closed-ion source and the instrument was always oriented along the satellite velocity vector. The idea behind this design was the hope that atomic oxygen would recombine to molecular oxygen inside the closed source and that one would be able to measure O quantitatively as $O_2$. Harpold and Horowitz (1971) have now reported on the observed ion source densities during perigee passes for this instrument. After the opening of the ion source in orbit, initially a large $CO_2$ peak was observed which reached a maximum concentration of approximately $2 \cdot 10^9 cm^{-3}$ nineteen hours after turn-on (even in this gold-plated source!). The observed $O_2$ density stabilized only after about 35 days in orbit at approximatley $4 \cdot 10^9 \, cm^{-3}$, when $n(CO_2)$ had fallen to $1 \cdot 10^8 \, cm^{-3}$. How the ambient O densities deduced from this instrument compared with present-day model atmospheres has not been published yet.

Let us take the atomic oxygen flux into the source of the OGO-6 spectrometer integrated over 35 days as a lower limit to stabilize the surface conditions with respect to O recombination. A rocket-borne spectrometer opened at 130 km will only be exposed to about 1% or less of the above mentioned stabilizing O flux during its time of flight (typically 5 min above 130 km). Only after long-time exposure of the ion source to atomic oxygen in the laboratory is there some hope that conventional ion sources may become capable of measuring O densities during the short flight times of sounding rockets.

### 3.3. Height of turbopause and eddy diffusion coefficients

The height of the turbopause has been determined from mass spectrometric determinations of the ratio $n(Ar)/n(N_2)$. Because measurements only start at about 115 km this ratio has been extrapolated downward under the assumption of diffusive equilibrium (Meadows and Townsend, 1958). The altitude at which the extrapolated ratio met the ground value for this ratio ($= 1.196 \cdot 10^{-2}$) was identified as the turbopause altitude. Table VII reviews the results of such extrapolations. The results are remarkably uniform and point strongly towards a turbopause altitude of 100 to 102 km. This implies an eddy diffusion coefficient of $K \simeq 1 \cdot 10^6 \, cm^2 \, sec^{-1}$, if one assumes $K$ to be independent of altitude. Unfortunately, no published atmospheric model has taken into account both this low $K$ value as well as argon densities. Colegrove et al. (1966) included argon but their lowest $K$ was $2.3 \cdot 10^6 \, cm^2$ $sec^{-1}$. Shimazaki and Laird (1970) and Keneshea and Zimmerman (1970) used low eddy diffusion coefficients in some of their models, but did not include argon in their calculations.

This shortcoming, however, should be removed in the near future and then interesting results can be expected from a comparison of atmospheric models including eddy diffusion and mass spectrometric $Ar/N_2$ determinations. In particular, when mass spectrometers become capable of measuring at altitudes as low as the mesopause, we hope that continuous profiles of K can be deduced over a limited altitude range about the turbopause. The first mass spectrometric measurement of this kind has been reported by Philbrick et al. (1971).

### 3.4. Diffusive equilibrium

The important question as to what extent diffusive equilibrium prevails in the upper atmosphere has not been answered conclusively, either by mass spectrometers or by other means. It has been clearly established that large scale diffusive

TABLE VII

Turbopause levels $h_T$ determined from Ar/N$_2$ separation ratios (from Von Zahn, 1970)

| Payload | Date of flight | Time | Latitude | $h_T$ (km) | Reference |
|---|---|---|---|---|---|
| **University of Michigan** | | | | | |
| NASA 14.08 UA | March 28, 1963 | 0255 | 38° N | 102 | Schaefer, 1969 |
| NASA 14.10 UA | Nov. 26, 1963 | 1316 | 38° N | 105 | Schaefer, 1969 |
| NASA 14.95 UA | Feb. 19, 1965 | 0317 | 59° N | 99 | Schaefer, 1969 |
| NASA 14.98 UA | March 11, 1965 | 0435 | 9° S | 103 | Schaefer, 1969 |
| NASA 14.99 UA | March 11, 1965 | 1507 | 10° S | 101 | Schaefer, 1969 |
| Average | | | | 102 | |
| **University of Minnesota** | | | | | |
| NASA 4.127 | April 15, 1965 | 0345 | 33° N | 99* | |
| NASA 4.181 | Nov. 30, 1966 | 0445 | 32° N | 102 | Kasprzak et al., 1968 |
| NASA 4.180 | Dec. 2, 1966 | 1409 | 32° N | 100 | Kasprzak et al., 1968 |
| NASA 4.179 | June 21, 1967 | 1249 | 32° N | 99 | Krankowsky et al., 1968 |
| NASA 4.212 | July 20, 1967 | 0200 | 32° N | 99 | Krankowsky et al., 1968 |
| NASA 4.211 | July 20, 1967 | 1224 | 32° N | 101 | Krankowsky et al., 1968 |
| NASA 4.272 | Feb. 4, 1969 | 0835 | 59° N | 102 | Hickman and Nier, 1970 |
| Average | | | | 100 | |
| **University of Bonn** | | | | | |
| ESRO CO7/2 | Dec. 11, 1965 | 0443 | 40° N | 103 | Von Zahn, 1967 |
| BB3/1 | Dec. 12, 1966 | 1305 | 59° N | 99 | Von Zahn and Gross, 1969 |
| BB3/2 | Dec. 4, 1968 | 1251 | 59° N | 102 | Gross and Von Zahn, 1971 |
| Average | | | | 101 | |

*Determined from data of Hedin and Nier (1966) by comparison with Ar/ N$_2$ separation ratios of Kasprzak et al. (1968) and Krankowsky et al. (1968).

separation sets in at altitudes above 100 km, but this is only a precursor to the more stringent condition of diffusive equilibrium. It is easier to give examples where evidently diffusive equilibrium did not hold, such as many helium measurements of the University of Minnesota group (see section 3.5) or the appearance of wave-like structures in the density profile (see section 3.6), than to give examples where diffusive equilibrium conditions were fulfilled. It appears rather straight-forward to check on the constancy of the ratio (15), but because of the doubtful O determinations and the limited altitude range over which Ar data is available, this check loses much of its significance. Even in composition measurements for which a close approach to diffusive equilibrium has been claimed (Hedin and Nier, 1965) the ratio (15) increases from a value $1.57 \cdot 10^3$ at 120 km to $1.84 \cdot 10^3$ at 155 km. It is gratifying to see that ratio (15) does not vary too much in this experiment, but there is certainly still room left for improvements.

*3.5. Helium and vertical transport*

Helium turns out to be a rather interesting constituent of the upper atmosphere.

Due to its low atomic weight, the scale height for helium is much greater than that of the major constituents O, $N_2$, and $O_2$. The ratio of $n(He)/n(N_2)$ therefore, is strongly altitude dependent, and any air mass can be characterized by its $n(He)/n(N_2)$ ratio. In this way helium can be used as a kind of tracer, especially for vertical motions, similar to ozone in the stratosphere and troposphere.

Figure 21 shows all available results obtained in the lower thermosphere so far.

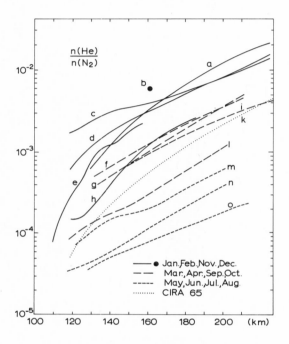

Fig. 21. Measured He/ $N_2$ profiles in lower thermosphere.

| Curve | Payload | Date | Time | Latit. | Reference |
|---|---|---|---|---|---|
| a | ESRO S61/ 1 | 25. 2.70 | 1723 CET | 70° N | Von Zahn and Schneppe (1971) |
| b | ESRO S61/ 2 | 26. 2.70 | 0547 CET | 70° N | Von Zahn and Schneppe (1971) |
| c | NASA 4.180 | 2.12.66 | 1409 MST | 32° N | Kasprzak et al. (1968) |
| d | NASA 4.181 | 30.11.66 | 0444 MST | 32° N | Kasprzak et al. (1968) |
| e | BB3/ 1 | 12.12.66 | 1305 CST | 59° N | Müller and Hartmann (1969) |
| f | ESRO S19/ 2 | 10.10.67 | 1533 CET | 40° N | Bitterberg et al. (1970) |
| g | ESRO S19/ 1 | 4.10.67 | 1423 CET | 40° N | Bitterberg et al. (1970) |
| h | NASA 4.272 | 4. 2.69 | 0835 CST | 59° N | Hickman and Nier (1970) |
| i | ESRO D34/ 1 | 4.10.67 | 1412 CET | 69° N | Pfeiffer and Von Zahn (1969) |
| k | CIRA 1965 | Mean Atmosphere | | | |
| l | NASA 4.127 | 15. 4.65 | 0345 MST | 32° N | Hedin and Nier (1966) |
| m | NASA 4.211 | 20. 7.67 | 1224 MST | 32° N | Krankowsky et al. (1968) |
| n | NASA 4.212 | 20. 7.67 | 0200 MST | 32° N | Krankowsky et al. (1968) |
| o | NASA 4.179 | 21. 6.67 | 1249 MST | 32° N | Krankowsky et al. (1968) |

There are two evident features in this plot. A strong seasonal variation exists in the helium content of the thermosphere, as the winter densities are much higher than the summer densities. There is, however, surprisingly no clear latitudinal variation evident, although the latitudes of the measurements spread out from 32°N to 70°N. As already discussed in secion 1.3.1.3, Johnson and Gottlieb (1970) interpret this seasonal variation as being due to a global circulation system which leads to rising air motions over the summer polar region and downward motions over the winter polar regions. A satisfying quantitative description of the temporal and spatial variations of the ratio $n(He)/n(N_2)$, however, has not yet been given.

A very interesting contribution to this subject was made by Reber et al. (1971) who studied the global helium distribution in June, 1969 by using data from the mass spectrometer on the satellite OGO-6. After normalizing for altitude effects by use of the Jacchia model atmosphere, the densities show an order of magnitude difference between the southern (winter) hemisphere and the northern (summer) hemisphere, with the maximum density occuring near −55° latitude. The exact location of the maximum varies between −40° and −70° geographic latitude, and is apparently correlated with the geomagnetic dipole latitude of −53°. This observation is not inconsistent with the idea that the helium distribution is quite sensitive to thermospheric winds. They in turn are controlled by ion concentrations and drift velocities ("ion drag") which are geomagnetically controlled.

Strong seasonal variations in the helium content of the lower exosphere were deduced from satellite drag analysis by Keating and Prior (1968) and Jacchia and Slowey (1968), and from optical observations of the 10830 Å emission line of helium (Christensen et al., 1971). In both instances the ratio of winter maximum density to summer minimum is found to be lower than in the case of the mass spectrometer measurements which need further study.

Apart from the seasonal variations, Fig. 21 indicates another notable feature. Quite a number of the altitude profiles show considerable deviations from a diffusive equilibrium distribution, which is given ideally by the CIRA (1965) profile. Kasprzak (1969) has reviewed the available profiles and concluded that the assumption of upward fluxes of helium, independent of altitude, is not inconsistent with the shape of the measured $n(He)$ profiles. The fluxes were calculated for seven different measured He profiles and ranged from $2 \cdot 10^8$ to $2.6 \cdot 10^{10}$ cm$^{-2}$sec$^{-1}$. These are rather high fluxes and "unfortunately" all are *upwards*, even at night and in winter time! Hodges, Jr. (1970) proposed a vertical transport mechanism for minor constituents through non-linear processes connected with gravity waves. Whether this mechanism can be effective enough to counteract the circulation system of Johnson and Gottlieb (1970), which calls for *downward* helium flow in winter time, remains to be seen.

*3.6. Wave phenomena and the geomagnetic activity effect*

A number of vertical number density profiles obtained by rocket-borne mass spectrometers exhibited wavelike structures even after careful correction for variations in payload attitude (Mauersberger et al., 1967; Von Zahn and Gross, 1969; Müller and Hartmann, 1969; flight NASA 4.211 of Krankowsky et al., 1968). Mauersberger et al. (1967) explained the wave structure by the presence of an internal gravity wave. But true identification of such a wave from the results of a single vertical probe will always be rather difficult. Futher investigations of these effects, their abundance and correlation with other geophysical phenomena would certainly be worthwhile.

The response of the neutral atmosphere to geomagnetic disturbances was studied for the first time with high temporal and spatial resolution by the mass spectrometer on the OGO-6 satellite. Taeusch et al. (1971) concluded from these exciting new data that the time lag between a given magnetic disturbance and an atmospheric response is indeed very short, appearing to be less than one hour. Furthermore, the major effect of a magnetic storm on the neutral atmosphere above 400 km altitude is quite localized in the polar regions, i.e., latitudes north of $+50°$ and south of $-50°$. The conclusions regarding time lag and latitudinal extension are not in agreement with respective results of Roemer (1971) deduced from satellite drag analysis. Again we have the well-known situation that new results open up more questions than they can answer. This seems to be the special fate of mass spectometric studies on the composition of the thermosphere.

## 4. THE OUTLOOK FOR FUTURE STUDIES

It is felt that a general survey of the atmosphere in the middle thermosphere has been fairly well accomplished. Many details of spatial and temporal variations still remain to be clarified, but it is expected that the main interest of mass spectrometric studies will shift in the future to lower altitudes, and to a lesser extent also to higher altitudes.

For the altitude range of 70 to 120 km, the main interest lies in measurements of the minor constituents $H_2O$, $NO$, $O_3$, $CO_2$ etc. But this poses challenging instrumental problems because of the rather low abundances of these trace gases, as well as for the rather high ambient pressure. The investigation of turbulent and transport processes is of major interest too, if only because processes acting at turbopause altitudes determine so many properties of the atmosphere much higher up.

For altitudes above 400 km, we have to consider only four gases: H, He, O, and $N_2$. Measurements of hydrogen densities with good time resolution would be very desirable, but again the necessary ion source technology is still inadequate. The escape processes for He and H atoms from the earth's gravitational field are only qualitatively understood, and good altitude profiles for both constituents would be of great interest. Another question concerns the possible existence of a summer oxygen bulge which has not been verified yet by experiment. Furthermore, mass spectrometers should be able to contribute to investigations of the mechanism which causes the diurnal temperature maximum to lag 2 to 3 h behind the density maximum in the thermosphere.

It is evident that the investigations above 300 km will almost exclusively be performed by satellite-borne instruments, whereas the altitudes below 120 km are safe havens for those of us who like sounding rocket work. There are still plenty of interesting problems left to be investigated by both kinds of experiments.

## REFERENCES

Ackerman, M., Ultraviolet solar radiation related to mesospheric processes. In: *Mesospheric Models and Related Experiments* (G. Fiocco, Editor) Reidel, Dordrecht, Holland, 298 pp., 1971.

Ackerman, M., Frimout, D. et Pastiels, R., Mesure du rayonnement ultraviolet solaire par ballon stratosphérique. *Ciel et Terre*, **84**:408, 1968.

Axford, W. I., The polar wind and the terrestrial helium budget. *J. Geophys. Res.*, **73**:6855, 1968.

Bailey, A.D. and Narcisi, R.S., Miniature mass spectrometers for upper atmosphere composition measurements. *Instrument. Pap.* 95. Air Force Cambridge Res.Labs., pp. 66–148, 1966.

Banks, P. M. and Holzer, T. H., The polar wind. *J. Geophys. Res.*, **73**:6846, 1968.

Bitterberg, W., Bruchhausen, K., Offermann, D. and Von Zahn, U., Lower thermosphere composition and density above Sardinia in October 1967. *J. Geophys. Res.*, **75**:5528, 1970.

Bonnet, R. M., Stigmatic spectra of the sun between 1800 Å and 2800 Å. *Space Res.*, **7**:458, 1968.

Chapman, S. and Cowling, T. G., *The Mathematical Theory of Non-Uniform Gases*. Cambridge University Press, 2nd ed., 1952; (third edition, 1970, 424 pp).

Christensen, A. B., Patterson, T. N. L. and Tinsley, B. A., Observations and computations of twilight helium 10, 830-Angstrom emission. *J. Geophys. Res.*, **76**:1764, 1971.

CIRA, 1965. *COSPAR International Reference Atmosphere 1965*, North-Holland, Amsterdam, 313 pp., 1965.

Colegrove, F. D., Hanson, W. B. and Johnson, F. S., Eddy diffusion and oxygen transport in the lower thermosphere. *J. Geophys. Res.*, **70**:4931, 1965.

Colegrove, F. D., Johnson, F. S. and Hanson, W. B., Atmospheric composition in the lower thermosphere. *J. Geophys. Res.*, **71**:2227, 1966.

Cooley, J. E. and Reber, C. A., Neutral atmosphere composition measurement between 133 and 533 kilometers from the Geoprobe rocket mass spectrometer. *NASA Goddard Space Flight Cent. Rep.* X-621-69-260, 1969.

Curtiss, C. F. and Hirschfelder, J. O., Transport properties of multicomponent gas mixtures. *J. Chem. Phys.*, **17**:550, 1949.

Detwiler, C. R., Garrett, D. L., Purcell, J. D. and Tousey, R., The intensity distribution in the ultraviolet solar spectrum. *Ann. Geophys.*, **17**:9, 1961.

Gross, J. Massenspektrometrische Untersuchung der Sauerstoffverluste an heissen Rheniumoberflächen im Druckbereich $10^{-8}$ bis $10^{-4}$ Torr. *Z. Naturforsch.*, **25a**:900, 1970.

Gross, J. and Von Zahn, U., Air density and composition in the lower thermosphere above Fort Churchill, In: *Space Research XI*. (K. Kondratyev, M. J. Rycroft and C. Sagan, Editors). Akademie-Verlag, Berlin, 875–885, 1971.

Hall, L. A., Schweizer, W. and Hinteregger, H. E., Improved extreme ultraviolet absorption measurements in the upper atmosphere. *J. Geophys. Res.*, **70**:105, 1965.

Hall, L. A., Chagnon, C. W. and Hinteregger, H. E., Daytime variations in the composition of the upper atmosphere. *J. Geophys. Res.*, **72**:3425, 1967.

Harpold, D. N. and Horowitz, R., Minor gases measured with the OGO-6 neutral mass spectrometer. *EOS, Trans. Am. Geophys. Union,* **52**:294, 1971.

Hartmann, G., Mauersberger, K. and Muller, D., Evaluation of the turbopause level from measurements of the helium and argon content of the lower thermosphere above Fort Churchill. In: *Space Research VIII* (A. P. Mitra, L. G. Jacchia and W. S. Newman, Editors) North-Holland, Amsterdam, pp. 940–946, 1968.

Hedin, A. E., Avery, C. P. and Tschetter, C. D., An analysis of spin modulation effects on data obtained with a rocket-borne mass spectrometer. *J. Geophys. Res.,* **69**:4637, 1964.

Hedin, A. E. and Nier, A. O., Diffusive separation in the upper atmosphere. *J. Geophys. Res.,* **70**:1273, 1965.

Hedin, A. E. and Nier, A. O., A determination of the neutral composition, number density, and temperature of the upper atmosphere from 120 to 200 kilometers with rocket-borne mass spectrometers. *J. Geophys. Res.,* **71**:4121, 1966.

Hesstvedt, E., On the effect of vertical eddy transport on atmospheric composition in the mesosphere and lower thermosphere. *Geofys. Publ. (Oslo),* **27**:1, 1968.

Hickman, D. R. and Nier, A. O., Neutral composition of polar atmosphere. *EOS, Trans. Am. Geophys. Union,* **51**:378 1970.

Hines, C. O., Paghis, I., Hartz, T. R. and Fejer, J. A. (Editors). *Physics of the Earth's Upper Atmosphere. Prentice-Hall, London,* 434 pp., 1965.

Hinteregger, H. E., Absorption spectrometric analysis of the upper atmosphere in the EUV region. *J. Atmos. Sci.,* **19**:351, 1962.

Hinton, B. B., Leite, R. J. and Mason, C. J., Comparison of water vapor measurements from two similar spacecraft. *EOS, Trans. Am. Geophys. Union,* **50**:267, 1969.

Hodges, Jr., R. R., Vertical transport of minor constituents in the lower thermosphere by nonlinear processes of gravity waves. *J. Geophys. Res.,* **75**:4842, 1970.

Jacchia, L. G., Revised static models of the thermosphere and exosphere with empirical temperature profiles. *Spec. Rep. 332, Smithsonian Astrophys. Obs.,* 115 pp., 1971.

Jacchia, L. G. and Slowey, J., Diurnal and seasonal-latitudinal variations in the upper atmosphere. *Planet. Space Sci.,* **16**:509, 1968.

Johnson, F. S. and Gottlieb, B., Eddy mixing and circulation at ionospheric levels. *Planet. Space Sci.,* **18**:1707, 1970.

Jursa, A. S., Nakamura, M. and Tanaka, Y., Molecular oxygen distribution in the upper atmosphere. *J. Geophys. Res.,* **68**:6145, 1963.

Kasprzak, W. T., Evidence for a helium flux in the lower thermosphere, *J. Geophys. Res.,* **74**:894, 1969.

Kasprzak, W. T., Krankowsky, D. and Nier, A. O., A study of day-night variations in the neutral composition of the lower thermosphere. *J. Geophys. Res.,* **73**:6765, 1968.

Keating, G. M. and Prior, E. J., The winter helium bulge. In: *Space Research VIII.* (A. P. Mitra, G. G. Jacchia and W. S. Newman, Editors). North-Holland, Amsterdam, 982 pp., 1968.

Keneshea, T. J. and Zimmerman, S. P., The effect of mixing upon atomic and molecular oxygen in the 70–170 km region of the atmosphere. *J. Atmos. Sci.,* **27**:831, 1970.

Kockarts, G., Helium and hydrogen distributions in the upper atmosphere. In: *Physics of the Upper Atmosphere.* (F. Verniani, Editor). Editrice Compositori, Bologna, 461 pp., 1971.

Krankowsky, D., Kasprzak, W. T. and Nier, A. O., Mass spectrometric studies of the composition of the lower thermosphere during summer 1967. *J. Geophys. Res.,* **73**:7291, 1968.

Lettau, H., Diffusion in the upper atmosphere. In: *Compendium of Meteorology* (T. F. Malone, Editor), Am. Meteorol. Soc., New York, pp. 320–333, 1951.

Mange, P., *J. Geophys. Res.,* **62**:279. 1957.

Mauersberger, K., Müller, D., Offermann, D. and Von Zahn, U., Neutral constituents of the upper atmosphere in the altitude range of 110 to 160 km above Sardinia. *Space Res.,* **7**:1150, 1967.

Mauersberger, K., Müller, D., Offermann, D. and Von Zahn, U., A mass spectrometric determination of the neutral constituents in the lower thermosphere above Sardinia. *J. Geophys. Res.,* **73**:1071, 1968a.

Mauersberger, K., Müller, D., Offermann, D. and Von Zahn, U., Eine Messung der Dichte und Zusammensetzung des Neutralgases in der unteren Thermosphäre. *Forschungsber.* W 68-05. Bundesministerium für wissenschaftliche Forschung, Bonn, 74 pp., 1968b.

Meadows, E. B., and Townsend, J. W., Diffusive separation in the winter night time arctic upper atmosphere 112 to 150 km. *Ann. Geophys.*, **14**:80, 1958.

Müller D., Ein einfaches Gaseinlaβsystem zur präzisen Herstellung niedriger Drucke. *Z. Angew. Phys.*, **23**:467, 1967.

Müller D. and Hartmann, G., A mass spectrometric investigation of the lower thermosphere above Fort Churchill with special emphasis on the helium content. *J. Geophys. Res.*, **74**:1287, 1969.

Nicolet, M., The aeronomic problem of helium. *Ann. Geophys.* **13**:1, 1957.

Nicolet, M., The properties and constitution of the upper atmosphere. In: *Physics of The Upper Atmosphere* (J. A. Ratcliffe, Editor). Academic Press, New York, London, 586 pp., 1960.

Niemann, H. B. O., An atomic oxygen beam system for the investigation of mass spectrometer response in the upper atmosphere. *Sci. Rep.* 07065-5-S. Space Phys. Res. Lab., The University of Michigan, 1969.

Niemann, H. B. and Kennedy, B. C., Omegatron mass spectrometer for partial pressure measurements in upper atmosphere. *Rev. Sci. Instr.*, **37**:722, 1966.

Nier, A. O., Mass spectrometry of the neutral constituents of the upper atmosphere. *Mass Spectrometry*, **15**:67, 1967.

O'Day, M. D., Upper air research by use of rockets in the U.S. Air Force. In: *Rocket Exploration of The Upper Atmosphere* (R. L. F. Boyd and M. J. Seaton, Editors), Pergamon, London, 1954.

Offermann, D. and Trinks, H., A rocket-borne mass spectrometer with helium cooled ion sources. *Rev. Sci. Instr.*, **42**:1836–1843, 1971.

Offermann, D. and Von Zahn, U., Atomic oxygen and carbon dioxide in the lower thermosphere. *J. Geophys. Res.*, **76**:2520, 1971.

Parkinson, W. H. and Reeves, E. M., Measurements in the solar spectrum between 1400 and 1875 Å with a rocket-borne spectrometer. *Solar Phys.*, **10**:342, 1969.

Paul, W., Reinhard, H. P. und Von Zahn, U., Das elektrische Massenfilter als Massenspektrometer und Isotopentrenner. *Z. Phys.*, **152**:143, 1958.

Pelz, D. T. and Newton, G. P. Midlatitude neutral thermosphere density and temperature measurements. *J. Geophys. Res.*, **74**:267, 1969.

Pfeiffer, G. and Von Zahn, U., Versuch einer Messung von Dichte und Zusammensetzung der arktischen Thermosphäre. *Forschungsber.* W 69-26. Bundesministerium für wissenschaftliche Forschung, Bonn, 1969.

Philbrick, C. R., Wlodyka, R. A. and Gardner, M. E., Neutral atmospheric composition measurements. *EOS, Trans. Am. Geophys. Union*, **51**:378, 1970.

Philbrick, C. R., Faucher, G. A. and Wlodyka, R. A., Neutral composition measurements of the mesosphere and lower thermosphere. *COSPAR Symposium on D- and E-Region Ion Chemistry*, Urbana, Ill., 1971 (presented paper).

Reber, C. A. and Cooley, J. E., Neutral atmosphere composition measurement between 180 and 420 km from the Geoprobe rocket mass spectrometer, *Trans. Am. Geophys. Union*, **48**:75, 1967.

Reber, C. A. and Harpold, D. N., Mass spectrometric measurements of atomic oxygen in the earth's upper atmosphere, *Trans. Am. Geophys. Union*, **48**:76, 1967.

Reber, C. A., Harpold, D. N., Horowitz, R. and Hedin, A. E., Horizontal distribution of helium in the earth's upper atmosphere. *J. Geophys. Res.*, **76**:1845, 1971.

Roemer, M., Structure of the thermosphere and its variations, deduced from satellite decay, In: *Physics of the Upper Atmosphere* (F. Verniani, Editor). Editrice Compositori, Bologna 461 pp., 1971.

Schaefer, E. J., The dissociation of oxygen measured by a rocket-borne mass spectrometer. *J. Geophys. Res.*, **68**:1175, 1963.

Schaefer, E. J., Composition and temperature of the neutral tropic lower thermosphere. *J. Geophys. Res.*, **74**:3488, 1969.

Schaefer, E. J. and Nichols, M. H., Mass spectrometer for upper air measurements. *Am. Rocket Soc. J.*, **31**:1773, 1961.

Schaefer, E. J. and Nichols, M. H., Neutral composition obtained from a rocket-borne mass spectrometer. In: *Space Research IV*. (P. Muller, Editor). North-Holland, Amsterdam, 205–234, 1964.

Schneppe G., Herstellung genau bekannter Drucke von Rein- und Mischgasen im Hochvakuumbereich. *Forschungsber.* W 69-05, Bundesministerium für wissenschaftliche Forschung, Bonn, 47 pp., 1969.

Schultz, F. V., Spencer, N. W. and Reifman, A., Atmospheric pressure and temperature measurements between the altitudes of 40 and 110 kilometers. *Upper Atmos. Rep.* 2. University of Michigan Research Institute, 1948.

Shimazaki, T., Dynamic effects on atomic and molecular oxygen density distributions in the upper atmosphere: A numerical solution to equations of motion and continuity. *J. Atmos. Terr. Phys.*, **29:**723, 1967.

Shimazaki, T. and Laird, A. R., A model calculation of the diurnal variation in minor neutral constituents in the mesosphere and lower thermosphere including transport effects. *J. Geophys. Res.*, **75:**3221, 1970.

Simmons, R. W., Carter, M. F. and Taeusch, D. R., NASA 18.50 thermosphere probe experiment. *Sounding Rocket Flight Rep.* 07065-8-R, Space Phys. Res. Lab., Univ. Mich., 48 pp., 1968.

Spencer, N. W., and Reber, C. A., A mass spectrometer for an aeronomy satellite. *Space Research III* (W. Priester, Editor) North-Holland, Amsterdam, 1151 pp., 1963.

Spencer, N. W., Newton, G. P., Carignan, G. R. and Taeusch, D.R., Thermosperic temperature and density variations with increasing solar activity. In: *Space Research, X* (T. M. Donahue, P. A. Smith and L. Thomas, Editors). North-Holland, Amsterdam, 389–412, 1970.

Taeusch, D. R., and Carignan, G. R., NASA 18.06 thermosphere probe experiment. *Sounding Rocket Flight Rep.* 07065-6-R. Space Phys. Res. Lab., Univ. Mich., 31 pp., 1968.

Taeusch, D. R., Carignan, G. R. and Reber, C. A., Response of the neutral atmosphere to geomagnetic disturbances. In: *Space Research XI* (K. Kondratyev, M. J. Rycroft and C. Sagan, Editors). Akademie-Verlag, Berlin, pp. 995–1002, 1971.

Taylor Jr., M. A., Brinton, H. C. and Smith, C. R., Instrumentation for atmospheric composition measurements. *Proc. 8th Ann. Aero-Space Symp.* Instrum. Soc. Am., Washington, D.C., pp.1–14, 1962.

Townsend Jr., J. W., Radiofrequency mass spectrometer for upper air research. *Rev. Sci. Instr.*, **23:**538, 1952.

Trinks, H., *Aufbau und Kalibrierung eines Neutralgasspektrometers mit offener, kathodenloser Elektronen-stoß*-Ionenquelle für Dichtemessungen in der oberen Atmosphäre. Thesis, Universität Bonn, 50 pp., 1971.

Von Zahn, U., Monopole spectrometer, a new electric field mass spectrometer. *Rev. Sc. Instrum.*, **34:**1–4, 1963.

Von Zahn, U., Mass spectrometric measurements of atomic oxygen in the upper atmosphere: a critical review. *J. Geophys. Res.*, **72:**5933, 1967.

Von Zahn, U., Space mass spectrometry. In: *Advances in Mass Spectrometry*, 4 (E. Kendrick, Editor). The Institute of Petroleum, London, pp. 869–895, 1968.

Von Zahn, U., Neutral air density and composition at 150 kilometers. *J. Geophys. Res.*, **75:**5517, 1970.

Von Zahn, U. and Gross, J., Mass spectrometric investigation of the thermosphere at high latitudes. *J. Geophys. Res.*, **74:**4055, 1969.

Von Zahn, U. and Schneppe, G., Neutral composition of polar thermosphere. *EOS, Trans. Am. Geophys. Union*, **52:**292,1971.

Weeks, L. H. and Smith, L. G., Molecular oxygen concentrations in the upper atmosphere by absorption spectroscopy. *J. Geophys. Res.*, **73:**4835, 1968.

Widing, K. G., Purcell, J. D. and Sandlin, G. D., The UV continuum 1450–2100 Å and the problem of the solar temperature minimum. *Solar Phys.*, **12:**52, 1970.

Wildman, P. J. L., Kerley, M. J. and Shaw, M. S., Molecular oxygen measurements from 100 to 150 km at Woomera, Australia. *J. Atmos. Terr. Phys.*, **31:**951, 1969.

Zimmerman, S. P., George, J. D. and Keneshea, T. J., Calculation of the latitudinal variation of major and minor neutral species using measured transport coefficients. *COSPAR Symp. on D- and E-Region Ion Chem.*, Urbana, Ill. (presented paper), 1971.

# STRUCTURE OF THE F REGION AND GLOBAL THERMOSPHERIC WINDS

HENRY RISHBETH

*Radio and Space Research Station, Slough (Great Britain)*

## SUMMARY

This paper is intended as background material for a course of lectures on the ionospheric F region and thermospheric winds. It covers this field very broadly, without much detail, and the references are necessarily only a small selection of the available literature.

The principal topics are as follows: Observations of the F region (Introduction, Experimental methods, Features of the undisturbed F layer, F-layer disturbances); A resumé of the F-region theory (Composition of the atmosphere and ionosphere, Production of ionization, F-region chemistry, Motions of ionization, Formation of the F2 peak); Global thermospheric winds (The global distribution of thermospheric temperature, The equations governing thermospheric wind velocities, The wind systems, Large-scale transport of air by winds, Effects of winds on the F2 layer, Some outstanding problems).

## 1. OBSERVATIONS OF THE F REGION

### 1.1. Introduction

#### 1.1.1. Historical note (Green, 1946)

Although Marconi could be said to have "discovered" the ionosphere when he transmitted radio signals across the Atlantic in 1901, the first real experimental observations of the ionosphere were those of Appleton and Barnett (1925), and Breit and Tuve (1925). Appleton and Barnett transmitted signals at around 1 MHz over a distance of 200 km, and showed that fading occurred when the signal frequency was varied, this being due to interference between the direct ground wave and the sky wave reflected from the ionosphere. Breit and Tuve transmitted pulsed signals and received echoes from the ionosphere. Appleton (1927) showed that at least two reflecting layers exist; these are now known as the E and F layers, the latter being sometimes divided into F1 and F2 components. The name "ionosphere", which fits the conventional terminology of the atmosphere (troposphere, stratosphere, etc.) was proposed by Watson-Watt in 1926 (Gardiner, 1969).

### 1.1.2. Important ionospheric parameters

The principal measurable parameter of the ionosphere is the electron concentration (or electron density), denoted by $N$; it is a function of height $h$ and time $t$. Since the ionosphere is almost exactly electrically neutral, and negative ions and doubly-charged positive ions are very scarce in the F region, the total positive ion concentration is also $N$. Other parameters include the temperatures of the electrons ($T_e$), ions ($T_i$) and neutral gas ($T_n$), which may all be different; and the plasma drift velocity $\vec{V}$. Since it is generally true that only small electric currents flow in the F region, it is often permissible to assume the electron and ion drift velocities to be equal; thus $\vec{V}_e = \vec{V}_i = \vec{V}$.

### 1.1.3. Basic symbols and units

Système International mksA units are used in these notes. The conventional symbols $c$, $e$, $m$, $k$, $\epsilon_o$, $\mu_o$ denote respectively the free-space velocity of light, the fundamental positive charge, electron mass, Boltzmann's constant, and the permittivity and permeability of free space.

## 1.2. Experimental methods (J. Atmos. Terr. Phys., April 1970)

### 1.2.1. Vertical incidence sounding (Piggott and Rawer, 1961; Davies, 1969)

The basic instrument for studying the ionosphere is the pulse sounder or "ionosonde", developed from the principle first used by Breit and Tuve. About 100 ionosondes are in routine use throughout the world, and at a few sites regular observations have been made for 35–40 years. The records, or "ionograms", contain a wealth of information and interpreting them has become a complex art and science.

Effectively, an ionosonde measures the "virtual height" of reflection ($h'$) as a function of radio frequency $f$: the out-and-back travel time of a pulse is simply $2h'/c$. As $f$ is varied the virtual height is recorded on film, a complete ionogram with $f$ ranging from, say, 1–20 MHz being obtained in about a minute. Reflection occurs where $f$ becomes equal to the plasma frequency:

$$f_N = \frac{1}{2\pi}(Ne^2/m\epsilon_o)^{1/2} \tag{1}$$

The frequency $f_o$ that will just penetrate an ionospheric layer, i.e., the "ordinary critical frequency" is related to the peak electron concentration $N_m$ by an equation like eq. 1. Inserting numerical values:

$$f_o = (80.6 N_m)^{1/2} = 8.98 N_m^{1/2} \quad [\text{Hz}] \tag{2}$$

In the presence of the geomagnetic induction $\vec{B}$, the ionosphere acts as a birefringent medium for radio waves, and there exists also the "extraordinary critical frequency" $f_x$. If:

$$f_H = Be/2\pi m \tag{3}$$

(called the electron "gyrofrequency"), then it may be shown that:

$$f_o = (f_x^2 - f_x f_H)^{1/2} \simeq f_x - \tfrac{1}{2}f_H \tag{4}$$

the approximation being valid if $f_o \gg f_H$, as is generally so in the F2 layer.

*1.2.2. Calculating the N(h) distribution from ionograms* (e.g., Budden, 1961 and Radio Science, October 1967)

The group velocity of a radio wave in the ionosphere is $c/\mu'$, the "group refractive index", $\mu'$ being a function of the wave frequency $f$ and the ambient electron concentration $N(h)$. For vertically incident waves, the virtual height $h'$ (always greater than the true height of reflection, $h_r$) is given by:

$$h'(f) = \int_0^{h_r} \mu'(N,f)\,\mathrm{d}h \tag{5}$$

This can be defined for either the "o" or "x" modes. Near the reflection height $\mu'$ becomes very large and, at frequencies close to a critical, $\mu'$ is large over an appreciable range of heights near a layer peak, so that large virtual heights are recorded on ionograms (Fig. 1).

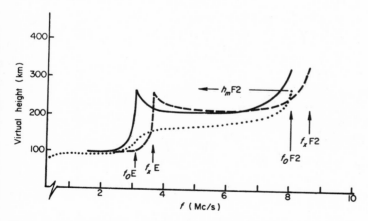

Fig. 1. An idealized daytime ionogram, showing virtual height versus frequency $f$. Only E and F2 layers are present. Full curve: "ordinary mode" echo. Broken curve: "extraordinary mode" echo. Dotted curve: a possible distribution of plasma frequency versus real height $f_N(h)$, which could produce these $h'(f)$ curves. (From Rishbeth and Garriott, 1969.)

In principle, the $N(h)$ distribution up to the F2 peak can be found by inverting eq. 5. In a "lamination" method this operation is carried out in a series of small steps. An alternative technique is to assume that the electron distribution can be represented by a polynomial, $h(f_N)$, of a given form and to determine its coefficients. Both approaches have merits and demerits. If the $N(h)$ profile is not monotonically increasing, there exists a so-called "valley ambiguity"; this difficulty can be mitigated by using the "extraordinary" curve of the ionogram as well as the "ordinary" $h'(f)$ curve (Titheridge, 1959).

### 1.2.3. Topside sounding (Proc. I.E.E.E., June 1969)

If a miniature ionosonde is carried in a satellite orbiting above the F2 peak, it can make soundings of the "topside" ionosphere below the satellite (Fig. 2). Topside ionograms can be analysed to give $N(h)$ profiles by methods analogous to those used for bottomside ionograms, though some particular problems are encountered.

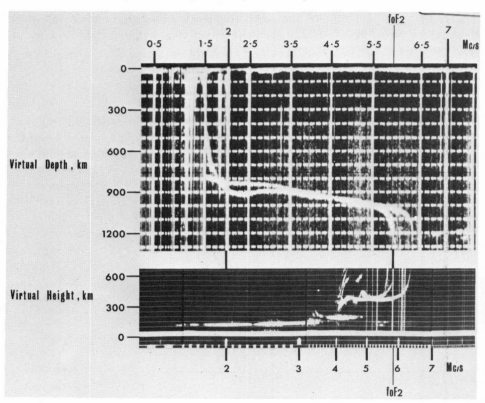

Fig. 2. Matching topside and bottomside ionograms recorded above Port Stanley, Falkland Islands (52°S) near 1100 local time on 2 Jan. 1963. The virtual height scales represent depth below the topside sounder satellite (Alouette I) for the upper ionogram, and height above ground for the lower ionogram. P. A. Smith, Radio and Space Research Station, Slough. (Crown Copyright.)

Because the topside sounder, unlike a ground-based sounder, is immersed in the ionospheric plasma, its transmitter excites a number of plasma resonances; by studying these phenomena, various parameters (such as $f_N$ and $f_H$) can be found.

### 1.2.4. Radio propagation experiments with space vehicles

The use of rockets enables $N(h)$ profiles to be determined by means of the "Faraday rotation" and "differential Doppler" methods. In effect, these measure the difference between the rocket-to-ground phase paths; for the "o" and "x" modes at a chosen r.f. in the Faraday method, and for two different r.f.'s in the Doppler method.

With satellite-borne transmitters, the Doppler and Faraday methods (or a "hybrid" method) can be used to obtain the integrated electron content along the path from the satellite to the observer, and (within certain assumptions) the "total electron content" $N_T = \int N \, dh$ below the satellite's altitude can be evaluated. For ordinary low-orbiting satellites, the spatial variation of $N_T$ along the satellite track can then be found. For geostationary satellites, on the other hand, Faraday rotation measurements can yield very accurate measurements of the time variation of $N_T$ at a given location.

### 1.2.5. Direct measurements using satellites and rockets

The most useful techniques for measuring F-region parameters include: (a) electron concentration and temperature. D.c. probes of specialized types (the simple Langmuir probe being of limited use); r.f. probes which measure plasma impedance and/or detect various resonance phenomena; (b) ion concentration and temperature. ion traps (probes with grids); (c) ion composition. Mass spectrometers of the "magnetic deflection", r.f. "time-of-flight" and "quadrupole mass filter" varieties; ion traps; (d) neutral gas composition. As for ions, but with provision for ionizing the incoming gas samples; and (e) neutral gas density. Principally derived from observations of the air drag on artificial satellites or the decay of their spin. From the density distributions rather indirect information on composition and temperature can be inferred.

### 1.2.6. Incoherent scatter (Evans, 1969)

If an intense radio beam (at an r.f. $\gg f_o F2$) is radiated vertically from the ground, a small signal is scattered back from the charged particles in the ionosphere. The scattering cross-section per particle is comparable to the Thomson cross-section of an electron, $10^{-28} \, m^2$; so, if $N_T \sim 10^{17} \, m^{-2}$, only 1 part in $10^{11}$ of the upgoing radiation is scattered. Consequently, large transmitter powers and sensitive receivers are required. Among the many parameters that can be measured with this technique are the following: $N(h)$, from the total power or from the Faraday rotation of the scattered signal; $T_e$ and $T_i$ (and sometimes the ion composition), from the spectrum of the scattered signal which is typically 50 kHz in width; components of the plasma drift velocity $\vec{V}$, from the Doppler shift (10 Hz or so) of the central frequency (a multistatic system is required to determine all components of $\vec{V}$).

A bistatic (or multistatic) system is necessary if c.w. signals are used, as the observing height has to be defined by the intersection of the transmitting and receiving aerial beams. With pulse systems, conventional "gating" procedures can be used to define the observing height. Most incoherent scatter systems use pulses, but better spectrum data are obtainable with c.w.

### 1.2.7. HF Doppler measurements (Davies and Baker, 1966)

Motions of the bottomside F layer can be studied by measuring the Doppler shift of signals (generally at a few MHz) reflected at vertical or oblique incidence. This technique is particularly suited to the study of wave motions.

### 1.2.8. Drift measurements (Kent and Wright, 1968; Kent, 1970)

The well-known "spaced receiver" method has given extensive data on the "drifts" of small-scale irregularities. In the F region, however, it is not clear how these "drifts" are related to the drift velocity $\vec{V}$ of the plasma as a whole.

### 1.2.9. Other techniques

In addition, it is possible to obtain important information about conditions in the F region (such as temperatures, and more recently winds) by observations of airglow (Chamberlain, 1961; Thomas, 1967; Armstrong, 1969). Experiments involving the release of artificial clouds have provided data on neutral-air winds and ion drifts (e.g., Haerendel et al., 1967).

Fig. 3 compares $N(h)$ profiles obtained by four different methods.

### 1.3. Features of the undisturbed F layer (Ratcliffe and Weekes, 1960; King, 1969; Rishbeth and Garriott, 1969)

#### 1.3.1. The F1 layer

When the F1 layer is observable as a separate layer, its critical frequency varies diurnally with solar zenith angle $\chi$, approximately as:

$$f_o F1 \propto (\cos \chi)^{0.2} \tag{6}$$

This is not very different from the $(\cos \chi)^{0.25}$ behaviour of the idealized Chapman layer (Chapman, 1931a). When corrected for variations of $\chi$, the variation with mean sunspot number $R$ is approximately given by:

$$(f_o F1)^4 \propto (N_m F1)^2 = (500 + 8R) \times 10^{20} \text{m}^{-6} \tag{7}$$

Largely because of the "valley ambiguity" problem (section 1.2.2), the height of the

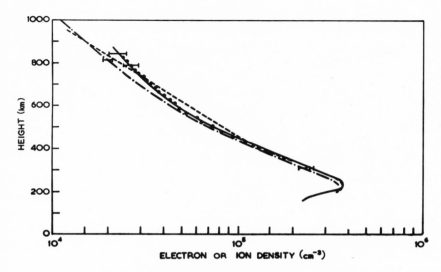

Fig. 3. Electron and ion concentration profiles in the F region and topside ionosphere, obtained nearly simultaneously by different techniques within the region 37–42°N, 69–74°W, on 2 July 1963. Dotted line = ion-trap experiment and continuous line = two-frequency c.w. propagation experiment, both carried by a NASA rocket (0921–0926 75°W meridian time); dashed-dotted line = Alouette topside sounder satellite (0922 h); dashed line = Lincoln Laboratory incoherent scatter (0930–1044 h). (After Bauer et al., 1964.)

F1 maximum, $h_m$F1, is difficult to compute accurately from ionograms, but it is normally between 160 and 200 km. However, the F1 layer is usually no more than a small "ledge" on the underside of the F2 layer, and often is not observable separately at all.

Very broadly, the F1 layer is observable: (a) only by day; (b) mainly at midlatitudes; (c) more frequently at sunspot minimum; and (d) in summer rather than winter. Sometimes extra F-layer stratifications, not necessarily related to the regular F1 layer and designated "F1$\frac{1}{2}$ layers", are seen. These often occur at low latitudes, where there is evidence of lunar control (McNish and Gautier, 1949) and during solar eclipses (section 1.4.4).

### 1.3.2. The peak electron concentration $N_m$F2

The worldwide distribution of the critical frequency $f_o$F2, which is of course related to $N_m$F2, is summarized by the "prediction maps" that are based on extensive ionosonde·data (Fig. 4 and 5). These show solstice conditions at sunspot minimum and maximum, and thus represent rather extreme conditions. Some leading features are: (a) normally $f_o$F2 is greater by day than by night, though the diurnal variation is smaller in the summer (southern hemisphere) than in winter (northern hemisphere); (b) solar control is more marked in winter than in summer;

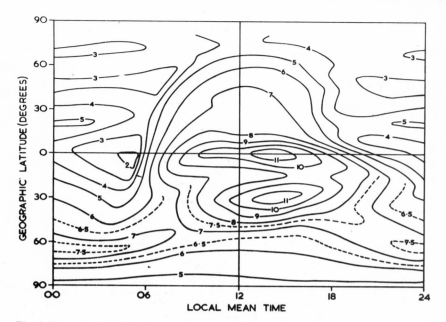

Fig. 4. Contours of predicted critical frequency (MHz) for the American zone, December 1954 (sunspot number $R = 10$). The map shows the "extraordinary" critical frequency $f_xF2$, but this is closely related to $f_oF2$ (section 1.2.1). (From the CRPL Series D Predictions, U.S. Dept. of Commerce.)

this is shown by the extent to which the contours follow the sunrise–sunset line. In the summer hemisphere, $f_oF2$ does not necessarily attain its greatest values near noon; (c) in general, $f_oF2$ decreases from the tropics towards the poles (but see d, e, g); (d) the "equatorial anomaly"; by day there exists a "trough" of $f_oF2$ centered on the magnetic equator. This "trough" extends over a greater range of local time at sunspot maximum than at sunspot minimum. Its width is about 30° in latitude; (e) at midlatitudes (35°–75°) and at sunspot maximum, $f_oF2$ is much greater in winter than in summer—the "seasonal anomaly". Although these maps do not show it, there is a marked semi-annual variation with maxima of $f_oF2$ at the equinoxes; this is particularly noticeable in the southern hemisphere; (f) at a given station, the variation with sunspot number $R$ is roughly like:

$$N_m F2 \propto (1 + R/50) \tag{8}$$

(g) these maps do not purport to give an accurate picture at high latitudes.

### 1.3.3. The height of the peak, $h_mF2$, and layer thickness, $Y_mF2$ (Thomas, 1963)

At midlatitudes (Slough, Watheroo), $h_mF2$ is higher at night than by day (Fig. 6). In low latitudes (Huancayo), this behaviour is almost reversed, though very large (and rather ill-determined) values occur after sunset. Intermediately (at Maui), the

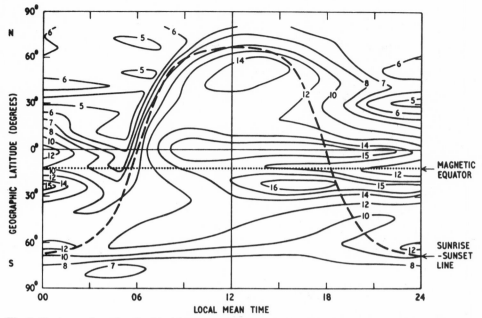

Fig. 5. Contours of predicted critical frequency (MHz) for the American zone, December, 1957 (sunspot number $R = 240$). The map shows the "extraordinary" critical frequency $f_xF2$, but this is closely related to $f_oF2$ (section 1.2.1). The dotted line shows the approximate position of the magnetic equator; the broken line shows where the solar zenith angle is 90°. (From the CRPL Series D Predictions, U.S. Dept. of Commerce.)

variation is more complex. In general $h_mF2$ is found to increase as the sunspot number increases. A marked semiannual variation of $h_mF2$ has been detected at midlatitudes, the greatest values being near equinox (Becker, 1967).

The sub-peak thickness of the F2 layer, $Y_mF2$, is greater in summer than in winter, and increases from sunspot minimum to sunspot maximum. It is thought to be related to the temperature of the F2 layer (Becker, 1967). By day, the thickness may be influenced by the presence of the F1 layer.

### 1.3.4. Variations of $N_mF2$ at high latitudes (Piggott and Shapley, 1962)

In recent years more detailed study has been made of the Antarctic region than the Arctic. It has been found that there exist: (a) seasonal variations, characterized by the different shapes of the diurnal variations of $f_oF2$; the transition between winter and summer behaviour often takes place rather abruptly, within the space of a few weeks (Fig. 7); (b) geographical variations; maps of $f_oF2$ often show enhancements in particular regions (Sato and Rourke, 1964); (c) Universal Time effects; taking the Antarctic as a whole, there is a maximum of F2 layer ionization at around 0600 UT (Duncan, 1962). In the Arctic the UT effect, if present, is much weaker (Challinor, 1970).

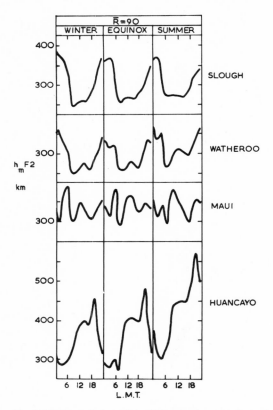

Fig. 6. Daily variation of $h_m$F2 for moderate solar activity at Slough ($51\frac{1}{2}$°N), Watheroo (30°S), Maui (21°N) and Huancayo (12°S, dip 2°). (From Thomas, 1963.)

*1.3.5. High-latitude structure revealed by satellite data* (Thomas and Andrews, 1969; Sato and Colin, 1969)

The coverage afforded by satellites in high-inclination orbits has enabled the construction of new models of the upper F2 layer at high latitudes (Fig. 8). There appears to be a fairly definite boundary, prominent at night and adjoining the so-called "trough", where the midlatitude F2 layer may be considered to end (Muldrew, 1965; Sharp, 1966). It is generally found at 50°–65° magnetic latitude, according to local time and magnetic activity. It can also be detected on ground-based ionograms (Bowman, 1969). It contains quite marked horizontal gradients of electron concentration.

The magnetospheric "plasmapause" or "knee" (Carpenter, 1966)—at which there is a rapid outward decrease of electron concentration—is connected by geomagnetic field lines to the F region in the vicinity of the "trough".

Poleward of the "trough", in the winter topside F layer, there is a "ring" of

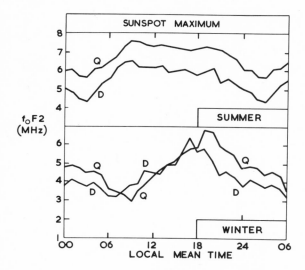

Fig. 7. Daily variation of $f_oF2$ for magnetically quiet ($Q$) and disturbed ($D$) days at Little America (78°S). (After Piggott and Shapley, 1962.)

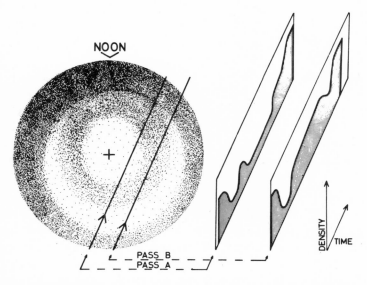

Fig. 8. Idealized distributions of electron density at 300–1000 km altitude in the winter polar ionosphere. The magnetic dip pole is at the centre of the diagram (+). The outer shaded region is the midlatitude F2 layer, whose poleward boundary adjoins the "trough" (and is linked with the magnetospheric plasmapause). The inner shaded region is the "auroral ring" (roughly coincident with the auroral oval) which surrounds the "polar cavity". In the noon sector, the "ring" almost merges with the midlatitude ionosphere so the "trough" is scarcely evident. On the right, two different cross-sections of the electron density distribution are shown, such as would be observed by satellites passing on tracks A and B. (From Andrews and Thomas, 1969.)

enhanced ionization surrounding the magnetic dip pole. It coincides, at least approximately, with the "auroral oval" defined by Feldstein and Starkov (1967). Within the "ring", a smaller electron concentration is found (Fig. 8).

The extent of the "trough" and "ring" depend on magnetic activity. They are closest to the poles during quiet times, and expand towards the equator during magnetic disturbances. The "stable arcs" of oxygen red airglow emission seem to be associated with the "trough" (Bowman, 1969).

### 1.4. F-layer disturbances

#### 1.4.1. Introduction

F-layer disturbance phenomena are extremely complex and any general statements must be treated with even more caution than is needed for general statements about quiet F-layer phenomena. Since a considerable day-to-day variation exists in the quiet F2 layer, e.g., of order 10% in $f_o$F2 (Wright, 1962), deviations from normal behaviour may occur when there are no specific solar or geomagnetic disturbances.

The gross effects of geomagnetic storms, solar eclipses and solar flares will be discussed in the following paragraphs. Travelling disturbances and other wavelike disturbances are not discussed, being dealt with elsewhere.

#### 1.4.2. A summary of storm phenomena in the F layer (Appleton and Piggott, 1952; Martyn, 1953b; King, 1962; Thomas, 1970)

The following are thought to be reasonably typical of what may happen (see Fig. 9): (a) in the early stages of a severe magnetic storm, large decreases of $N_m$F2 (up to 50%) and large increases of $h_m$F2 ($\sim$ 100 km) may occur and last for 2–3 hours (Thomas, 1968). These are seen simultaneously at widely separated places; (b) within a few hours of the "onset of the main phase" of the magnetic storm (i.e., the time at which the $D$st "stormtime" component of the horizontal geomagnetic field becomes negative), decreases of $N_m$F2 may set in and last for 1–3 days. These "negative disturbances", which may appear to be continuations of the decreases (a), are most likely if the "main phase onset" takes place at night. Otherwise, the storm may produce increases of $N_m$F2, or have no effect; (c) around the time of onset of the main phase of the magnetic storm, increases of $N_T$ and $N_m$F2 may occur at midlatitudes, particularly if the time when magnetic $D$st decreases rapidly occurs during daylight (Jones, 1971). These increases in $N_T$ may exceed 100%; (d) sometimes these increases persist for a day or so, and the storm is then termed "positive". At midlatitudes, "positive" storms are commonest in winter; (e) the behaviour of $N_m$F2 in a given storm can be very different at magnetically conjugate stations (Thomas, 1970); (f) except during the fluctuations (a) early in the storm, $h_m$F2 does not generally vary by more than about ±20 km from its quiet day values (ionograms may show large increases of virtual height but they are misleading); (g)

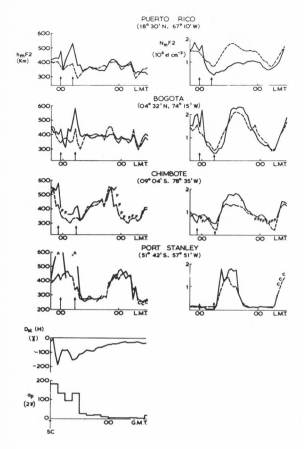

Fig. 9. Variations of $h_mF2$ and $N_mF2$ at stations in the American sector, and of geomagnetic parameters, during the storm of 7 June 1958. In the plots of F2 parameters, the broken curves represent the quiet-day variations. The arrows show the times of the maximum increases of $h_mF2$. (From Thomas, 1970.)

the F1 layer often becomes more prominent during storms. In extreme cases a "G-condition", in which $N_mF2 < N_mF1$, may occur; (h) at low latitudes, positive storms are very common, though very large storms tend to produce negative effects in $N_mF2$. King et al. (1967) show that the equatorial "trough" tends to fill up during a storm; (i) few storms are truly world-wide as regards the F2-layer effects produced during the main phase of the magnetic storm. Sometimes a "storm center", where the effects are strongest, can be located (Appleton and Piggott, 1952); (j) F2-layer storm effects in high latitudes are extremely complex. The "trough" and "ring" features (section 1.3.5) expand equatorwards during storms; (k) satellite drag data show that the neutral atmospheric temperature increases during storms (Jacchia et al., 1967). The electron temperature in the F2 layer does not appear to increase markedly by day, though some increases occur at night (Evans, 1970a, b); and (1) the magnetic disturbance parameters ($D$st and $Kp$ or $ap$) tend to peak earlier than the F2 layer effects (by up to a day).

### 1.4.3. Solar flare effects

During a solar flare, the sun's output of ionizing radiation is enhanced, causing short-lived increases of ionospheric electron concentration. The lower ionosphere perturbations (SID's) are well-k..own but, owing to their short duration of 10–20 min, the comparatively small F-layer effects are not readily measurable by iono-sondes operating on normal schedules. Continuous $N_T$ measurements using geostationary satellites (section 1.2.4), and HF Doppler measurements (section 1.2.7) of the lower F layer, have made possible detailed analysis of the phenomena (Garriott et al., 1969; Davies and Baker, 1966). The F1-layer ion production rate may increase by 30% during a big flare.

### 1.4.4. Eclipse effects (Ratcliffe, 1956b; Rishbeth, 1968b)

The ionospheric effects of solar eclipses have been studied for decades, in the hope that they might provide a means of determining production and loss rates. In fact, it has proved very difficult to obtain accurate results from eclipse observations of the F layer. For various reasons, the problems seem least severe in low latitudes. A typical eclipse observation is illustrated in Fig. 10.

In principle, the following information might be obtainable: (a) production and loss rates. The results are difficult to interpret, largely because the emission of radiation from the solar disc is very "patchy"; moreover, the results depend strongly on how much residual radiation is received from the totally-eclipsed sun; (b) F-region drift velocities can sometimes be estimated from eclipse data; (c) temperature changes have been measured with the aid of space vehicles and incoherent scatter observations.

### 1.4.5. Lunar perturbations of the F2 layer

In low latitudes, some small perturbations of the F2 layer can be attributable to lunar tides (McNish and Gautier, 1949). Smaller variations, having a different phase in lunar time, exist at midlatitudes (Bossolasco and Elena, 1960; Rastogi, 1961).

## 2. A RESUME OF THE F-REGION THEORY

### 2.1. The composition of the atmosphere and ionosphere (Johnson, 1969)

#### 2.1.1. Basic equations

As a preliminary to discussing how the F layers are formed, it is necessary to have some knowledge of the structure of the neutral atmosphere. Above 100 km altitude,

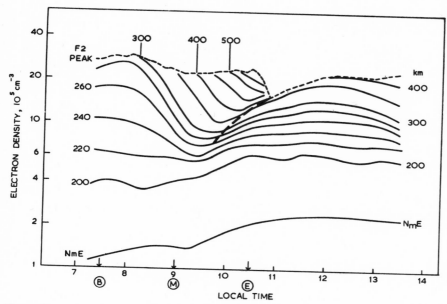

Fig. 10. Electron density variations at fixed heights, and at the F2 and E layer peaks, at Singapore (1°N, dip −18°) during the partial eclipse of 19 April 1958. The broken line refers to the "Fl$\frac{1}{2}$ ledge" (section 1.3.1). Letters $B$, $M$, $E$ mark the beginning, middle and end of the eclipse. (Radio and Space Research Station, Slough. Crown Copyright.)

the distributions of all neutral gases (except possibly H and He) conform closely to the hydrostatic formula. Let $p$, $\rho$, $n$, $m$ denote respectively the pressure, density, concentration and particle mass for a single gas; $T$ is the absolute temperature, and $g$ the acceleration due to gravity. Then the perfect gas law:

$$p = nkT \tag{9}$$

and the barometric equation:

$$dp/dh = -\rho g \tag{10}$$

can be used to derive the "scale height" (noting also that $\rho = nm$)

$$H = -\{d(\ln p)/dh\}^{-1} = kT/mg \tag{11}$$

The scale height is a measure of how the pressure varies with height.

If several gases are present and each conforms independently to the barometric eq.10, then eq.11 holds for the scale height of the atmosphere as a whole, provided the mean particle mass:

$$\overline{m} = (\Sigma n_j m_j)/\Sigma n_j \tag{12}$$

(where $j = 1, 2, \ldots$ refers to the individual gases) is used in the equation.

### 2.1.2. Neutral gas composition

Typical data on gas composition are shown in Fig. 11. The major neutral atomic gas is O, and the most important molecular gases are $N_2$ and $O_2$. The atomic/molecular ratio is hard to determine accurately; it is still uncertain and may well be variable. The presence of H and He is insignificant, except at great heights.

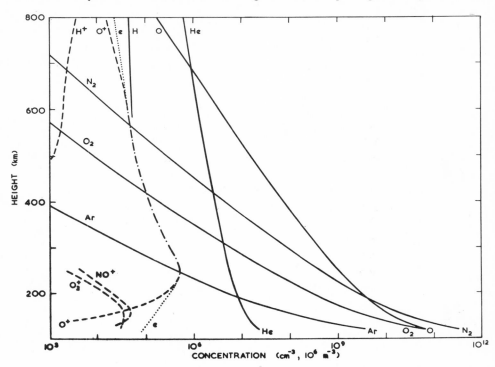

Fig. 11. Day-time ion and neutral composition at 100–800 km altitude for sunspot minimum. The positive ion and electron concentrations are from Johnson (1969). The neutral gas concentrations are from the COSPAR International Reference Atmosphere, model 2 for noon, North-Holland, Amsterdam, 1965. (Radio and Space Research Station, Slough. Crown Copyright.)

### 2.1.3. Positive ion composition

Fig. 11 shows that at the F2 peak, around 300 km altitude, the electrons and ions represent only about 0.1% of the total gas concentration. Even at 800 km, they are only about 1% of the total. Throughout almost the entire height range shown in Fig. 11, $O^+$ is the dominant ion. Below 200 km, however, $NO^+$ and $O_2^+$ ions become abundant, and below 150 km they are more abundant than $O^+$. Since neutral NO is only a trace constituent of the F region (and is too scarce to be shown in the figure), the $NO^+$ must be formed from other ions, as discussed in section 2.3.2.

## 2.2. Production of ionization (Allen, 1965)

### 2.2.1. The production function (Chapman, 1931a)

To begin with, a very simple situation will be discussed. Consider a spherically symmetric atmosphere acted upon by monochromatic solar ionizing radiation. The rate of production or ionization $q(h, t)$ is proportional to the intensity $I(h, t)$ of the radiation, the concentration $n(h)$ of ionizable gas, the absorption cross-section $\sigma$, and an efficiency factor $\eta$ (of order unity). The downcoming radiation is attenuated as it travels through the atmosphere; if $I_\infty$ is its original intensity at the top of the atmosphere, then at any height $h$ the "optical depth" $\tau$ is defined by the equation:

$$I(h, t) = I_\infty e^{-\tau(h,\chi)} \tag{13}$$

The optical depth depends on the total amount of gas along the path of the radiation from the top of the atmosphere to height $h$. It depends on the solar zenith angle $\chi$, which of course varies with local time. For a horizontally stratified plane atmosphere, $\tau$ is proportional to $\sec \chi$. In practice, the earth's curvature must be taken into account if $\chi > 80°$, and then $\sec \chi$ is replaced by the "grazing incidence integral" defined by Chapman (1931b), namely $Ch(\chi)$. Some production occurs for zenith angles up to about $100°$.

If there is only one ionizable gas, then:

$$q(h, t) = I(h, t)\eta\sigma n(h) = I_\infty e^{-\tau(h,\chi)}\eta\sigma n(h) \tag{14}$$

$$\tau(h, \chi) = \int_h^\infty \sigma n(h)\, \mathrm{d}h Ch(\chi) = \sigma n(h) H(h) Ch(\chi) \tag{15}$$

The integral in eq.15 has been evaluated by using a well-known result, derivable from the equations of section 2.1.1, that the total columnar content of gas above any given level is the product of the gas concentration and the scale height at that level. This result holds strictly for a plane atmosphere in which $g$ is independent of height.

These equations lead to the famous Chapman production formula. Let $z$ denote "reduced height", measured in units of $H$; the level $z = 0$ is taken to be the level of unit optical depth for $\chi = 0$; thus

$$z(h) = \int_{h_o}^h (\mathrm{d}h/H) = -\ln(p/p_o) \tag{16}$$

(and it is easily shown that $z$ varies linearly with pressure, $p_o$ being the pressure at the zero level $h_o$ where $z = 0$). Using the fact that $(nH/n_o H_o) = e^{-z}$, it follows that $\tau = e^{-z} \sec \chi$ whence, from eq. 14 and 15, it is not difficult to show that:

$$q(z, \chi) = (\eta I_\infty/eH)\exp(1 - z - e^{-z} \sec \chi) \tag{17}$$

which is one version of Chapman's formula. If, furthermore, $H$ is assumed independent of height, it is easily shown that the peak value of $q$ is $(\eta I_\infty / eH)\cos\chi$, and is found at the level $z = \ln\sec\chi$; i.e., at $z = 0$ if $\chi = 0$. Slightly more complex formulas apply if $H$ varies with height.

### 2.2.2. The production function for a mixture of gases

If different gases (distinguished by $j = 1, 2, \ldots$) are present, then $\eta$, $\sigma$, $n$, $H$ and (if $\chi$ is large) $Ch(\chi)$ are different for each. Then eq.14 and eq.15 must be replaced by composite equations:

$$q(h, t) = I_\infty e^{-\tau(h,\chi)} \Sigma \eta_j \sigma_j n_j(h) Ch_j(\chi) \tag{18}$$

$$\tau(h, \chi) = \Sigma \sigma_j n_j(h) H_j(h) Ch_j(\chi) \tag{19}$$

Lastly, the $\eta_j$, $\sigma_j$ and $I_\infty$ are functions of the optical wavelength $\lambda$, so that $q$ ought really to be determined from a more complex equation which contains an integration with respect to $\lambda$. In practice, it suffices to calculate $q$ from a summation over a number of discrete bands of wavelength, so chosen that $\eta$ and $\sigma$ do not vary too much within any one band. At the F2 peak and above, $\tau \ll 1$ for much of the day, and the equations for $q$ are then simplified.

In reality, a broad spectrum of solar radiation is responsible for producing the F-layer ionization. Fig. 12 shows how the "height of unit optical depth" for overhead sun varies with wavelength. This height lies above 140 km, i.e., in the F1 layer, for

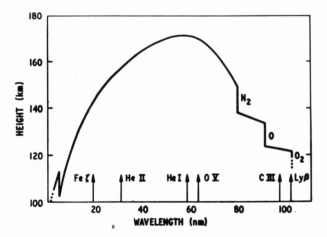

Fig. 12. The level of unit optical depth, at which vertically incident radiation is attenuated to a fraction $e^{-1}$ of its original intensity, for extreme-UV and X-radiations that ionize the E and F layers. Each radiation makes its maximum contribution to ion production in the vicinity of this level, when $\chi = 0$ (section 2.2.1). The break near 3 nm is due to the K-absorption limit of $N_2$; the breaks at about 80, 91, 103 nm are respectively at the ionization limits of $N_2$, O and $O_2$. The arrows mark the wavelengths of some important solar emissions. (Computed from Norton et al., 1963.) (Radio and Space Research Station, Slough. Crown Copyright.)

wavelengths between about 17 and 80 nm; the ionization is produced partly by a few strong line emissions (as shown), and partly by continuum radiation. Wavelengths outside this range have cross-sections $\sigma$ that are so small that the level of unit optical depth lies in the E layer, or even lower.

The total input of solar energy in the 17–80 nm range is 1.8 mW m$^{-2}$ (from Allen, 1965), for low sunspot activity. This represents about a millionth of the whole "solar constant", which is 1.4 kW m$^{-2}$.

### 2.2.3. Corpuscular ionization (Torr and Torr, 1970)

Ionization can be produced in the F layer by charged particles; the extent to which this might contribute to the normal F2 layer is a matter of current discussion. The absorption cross-sections for protons and electrons are such that only particles with energy $\lesssim 1000$ eV will be stopped in the F layer, if incident from above. The motions of such particles are strongly controlled by the geomagnetic field. Particles will cause heating, particularly of the electron gas, as well as ionization; they also cause excitation of neutral particles, sometimes leading to airglow emission. Thus airglow measurements can give information on the particle influx.

At middle and low latitudes, where the geomagnetic field lines are closed, the principal source of charged particles is likely to be photoelectrons from the opposite hemisphere. These have energies $\lesssim 100$ eV. This source is most intense by day, but its most interesting effects occur when one end of a field line is sunlit and the other in darkness (e.g., Carlson, 1966). Essentially, the photoelectrons just redistribute energy supplied by solar radiation.

At higher latitudes, effects may be produced by auroral particles, which may be responsible for some of the phenomena described in section 1.3.5; and by magnetospheric particles entering the ionosphere along open field lines.

## 2.3. F-region chemistry (Ferguson, 1969; Johnson, 1969)

### 2.3.1. Production and loss reactions

Since atomic oxygen is an abundant neutral constituent, $O^+$ ions are produced copiously. They recombine with electrons only very slowly as the relevant process, i.e., radiative recombination ($O^+ + e \rightarrow O + $ photon), is very much slower than the dissociative recombination process which applies only to molecular ions (e.g., $O_2^+ + e \rightarrow O + O$). Instead, $O^+$ ions are lost via ion-atom interchange reactions in which molecular ions (notably $NO^+$ and $O_2^+$) are produced. Important interchange reactions are:

$$O^+ + O_2 \rightarrow O_2^+ + O \tag{X}$$

$$O^+ + N_2 \rightarrow NO^+ + N \tag{Y}$$

$$N_2^+ + O \rightarrow NO^+ + N \tag{Z}$$

Fig. 11 shows that O and $N_2$ are major constituents in the F region, so that $O^+$ and $N_2^+$ are the ions principally produced. The $O^+$ ions are lost via reactions (X) and (Y), and the molecular ions are subsequently lost by dissociative recombination. The $N_2^+$ ions can also disappear by dissociative recombination, but this process is not thought to be rapid enough to account for the almost total absence of $N_2^+$ ions (there are too few to be shown in Fig. 11), so reaction (Z) is invoked to remove $N_2^+$ ions (Norton et al., 1963). Since there is very little neutral NO in the E and F regions, reactions like (Y) and (Z) must be the source of the $NO^+$ ions that are observed. Some $O_2^+$ is produced by photoionization of $O_2$, but (X) is also a source of this ion.

### 2.3.2. The linear and square-law loss coefficients

For simplicity, consider just the following sequence of ion reactions (it is probably the most important in practice):

$$O \xrightarrow[\substack{\text{ionization} \\ \text{(rate } q)}]{\text{photo-}} O^+ \xrightarrow[\substack{\text{reaction (Y)} \\ \text{(coefficient } \gamma)}]{\text{interchange}} NO^+ \xrightarrow[\substack{\text{recombination} \\ \text{(coefficient } \alpha)}]{\text{dissociative}} N, O$$

The balance equations for this scheme of reactions at equilibrium are:

$$q = \gamma[O^+][N_2] = \alpha[NO^+][e] \tag{20}$$

in which the square brackets represent concentrations of the relevant constituent. There are two special cases:

(1) If neutral $N_2$ is so abundant that the $O^+$ ions are converted to $NO^+$ via reaction (Y), as soon as they are formed, then almost all the ions are $NO^+$; i.e., $[NO^+] \simeq [e]$ so that in eq.20 $q \simeq \alpha[e]^2$.

(2) If neutral $N_2$ is so scarce that reaction (Y) can only proceed slowly, but the $NO^+$ ions recombine with electrons very quickly once they are formed, then almost all the ions are $O^+$; i.e., $[O^+] \simeq [e]$ so that in eq.20: $q \simeq \gamma[N_2][e]$. Let $\beta = \gamma[N_2]$.

Case (1) applies at lower heights, and case (2) at higher heights in the F region; the transition between (1) and (2) occurs where $\alpha[e] \sim \gamma[N_2]$. Since there must be electrical neutrality:

$$[e] = [O^+] + [NO^+] \tag{21}$$

By using eq.21 it is possible to eliminate both $[O^+]$ and $[NO^+]$ from eq.20. Reverting to the symbol $N$ for electron concentration, instead of $[e]$, one obtains in the general case:

$$\frac{1}{q} = \frac{1}{\beta N} + \frac{1}{\alpha N^2} \tag{22}$$

and in the special cases mentioned above:

$$\text{Case (1):} \qquad q = \alpha N^2 \qquad \text{or:} \qquad N = N_\alpha = (q/\alpha)^{1/2} \qquad (23)$$

$$\text{Case (2):} \qquad q = \beta N \qquad \text{or:} \qquad N = N_\beta = q/\beta \qquad (24)$$

Note that $\beta$, as defined above, decreases rapidly upwards as it is proportional to the concentration of molecular gas.

### 2.3.3. F1/F2 splitting (Ratcliffe, 1956a; King, 1961; Stubbe, 1970)

Even when the analysis is generalized to include other constituents ($O_2$, N, $O_2^+$, $N_2^+$, etc.), the situation is basically similar to that described in section 2.3.2. As shown in Fig. 11, there is a lower (F1) layer of $NO^+$ and $O_2^+$ ions and electrons, and an upper (F2) layer of $O^+$ ions and electrons. Whether the lower layer is prominent enough to form a separate "bump" in the $N(h)$ profile, thus giving an observable F1 layer, can be shown to depend upon a "shape parameter" given by:

$$G = \beta^2/\alpha q = N_\alpha^2/N_\beta^2 \qquad (25)$$

The larger is $G$, the more prominent is the F1 layer. Many features of F1-layer behaviour (section 1.3.1) can be accounted for in terms of variations of $G$. The two ionograms shown in Fig. 13 have different values of $G$, which may be attributable

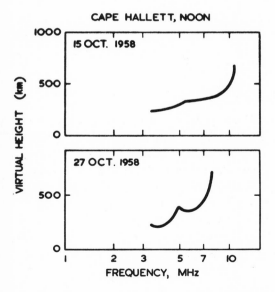

Fig. 13. Ordinary mode echoes [$h'(f)$ curves] from noon ionograms at Cape Hallett (72°S), before (15 October 1958) and after (27 October 1958) the onset of a polar stratospheric warming, which is thought to influence the atmospheric composition. The shape of the lower ionogram, with a more prominent F1 layer, is consistent with an enhanced loss coefficient $\beta$ (section 2.3.3). (After Bullen, 1964.)

to changes in atmospheric composition that affected $\beta$, and perhaps connected with a stratospheric warming that occurred in the meantime (Bullen, 1964). However, the visibility of the F1 layer also depends on the height of the F2 peak, which is largely determined by transport processes, as will now be considered.

## 2.4. Motions of ionization (Rishbeth and Garriott, 1969)

### 2.4.1. The continuity equation

In general the ionization in the F2 layer is neither in equilibrium nor stationary. Its behaviour must therefore be described with the aid of the full continuity equation:

$$\partial N/\partial t = q - \beta N - \text{div}(N\vec{V}) \tag{26}$$

in which the linear loss coefficient $\beta$, defined in section 2.3.2, is used, and $\vec{V}$ is the "plasma drift velocity". On expanding the "movement term":

$$\partial N/\partial t = q - \beta N - N \text{ div } \vec{V} - \vec{V} \cdot \text{grad } N \tag{27}$$

In its lifetime (of order $\beta^{-1}$) the ionization travels a distance $\sim V/\beta$. The drift will have an appreciable effect on the distribution of ionization if either (a) $N$ varies significantly within a distance $V/\beta$ in the direction parallel to $\vec{V}$; i.e. if:

$$|\vec{V} \cdot \text{grad } N|/N \gtrsim \beta \tag{28}$$

or (b) if the spatial variation of $\vec{V}$ is sufficiently rapid; i.e., if:

$$|\text{div } \vec{V}| \gtrsim \beta \tag{29}$$

Typically, at the F2 peak $V \sim 30$ m sec$^{-1}$ and $\beta \sim 10^{-4}$sec$^{-1}$, so that $V/\beta \sim 300$ km. This is smaller than a typical scale length for horizontal variations of $N$ ($\sim 1000$ km), but larger than the typical scale length for vertical variations (one scale height, $\sim 50$ km). Thus vertical drifts are more important than horizontal drifts, in general. In much the same way, the condition (29) is only likely to be fulfilled for vertical drifts. But in particular places, where the horizontal gradients of either $N$ or $\vec{V}$ are unusually large, such as near sunrise and sunset or in the vicinity of the plasma-pause, horizontal motions may produce important effects.

In the F1 layer, where the lifetime of the ionization is short ($\sim 100$ sec), movements produce smaller perturbations of $N$ than they do in the F2 layer.

### 2.4.2. Forces acting on the plasma

Provided ions and electrons move together ($\vec{V}_e = \vec{V}_i = \vec{V}$), as is very nearly the case in the F region, a single equation of motion suffices for the plasma as a whole.

The forces acting on the plasma include (a) gravity; (b) gradients of the partial pressure $(NkT_e + NkT_i)$ of the electron-ion plasma; (c) collisions with the neutral air, which moves with velocity $\vec{U}$ (the wind velocity); and (d) electric fields.

### 2.4.3. The plasma diffusion equation

Under the conditions prevailing in the F region (magnetic gyrofrequency $\gg$ collision frequency with neutral particles, both for ions and electrons), the forces (a) (b) (c) listed above produce only motion parallel to the geomagnetic field $\vec{B}$. It may be shown that only the ions make a significant contribution, so far as the gravitational and collisional terms in the equation of motion are concerned. This equation is:

$$0 = m_i \vec{g} - N^{-1}\vec{\nabla}[Nk(T_i + T_e)] - m_i \nu_{in}(\vec{V} - \vec{U}) \tag{30}$$

in which $m_i$ is the ion mass, $\nu_{in}$ the ion-neutral collision frequency for momentum transfer, and $\vec{g}$ the acceleration due to gravity. Strictly speaking, the left-hand side should contain the inertial term, $m_i\,d\vec{V}/dt$, but this is negligible in practice. If more than one ion species is present, a separate equation should be written for each; but in the F2 layer it is usually legitimate to assume all the ions to be $O^+$ in writing the equation of motion.

If there is no magnetic field, or if the field is vertical, and the various quantities are assumed to vary in the vertical direction only, then eq. 30 reduces to an equation for $W_p$, the vertical diffusion velocity. Taking the air to be stationary ($\vec{U} = 0$), and replacing $\vec{\nabla}$ by $\partial/\partial h$ and $\vec{g}$ by $-g$:

$$W_p = -\frac{1}{m_i \nu_{in}}\left\{\frac{1}{N}\frac{\partial}{\partial h}[Nk(T_i + T_e)] + m_i g\right\} \tag{31}$$

Now let:

$$T_p = \tfrac{1}{2}(T_i + T_e) \qquad \text{``plasma temperature''} \tag{32}$$

$$H_p = 2kT_p/m_i g \qquad \text{``plasma scale height''} \tag{33}$$

$$D_p = 2kT_p/m_i \nu_{in} \qquad \text{``plasma diffusion coefficient''} \tag{34}$$

Then (31) becomes:

$$W_p = -D_p\left\{\frac{1}{NT_p}\frac{\partial}{\partial h}(NT_p) + \frac{1}{H_p}\right\} \tag{35}$$

showing that $W_p$ depends on gravity and on the vertical plasma pressure gradient.

### 2.4.4. Diffusive equilibrium

The vertical diffusion velocity $W_p$ given by Eq.35 vanishes if the plasma pressure varies exponentially with height such that:

$$(NT_p) \propto \exp(-h/H_p) \propto \exp(-m_i gh/2kT_p) \qquad (36)$$

The scale height $H_p$ of this distribution is twice that of a neutral gas of molecular mass $m_i$ at temperature $T_p$. Eq.36 may be expected to apply at heights well above the F2 peak, where $v_{in}$ is very small and $D_p$, therefore, large.

### 2.4.5. Diffusion in an inclined magnetic field

To a good approximation plasma diffusion takes place only in the direction of the magnetic field. If $I$ is the magnetic dip angle, then from simple geometrical considerations the field-aligned diffusion velocity is (sin $I$) times as large, and the vertical velocity (sin$^2$ $I$) times as large, as the velocity $W_p$ given by eq.35. But Dougherty (1961) has shown that, if the neutral air is free to move horizontally, it is accelerated in such a way that the vertical diffusion velocity of the plasma is, after all, simply given by eq.35.

Near the magnetic equator, the curvature of the magnetic field lines must be taken into account, and the equations become more complex (Kendall, 1962).

### 2.4.6. Effect of an electric field

In the F region an electric field $\vec{E}$, normal to $\vec{B}$, exists as a result of dynamo action in the E region (Martyn, 1953a). Other processes may contribute to $\vec{E}$, particularly during magnetic disturbances. The electric field produces a drift velocity normal to $\vec{B}$, viz:

$$\vec{V}_E = \vec{E} \times \vec{B}/B^2 \qquad (37)$$

It may be shown that in general div $\vec{V}_E$ is very small. Thus any effective contribution of the "electromagnetic drift" $\vec{V}_E$ to eq. 27 is likely to depend mainly on the term $\vec{V}_E \cdot$ grad $N$; this is largest when $\vec{V}_E$ has a substantial vertical component, as happens for a zonal electric field near the magnetic equator. At midlatitudes, the effects of electromagnetic drift are smaller, and are further reduced by the "ion-drag effect" (Dougherty, 1961); as a result of collisions between the drifting plasma and the air, the air is accelerated horizontally, which produces an additional component of ion drift. This additional drift, when added to $\vec{V}_E$, gives an almost horizontal resultant drift which makes very little contribution to the continuity equation 26.

### 2.4.7. Effect of a horizontal wind

A wind $U_y$ blowing northwards in the magnetic meridian produces a field-aligned drift of the plasma ($U_y$ cos $I$). The vertical component of this is ($-U_y$ cos $I$ sin $I$), reckoned positive upwards. Winds also produce a *very* small ion drift normal to $\vec{B}$ (Rishbeth, 1971).

## 2.5. *Formation of the F2 peak*

### 2.5.1. *Day equilibrium* (Rishbeth and Barron, 1960; Yonezawa, 1970)

The mid-latitude F2 peak forms at a height $h_m$ where the effects of plasma diffusion and loss are balanced. This is expressed by an equation:

$$\beta(h_m) \sim D(h_m)/H^2 \tag{38}$$

where $H$ is the scale height of the ionizable gas. Note that $D$ increases upwards, being inversely proportional to $\nu_{in}$, while $\beta$ decreases upwards. Well below the peak, the electron distribution approximates to photochemical equilibrium which, for the F2 layer, is given by $N = q/\beta$ as in (24). Well above the peak, the diffusive equilibrium formula (36) applies and, if $T_p$ does not vary much with height, $N \sim \exp(-h/H_p)$. (If also $T_e = T_i$ then $H_p = 2H$ so that, defining reduced height $z$ as in (16), $N \sim \exp(-z/2)$ as in Fig. 14).

The thickness of the peak is of the order of the scale height $H$.

### 2.5.2. *Effect of vertical drift*

A vertical drift velocity $W$ due to a wind, or to an electric field, displaces the height of the peak by an amount of order $WH/D(h_m)$ scale heights; i.e., $h_m$ is

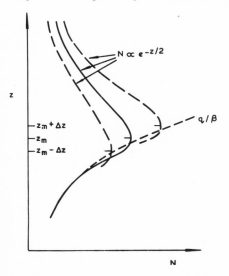

Fig. 14. Equilibrium electron distributions $N(z)$ for the F2 layer, $z$ being reduced height (measured in units of the scale height of the ionizable gas). In the absence of vertical drift, the peak lies at a level where determined by diffusion and loss (full curve). A vertical drift raises the peak if upwards, and lowers it if downwards (broken curves), the displacements $\Delta z$ being roughly proportional to the drift. All curves conform to the photochemical equilibrium curve $(q/\beta)$ at heights well below the peak, and to the diffusive equilibrium curve $(N \propto e^{-z/2})$ at heights well above the peak. (From Rishbeth and Garriott, 1969.)

changed by:

$$\Delta h_m \sim WH^2/D(h_m) \tag{39}$$

as shown in Fig. 14. An upward drift raises $h_m$ and increases the equilibrium value of $N_m$; a downward drift lowers $h_m$ and decreases the equilibrium $N_m$.

### 2.5.3. Night-time layers (Duncan, 1956; Dungey, 1956)

At night the F2 layer develops into a "shape-preserving" form, which decays as a whole with a decay coefficient $\beta'$, of the order of $\beta(h_m)$. In the absence of vertical drift, the peak is near the level given by eq.38; but an upward drift raises the layer, according to eq.39, and thereby reduces $\beta'$.

### 2.5.4. High latitudes

The general principles just stated should apply but there are some special factors: (a) ion production by energetic particles (section 2.2.4) is important; (b) the dip angle $I$ is large, which limits the effectiveness of winds and electric fields in producing vertical drift; (c) the $N(h)$ distribution can be influenced by fluxes of ions and electrons along open geomagnetic field lines (i.e., the polar wind (Banks and Holzer, 1969), which probably causes the "trough" described in section 1.3.5).

### 2.5.5. Low latitudes.

Diffusion is strongly affected by the magnetic field geometry (section 2.4.5), but electric fields can produce vertical drifts. The equatorial anomaly (section 1.3.2) results from the combined effects of electric fields and diffusion along magnetic field lines (Duncan, 1960).

### 2.5.6. The actual field-aligned drift velocity at the midlatitude F2 peak

At equilibrium the plasma drift velocity, due to all causes, must always be downwards so as to offset the imbalance between production and loss above the F2 peak, where $q > \beta N$. The field-aligned drift velocity $V_B$ can be found by resolving eq.30 in the direction of $\vec{B}$. At the F2 peak, the plasma pressure is a maximum, so its gradient is negligible. Using section 2.4.7 also, the *vertical* component of $\vec{V}_B$ (positive upwards) is found to be:

$$V_{Bz} = -V_B \sin I = -U_y \cos I \sin I - (g/\nu_{in})\sin^2 I \tag{40}$$

If the wind is poleward, $V_{Bz}$ is clearly downwards as then $U_y \sin I$ is positive in both hemispheres ($U_y$ being positive northwards). But $V_{Bz}$ is also downwards if the wind is equatorwards, because then the F2 peak is pushed up to a height where $\nu_{in}$ is so small that the downward "gravity" term in eq.40 slightly exceeds the upward "wind" term. If there is also an electric field, the situation is somewhat more complicated.

## 3. GLOBAL THERMOSPHERIC WINDS

### 3.1. The global distribution of thermospheric temperature

#### 3.1.1. The diurnal bulge (Jacchia, 1965)

The solar UV heating of the thermosphere produces large day-to- night variations of temperature, of order 30%. The thermal expansion due to the heating produces the "diurnal bulge" on the day side of the earth. As a result the thermosphere is not horizontally stratified and so there are horizontal pressure gradients, which drive the thermospheric winds. As discussed subsequently in this chapter, the wind velocities are limited by Coriolis force due to the earth's rotation, ion-drag due to collisions between the air and F-region ions, and the molecular viscosity of the air.

#### 3.1.2. Thermospheric models based on satellite drag data

The diurnal bulge was discovered, and its form determined, from observations of the drag on artificial satellites (Jacchia, 1959). The satellite data actually give values of $\rho H^{1/2}$ ($\rho$ = air density, $H$ = scale height); by use of the perfect gas law and barometric equation (eq.9 and 10), with suitable assumptions about atmospheric composition, it is possible to derive density profiles $\rho(h)$ to fit the data.

To construct global models giving the distribution of various atmospheric parameters, it is usual to: (a) assume fixed boundary conditions (temperature, gas concentrations) at a lower boundary level, often taken as 120 km; (b) assume that the temperature profile $T(h)$ has a certain shape, usually one that can be specified by one or two parameters. At great heights, $T$ is generally assumed to tend to a limiting value, the "exospheric" temperature $T_\infty$, which is a function of latitude, local time, etc.; (c) using (a) and (b), compute the vertical distribution of each constituent gas, and thus obtain the density profile $\rho(h)$, for any given $T_\infty$; (d) construct empirical formulas to describe the dependence of $T_\infty$ on latitude, local time, season, solar activity and magnetic disturbance, using the observational data as a basis.

Figure 15 shows the global distribution of $T_\infty$ for Jacchia's model at equinox.

An alternative procedure (cf. Harris and Priester, 1962) is to assume lower boundary conditions as in (a); assume an appropriate function to represent the time-varying input of energy from the sun (or other sources); and compute the temperature as a function of height and local time, $T(h, t)$, using the heat-balance equation. The distributions of individual gases are then obtained with the aid of the gas laws.

#### 3.1.3. Difficulties with existing thermospheric models

(a) The observational data consist of *density* measurements. However accurate the density values may be, there are uncertainties in the other parameters derived from them, such as pressure and temperature.

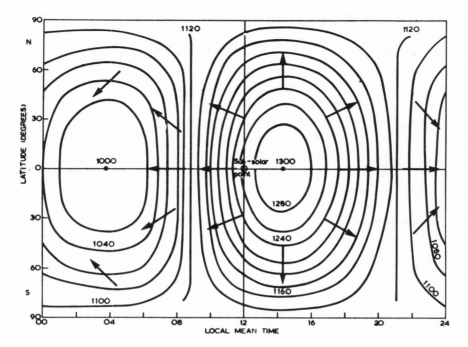

Fig. 15. Global distribution of exospheric temperature for equinox, moderate solar activity, according to the model of Jacchia (1965). The minimum and maximum values are respectively 1000°K and 1300°K; isotherms at intervals of 20°K are plotted on a latitude–local-time map. The arrows show approximate directions of thermospheric winds; they are normal to the isotherms by day, when the winds are controlled by ion-drag, and inclined to the isotherms at night, when ion-drag and Coriolis force are both important (section 3.3.2). (Radio and Space Research Station, Slough, Crown Copyright.)

(b) Indeed the temperature variations of the models disagree with other data, such as incoherent scatter data on temperature (Waldteufel and McClure, 1969). For example, the model temperatures peak around 1400 h local time, while the measured temperatures peak around 1600 h; the model places the global minimum of temperature at night at low latitudes, whereas the incoherent scatter data suggest that the minimum is located at high latitudes.

(c) The models are highly averaged; there may well be much fine-scale variation of atmospheric structure that the models cannot show.

(d) The fixed lower boundary conditions are unrealistic. Conditions may well vary diurnally and seasonally at 120 km (or at any other height that might be chosen as a lower boundary), and this may influence the values of parameters at greater heights (e.g., Chandra and Stubbe, 1970).

## 3.2. The equations governing thermospheric wind velocities

### 3.2.1. Forces acting on the air

The various forces acting on the air each contribute a term to the equation of motion. The quantities given below are all in terms of force per unit mass, i.e., acceleration:

*Gravity.* Though this is one of the largest forces acting on the air, it is not effective in producing horizontal winds.

*Pressure-gradient force* (given by $-\rho^{-1}\vec{\nabla}p$). The vertical component of this force is by far the largest but, since hydrostatic equilibrium prevails to a very good approximation, this component is almost exactly balanced by gravity (cf. eq.10). The horizontal components of the pressure-gradient force, (which are only about $10^{-3}$ of the vertical component) provide the driving force for thermospheric winds, the zonal x-component (positive eastward) and the meridional y-component (positive northward) being respectively:

$$F_x = -\rho^{-1}\partial p/\partial x; \; F_y = -\rho^{-1}\partial p/\partial y \tag{41}$$

*Viscous force.* From gas-kinetic theory this is:

$$\vec{F}_v = (\mu/\rho)\nabla^2\vec{U} \simeq (\mu/\rho)\partial^2\vec{U}/\partial h^2 \tag{42}$$

where $\mu$ is the coefficient of molecular viscosity for the air, and the approximation holds if only the vertical variations of $\vec{U}$ are considered.

*Ion-drag.* Due to collisions between neutral particles and ions (velocities respectively $\vec{U}$, $\vec{V}$). This may be written:

$$F_i = \nu_{ni}(\vec{V} - \vec{U}) = KN(\vec{V} - \vec{U}) \tag{43}$$

where $\nu_{ni} = KN$ is the frequency of collision of neutral particles with ions (which is related to the ion-neutral collision frequency of section 2.4.3 by

$$\nu_{in}/\rho = \nu_{ni}/Nm_i \simeq 2.7 \cdot 10^{10}\,\text{kg}\,\text{m}^3\,\text{sec}^{-1} \tag{44}$$

and the numerical value is for $O^+$ ions and O atoms at $1000°K$).

### 3.2.2. Acceleration of the air

The sum of the forces listed above equals the total acceleration of the air:

$$\vec{g} - \rho^{-1}\vec{\nabla}p + \vec{F}_v + \vec{F}_i = d\vec{U}/dt + 2\vec{\Omega} \times \vec{U} + \vec{\Omega} \times \vec{\Omega} \times \vec{R} \tag{45}$$

The right-hand side of this equation gives the acceleration in a frame of reference rotating with the earth's angular velocity $\vec{\Omega}$, at a point whose position with respect to the earth's centre is $\vec{R}$. The centripetal acceleration $\vec{\Omega} \times \vec{\Omega} \times \vec{R}$ is generally neglected, being constant at a given point and less than 1% of $\vec{g}$ in magnitude. But the Coriolis term $2\vec{\Omega} \times \vec{U}$ is important.

According to the equations of hydrodynamics:

$$d\vec{U}/dt = \partial\vec{U}/\partial t + (\vec{U} \cdot \vec{\nabla})\vec{U} \tag{46}$$

The nonlinear term $(\vec{U} \cdot \vec{\nabla})\vec{U}$ may be important if the wind speed $U$ is comparable with the peripheral speed of rotation of the earth, $R\Omega$; but it is frequently omitted from calculations to avoid mathematical difficulties.

### 3.2.3. Equations for horizontal winds

Taking the eastward (x) and northward (y) components of the equation of motion, eq.45, and substituting from (41–43), one obtains the equations for the horizontal winds:

$$dU_x/dt = F_x - KN(U_x - V_x) + fU_y + (\mu/\rho)\partial^2 U_x/\partial h^2 \tag{47}$$

$$dU_y/dt = F_y - KN(U_y - V_y) - fU_x + (\mu/\rho)\partial^2 U_y/\partial h^2 \tag{48}$$

where the "Coriolis parameter" is given for a geographic latitude $\phi$ by:

$$f = 2\Omega \sin \phi \tag{49}$$

In eq.47 and 48 centripetal accelerations, and the very small contribution of the vertical air velocity $U_z$ to the Coriolis term, have been ignored.

### 3.2.4. Lower boundary conditions

Many atmospheric models used for wind calculations assume that the atmospheric pressure, density and temperature are fixed at some lower boundary, often taken as 120 km (cf. section 3.1.2). In this case the pressure-gradient force components $F_x$, $F_y$ vanish at this boundary, so it is generally assumed that $U_x = U_y = 0$ at this level. According to Lindzen (1967), the choice of any other lower boundary values of $U_x$ and $U_y$ would make little difference to the values computed at levels more than one or two scale heights above the boundary. But if, instead, the atmospheric parameters $p$, $\rho$, $T$ were assumed to vary at the lower boundary (as they probably do in reality), this would have some effect on the computed winds at greater heights. See Chandra and Stubbe (1970).

### 3.2.5. Upper boundary conditions

At great heights, the quantity $(\mu/\rho)$, known as the "kinematic viscosity", becomes very large. Consequently, in order that the viscosity terms in eq. 47 and 48, should not become overwhelmingly large, the derivatives $\partial^2 U_x/\partial h^2$, $\partial^2 U_y/\partial h^2$ must become small at great heights. This implies that the "wind shears" $\partial U_x/\partial h$, $\partial U_y/\partial h$ must become constant. However there does not exist at great heights any force capable of maintaining a vertical wind shear, so that the more stringent condition

$\partial U_x/\partial h$, $\partial U_y/\partial h \rightarrow 0$ must hold at great heights; i.e., $U_x$, $U_y$ become independent of height.

### 3.3. The wind systems (Geisler, 1966, and 1967; Kohl and King, 1967)

#### 3.3.1. Steady-state conditions

To understand the factors that control the wind speed and direction, it is helpful to simplify the equations 47 and 48 as follows: neglect viscosity; assume that the horizontal ion motions $V_x = V_y = 0$, as would be the case if the magnetic field were vertical and there were no horizontal electric fields; assume a steady state exists, with $d/dt = 0$. Then:

$$F_x - KNU_x + fU_y = 0 \tag{50}$$

$$F_y - KNU_y - fU_x = 0 \tag{51}$$

The two special cases of interest are:
Ion-drag dominant $(KN \gg |f|)$:   the wind is parallel to the driving force, and:

$$U_x = F_x/KN; \; U_y = F_y/KN \tag{52}$$

Coriolis force dominant $(KN \ll |f|)$;   the wind is perpendicular to the driving force, its direction being given by the Buys Ballot law of meteorology, such that:

$$U_x = F_y/f; \; U_y = -F_x/f \tag{53}$$

Usually, in the F2 layer the situation lies between these extremes. Let $F$ and $U$ be the magnitudes of the horizontal pressure-gradient force and wind velocity, respectively, so that:

$$F = (F_x^2 + F_y^2)^{1/2}; \quad U = (U_x^2 + U_y^2)^{1/2} \tag{54}$$

Then, from eq.50 and eq.51 it is found that:

$$U = F(K^2N^2 + f^2)^{-1/2} \tag{55}$$

If azimuth is measured clockwise from geographic north, then the azimuths of the pressure-gradient force and the wind are respectively arc tan $(F_x/F_y)$, arc tan $(U_x/U_y)$. In a steady state, the angle $A$ between them is given by:

$$\tan A = f/KN \tag{56}$$

which has the same sign as $f$ (positive in the northern, negative in the southern hemisphere).

### 3.3.2. Global wind patterns

The arrows in Fig. 15 show the directions of the winds produced by the pressure-gradient forces of Jacchia's model. It is assumed that steady-state conditions apply, and that ion-drag is dominant by day ($KN \gg |f|$) so that the wind blows across the isotherms (which, in Jacchia's model, are also iosbars). At night it is assumed that ion-drag and Coriolis force are equally important ($KN = |f|$), so that the winds are inclined at 45° to the isotherms, which at midlatitudes occurs for a plasma frequency of 3–4 MHz.

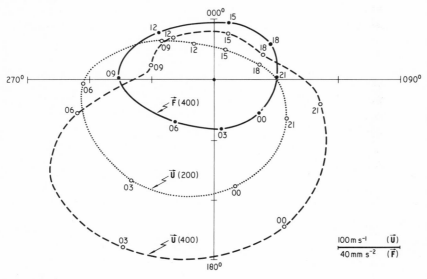

Fig. 16. Vector diagram showing how the horizontal wind velocity $\vec{U}$ and pressure-gradient force $\vec{F}$ vary with local time, according to calculations for latitude 51°N, equinox, solar activity 160 flux units (kindly supplied by Dr. R. Rüster). Open circles show U at 3-h intervals of local time at heights of 200 km (dotted line) and 400 km (dashed line). Filled circles and the curve (continuous line) show $\vec{F}$ at 400 km; the variation of $\vec{F}$ at 200 km is not shown, but at any local time $\vec{F}(200)$ is in the same direction as $\vec{F}(400)$ and its magnitude is about 23% of that of $\vec{F}(400)$. A scale giving the magnitudes of $\vec{U}$ and $\vec{F}$ is shown; azimuths are in degrees east of north. (Radio and Space Research Station, Slough. Crown Copyright.)

### 3.3.3. Daily variation of wind speed and direction

Figure 16 is a vector diagram showing how the pressure-gradient force $\vec{F}$ and wind velocity $\vec{U}$ vary with local time, at equinox in latitude 51°N for moderate solar activity, according to calculations by Rüster (1971) based on Jacchia's model. The force vector $\vec{F}$ rotates through 360° in 24 h (full curve); the wind at 400 km (broken curve) is in virtually the same direction as $\vec{F}$ during the day when ion-drag is dominant. At night the electron density decreases so that ion-drag becomes smaller; as a result, the wind speed increases, and moreover Coriolis force becomes fairly important so that the wind direction "leads" the direction of $\vec{F}$ by approximately

30°. The wind direction does not accurately conform to the steady-state formula (eq.56) because of the inertial terms in eq. 47–48, the effects of which may alter the wind direction by 15° or so.

### 3.3.4. The effect of viscosity on the height variation of wind velocity

A simplified analysis which neglects viscosity, based on eq. 50–51, can give a reasonably good picture of the winds in the neighbourhood of the F2 peak. At heights both above and below the F2 peak, ion-drag is weaker and Coriolis force would be expected to be important. It is found, however, that viscosity exerts a strong control, particularly at great heights (section 3.2.5), and as a result the wind speed and direction is practically independent of height above the F2 peak, according to calculation.

Below the peak there is an appreciable height variation, as may be seen by comparing the variations of $\vec{U}$ at 200 and 400 km in Fig. 16. Nevertheless, viscosity does have some effect as low as 200 km. At night the electron density at 200 km is so small that Coriolis force should be completely dominant, so that the wind direction should be about 90° from that of $\vec{F}$. In fact the difference is never more than about 50° (the direction of $\vec{F}$ being just the same at 200 km as at 400 km), and this is mainly because viscosity prevents the wind velocity at 200 km from being very different from the velocity at greater heights.

### 3.3.5. Proper account of ion motions

For accurate calculations the assumptions $V_x = V_y = 0$ (section 3.3.1) are inadequate, and all causes of ion motion (section 2.4.2) must be taken into account. When this is done, it is found that the electromagnetic drift (eq. 37) has some effect on the wind velocity, but the winds are however not *grossly* different from those calculated with the assumption $V_x = V_y = 0$.

## 3.4. Large-scale transport of air by winds

### 3.4.1. Calculated prevailing winds

Wind calculations based on the Jacchia or similar atmospheric models give a prevailing equatorward wind. This is because the reduced ion-drag at night leads to greater wind speeds at night (when the meridional wind is equatorward) than by day (when the meridional wind is poleward); (see Fig. 16). According to Dickinson and Geisler (1968), this equatorward motion, together with the upflow of air in high latitudes and downflow in low latitudes that would accompany it, tends to transport heat from the poles towards the equator. One would expect the heat flow to go the other way. This equatorward transport of heat and air can be suppressed by modifying the atmospheric model so that the mean temperature at high latitudes is reduced.

In contrast, the calculations give very little prevailing zonal wind. We must bear in mind that, because of the upward decrease of density, the winds in the lower thermosphere are more important than those at greater heights, as regards the transport of air mass. Unfortunately, the existing calculations are unreliable at lower heights, largely because of the problem of the lower boundary conditions (section 3.1.3).

### 3.4.2. Atmospheric superrotation (King-Hele, 1970)

Notwithstanding the small zonal prevailing winds obtained from calculations, there is strong observational evidence that the thermosphere rotates 25% faster than the earth; i.e., there is a prevailing eastward (west to east) wind of around 100 m sec$^{-1}$. Several theories of this superrotation have been advanced: (a) a periodic thermal forcing mechanism, which may be responsible for the general atmospheric circulation on Venus (Schubert and Young, 1970), might also operate in the earth's atmosphere; (b) a nonlinear instability of convection motions due to solar heating has been proposed as a cause of the Venus circulation (Thompson, 1970); (c) Cole (1971) suggests that electric fields associated with auroral substorms might impart a mean eastward motion to the neutral air; (d) Challinor (1969) shows that a large day-to-night variation of ion-drag—due to the diurnal variation of electron density in the F region—could cause a mean eastward transport of air momentum, the night-time winds (which are eastward) being faster than the daytime winds (which are westward). His results are not necessarily compatible with those of the detailed wind calculations mentioned in section 3.4.1; (e) Rishbeth (1971) proposes that polarization fields, that are produced by zonal winds, reduce the ion-drag at night and thereby enhance the night-time eastward wind; and (f) Matuura (1968) attributes superrotation to motions produced by plasma·diffusion in the F2 layer, combined with Coriolis force. Several of the above mechanisms may contribute to the superrotation phenomenon, though (f) does not seem very probable.

### 3.4.3. Continuity of air motion (Dickinson and Geisler, 1968; Rishbeth et al., 1969)

Quite apart from the prevailing motions (section 3.4.1), there are substantial horizontal air motions, due to thermospheric winds, on time-scales of several hours, because the winds remove air from the day side of the earth and transport it to the night side. This transport must be accompanied by vertical air motion, two components of which can be identified: (a) *the barometric velocity* $W_B$, which is associated with thermal expansion and contraction of the atmosphere. It is the vertical velocity of the levels of constant pressure; (b) *the divergence velocity* $W_D$, which is needed to balance the transport of air by the horizontal winds, i.e., the "divergence" by day and "convergence" by night that are implied by the arrows in Fig. 15.

These velocities appear in the continuity equation for the air:

$$-\frac{\partial n}{\partial t} = \text{div}(n\vec{U}) = \frac{\partial}{\partial x}(nU_x) + \frac{\partial}{\partial y}(nU_y) + \frac{\partial}{\partial h}(nW_B + nW_D) \tag{57}$$

In an atmospheric model, such as Jacchia's, that neglects horizontal transport of air but satisfies the barometric equation, it can be shown that:

$$-\partial n/\partial t = \frac{\partial}{\partial h}(nW_B) \tag{58}$$

so that the barometric velocity can be eliminated from eq. 57 and the equation then used to compute $W_D$. At midlatitudes, $W_D$ is found to be upwards by day and downwards by night, with velocities of order 2 m sec$^{-1}$ (comparable to $W_B$). Though small in comparison to the horizontal wind velocity (50–100 m sec$^{-1}$), this vertical motion may contribute significantly to the energy balance of the thermosphere, owing to the adiabatic heating or cooling associated with vertical air motion (Johnson and Gottlieb, 1969).

### 3.5. Effects of winds on the F2 layer

#### 3.5.1. Vertical drift produced by a horizontal wind

A horizontal wind $U$, blowing in a direction $\theta$ (reckoned east from geographic north), has a component in a magnetic meridian given by $U \cos(\theta - D)$. The vertical ion drift produced by this wind is, by simple geometrical considerations (section 2.4.7)

$$W = -U \cos(\theta - D)\cos I \sin I \tag{59}$$

where $D$ = magnetic declination, $I$ = dip. The effect of this drift on the F2 layer can be estimated by using the considerations of sections 2.5.2–2.5.3.

A poleward wind ($\theta \sim 0°$ in the northern hemisphere, $\theta \sim 180°$ in the southern), such as occurs by day (Fig. 15), produces downward drift. An equatorward wind ($\theta \sim 180°$ in the northern hemisphere, $\theta \sim 0°$ in the southern), such as occurs at night, produces upward drift.

#### 3.5.2. The diurnal variation of $N_mF2$ and $h_m F2$ (Kohl et al., 1968)

Figure 17 shows how winds affect the diurnal variations of the F2 layer. Without winds (broken curves), the calculations give a simple daily variation; at night the layer decays in the way described in section 2.5.3. Winds tend to depress $h_mF2$ by day (full curves), thereby reducing $N_mF2$ (i.e., $f_oF2$); the summer (December) curve shows an extreme case in which $f_oF2$ is smaller at noon than at midnight; this is because at the latitude of Port Lockroy production of ionization occurs throughout the night at midsummer.

The extent to which the layer is preserved at night depends on the phase of the

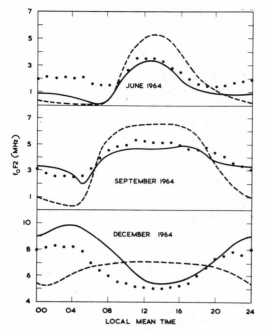

Fig. 17. Diurnal variations of $f_oF2$ at Port Lockroy (65°S). Dots are mean observed values; the dashed curves are obtained from calculations including production, loss and diffusion, but excluding winds; the full curves are obtained from calculations that include winds also. (After Kohl et al., 1968.)

local-time variation of drift, i.e., of $(\theta - D)$ in eq. 59. If upward drift starts before sunset, the layer is preserved, and there may even be a slight increase of electron density before sunset (September curve, Fig. 17). But if the drift remains downward till after sunset (June curve, i.e., winter), the ionization decays rapidly when production ceases and not much remains during night. Other mechanisms, such as influx of ionization from the protonosphere, are apparently needed to maintain the observed nighttime values of $f_oF2$ (dots in Fig. 17).

### 3.5.3. Longitude variations (Challinor and Eccles, 1971; Eccles et al., 1971)

At any given latitude and local time, the vertical drift $W$ varies with longitude, mainly because in eq. 59 the magnetic parameters $D$ and $I$ are functions of longitude. A contributory, though minor, cause is that the wind direction $\theta$ varies slightly with longitude at fixed local time, owing to variations of ion-drag. Fig. 18 shows maps of vertical drift derived from simplified wind calculations. One might expect the greatest electron densities to occur where the drift $W$ is most positive (upward), and the smallest electron densities where $W$ is most negative; this is largely borne out by the satellite data shown. These data were obtained at heights well above the F2 peak, and the variations of $N$ with longitude are sometimes 3 : 1

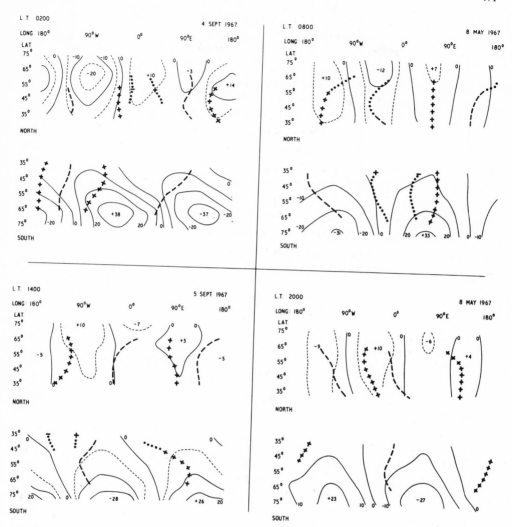

Fig. 18. Maps showing the distribution in geographic coordinates of vertical drift due to winds, computed from equations 50, 51, 59, using as a basis Jacchia's model for equinox, medium solar activity, and observational values of $N_mF2$ for computing ion-drag. Contours are plotted at intervals of 10 m sec$^{-1}$ (full curves), some intervening contours (intervals of 5 m sec$^{-1}$) being shown dashed. Maxima and minima are shown as spot values. The heavy curves are lines of maximum (+ + +) and minimum (− − −) electron density observed at 500–600 km altitude by the University of Birmingham experiment aboard the Ariel III satellite on the dates shown; some indistinct extrema are shown dotted. The maps are drawn for four different local times: 0200, 0800, 1400 and 2000. (From Rishbeth and Kelley, 1971.)

(at the F2 peak the variations might be 1.5 : 1). The extrema shift progressively in longitude as local time advances.

### 3.5.4. Declination effects

Another consequence of the dependence of $W$ on $D$ (as in eq. 59), is that the phase of the local-time variation of $W$ is different at places with a different declination. This leads to systematic variations of $N_mF2$ (Kohl et al., 1969), which accounts for the "declination effect" noticed by Eyfrig (1963).

The way in which the drift varies with declination, and hence with longitude, can be studied with the aid of Fig. 19, which shows for latitudes 45°N and 45°S : (a) the approximate variation of wind direction $\theta$ with local time; (b) the declination; and (c) the magnitude of (cos $I$ sin $I$) as functions of longitude. It may be shown that places where the declination has an extreme value (two in the southern hemisphere, $P$ and $Q$; and four in the northern hemisphere, $A$, $B$, $A'$, $B'$) are often the places where the drift, and hence the electron density, has a maximum or minimum with respect to longitude.

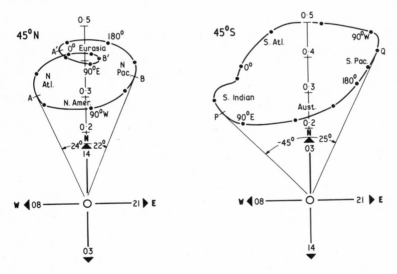

Fig. 19. To each place at latitude 45°N (left) or 45°S (right), there corresponds a point on the closed loops, longitudes being shown by solid circles at intervals of 30°, labelled at intervals of 90°. Oceanic and continental regions are indicated. The vector from origin O to any point on a closed loop represents, by its orientation, the declination at the corresponding place and, by its length, the values of (cos $I$ sin $I$), according to the scales marked along the "northward" axes. For example at 45°S, cos $I$ sin $I$ is about 0.22–0.25 in the Australian sector and 0.45–0.5 in the South Atlantic. Extreme values of declination at 45°S occur at $P$, $Q$ (longitudes 75°E, 120°W respectively) and at 45°N at $A$, $B$, $A'$, $B'$ (longitudes 50°W, 135°W, 135°E, 60°E respectively). The wind blows in the cardinal directions at approximately 03, 08, 14, 21 h local time, as shown; the vertical drift it produces at any place is proportional to the projection of the wind vector on the vector representing cos $I$ sin $I$ and $D$ at that place. (Radio and Space Research Station, Slough. Crown Copyright.)

### 3.5.5. Seasonal variations

To some extent, the seasonal changes of *shape* of the diurnal variations of $N_mF2$ can be accounted for by winds (Fig. 17), but the seasonal anomaly in the values of noon $N_mF2$ (section 1.3.2) cannot be. Other causes, such as seasonal variations of atmospheric composition affecting the production and loss rates, must be sought (Rishbeth, 1968a). The semiannual variations (Becker, 1967), in phase with the semiannual temperature variation of the neutral atmosphere, may partly be due to winds.

### 3.5.6. Wind effects at high latitudes (King et al., 1968)

The winds blowing across polar regions cause diurnal variations of $h_mF2$ and $N_mF2$, much as they do at midlatitudes. Near the dip pole, wind effects become small because cos $I$ is small in eq. 59. In the Antarctic there exists a Universal Time effect, because at around 0600 UT the wind direction is such as to produce upward drift all over Antarctica, giving rise to maxima of $h_mF2$ and $N_mF2$ (Fig. 20). In the Arctic there is no real Universal Time effect due to winds (Challinor, 1970).

Since winds do not transport ionization across field lines, they cannot bring ionization to high latitudes from middle latitudes, and so are not the mechanism that maintains the F2 layer during the polar winter. Other processes must be invoked to account for the geographical distributions of ionization (e.g., Duncan, 1969).

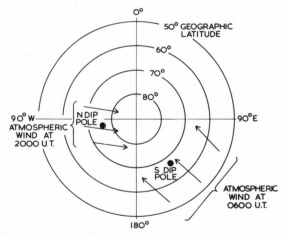

Fig. 20. Location of the Arctic and Antarctic dip poles with respect to the geographic poles. The south dip pole is on the edge of the Antarctic continent, and all the ionospheric observatories are so placed that at 0600 Universal Time, when the wind blows from the dip pole towards the geographic pole, upward drifts occur in the F2 layer and maxima of $h_mF2$ and $N_mF2$ are observed. But in the Arctic the dip pole is closer to the geographic pole, and the distribution of ionospheric observatories is such that upward drifts do not occur at all stations at 2000 Universal Time, even though the wind at that time does produce upward drift at the geographic pole. (J. W. King, Radio and Space Research Station, Slough. Crown Copyright.)

### 3.5.7. Wind effects at low latitudes

Although electric fields are thought to be the basic cause of the equatorial anomaly (section 2.5.5), winds may influence its shape. A meridional wind blowing across the equator will transport ionization from one hemisphere to the other, thereby making the distribution asymmetrical about the equator (Bramley and Young, 1968). Actually the wind causes a reduction of $N_mF2$ on *both* sides of the equator, the reduction being greatest in the "upwind" hemisphere by day; at night the effect is reversed, and $N_mF2$ decays more rapidly in the "downwind" hemisphere (Abur-Robb and Windle, 1969). The wind pattern shown in Fig. 15, however, is symmetrical about the equator, and there are no meridional winds blowing across the equator. There are winds blowing *away* from the equator in both hemispheres by day, which will remove ionization from the equator. Conversely, the winds blowing towards the equator at night will tend to maintain the equatorial F2 layer. See Abur-Robb (1969) and Sterling et al. (1969).

### 3.5.8. Storm effects

The heating of the thermosphere during storms (e.g., Jacchia et al., 1967) will affect the wind pattern (Kohl and King, 1967). The extent to which this may cause F2 layer storm phenomena has not been fully worked out. It does seem that equatorward winds, caused by strong heating in the auroral zone, could produce some "positive storm effects" that are observed at midlatitudes (section 1.4.2); see Jones and Rishbeth (1971).

## 3.6. Some outstanding problems

Various matters needing further investigation are mentioned below; some (but not all) have been referred to already in these notes, and text references are given where appropriate.

### 3.6.1. Matters relating to the neutral thermosphere

(a) More needs to be known about atmospheric composition and its variations (section 2.1.2); could variations be produced by winds?

(b) The temperature distribution (3.1.2) needs confirmation, in view of some conflicting experimental data (3.1.3), and energy- balance problems associated with the calculated prevailing air motions (3.4.1).

(c) Localized variations of atmospheric parameters (e.g., in the auroral zone, but probably elsewhere too) might affect the wind patterns, but knowledge of them is very scanty.

(d) The lower boundary conditions may vary diurnally, seasonally, and geographically; this would affect density, temperature, etc. and the wind patterns at greater heights (3.1.3), and might be relevant to the problem of the "phase of the diurnal bulge".

(e) The thermosphere is known to be acted upon by tidal forces transmitted from lower levels (e.g., Volland, 1969). These may well affect the lower boundary conditions, (d). What is the relative importance of tidal energy and solar XUV radiation as regards the thermospheric energy balance?

(f) In relation to (e), there is need for a full three-dimensional theoretical computation of thermospheric motions, using the continuity, momentum and energy equations.

(g) The non-linear acceleration term (3.2.2) is often neglected; does this matter?

(h) What is the most important cause of superrotation (3.4.2)?

### 3.6.2. Matters relating to the F layer

(a) Storms present the most outstanding problem. What parts are played by winds, composition changes and electric fields (1.4.2, 3.5.8)?

(b) More needs to be known about the high-latitude F layer and its structure (1.3.5), and its relation to auroral phenomena, particle precipitation, airglow emission. How is the polar winter F2 layer maintained?

(c) There is an interesting "feedback" effect that might merit further study: the wind velocity is largely controlled by ion-drag, and hence depends on the electron concentration (3.3.1), but the latter is strongly influenced by the drift due to the winds (3.5.1).

(d) The diurnal variations of $N_mF2$ are largely controlled by winds (3.5.2), but the sunset increases could also be due to thermal contraction. The relative importance of different mechanisms requires clarification.

(e) Winds help to maintain the F2 layer at night (3.5.2), but in addition some input of ionization seems to be required. Does it come from the protonosphere or from energetic particles?

(f) The seasonal and semiannual variations (3.5.5) appear to be due to variations of atmospheric temperature and composition, but are nevertheless not yet fully explained.

(g) Besides the longitude variations (3.5.3), there appear to be some fairly localized variations of F2 layer parameters. Are they connected with variations of the neutral atmospheric structure?

(h) Although the basic form of the equatorial ionosphere seems to be explicable in terms of electric fields, modifications due to winds probably exist, and require further investigation (3.5.7). There is scope for detailed studies of storm effects at low latitudes.

## ACKNOWLEDGEMENTS

Acknowledgment is made to the Academic Press for Fig. 1 and 14, the American Geophysical Union for Fig. 3 and 7, Macmillan Journals Ltd. for Fig. 8, and Pergamon Press Ltd. for Fig. 6, 9, 13, 17 and 18.

This paper is published by permission of the Director of the Radio and Space Research Station in the U.K. Science Research Council.

## REFERENCES

Abur-Robb, M. F. K., Combined world-wide neutral air wind and electrodynamic drift effects on the F2-layer. *Planet. Space Sci.*, **17**:1269–1279, 1969.

Abur-Robb, M. F. K. and Windle, D. W., On the day and night reversal in NmF2 north-south asymmetry. *Planet. Space Sci.*, **17**:97–106, 1969.

Allen, C. W., The interpretation of the XUV solar spectrum. *Space Sci. Rev.*, **4**:91–122, 1965.

Andrews, M. K. and Thomas, J. O., Electron density distribution above the winter pole. *Nature*, **221**:223–227, 1969.

Appleton, E. V., The existence of more than one ionized layer in the upper atmosphere. *Nature*, **120**:330, 1927.

Appleton, E. V. and Barnett, M. A. F., Local reflection of wireless waves from the upper atmosphere. *Nature*, **115**:333–334, 1925.

Appleton, E. V. and Piggott, W. R., The morphology of storms in the F2-layer of the ionosphere, 1: Some statistical relationships. *J. Atmos. Terr. Phys.*, **2**:236–252, 1952.

Armstrong, E. B., Doppler shifts in the wavelength of the OI λ6300 line in the night airglow. *Planet. Space Sci.*, **17**:957–974, 1969.

Banks, P. M. and Holzer, T. E., Features of plasma transport in the upper atmosphere. *J. Geophys. Res.*, **74**:6304–6316, 1969.

Bauer, S. J., Blumle, L. J., Donley, J. L., Fitzenreiter, R. J. and Jackson, J. E., Simultaneous rocket and satellite measurements of the topside ionosphere. *J. Geophys. Res.*, **69**:186–189, 1964.

Becker, W., The temperature of the F region deduced from electron number-density profiles. *J. Geophys. Res.*, **72**:2001–2006, 1967.

Bossolasco, M. and Elena, A., On the lunar semidiurnal variations of the D and F2 layers. *Geofis. Pura e Appl.*, **46**:167–172, 1960.

Bowman, G. G., Ionization troughs below the F2-layer maximum. *Planet. Space Sci.*, **17**:777–796, 1969.

Bramley, E. N. and Young, M., Winds and electromagnetic drifts in the equatorial F2-region. *J. Atmos. Terr. Phys.*, **30**:99–111, 1968.

Breit, G. and Tuve, M. A., A radio method of estimating the height of the conducting layer. *Nature*, **116**:357, 1925.

Budden, K. G., *Radio Waves in the Ionosphere*. Cambridge University Press, 542 pp., 1961.

Bullen, J. M., Ionospheric recombination and the polar stratospheric warming. *J. Atmos. Terr. Phys.*, **26**:559–568, 1964.

Carlson, H. C., Ionospheric heating by magnetic conjugate-point photoelectrons. *J. Geophys. Res.*, **71**:195–199, 1966.

Carpenter, D. L., Whistler studies of the plasmapause in the magnetosphere, 1: Temporal variations in the position of the knee and some evidence on plasma motions near the knee. *J. Geophys. Res.*, **71**:693–709, 1966.

Challinor, R. A., Neutral-air winds in the ionospheric F-region for an asymmetric global pressure system. *Planet. Space Sci.*, **17**:1097–1106, 1969.

Challinor, R. A., The behaviour of the Arctic F-region in winter. *J. Atmos. Terr. Phys.*, **32**:1959–1965, 1970.

Challinor, R. A. and Eccles, D., Longitudinal variations of the mid-latitude ionosphere produced by neutral-air winds, 1: Neutral-air winds and ionospheric drifts in the northern and southern hemispheres. *J. Atmos. Terr. Phys.*, **33**:363–369, 1971.

Chamberlain, J. W., *Physics of Aurora and Airglow*. Academic Press, New York, N. Y., 704 pp., 1961.

Chandra, S., and Stubbe, P., The diurnal phase anomaly in the upper atmospheric density and temperature. *Planet. Space Sci.*, **18**:1021–1033, 1970.

Chapman, S., The absorption and dissociative or ionizing effect of monochromatic radiation in an atmosphere on a rotating earth. *Proc. Phys. Soc. Lond.*, **43**:26–45, 1931a.

Chapman, S., The absorption and dissociative or ionizing effect of monochromatic radiation in an atmosphere on a rotating earth, II: Grazing incidence. *Proc. Phys. Soc. Lond.*, **43**:483–501, 1931b.

Cole, K. D., Electrodynamic heating and movement of the thermosphere. *Planet. Space Sci.*, **19**:59–75, 1971.

Davies, K., *Ionospheric Radio Waves*. Blaisdell, Waltham, Mass., 460 pp., 1969.

Davies, K. and Baker, D. M., On frequency variations of ionospherically propagated HF radio signals. *Radio Sci.*, **1**:545–556, 1966.

Dickinson, R. E. and Geisler, J. E., Vertical motion field in the middle thermosphere from satellite drag densities. *Monthl. Weather Rev.*, **96**:606–616, 1968.

Dougherty, J. P., On the influences of horizontal motion of the neutral air on the diffusion equation of the F region. *J. Atmos. Terr. Phys.*, **20**:167–176, 1961.

Duncan, R. A., The behaviour of a Chapman layer in the night F2 region of the ionosphere, under the influence of gravity, diffusion and attachment. *Aust. J. Phys.*, **9**:436–439, 1956.

Duncan, R. A., The equatorial F-region of the ionosphere. *J. Atmos. Terr. Phys.*, **18**:89–100, 1960.

Duncan, R. A., Universal time control of the Arctic and Antarctic F region. *J. Geophys. Res.*, **67**:1823–1830, 1962.

Duncan, R. A., F-region seasonal and magnetic storm behaviour. *J. Atmos. Terr. Phys.*, **31**:59–70, 1969.

Dungey, J. W., The effect of ambipolar diffusion in the night-time F layer. *J. Atmos. Terr. Phys.*, **9**:90–102, 1956.

Eccles, D., King, J. W. and Rothwell, P., Longitudinal variation of the mid-latitude ionosphere produced by neutral-air winds, II: Comparisons of the calculated variations of electron concentration with data obtained from the Ariel I and Ariel III satellites. *J. Atmos. Terr. Phys.*, **33**:371–377, 1971.

Evans, J. V., Theory and practice of ionosphere study by Thomson scatter radar. *Proc. I.E.E.E.*, **57**:496–530, 1969.

Evans, J. V., Midlatitude ionospheric temperatures during three magnetic storms in 1965. *J. Geophys. Res.*, **75**:4803–4813, 1970a.

Evans, J. V., F-region heating observed during the main phase of magnetic storms. *J. Geophys. Res.*, **75**:4815–4823, 1970b.

Eyfrig, R. W., The effect of the magnetic declination on the F2 layer. *Ann. Géophys.*, **19**:102–117, 1963.

Feldstein, Y. I. and Starkov, G. V., Dynamics of auroral belt and polar geomagnetic disturbances. *Planet. Space Sci.*, **15**:209–229, 1967.

Ferguson, E. E., Laboratory measurements of F-region reaction rates. *Ann. Géophys.*, **25**:819–823, 1969.

Gardiner, G. W., Origin of the term 'ionosphere'. *Nature*, **224**:1096, 1969.

Garriott, O. K., da Rosa, A. V., Davies, M. J., Wagner, L. S. and Thome, G. D., Enhancement of ionizing radiation during a solar flare. *Solar Phys.*, **8**:226–239, 1969.

Geisler, J. E., Atmospheric winds in the middle latitude F-region. *J. Atmos. Terr. Phys.*, **28**:703–720, 1966.

Geisler, J. E., A numerical study of the wind system in the middle thermosphere. *J. Atmos. Terr. Phys.*, **29**:1469–1482, 1967.

Green, A. L., Early history of the ionosphere. *Amalgamated Wireless of Australasia Tech. Rev.*, **7**:177–228, 1946.

Haerendel, G., Lüst, R. and Rieger, E., Motion of artificial ion clouds in the upper atmosphere. *Planet. Space Sci.*, **15**:1–18, 1967.

Harris, I. and Priester, W., Time-dependent structure of the upper atmosphere. *J. Atmos. Sci.*, **18**:286–301, 1962.

Jacchia, L. G., Two atmospheric effects in the orbital acceleration of artificial satellites. *Nature*, **183**:526–527, 1959.

Jacchia, L. G., Static diffusion models of the upper atmosphere with empirical temperature profiles. *Smithsonian Contrib. Astrophys.*, **8**:215–257, 1965.

Jacchia, L. G., Slowey, J. and Verniani, F., Geomagnetic perturbations and upper-atmosphere heating. *J. Geophys. Res.*, **72**:1423–1434, 1967.

Johnson, C. Y., Ion and neutral composition of the ionosphere. *Ann. Int. Quiet Sun Years*, **5**:197–213, 1969.

Johnson, F. S. and Gottlieb, B., Composition changes in the lower thermosphere. In: *Space Research IX* (K. S. W. Champion, P. A. Smith, R. L. Smith-Rose, Editors). North-Holland, Amsterdam, 442–446, 1969.

Jones, K. L., Storm time variation of F2-layer electron concentration. *J. Atmos. Terr. Phys.*, **33**:379–389, 1971.

Jones, K. L. and Rishbeth, H., The origin of storm increases of mid-latitude F-layer electron concentration. *J. Atmos. Terr. Phys.*, **33**:391–401, 1971.

Kendall, P. C., Geomagnetic control of diffusion in the F2-region of the ionosphere, I. The form of the diffusion operator. *J. Atmos. Terr. Phys.*, **24**:805–811, 1962.

Kent, G. S., Measurement of ionospheric movements. *Rev. Geophys.*, **8**:229–288, 1970.

Kent, G. S. and Wright, R. W. H., Movements of ionospheric irregularities and atmospheric winds. *J. Atmos. Terr. Phys.*, **30**:657–691, 1968.

King, G. A. M., Analysis of the F1-F2 transition region. *J. Geophys. Res.*, **66**:2757–2762, 1961.

King, G. A. M., The ionospheric F region during a storm. *Planet. Space Sci.*, **9**:95–100, 1962.

King, J. W., A review of the large-scale structure of the ionospheric F layer. *Ann. Int. Quiet Sun Years*, **5**:131–165, 1969.

King, J. W., Reed, K. C., Olatunji, E. O. and Legg, A. J., The behaviour of the topside ionosphere during storm conditions. *J. Atmos. Terr. Phys.*, **29**:1355–1363, 1967.

King, J. W., Kohl, H., Preece, D. M. and Seabrook, C., An explanation of phenomena occurring in the high-latitude ionosphere at certain Universal Times. *J. Atmos. Terr. Phys.*, **30**:11–23, 1968.

King-Hele, D. G., "Super-rotation" of the upper atmosphere at heights of 150-170 km. *Nature*, **226**:439–440, 1970.

Kohl, H. and King, J. W., Atmospheric winds between 100 and 700 km and their effects on the ionosphere. *J. Atmos. Terr. Phys.*, **29**:1045–1062, 1967.

Kohl, H., King, J. W. and Eccles, D., Some effects of neutral air winds on the ionospheric F-layer. *J. Atmos. Terr. Phys.*, **30**:1733–1744, 1968.

Kohl, H., King, J. W., and Eccles, D., An explanation of the magnetic declination effect in the ionospheric F2-layer. *J. Atmos. Terr. Phys.*, **31**:1011–1016, 1969.

Lindzen, R. S., Reconsideration of diurnal velocity oscillation in the thermosphere. *J. Geophys. Res.*, **72**:1591–1598, 1967.

Martyn, D. F., Electric currents in the ionosphere III. Ionization drift due to winds and electric fields. *Philos. Trans. R. Soc. Lond.*, Ser. A, **246**:306–320, 1953a.

Martyn, D. F., The morphology of the ionospheric variations associated with magnetic disturbance, 1: Variations at moderately low latitudes. *Proc. R. Soc. Lond.*, Ser. A, **218**:1–18, 1953b.

Matuura, N., Effect of the ionosphere on the upper atmosphere rotation. *J. Atmos. Terr. Phys.*, **30**:763–778, 1968.

McNish, A. G. and Gautier, T. N., Theory of lunar effects and midday decrease in F2 ion-density at Huancayo, Peru. *J. Geophys. Res.*, **54**:181–185, 1949.

Muldrew, D. B., F-layer ionization troughs deduced from Alouette data. *J. Geophys. Res.*, **70**:2635–2650, 1965.

Norton, R. B., Van Zandt, T. E. and Denison, J. S., A model of the atmosphere and ionosphere in the E and F1 regions. In: *Proc. Int. Conf. Ionosphere*. Institute of Physics and Physical Society, London, pp. 26–34, 1963.

Piggott, W. R. and Rawer, K., *URSI Handbook of Ionogram Interpretation and Reduction*. Elsevier, Amsterdam, 192 pp., 1961.

Piggott, W. R. and Shapley, A. H., The ionosphere over Antarctica. In: *Antarctic Research, Geophys. Monogr. 7*. Am. Geophys. Union, Washington, pp. 111–126, 1962.

Rastogi, R. G., The morphology of lunar semi-diurnal variation in $f_oF2$ near solar noon. *J. Atmos. Terr. Phys.*, **22**:290–297, 1961.

Ratcliffe, J. A., The formation of the ionospheric layers F-1 and F-2. *J. Atmos. Terr. Phys.*, **8**:260–269, 1956a.

Ratcliffe, J. A., A survey of solar eclipses and the ionosphere. In: *Solar Eclipses and the Ionosphere* (W. J. G. Beynon and G. M. Brown, Editors). Pergamon Press, London, pp. 1–13; 306–307, 1956b.

Ratcliffe, J. A. and Weekes, K., The ionosphere. In: *Physics of the Upper Atmosphere* (J. A. Ratcliffe, Editor). Academic Press, New York, pp. 377–470, 1960.

Rishbeth, H., On explaining the behavior of the ionospheric F region. *Rev. Geophys.*, 6:33–71, 1968a.

Rishbeth, H., Solar eclipses and ionospheric theory. *Space Sci. Rev.*, 8:543–554, 1968b.

Rishbeth, H., Polarization fields produced by winds in the equatorial F-region. *Planet. Space Sci.*, 19:357–369, 1971.

Rishbeth, H. and Barron, D. W, Equilibrium electron distributions in the ionospheric F2 layer. *J. Atmos. Terr. Phys.*, 18:234–252, 1960.

Rishbeth, H. and Garriott, O. K., *Introduction to Ionospheric Physics*. Academic Press, New York, 331 pp., 1969.

Rishbeth, H. and Kelley, D. M., Maps of the vertical F-layer drifts caused by horizontal winds at midlatitudes. *J. Atmos. Terr. Phys.*, 33:539–545, 1971.

Rishbeth, H., Moffett, R. J. and Bailey, G. J., Continuity of air motion in the mid-latitude thermosphere. *J. Atmos. Terr. Phys.*, 31:1035–1047, 1969.

Rüster, R., Solution of the coupled ionospheric continuity equations and the equations of motion for the ions, electrons and neutral particles. *J. Atmos. Terr. Phys.*, 33:137–147, 1971.

Sato, T. and Colin, L., Morphology of electron concentration enhancement at a height of 1000 kilometers at polar latitudes. *J. Geophys. Res.*, 74:2193–2207, 1969.

Sato, T. and Rourke, G. F., F-region enhancements in the Antarctic. *J. Geophys. Res.*, 69:4591–4607, 1964.

Schubert, G. and Young, R. E., The 4-day Venus circulation driven by periodic thermal forcing. *J. Atmos. Sci.*, 27:523–528, 1970.

Sharp, G. W., Midlatitude trough in the night ionosphere. *J. Geophys. Res.*, 71:1345–1356, 1966.

Sterling, D. L., Hanson, W. B. , Moffett, R. J. and Baxter, R. G. Influence of electromagnetic drifts and neutral air winds on some features of the F2 region. *Radio Sci.*, 4:1005–1023, 1969.

Stubbe, P., Simultaneous solution of the time dependent coupled continuity equations, heat conduction equations, and equations of motion for a system consisting of neutral gas, an electron gas and a four component ion gas. *J. Atmos. Terr. Phys.*, 32:865–903, 1970.

Thomas, J. O., The ionized layers in the E- and F-regions of the upper atmosphere. In: *Meteorological and Astronomical Influences on Radio Wave Propagation* (B. Landmark, Editor). Pergamon, London, pp. 43–66, 1963.

Thomas, J. O. and Andrews, M. K., The trans-polar exospheric plasma, 3: A unified picture. *Planet. Space Sci.*, 17:433–446, 1969.

Thomas, L., Ionospheric implications of aurora and airglow studies. In: *Aurora and Airglow* (B. M. McCormac, Editor). Reinhold, New York, N. Y., pp. 93–106, 1967.

Thomas, L., World-wide disturbances in the F-region accompanying the onset of the main phase of severe magnetic storms. *J. Atmos. Terr. Phys.*, 30:1623–1630, 1968.

Thomas, L., F2-region disturbances associated with major magnetic storms. *Planet. Space Sci.*, 18:917–928, 1970.

Thompson, R., Venus's general circulation is a merry-go-round. *J. Atmos. Sci.*, 27:1107–1116, 1970.

Titheridge, J. E., The use of the extraordinary ray in the analysis of ionospheric records. *J. Atmos. Terr. Phys.*, 17:110–125, 1959.

Torr, D. G. and Torr, M. R., A theoretical investigation of corpuscular radiation effects on the F-region of the ionosphere. *J. Atmos. Terr. Phys.*, 32:15–34, 1970.

Volland, H., A theory of thermospheric dynamics, I. Diurnal and solar-cycle variations. *Planet. Space Sci.*, 17:1581–1597, 1969.

Waldteufel, P. and McClure, J. P., Preliminary comparisons of middle and low latitude Thomson scatter data. *Ann. Géophys.*, 25:785–793, 1969.

Wright, J. W., Diurnal and seasonal changes in structure of the mid-latitude quiet ionosphere. *J. Res. Nat. Bureau Standards*, 66D:297–312, 1962.

Yonezawa, T., The vertical distribution of ionization at the F2 peak related to ionic production and transport processes. *Ann. Géophys.*, 26:581–588, 1970.

# MID-LATITUDE IONOSPHERIC IRREGULARITIES

MARIO BOSSOLASCO AND ANTONIO ELENA

*Istituto Geofisico e Geodetico, Università di Genova, Genoa (Italy)*

## SUMMARY

Irregular fluctuations of the F2-region plasma frequency at a middle latitude (Genoa–Mt. Capellino) are investigated, in relation to neutral winds, drifts and gravity waves. Thundery activity appears to contribute to the energy of gravity waves propagating upwards in the ionosphere. Some coupling between ionospheric and tropospheric irregularities is pointed out.

## 1. INTRODUCTION

At every level the ionosphere exhibits ionization irregularities variable in time and space. Turbulence and instability are also related to the motion of neutral air and play a role in producing small-scale irregularities in the lower ionosphere; conversely, atmospheric waves propagating to ionospheric heights from the stratosphere produce large-scale travelling ionospheric disturbances. Large-scale irregularities, exceeding about 50 km in horizontal extent, generally form a sequence, suggesting that such disturbances are wavelike.

Some irregularities of the F2-region plasma frequency over Genoa–Mt. Capellino will be investigated in this paper, with the aim of contributing to the explanation of their generative processes.

## 2. THE PRE-SUNRISE MAXIMUM IN THE F2 REGION

It is well known that the maximum plasma frequency of the F2 region shows a local time variation with a maximum near sunset (evening concentration) in summer and a secondary shallow maximum before sunrise in winter; at equinoxes the daily variation is of a somewhat mixed type (Bossolasco et al., 1965; Evans, 1970).

We shall investigate the behaviour of the pre-sunrise maximum (P.S.M.) of the F2-region plasma frequency at a middle latitude, with reference to the continuous recording of MUF(3000)F2 taken at Genoa–Mt. Capellino.

As it is known, MUF(3000)F2 is the highest frequency which can be propagated by rays refracted in the ionosphere to a distance, on the earth's surface, of 3000 km. For a parabolic layer, MUF(3000)F2 is related to the height $h_M$ (in km) of the maximum electron density by the following empirical relationship:

$$h_M = -176 + 1490\frac{f_o F2}{\mathrm{MUF}(3000)F2} \tag{1}$$

from which we obtain:

$$\frac{\delta \mathrm{MUF}}{\mathrm{MUF}} = \frac{\delta f_o F2}{f_o F2} - \frac{\delta h_M}{h_M + 176} \tag{2}$$

An increase of MUF(3000)F2 is generally due to a simultaneous increase of $f_o F2$ and decrease of $h_M$. For instance, a relative variation of +40% in $f_o F2$ and of −10% in $h_M$ leads to a relative variation of +50% in MUF(3000)F2. The variation of the plasma frequency plays the main role in the changes of MUF; the latter, however, are larger than the corresponding changes of $f_o F2$.

The mean daily variations of MUF(3000)F2 and $f_o F2$ at Genoa-Mt.Capellino from January to December 1964, i.e., in a period of low solar activity, are shown in Fig. 1. Here MUF(3000)F2, rather than $f_o F2$, shows the occurrence of P.S.M. in winter months.

The occurrence of P.S.M. at Genoa–Mt. Capellino from February 1964 to January 1965, is given in Table I where the number of cases in which P.S.M. was surely present or absent is indicated. P.S.M. is frequent in winter and very rare in summer.

Table II shows the monthly mean time of the maximum of MUF(3000)F2 during P.S.M. and of the minimum MUF(3000)F2 at the end of the phenomenon, respectively. The difference between these times is related to the duration of the

TABLE I

The occurrence of P.S.M. of MUF(3000)F2 at Genoa–Mt. Capellino (February 1964–January 1965)

| Date | | Present P.S.M. | Absent P.S.M. | Doubtful P.S.M. |
|------|------|------|------|------|
| 1964 | February | 17 | 7 | 5 |
| | March | 5 | 12 | 8 |
| | April | 8 | 12 | 8 |
| | May | 5 | 14 | 6 |
| | June | 3 | 19 | 6 |
| | July | 5 | 20 | 5 |
| | August | 4 | 17 | 5 |
| | September | 12 | 10 | 8 |
| | October | 18 | 6 | 4 |
| | November | 10 | 4 | 10 |
| | December | 15 | 6 | 7 |
| 1965 | January | 10 | 4 | 10 |

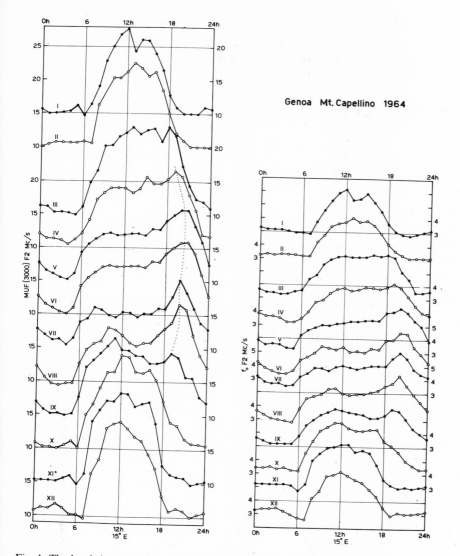

Fig. 1. The local time variation of $f_oF2$ and MUF(3000)F2 at Genoa–Mt.Capellino in a period of low solar activity (January–December 1964).

phenomenon, which becomes maximum in winter, being 60 min in December and 55 min in January.

These findings agree with the corresponding behaviour of the F2 region over Freiburg, Germany. In Fig. 2 we plot the mean daily variation, for winter, equinoctial and summer months, of the ratio $\Delta MUF/MUF$ between the interquartile monthly range of MUF(3000)F2 and monthly median MUF(3000)F2 at Freiburg.

TABLE II

The monthly mean time (15°E) of the MUF (3000)F2 maximum and
the corresponding mean time of the MUF (3000)F2 minimum at the
end of P.S.M., at Genoa–Mt. Capellino (February 1964–January
1965)

| Date | | max.MUF | min.MUF |
|---|---|---|---|
| 1964 | February | 0535h | 0625h |
| | March | 0455h | 0535h |
| | April | 0505h | 0525h |
| | May | 0410h | 0455h |
| | June | 0345h | 0445h |
| | July | 0400h | 0445h |
| | August | 0415h | 0500h |
| | September | 0500h | 0540h |
| | October | 0525h | 0605h |
| | November | 0535h | 0615h |
| | December | 0525h | 0625h |
| 1965 | January | 0530h | 0625h |

This ratio, which is indicative of the scatter of individual values of MUF(3000)F2
with reference to the monthly median values, becomes maximum at the time of
P.S.M. in winter and at the time of the evening concentration in summer.

As far as the explanation of P.S.M. is concerned, we remember first that, because
of the displacement of the geomagnetic North pole with respect to the geographic

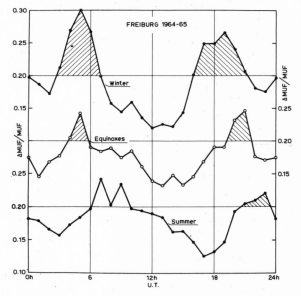

Fig. 2. The local time variation of the hourly percentage interquartile range of MUF(3000)F2 at
Freiburg, in a period of low solar activity (January 1964–December 1965).

North pole, photoelectrons streaming from the magnetically conjugate ionosphere (lying somewhat midway between Johannesburg and Capetown) will produce significant heating over Genoa–Mt.Capellino well before the local sunrise. Such exchange of electrons between conjugate ionospheres should also influence the electron density by secondary effects of heating, excitation and impact ionization; this additional ionization could explain the occurrence of P.S.M. in winter, at least in part (Evans, 1970).

The P.S.M., however, does not occur on all winter days. On the other hand, the ionospheric vertical drift, caused by the difference between the ion mass density and the neutral particle density, increases during night-time and can contribute to the genesis of P.S.M.. Because of the vertical motion of ionization, due to the buoyancy, the ionospheric layers are somewhat displaced and an enhancement of $f_oF2$ as well as of MUF(3000)F2 arises. A downward motion of the F2 region, increasing from midnight to just before sunrise, affording higher values of $f_oF2$ and MUF(3000)F2, appears also from measurements of vertical drift (Srivastava et al., 1970).

## 3. BAYLIKE DISTURBANCES IN THE F2 REGION

Many perturbations of the F2 region, not linked to geomagnetic activity, occur in winter months: e.g., F-layer spreading and anomalous day-to-day variations of $f_oF2$.

In Fig. 3 we have reported the ratio $q$ between the monthly means of hourly percentage interquartile range of $\Delta$MUF/MUF at Freiburg and the corresponding monthly mean of the geomagnetic index $Ap$ for two years of low solar activity (1964–1965) and of enhanced solar activity (1968–1969), respectively. The behaviour of $q$, although somewhat irregular, shows maxima in winter months, thus supporting evidence that in this season ionospheric irregularities can occur even with quiet geomagnetic conditions.

Many fluctuations of MUF(3000)F2 at Genoa-Mt.Capellino have a quasi-sinusoidal form with quasi-periods ranging from several minutes to about three hours. Such fluctuations will be called "baylike disturbances" (B.D.).

The B.D. are observed in the records of MUF(3000)F2 at all local times, but with greatest frequency near sunset, during the winter decrease of MUF(3000)F2 or during the summer evening concentration. The occurrence of B.D. becomes minimum in the first two or three hours after midnight; the annual variation of the occurrence of B.D. shows a maximum in winter and a minimum in summer. In winter, at the time of the decrease of MUF(3000)F2 near sunset, B.D. almost always occur (Table III). Such results are somewhat confirmed by the behaviour of the interquartile range of $\Delta$MUF/MUF at Freiburg (Fig. 2) which shows a maximum near sunset, related to the corresponding fluctuations of MUF(3000)F2.

The quasi-period $T$ of B.D. depends on local time, becoming maximum from

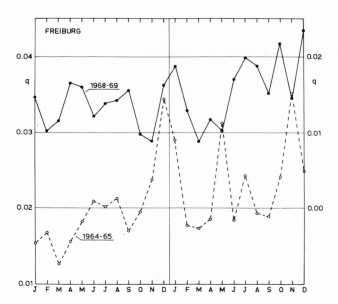

Fig. 3. The ratio $q$ of the monthly means of hourly percentage interquartile range of MUF(3000)F2 at Freiburg to the corresponding monthly mean $Ap$, in periods of low and enhanced solar activity (January 1964–December 1965 and January 1968–December 1969, respectively).

TABLE III

The occurrence of baylike disturbances (B.D.) during the decrease of MUF (3000)F2 at Genoa–Mt. Capellino at sunset in winter, during the evening concentration at equinoxes and in summer

| Date | | Present B.D. | Absent B.D. | Doubtful B.D. |
|------|------|------|------|------|
| 1964 | February | 22 | 2 | 3 |
| | March | 13 | 4 | 10 |
| | April | 6 | 12 | 4 |
| | May | 6 | 10 | 6 |
| | June | 5 | 11 | 5 |
| | July | 5 | 8 | 10 |
| | August | 7 | 10 | 7 |
| | September | 3 | 8 | 9 |
| | October | 7 | 11 | 7 |
| | November | 18 | 6 | 6 |
| | December | 23 | 5 | 2 |
| 1965 | January | 17 | 1 | 3 |

sunset to the first two or three hours after midnight. From the sunrise till about 1400 h, quasi periods $T$ of about 15 min and 30 min strongly prevail; $T$ shifts towards higher values in the afternoon (Table IV).

The very rare occurrence of B.D. in the first hours after midnight appears to be

TABLE IV

The distribution of the "quasi-period" $T$ of baylike disturbances in MUF (3000)F2 at Genoa–Mt. Capellino, from February 1964 to January 1965 included (Time: 15°E)

| Quasi-period $T$ (min) | Number of occurrences | |
|---|---|---|
| | from sunrise to 1400 h | from 1400 to 2400 h |
| $\leqslant 15$ | 42 | 22 |
| 15–30 | 46 | 31 |
| 30–45 | 25 | 19 |
| 45–60 | 39 | 30 |
| 60–90 | 8 | 18 |
| 90–120 | 0 | 7 |

independent of geomagnetic activity. Some instances when B.D. occurred from 0000 to 0300 h (15°E), under quiet geomagnetic conditions, are listed in Table V, thus further confirming the "meteorological" character of the wintertime F2-region irregularities.

TABLE V

Some cases of after-midnight baylike fluctuations of great period (2–3 h), under quiet geomagnetic conditions (see the corresponding $Ap$)

| Date | | | $Ap$ | Date | | | $Ap$ |
|---|---|---|---|---|---|---|---|
| 1964 | February | 21 | 4 | 1964 | December | 21 | 4 |
| 1964 | November | 3 | 4 | 1964 | December | 26 | 5 |
| 1964 | November | 16 | 10 | 1964 | December | 30 | 2 |
| 1964 | December | 2 | 3 | 1968 | November | 15 | 3 |
| 1964 | December | 3 | 5 | 1969 | January | 6 | 1 |
| 1964 | December | 11 | 3 | 1969 | February | 4 | 13 |
| 1964 | December | 12 | 2 | | | | |

## 4. DISCUSSION

The lower ionosphere and the mesosphere in winter are subjected to irregular changes of "meteorological" character, such as the winter anomaly in ionospheric absorption. Conversely, for the F2 region, only indirect evidence of a coupling with the lower atmosphere has been attained at present (Davies and Jones, 1971). At the height of the F2-region maximum plasma frequency, the terms of movement, diffusion and electrodynamical drift are comparable with the production and recombination terms. Variations of the F2-region plasma frequency can be due either to changes in electron production rate and temperature, or to movements of ionization because of drifts and neutral winds.

As far as the local time variation of B.D. is concerned, the maximum occurrence near sunset may be linked to a corresponding maximum of vertical drift in the F2 region; whereas the infrequent occurrence of baylike disturbances after midnight can be coupled to a corresponding minimum of the vertical drift velocity (Srivastava et al., 1970).

B.D., observed in the records of MUF(3000)F2 at Genoa–Mt.Capellino, are excited by gravity waves propagating from lower levels in the atmosphere. Direct evidence of gravity waves in the F2 region was found by Dyson et al. (1970). The characteristics of these waves depend on the atmospheric conditions, as well as on the nature of the source; neutral winds, moreover, provide a directional filter effect (Cowling et al., 1971).

During the daytime, at a height of about 250 km over Westwood, N. J., Tolstoy and Montes (1971) found power spectra of phase height fluctuations with maxima around 15 min. Moreover, the height variation of power spectra suggested a trend similar to the Vaisälä frequency profile, according to the formula:

$$N^2 = \frac{g}{T}\left[\frac{g}{R'}\left(1 - \frac{1}{\sigma}\right) + \frac{dT}{dz}\right] \tag{3}$$

where $N$ is the local value of the Vaisälä frequency; $R' = R/M_o$, $R$ being the perfect gas constant and $M_o$ the molecular weight at sea level; $\sigma$ the ratio of the specific heats at constant pressure and constant volume $c_p/c_v$; $T$ the molecular scale temperature.

Such findings of Tolstoy and Montes (1971) appear to agree with our results of Table IV concerning the quasi-period $T$ of gravity waves at the height of the daytime F2 region.

We can remember that, from a thunderstorm area, spherics are generally emitted in groups having a mean duration of about 15 min or 30 min or 45 min (Bossolasco et al., 1969). With reference to Table IV, we can suggest that thundery activity may contribute to the energy of the propagating acoustic gravity waves, responsible for baylike F2-region disturbances. Of course, this does not mean that F2-region winter irregularities are due to thundery activity at the same place; but it is probable that the energy of gravity waves in the ionosphere can be supplied by global thundery activity. Some evidence of infrasonic waves in the ionospheric F2 region, associated with severe thundery activity, was recently found by Davies and Jones (1971); following these authors, mechanical motion of thunderstorms should be responsible for the origin of infrasonic waves.

Because of the very large frequency spectrum of atmospheric perturbations, many ionospheric irregularities are likely to be gravity waves due to tropospheric phenomena. In this respect, the highest occurrence of baylike disturbances in the winter F2 region could be ascribed to the contemporaneous maximum of atmospheric perturbations.

## REFERENCES

Bossolasco, M., Dagnino, I. and Elena, A., Ricerche di Aeronomia III. *Geofis. Meteor.*, **XIV**:139–153, 1965.

Bossolasco, M., Dagnino, I. and Flocchini, G., Primi risultati sulla registrazione rapida delle scariche elettroatmosferiche. *Geofis. Meteor.*, **XVIII**:90–97, 1969.

Cowling, D. H., Webb, H. D. and Yeh, K. C., Group rays of internal gravity waves in a wind-stratified atmosphere. *J. Geophys. Res.*, **76**:213–220, 1971.

Davies, K. and Jones, J. E., Ionospheric disturbances in the F2 region associated with severe thunderstorms. *J. Atmos. Sci.*, **28**:254–262, 1971.

Dyson, P. L., Newton, G. P. and Brace, L. H., In situ measurements of neutral and electron density wave structure from the Explorer 32 satellite. *J. Geophys. Res.*, **75**:3200–3210, 1970.

Evans, J. V., Millstone Hill Thomson scatter results for 1965. *Planet. Space Sci.*, **18**:1225–1253, 1970.

Srivastava, S. K., Pradhan, S. M. and Tantry, B. A. P., Vertical drifts in F region of the ionosphere. *Ann. Géophys.*, **26**:881–892, 1970.

Tolstoy, I. and Montes, H., Phase height fluctuations in the ionosphere between 130 and 250 km. *J. Atmos. Terr. Phys.*, **33**:775–781, 1971.

# ELECTRONS PRECIPITATING INTO THE D REGION FROM THE INNER AND OUTER VAN ALLEN ZONES

GIOVANNI E. PERONA

*Istituto di Elettronica e Telecomunicazioni, Politecnico di Torino, Turin (Italy)*

## SUMMARY

Estimated fluxes of precipitating electrons from the magnetosphere into the lower ionosphere as a function of magnetic latitude are briefly reviewed. Their importance as sources of ionization in the D region is recognized, both during magnetically quiet and disturbed periods, even at middle latitudes.

## 1. INTRODUCTION

The effect of magnetospheric electrons precipitating into the ionosphere has been widely studied at high latitudes and in the polar regions, both theoretically and experimentally (McCormac, 1970). Much less attention has been given to electron precipitation in the middle and low latitude regions where the phenomenon, though less conspicuous, is not of secondary importance. In recent years, evidence has been growing that low energy electrons produce a pronounced effect on the F region (Mariani, 1963; Gledhill et al., 1967; Torr and Torr, 1970). In particular, Torr and Torr (1969, 1970) found very large precipitations of 0.05–2.00 keV electrons into the F region at latitudes lower than the auroral ones and at Johannesburg (26°S, 28°E). Recently, attempts have been made to describe the winter anomaly in terms of electron precipitation from the magnetosphere (Maehlum, 1967). An alternative explanation of such a phenomenon in terms of meteorological influences finds support in a large body of indirect experimental evidence (Bossolasco and Elena, 1963; Gregory and Manson, 1969) and in a few direct measurements (Sechrist et al., 1969). However, both meteorological factors and electron precipitation seem to be important as generating factors of the winter anomaly (Manson and Merry, 1970).

In this paper, only the precipitation of "natural" electrons from the inner and outer Van Allen zones will be considered, excluding polar cap and auroral phenomena. Furthermore, we will examine only electrons capable of penetrating below 100 km. The first constraint sets an upper limit on the magnetic latitude range to be studied, so that it will not be larger than 70°; the second constraint sets a

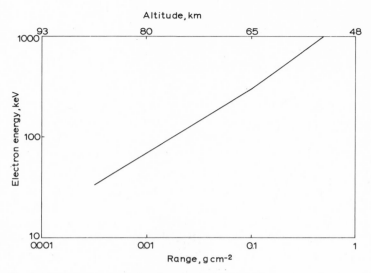

Fig. 1. Range-energy curve for electrons in the atmosphere.

lower limit on the electron energies of interest. Indeed, the range of electrons as a function of their energy is shown in Fig. 1; the upper scale gives the corresponding altitude in the ionosphere as deduced from CIRA (1965): only energies larger than $\sim$ 10 keV can be of interest in the present study. However, a comparison of the range $R$ of the particles with the amount of material $D$ necessary for Coulomb scattering shows the importance of this second phenomenon. Quantity $D$ can be obtained using the following formula (Hess, 1968):

$$\theta^2 = 7000 \; D/E^2$$

where $E$ is the energy of the particles (keV), $\theta$ the mean angle of scattering (rad), and $D$ is the path length (cm) of traversed air (at standard pressure and temperature). For 1 MeV electrons and $\theta = 0.5$ rad, $D$ is 35 cm, that is equivalent to $\sim$ 0.04 g cm$^{-2}$. Consequently, 1 MeV precipitating electrons can be scattered back before reaching their lower altitude limit of 50 km, even if their pitch angles were initially very small, since their range $R$ is more than 10 times larger than $D$. Conversely, electrons having relatively large pitch angles and reflection points well above 50 km, could be scattered into the loss cone. Therefore, measurements of electron fluxes in the loss cone made by satellite well above the ionosphere should be considered with some caution before interpreting them as fluxes of particles really precipitating into the lower part of the ionosphere. A detailed analysis of the behaviour of electrons interacting with the atmosphere has been given by Walt (1964).

This presentation is divided into three sections: in the first, fluxes and spectra of precipitated electrons are examined as a funcion of magnetic latitude $\Lambda$; in the second, the effects of such electrons on the lower ionosphere are briefly considered and finally, in the last section, the precipitation during disturbed times is evaluated.

## 2. PRECIPITATION OF ELECTRONS DURING MAGNETICALLY QUIET PERIODS

Only in a few cases has the precipitation of electrons over low and middle latitude regions been measured. Potemra and Zmuda (1970) give a review of some data supplied by satellites, relative to precipitating electrons with energies $E$ larger than 40 keV. The concerned satellites were: Injun 1 (O'Brien, 1962), Injun 3 (O'Brien, 1964), Explorer 12 and Alouette. Approximate upper and lower limits of the fluxes are given as a function of magnetic latitude in Fig. 2. On the latter, crosses indicate values deduced from Injun 1 and Injun 3 data and used by Manson and Merry (1970) in their analysis of the winter anomaly. To have an approximate idea of the ionospheric effect of such fluxes, let us point out that Gledhill et al. (1967) experimentally deduced a critical value of $\sim 2 \cdot 10^4$ el sec$^{-1}$ cm$^{-2}$ (or $\sim 3 \cdot 10^3$ el sec$^{-1}$ ster$^{-1}$ cm$^{-2}$) as the minimum flux of electrons with $E > 40$ keV, capable of producing ionospheric changes at $\Lambda \simeq 60°$. Measurements by Injun 3, reported by Maehlum (1967, see his fig.7) give a wide range of fluxes, from as low as 20 to $10^4$ el sec$^{-1}$ ster$^{-1}$ cm$^{-2}$, at $\Lambda \simeq 55°$. The same satellite showed that fluxes larger than $3 \cdot 10^3$ el sec$^{-1}$ ster$^{-1}$ cm$^{-2}$ were present for more than 50% of the time at local noon, and $\Lambda \simeq 55°$ (Maehlum, 1967).

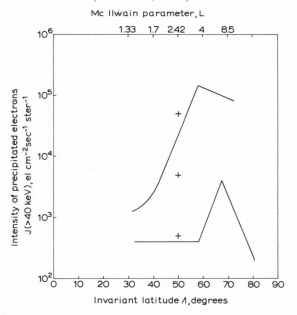

Fig. 2. Minimum and maximum intensities, at altitudes $\leqslant 1100$ km, of precipitated electrons with energies larger than 40 keV, as a function of magnetic latitude $\Lambda$.

Figure 1 shows that precipitating electrons may influence the lower D region only if their energy is larger than a few hundred keV. Therefore, it is important to derive the energy spectra of the electron fluxes. Unfortunately, there are very few measurements of fluxes at energies larger than, say, 300 keV. Potemra and Zmuda (1970) collected all available direct measurements and built three model spectra that they used in calculations relative to the mid-latitude D region (curves *A*, *B* and *C* in Fig. 3). Manson and Merry (1970) hypothesized the spectrum plotted as curve *D* in Fig. 3, where the e-folding energy has been assumed equal to 15 keV. All data refer to a midlatitude with $\Lambda \simeq 50°$. Curve *D* is very significantly different from the other model spectra and requires some comments. Indeed an e-folding energy of 15 keV seems too low, especially in the high energy range. Measurements by Imhof et al. (1970) show that the e-folding energy in the range 100–1000 keV is of the order of 100 keV. Such results refer to fluxes detected at $\Lambda \simeq 60°$ by a polar satellite in an orbit at 200 km altitude. In particular, one of the spectra presented corresponds to a *B–L* point which has a conjugate at a height as low as 69 km (Imhof et al., 1970; see their fig.7). Therefore, such a flux can be considered as a measure of the intensity

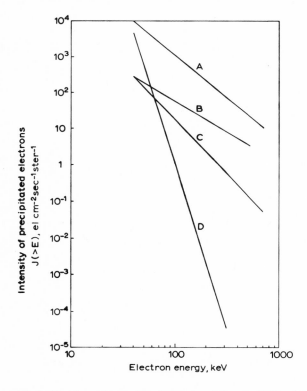

Fig. 3. Model spectra used by Potemra and Zmuda (1970; *A,B,C*) and by Manson and Merry (1970; *D*) representing precipitated electrons at middle latitudes.

of precipitated electrons and, in the units of Fig. 3, it is equal to $J(E > 100 \text{ keV}) \simeq$ 40 el sec$^{-1}$ ster$^{-1}$ cm$^{-2}$. This flux refers to a magnetic latitude larger than the latitude of Fig. 3; nevertheless, it is interesting to point out that its value is intermediate between curves $B$ and $C$ and much larger than $D$. Another important feature of the fluxes measured by Imhof et al. (1970) is that their e-folding energy is almost independent of the altitude of the conjugate point, be it lower or higher than 100 km, thus suggesting the possibility that trapped, quasi-trapped and precipitating electrons have a not too different spectral shape. The e-folding energy of the electron spectrum, is not too sensitive a function of the $B$–$L$ points, even at a much lower latitude, in the inner zone. Its value ranges between 1000 keV and 300 keV in the region 0.2 gauss $< B < 0.26$ gauss and $1.20 < L < 1.8$ for electron energies larger than 500 keV (Rosen et al., 1968).

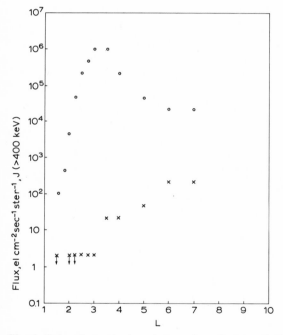

Fig. 4. Order-of-magnitude estimates of maximum and minimum electron fluxes with $E > 400$ keV, measured at $\sim 300$ km (Imhof, 1968).

The current literature presents a large number of data concerning quasi-trapped particles that are lost over the South Atlantic anomaly. Fig. 4 shows order-of-magnitude estimates of maxima and minima of electron fluxes with $E > 400$ keV— measured by a satellite in a polar orbit (Imhof, 1968)—as a function of the McIlwain (1961) parameter $L$. The general behaviour of the maxima is confirmed by other measurements (e.g., Williams and Kohl, 1965). The minima were detected just after the electrons had gone through the South Atlantic anomaly and can be used to give an approximate idea of the relative distribution of the precipitating electrons as a function of the magnetic latitude. Their absolute values can be considered as an upper estimate of the precipitating electrons and are consistent with curve $B$ of Fig. 3, but are many orders of magnitude larger than curve $D$. The e-folding energy measured by Imhof (1968) is of the order of 200 keV and does not change too much with the distance from the South Atlantic anomaly or the McIlwain parameter $L$ in the range $2 < L < 7$. In conclusion, it seems reasonable to assume that experimental, though indirect, evidence supports the model spectra used by Potemra and Zmuda (1970).

## 3. D-REGION PERTURBATIONS FROM PRECIPITATING ELECTRONS

One way of assessing the relative importance of different ionizing agents is to plot their corresponding electron production rate $q(h)$ as a function of altitude. Fig. 5 gives the computed $q(h)$ for various sources. The curve labelled C.R. refers to galactic cosmic rays during summer at solar minimum, and $\Lambda = 41°$ (Velinov, 1968): curve $X$ refers to the maximum of the X-ray source SCO XR1 at $0°$ zenith angle (Francey, 1970); curve $L\beta$ and $L\alpha$-Pearce are estimates of $q(h)$ due to Lyman-$\beta$ and Lyman-$\alpha$ in the night-time sky (Francey, 1970; Potemra and Zmuda, 1970). In computing the values of $q(h)$ due to Lyman-$\alpha$, the nitric oxide concentration measured by Pearce (1969) has been used. However, if the nitric oxide intensity profile derived by Mitra (1968) is adopted, a much lower production rate by Lyman-$\alpha$ is computed (Francey, 1970). Indeed, recent results (Meira, 1971) do not

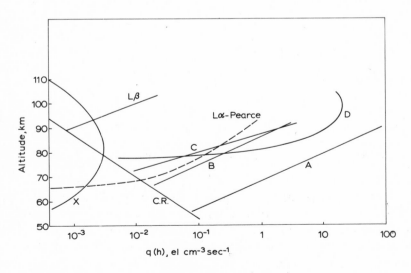

Fig. 5. Ionization rates in the night-time D region (see text).

seem to confirm Pearce's data. Curves $A$, $B$, $C$ and $D$ refer to the model spectra of precipitating electrons given in Fig. 3. In the high altitude region $h \simeq 90$–$100$ km the four models give consistent results. Indeed, such a height range is interested by the low energy part of the precipitating electrons (see Fig. 1), and the four models assume comparable fluxes in that part of the spectrum. At lower heights, model $D$ exhibits a very small electron production rate as a consequence of the small fluxes assumed at energies larger than $\sim 100$ keV. It is noteworthy to point out that the small e-folding energy (15 keV) used by Manson and Merry (1970) was not supported by any experimental results. The working criterion followed by the authors in their choice was "that the electron density increase at 80 km due to an average flux ($5 \cdot 10^3$ el sec$^{-1}$ ster$^{-1}$ cm$^{-2}$) of precipitated electrons at $L = 2.5$–$3.5$ be smaller than the electron density due to the photoionization under quiet solar conditions during summer months". It is unfortunate that the models give such different results in the height range 70–80 km, because the total electron production rate, due to all other natural sources in the night sky, presents a minimum there. Only further measurements of precipitating electrons with energies larger than 100 keV can definitely clarify the problem.

We must realize that curves $A$, $B$, $C$ and $D$ in Fig. 5 can only give order-of-magnitude estimates of the importance of precipitating electrons in the night sky and have to be considered with great caution. Indeed, in preparing the model spectra of Fig.3, day and night data have been indiscriminately used. However, diurnal variations in the particle fluxes can be very great, especially at high latitudes. At $L = 4$, there is an average day-to-night ratio of about 20 in the flux of precipitated electrons with energies larger than 40 keV (Frank et al., 1964). McDiarmid and Burrows (1964) indicate a diurnal modulation even at lower $L$ values. These local time changes have been shown to be statistically consistent with the diurnal variation of the winter anomaly (Maehlum, 1967).

## 4. PRECIPITATION DURING MAGNETICALLY DISTURBED PERIODS

The flux of precipitating electrons with $E > 40$ keV during magnetically disturbed periods can be represented by the upper curve in Fig. 2. At higher energies, measurements of quasi-trapped particles may be used to give the relative latitude distribution of the precipitating flux. Rosen et al. (1968) presented data on the temporal behaviour of inner zone electrons with $E > 500$ keV, obtained with Pegasus 1 orbiting at a height of $\sim 600$ km. Their results show an up to 5-fold increase in the electron flux at $L \simeq 1.18$, depending on the frequency and severity of magnetic storm activity. Bostrom et al. (1970) report increases by a factor of 10 in the electron fluxes with energies larger than 280 keV, at $L \simeq 1.2$ during the November 1968 storm. Their data, collected with the 1963 38C satellite at an altitude of 1100 km, present a 100 fold increase in the electron flux at $L = 2.2$. A very detailed picture of quasi-trapped electron fluxes with energies larger than 300 keV is given by Rosen and Sanders (1971), who used data from a Pegasus satellite orbiting at $\sim$ 500 km. They showed up to a 1000 fold increase in the flux of quasi-trapped particles at $1.8 < L < 2.3$ towards the end of May, 1967 when the $Ap$ index was above 120.

Electron fluxes in the outer Van Allen zone are much more sensitive to magnetic disturbances and have been thoroughly studied (Hess, 1968; McCormac, 1970).

From the above brief discussion, it appears that during magnetically disturbed periods, even at middle and low latitudes, an order of magnitude increase in the precipitated flux is plausible if such a flux follows the behaviour of the quasi-trapped particles. Consequently, the importance of the precipitation into the D region is stressed and at middle latitudes, the model spectrum $A$ of Fig. 3 can be considered realistic.

At higher latitudes, precipitation events are often detected through cosmic noise absorption by means of riometers (Hartz, 1963), or the effects on VLF propagation (Lauter and Knuth, 1967; Belrose and Thomas, 1968), or X-ray balloon measurements (Ullaland et al., 1970). In particular, a typical and well-characterized phenomenon, i.e., the precipitation of electrons at the time of a Sudden Commencement (S.C.), has been extensively studied.

It has been definitely shown that precipitation during an S.C. presents a very pronounced latitude effect which is revealed by simultaneous analysis of riometric absorption measurements and geomagnetic field records (e.g., Ortner et al., 1962). In Fig. 6 adapted from Hartz (1963), the probability and the amount of cosmic noise absorption that is observed during an S.C. are plotted as a function of the geomagnetic latitude. The two curves have a broad maximum between 65° and 70° and fall down to very low values at 60° and 75°. These results show a spread in latitude, which is in part an effect of the statistical analysis that was performed over many precipitation events. Ortner et al. (1962) show that cosmic noise absorption is maximum just to the south of the auroral zone, and is negligible 500 km south of this maximum.

The precipitating electrons show a well-defined local time dependence. At Churchill (Hartz, 1963), the probability of Sudden Commencement Absorption (S.C.A.), as a function of local mean solar time, has a relatively broad maximum at noon, falling down rapidly to a very low minimum in the midnight region.

Estimates of electron precipitation fluxes during an S.C. were presented by Brown et al. (1961), and give a maximum value of 2 to $6 \cdot 10^7$ el cm$^{-2}$ sec$^{-1}$. For a different case, Hofmann and Winckler (1963) deduced a flux of $10^9$ el cm$^{-2}$ sec$^{-1}$. However, this flux was obtained through X-ray measurements, and possibly a part of the X-rays was due to protons; therefore, the electron flux could be lower than the value given by these authors. From these data it seems reasonable to assume that at the peak of an S.C.A. event the number of electrons precipitating per second is less than

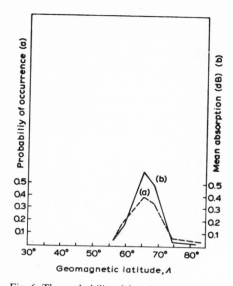

Fig. 6. The probability, (a) and the mean magnitude of cosmic noise absorption (b) as a function of the magnetic latitude Λ (Hartz, 1963).

0.01–0.001 of the total electron content of the tube of flux. Hence, their lifetime is 100–1000 sec, which is comparable to the lifetime expected when strong pitch-angle diffusion regime is reached (Kennel and Petschek, 1966). Furthermore, some data of Brown et al. (1961; see their Fig. 1 and 2) show that the tube of flux, which is not emptied of its electrons at the onset of an S.C., can supply a second burst of electron precipitation, due to a second increase in the magnetic field following the main peak. Recently, a theory was put forward (Perona, 1972) whereby VLF waves in the magnetosphere at the time of an S.C. can generate a pitch-angle diffusion of local electrons, and the latter, in turn, produce the kind of precipitation that is observed in the lower ionosphere. More precisely, it was shown that in the sunlit hemisphere between 65° and 70° of geomagnetic latitude, the compression of the geomagnetic field during an S.C. is often strong enough to increase the VLF level up to the point where strong pitch-angle diffusion will take place.

## 5. CONCLUSIONS

At high latitudes, precipitation of magnetospheric electrons into the lower ionosphere has been recognized for a long time as an important source of ionization in the D region. Recently, more weight has been given to precipitation at lower magnetic latitudes. However, it is very difficult to assess its importance because of a lack of quantitative information on the precipitating fluxes. Some preliminary analysis seems to confirm that these fluxes are a noticeable source of ionization, both during day and night, at middle latitudes.

## REFERENCES

Belrose, J. S. and Thomas, L., Ionization changes in the middle latitude D region, associated with geomagnetic storms, *J. Atmos. Terr. Phys.*, **30**:1397–1413, 1968.

Bossolasco, M. and Elena, A., Absorption in the D layer and the temperature of the magnetosphere. *C.R.Hebd. Seanc. Acad. Sci.*, **256**:4491–4493, 1963.

Bostrom, C. O,, Beall, D. S. and Armstrong, J. C., Time history of the inner radiation zone, October 1963 to December 1968. *J. Geophys. Res.*, **75**:1246–1256, 1970

Brown, R. R., Hartz, T. R., Landmark, B., Leinbach, H. and Ortner, J., Large scale electron bombardment of the atmosphere at the Sudden Commencement of a geomagnetic storm. *J. Geophys. Res.*, **66**:1035–1041, 1961.

CIRA 1965, *Cospar International Reference Atmosphere*. North Holland Publishing Company, Amsterdam, 1965.

Francey, R. J., Electron production in the ionospheric D region by cosmic X rays. *J. Geophys. Res.*, **75**:4849–4862, 1970.

Frank, L. A., Van Allen, J. A. and Craven, J. D., Large diurnal variations of geomagnetically trapped and of precipitating electrons observed at low altitudes. *J. Geophys. Res.*, **69**:3155–3167, 1964.

Gledhill, J. A., Torr, D. G. and Torr, M. R., Ionospheric disturbance and electron precipitation from the outer radiation belt. *J. Geophys. Res.*, **72**:209–214, 1967.

Gregory, J. B. and Manson, A. H., Seasonal variations of electron densities below 100 km at midlatitudes, II. *J. Atmos. Terr. Phys.*, **31**:703–729, 1969.

Hartz, T. R., Multi-station riometer observations. In: *Radio Astronomical and Satellite Studies of the Atmosphere*. (J. Aarons, Editor). North-Holland, Amsterdam, p. 220, 1963.

Hess, W. N., *The Radiation Belt and Magnetosphere*. Blaisdell, Waltham, Mass., 1968.

Hofmann, D. J. and Winckler, J. R., Simultaneous balloon observations at Forth Churchill and Minneapolis during the solar cosmic-ray events of July 1961. *J. Geophys. Res.*, **68**:2067–2097, 1963.

Imhof, W. L., Electron precipitation in the radiation belts. *J. Geophys. Res.*, **73**: 4167–4184, 1968.

Imhof, W. L., Gaines, E. E. and Reagan, J. B., High-resolution studies of electrons 25 KeV to 2.7 MeV precipitating at high latitudes. *J. Geophys. Res.*, **75**:776–782, 1970.

Kennel, C. F. and Petschek, H. E., Limit on stably trapped particle fluxes. *J. Geophys. Res.*, **71**:1–28, 1966.

Lauter, E. A. and Knuth, R., Precipitation of high energy particles into the upper atmosphere at medium latitudes after magnetic storms. *J. Atmos. Terr. Phys.*, **29**:411–417, 1967.

Maehlum, B., On the winter anomaly in the midlatitude D region. *J. Geophys. Res.*, **72**:2287–2299, 1967.

Manson, A. H. and Merry, M. W. J., Particle influx and the winter anomaly in the mid-latitude ($L = 2.5$–$3.5$) lower ionosphere. *J. Atmos. Terr. Phys.*, **32**:1169–1181, 1970.

Mariani, F., Evidence for the effect of corpuscular radiation on the ionosphere. *J. Atmos. Sci.*, **20**:479–491, 1963.

McCormac, B. (Editor), *Particles and Fields in the Magnetosphere*. Reidel, Dordrecht, 450 pp., 1970.

McDiarmid, I. B. and Burrows, J. R., Diurnal intensity variations in the outer radiation zone at 1000 km. *Can. J. Phys.*, **42**:1135–1148, 1964.

McIlwain, C., Coordinates for mapping the distribution of magnetically trapped particles. *J. Geophys. Res.*, **66**:3681–3691, 1961.

Meira, L. G., Rocket measurements of upper atmospheric nitric oxide and their consequences to the lower ionosphere. *J. Geophys. Res.*, **76**:202–212, 1971.

Mitra, A. P., A review of D region processes in non polar latitudes. *J. Atmos. Terr. Phys.*, **30**:1065–1114, 1968.

O'Brien, B. J., Lifetimes of outer-zone electrons and their precipitation into the atmosphere. *J. Geophys. Res.*, **67**:3697–3706, 1962.

O'Brien, B. J., High latitude geophysical studies with satellite Injun 3. *J. Geophys. Res.*, **69**:13–43, 1964.

Ortner, J., Hultquist, B., Brown, R. R., Hartz, T. R., Holt, O., Landmark, B., Hook, J. L., and Leinbach, H., Cosmic noise absorption accompanying geomagnetic storm Sudden Commencement. *J. Geophys. Res.*, **67**:4169–4186, 1962.

Pearce, J. B., Rocket measurements of nitric oxide between 60 and 96 km, *J. Geophys. Res.*, **74**:853–861, 1969.

Perona, G. E., A theory on the precipitation of magnetospheric electrons at the time of a Sudden Commencement. *J. Geophys. Res.*, **77**:101–111, 1972.

Potemra, T. A. and Zmuda, A. J., Precipitating energetic electrons as an ionization source in the midlatitude night-time D region. *J. Geophys. Res.*, **75**:7161–7167, 1970.

Rosen, A. and Sanders, N. L., Loss and replenishment of electrons in the inner radiation zone during 1965–1967. *J. Geophys. Res.*, **76**:110–121, 1971.

Rosen, A., Sanders, N. L., Shelton, R., Potter, R. and Urban, E., Pegasus 1: Observations of the temporal behaviour of the inner zone electrons, 1965–1966. *J. Geophys. Res.*, **73**:1019–1033, 1968.

Sechrist, C. F., Mechtly, E. A. and Shirke, J. S., Coordinated rocket measurements on the D-region winter anomaly, I: Experimental results. *J. Atmos. Terr. Phys.*, **31**:145–153, 1969.

Torr, D. G. and Torr, M. R., A theoretical investigation of corpuscular radiation effects on the F region of the ionosphere. *J. Atmos. Terr. Phys.*, **32**:15–34, 1970.

Torr, M. R. and Torr, D. G., The inclusion of a particle source of ionization in the ionspheric continuity equation. *J. Atmos. Terr. Phys.*, **31**:611–614, 1969.

Ullaland, S. L., Wilhelm, K., Kangas, J. and Riedler, W., Electron precipitation associated with a Sudden Commencement of a geomagnetic storm. *J. Atmos. Terr. Phys.*, **32**:1545–1552, 1970.

Velinov, P., On ionization in the ionospheric D region by galactic and solar cosmic rays. *J. Atmos. Terr. Phys.*, **30**:1891–1905, 1968.

Walt, M., The effects of atmospheric collisions on geomagnetically trapped electrons. *J. Geophys. Res.*, **69**:3947–3958, 1964.

Williams, D. J. and Kohl, J. W., Loss and replenishment of electrons at middle latitudes and high B values. *J. Geophys. Res.*, **70**:4139–4150, 1965.

# CORPUSCULAR EFFECTS IN THE IONOSPHERE

FRANCO MARIANI

*Istituto di Fisica dell'Università de L'Aquila, L'Aquila (Italy)*

## SUMMARY

Ionospheric effects attributed to particles at middle and low latitudes are presented. Particle contribution to the ionospheric ionization is suggested by several kinds of observations, namely maintenance of nocturnal ionization, temperature altitude distribution of neutral and charged particles, correlations between ionospheric electron density and solar activity, etc. Low energy particles are actually observed at ionospheric latitudes.

The author demonstrates the plausibility of attributing ionospheric effects which hitherto could not be explained only in terms of electromagnetic radiation to particle bombardment. As regards effects attributed to particles in the energy range above several tens of eV up to tens of keV, the evidence is good although not completely conclusive. On the very low energy side, heating of the ambient electrons by the photoelectrons produced by UV solar radiation is well established.

A summary of unpublished results on the last topic is included.

## 1. INTRODUCTION

Corpuscular radiation effects in the ionosphere are essentially due to particles: (a) coming from interplanetary space, like galactic or solar cosmic rays, particles from the radiation belts, solar wind; (b) produced or accelerated in the upper atmosphere, like the electrons, produced with energies of a few eV in the photoionization of the atmospheric constituents.

We will concentrate our attention on the low and middle latitude region of the magnetosphere, without considering the auroral phenomena where the particle effects have been well studied and established. The main experimental facts, partially or completely attributable to particle penetration at low and middle latitudes are: (a) the ionization at altitudes of 70–90 km and the maintenance of the nocturnal ionosphere; (b) long-term variations of the ionospheric electron density and correlations with the solar activity; (c) airglow; (d) temperature effects.

The early suggestion that a corpuscular radiation may be responsible for some ionospheric effects was recently revived as a result of several observational facts. The maintenance of the nocturnal ionization is one of these (Antonova and Ivanov-Kholodny, 1961). Another fact is the correlation between magnetic activity and the so-called winter anomaly found by Thomas (1962). A correlation between noon-time F2-layer maximum electron densities and solar activity peaked at latitudes between 55° and 65° was attributed by Mariani (1963) to a particle influx, possibly dumped from the radiation belts. Precipitating electrons during the winter anomalous periods were correlated with the absorption of electromagnetic waves in the 2 to 4 MHz range by Maehlum (1967). Ionization enhancements and other effects on VLF, LF and MF, associated with or following magnetic storms, have been attributed to particles by Lauter and Knuth (1967) and by Belrose and Thomas (1968). Moreover, electron density enhancements over the South Pacific and Africa at the 300-km level were attributed to precipitating particles (Knudsen and Sharp, 1968). No matter what the origin of the particles may be, emission of airglow by interaction with the ambient atmospheric particles can be predicted (Dalgarno, 1964). So, if this emission is above the instrumental background, indirect information on the incoming fluxes of electrons and/or protons is obtained. Fast electrons, with energies up to several hundreds eV, can excite several bands of the first negative system of molecular nitrogen $N_2^+$, corresponding to the wavelengths 3914, 4278 and 4709 Å. The cross sections for excitation of these bands are respectively 2, 0.7 and 0.2 times $10^{-3}$ of the total ionization cross section. Protons also are able to excite the same lines with cross sections essentially similar to those of the fast electrons with the same velocity. Upper limits for the particle fluxes can then be estimated from airglow intensity measurements. For example, an electron energy flux of 0.1 erg $cm^{-2}$ $sec^{-1}$ corresponds approximately to 20 rayleighs of 3914 Å radiation. A systematic study of the airglow can then be very helpful in looking into this matter.

## 2. THE MAINTENANCE OF THE NIGHT IONOSPHERE

Nocturnal observations are particularly suitable for studying corpuscular effects because of the absence of direct electromagnetic radiation. Two approaches can be taken, a priori: one is to consider the night ionosphere as a remnant of the daytime ionosphere, because of a dynamic redistribution of the particles, of long recombination times, etc.; the other is to think in terms of a corpuscular flux as an additional source of ionized particles.

Let us first give our attention to the D region, which is the lowest atmospheric region we are interested in. Up to approximately 70–80 km from the ground it is possible to balance the loss of charged particles by the ionization produced by the cosmic rays. At higher levels, if one takes the usual equation describing the time variation of the electron density, the loss term $-\alpha N_e^2$ (where $\alpha$ is the "effective" recombination coefficient and $N_e$ is the electron density) becomes, according to Ivanov-Kholodny (1965), too high, so that maintenance of the ionization requires an additional source, possibly a corpuscular source. Particles able to produce ionization at the D region altitude, i.e., appreciably below 100 km, must have primary energies greater than 40 keV. An energetic flux of the order of a few $10^{-5}$ erg cm$^{-2}$ sec$^{-1}$ is sufficient. Electrons with energy in this range have actually been observed on many occasions with much higher energetic fluxes, as can be derived from the intensities shown in Fig. 1 (O'Brien, 1962, 1964; McDiarmid et al., 1963; O'Brien and Laughlin, 1963; Fritz, 1968).

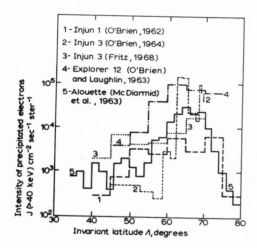

Fig. 1. Average intensities of electrons with energy > 40 keV, as observed by several satellites at altitudes below 1100 km, in the years 1961 to 1963 (redrawn from Potemra and Zmuda, 1970).

A remarkable maximum of the flux appears at invariant latitudes between 60° and 70°. The height distribution of electrons and positive ions of corpuscular origin can be computed for a given model of neutral atmosphere. The results for three different integral energy spectra of primary electrons $kE^{-\gamma}$ with $\gamma = 3$ to 5 (Potemra and Zmuda, 1970) are summarized in Fig. 2, where the observed concentrations are also shown, as well as the expected concentrations computed on the basis of the alternative non-corpuscular hypothesis that the D region is produced by photoionization of nitric oxide, NO, by scattered Lyman $\alpha$ radiation (Ogawa and Tohmatsu, 1966). Although the details of measured profiles cannot be reproduced by any smoothed theoretical profile, it is clear that the measured electron profiles by Mechtly and Smith (1968) in the altitude range 75–95 km may well be interpreted as being produced by primary electrons.

The observational results by Hale (1967) below 75 km are also shown in the

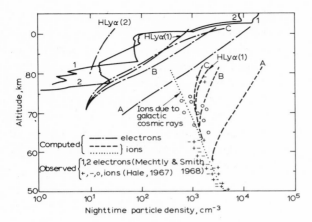

Fig. 2. Measured and computed densities of electrons, positive and negative ions. The results obtained using three energy spectra of primary electrons are labeled *A*, *B* and *C*. The curves labeled *HLyα*(2) and *HLyα*(1) are based on the computations by Ogawa and Tohmatsu (1966). (After Potemra and Zmuda, 1970.)

figure: an additional source of ionization actually seems necessary at altitudes above 70 km.

It is obvious that a special profile of the NO distribution can also lead to a curve of the ionization density that fits the experimental profile. The only way to solve the ambiguity between electromagnetic radiation and the particles is to carry out a coordinated experiment of particle and ionospheric measurements, ideally at points along the same field line.

At upper altitudes, i.e., above the 100 km level up to the basis of the F region, nocturnal electron density profiles show a very detailed structure with several narrow layers having peaks 3 to 7 times greater than the density of surrounding regions (Sagalyn and Smiddy, 1964; Ivanov-Kholodny, 1965). This compares with the rather smooth profiles at higher levels, near and above the F-region maximum. The presence of the sporadic Es during the night is an additional almost constant feature of the night ionosphere. The local narrow peaks of the electron densities are to be attributed to localized ionization sources, capable of producing electrons at a rate of the order of 10 cm$^{-3}$ sec$^{-1}$ which is 100 to 1000 times less than the daytime ultraviolet production rate. The total power flux can then be estimated in the range 10$^{-3}$ to 10$^{-2}$ erg cm$^{-2}$ sec$^{-1}$. The total absorption cross section for an ionizing agent penetrating the atmosphere to the E-layer altitude is within the 2 · 10$^{-17}$ to 10$^{-19}$ cm$^2$ range; several discrete values of this cross section are then required to explain the observed localized peaks. Rees (1963) studied the penetration of monochromatic steep energy spectrum electrons. These ionization curves are not as narrow as those experimentally observed. The expected peak occurs within the altitude range 100–200 km, when the energy of the electrons is in the range 0.5–40 keV. The altitude range of the energy deposition increases strongly when the energy increases. The evidence for the corpuscular origin of the peaks is not conclusive, although it has to be very seriously taken into account. At the F-region level, i.e., at altitudes above 200 km, it is well established that a diffusive equilibrium exists. After sunset the electron density $N_e$ is expected to decay according to an exponential law:

$$N_e \sim \exp(-\beta t)$$

where $\beta$ is the attachment coefficient and $t$ is the time measured from sunset. Due to the rather high measured values of $\beta$ ($\simeq 10^{-4}$ sec$^{-1}$), the electron density should then decrease by a factor of 10 in 6.5 h, and 100 in 13 h. This contrasts with the much lower excursion of the electron densities. Although more refined values for the parameters involved in the maintenance and time evolution of ionosphere density can lead to different values of the above factors, it seems difficult to avoid the catastrophic destruction of the ionosphere at night, unless an additional source of electrons exists. With $N_e \approx 10^5$ el cm$^{-3}$, $\beta \approx 4 \cdot 10^{-4}$ sec$^{-1}$ and a thickness $\approx 100$ km, the power required to maintain the nocturnal ionization in the diffusion controlled region was estimated by Ivanov-Kholodny (1965) to be around $2 \cdot 10^{-2}$ erg cm$^{-2}$ sec$^{-1}$.

The actual problem is to establish the nature and the real power flux of this possible corpuscular source. On the other hand, we must admit that alternative interpretations of the maintenance of the nocturnal ionosphere at high levels are possible, based on the horizontal transport of ionization, on the diffusion from upper altitudes, etc. Therefore, the problem is still open to further study.

## 3. LONG-TERM VARIATIONS OF THE IONOSPHERIC ELECTRON DENSITY AND CORRELATIONS WITH SOLAR ACTIVITY

Several effects observed in the daytime have been considered as evidence of corpuscular radiation at ionospheric altitudes. An early statistical study of seasonal and non-seasonal variations of the peak electron density in the F2 layer led Yonezawa and Arima (1959) to suggest that a contribution of the order of 20–30% of the variations is attributable to non-seasonal variations, i.e., to variations having the same phase both in the northern and southern hemisphere, exhibiting a maximum when the sun–earth distance is at a minimum. Now, an annual distance effect can be expected in the ultraviolet radiation. However, the intensity variation, as derived from the geometry of the earth's orbit seems insufficient to explain the observational facts. Yonezawa and Arima (1959) suggested the existence of a very low energy solar corpuscular radiation undergoing Coulombian scattering by interplanetary matter. A strong argument against this suggestion is how such radiation could directly penetrate to F-region altitudes under the deviating effect of the magnetic field.

508

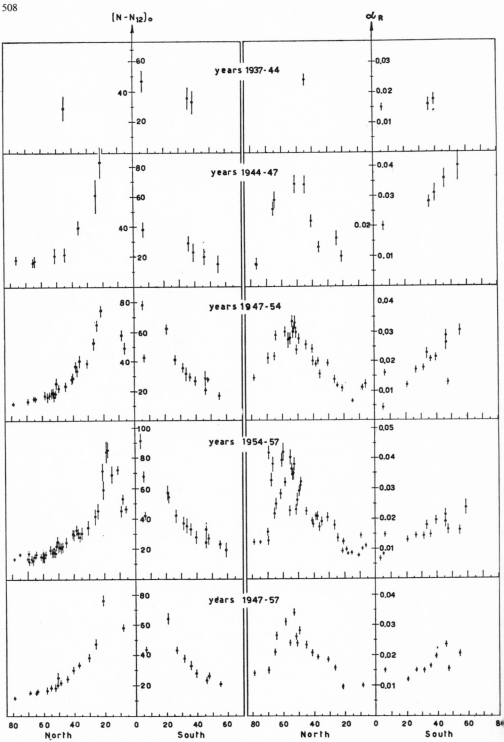

Fig. 3. Correlation of month to month variation $(N - N_{12})$ and $R$. (After Mariani, 1963.)

A statistical investigation of the correlation between peak electron density in the F2 layer at noon (when steady conditions $I(z) = \alpha N_e^2$ occur) and solar activity at all existing observatories for the years 1937–1957 was made by Mariani (1963). If a linear relationship $N - N_{12} = (N - N_{12})_0(1 + \alpha_R R)$ is considered, where $N$ is the monthly average electron density, $N_{12}$ is the seasonal variation and $R$ is the monthly average Wolf's number, the regression coefficient $\alpha_R$, rather than being a constant at each latitude, shows (Fig. 3) a typical behaviour with an identical minimum at the equator and at auroral latitudes, 55° to 65° north and south. Similar results are also obtained from the correlation with other parameters of the solar activity, such as the sunspot area, flocculi, hydrogen filaments, etc. It was suggested that this latitudinal effect can be interpreted as a corpuscular effect. The lack of any effect of this kind at lower altitudes, at the F1 and the E layer level, puts a limit to the energy of the particle stream in the order of keV's. Assuming that the rate of percentage increase of the electron density with the solar activity produced by the ultraviolet radiation does not depend upon the latitude and that the corpuscular effect does not exist at equatorial and polar latitudes, the ionization rate by corpuscular radiation was estimated at about 50 ion pairs $cm^{-3}$ $sec^{-1}$ at maximum solar activity. The corpuscular contribution should decrease with solar activity.

The energy flux suggested by Mariani (1963) is then of the order of 0.1 erg $cm^{-2}$ $sec^{-1}$. This value would give an airglow emission of about 20 rayleighs at 3914 Å, i.e., not far from the minimum detectable intensity. It is, anyway, important to notice that the maximum of the latitude distribution of the corpuscular flux found by Mariani (1963) and that of the electrons with energy above 40 keV shown in Fig. 1 are almost coincident. Meredith et al. (1955) at an early date observed a permanent flux of energy in the range 10–100 keV at low altitudes corresponding to energy fluxes of 0.1–1 erg $cm^{-2}$ $sec^{-1}$. The energy spectrum of the particles is consistent with the suggestion that there should be no appreciable contribution at energies below approximately 1 keV. Although it is difficult to attribute the latitudinal effect to the ultraviolet radiation, the evidence for a corpuscular contribution is, however, not direct.

## 4. TEMPERATURE EFFECTS

### 4.1. Neutral and charged particle temperatures

Other facts for which a corpuscular interpretation has been suggested are related to the thermal behaviour of the upper atmosphere. According to Harris and Priester (1962) the diurnal variation of density and temperature of the neutral particles in the thermosphere cannot be explained only by considering the energetic flux of the ultraviolet radiation. The authors performed a simultaneous integration of the heat-conduction equation and of the hydrostatic equation to produce a self-consistent time variable model of the exosphere. When using only the ultraviolet heat source they got a too large and wrongly phased diurnal temperature variation. After introducing an "ad hoc" source having a maximum intensity at 9 a.m. about 10% greater than the maximum ultraviolet intensity, they were able to fit the computed temperatures to the measured ones. This additional heat source represents, according to the authors, the "corpuscular" heating component. It is, however, hard to understand the physical meaning of such an intense and strongly time variable source.

Jacchia (1962) and Paetzold (1962) found an 11-year variation of the exospheric neutral particle temperature, which is maximum when solar activity is also maximum. This effect, which was named "solar wind effect", was believed to indicate a contribution of the solar wind to the heating of the upper atmosphere, up to 30–40% of the total heating flux to the exosphere. A possible mechanism could be heating by hydromagnetic waves in the ionosphere. A strong solar cycle variation of this effect would be required to fit the observational temperature variation: in particular a variation of a factor of 3 in the solar wind density.

A difficulty in the corpuscular interpretation arises because the observed solar cycle variation is not too large and contrary to the one suggested by Jacchia and Paetzold: the plasma density was higher in 1964, at minimum solar activity, than in 1968–69 near the maximum. Another difficulty arises from the fact that the fluxes required by the above authors are much higher than those suggested by the airglow observations (Galperin, 1962). The latter finds that the intensity of the Doppler shifted $H\alpha$ emission line does not exceed 3 rayleighs at middle and low latitudes, which corresponds to an extremely low energy flux of $6 \cdot 10^{-4}$ erg cm$^{-2}$ sec$^{-1}$. If one thinks in terms of fast electrons, a reasonable value of the conversion efficiency of kinetic energy to thermal energy 20 to 50% can be used. Then 10 rayleighs of 3914 Å radiation would be associated with $5 \cdot 10^{-2}$ erg cm$^{-2}$ sec$^{-1}$, i.e., to a heating flux of the order of $10^{-2}$ erg cm$^{-2}$ sec$^{-1}$. The consequence is that, should fast electrons be responsible for the thermal effects observed by Jacchia and Paetzold, then their flux should be sufficiently high to be detected, in contrast with the finding from the airglow measurements.

Early measurements of the electron temperature at altitudes between 400 and 1200 km made onboard the Ariel I satellite (Bowen et al., 1964) have shown an increase with the altitude and the latitude which can be accounted for as a heating effect by precipitating electrons.

The large diurnal variation of the temperature and its latitude variation, which is also maintained during the night would imply, according to the authors, the existence of an additional heat source. Since this source is required to give energy preferentially to the electrons and, on the other hand, it must do so in the presence of the geomagnetic field, the simplest suggestion is that an inward flux of relatively high energy electrons is at work. A flux of the order of $10^8$ el cm$^{-2}$ sec$^{-1}$ with individual energies of a few tens of eV is sufficient. The corresponding energy flux would then be of the order of 0.01–0.1 erg cm$^{-2}$ sec$^{-1}$. Actually, soft radiation in the equatorial region at altitudes of 320 km was already observed by the Soviet Sputniks (Savenko et al., 1963). Fluxes as high as $5 \cdot 10^9$ particles cm$^{-2}$ sec$^{-1}$ sterad$^{-1}$ were estimated, if interpreted as 10 keV electrons.

## 4.2. Photoelectron drift and conjugate effects*

The upward escape of a fraction of the photoelectrons produced by the ultraviolet radiation in the sunlit atmosphere was first suggested by Hanson (1963). Much indirect evidence has been accumulated since then: i.e., pre-sunrise effects in the upper ionosphere, heating and high positive gradients of the electron temperature etc. Recently, observational evidence on these photoelectrons has also been obtained by several authors (Rao and Donley, 1969; Heikkila, 1970; Rao and Maier, 1970). A series of electron energetic spectra obtained at the altitude of 3500 km at middle latitudes by Heikkila from the satellite ISIS-I is shown in Fig. 4.

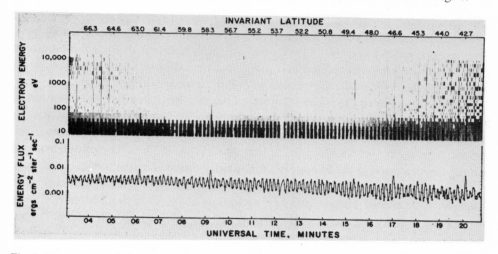

Fig. 4. An experimental electron energy spectrogram (above) and the total energy carried by electrons in the range 8 to 100 eV. The time is day February 9, 1969 at 01 h UT plus the minutes shown. The geographic longitude is 132°E. (After Heikkila, 1970.)

An intense flux peaked at about 20 eV appears with two maxima per spin period of the spacecraft ($Ts = 20.4$ sec). These maxima have almost the same amplitude and occur when the detector points into the geomagnetic plane. The total fluxes are $2.5 \cdot 10^8$ el cm$^{-2}$ sec$^{-1}$ and $8 \cdot 10^{-3}$ erg cm$^{-3}$ sec$^{-1}$. Extrapolation to 1000 km

---

* Most of the results given in this section are based on unpublished work done by the author as a NAS-NASA Senior Postdoctoral Associate with the NASA Goddard Space Flight Center, Greenbelt, Md. in 1963.

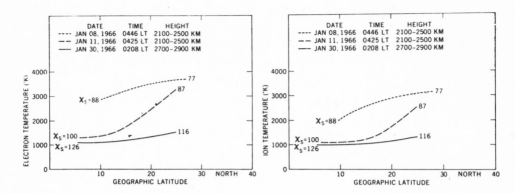

Fig. 5. Electron and ion temperatures during three passes of Explorer 31 in the pre-dawn period in the northern hemisphere. The solar zenith angles $\chi_s$ at the southern conjugate point are indicated; the angles at the northern conjugate point were always greater than 120°. (After Rao and Maier, 1970.)

would increase the above figure by a factor of 3; an additional factor of 2 would come to the particle flux taking into account the electrons with energy below 8 eV, although the energy flux is not significantly changed. Rao and Maier (1970) have observed on board Explorer 31 an increase of both electron and ion temperatures $T_e$ and $T_i$ in the pre-dawn period at a given location associated with the sunrise in the opposite hemisphere in the magnetically conjugate location (Fig. 5).

Carlson (1966), from the interpretation of the Arecibo incoherent back scattering measurements, derived a heating of the upper ionosphere coincident with sunrise at the conjugate location, as well as a continued higher temperature until sunset at the same location (Fig. 6). This applies to altitudes above 300 km. The photoelectrons coming down along the field lines are then rapidly thermalized below a critical level $z_e$ at which their mean free path becomes smaller than the scale height. Inelastic collisions with atomic oxygen produce excitation of its $^1D$ state. This may be the cause of the pre-dawn increase of the 6300 Å airglow, observed by Cole (1965), who, however, attributed it to excitation by thermal electrons. Duboin et al. (1968) proved that the interpretation was correct, since there is synchronism between the increases of the electron temperature and of the 6300 Å airglow at St. Santin de Maurs with the sunrise at the geomagnetically conjugate location (Fig. 7). On the other hand, a daytime enhancement of the plasma resonance lines in the incoherent backscatter has been observed at Arecibo, just as predicted by Perkins and Salpeter (1965).

The basic ideas on the escape of photoelectrons and the heat deposition along their trajectory from one hemisphere to the other are very simple. First of all, cross sections of all processes which slow down the electron motion are required. This is possible on the basis of the work by Hanson and Johnson (1961). Fig. 8 shows the altitude behaviour of the time constants for five different processes, for noon

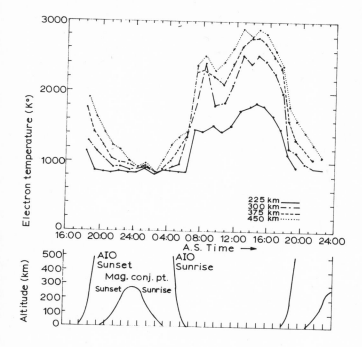

Fig. 6. Diurnal variation of $T_e$ at different altitudes. The bottom part shows the time of sunrise at Arecibo (labeled AIO) and at the southern conjugate point, at different altitudes. (After Carlson, 1966.)

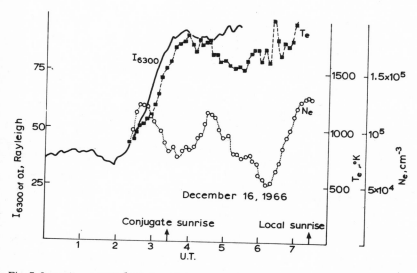

Fig. 7. Intensity of 6300 Å OI line, electron temperature $T_e$ and electron density $N_e$ observed by Duboin et al. (1968).

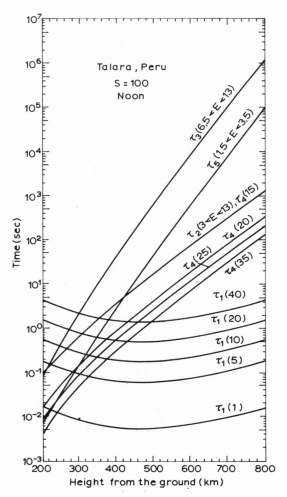

Fig. 8. Time constants for five different processes (see text) and for different energies at low latitudes. Noon conditions and medium solar activity are assumed. (F. Mariani, unpublished data, 1963.)

conditions: $\tau_1$, for elastic collisions with ambient electrons (kinetic energy $3 < E < 13$ eV); $\tau_2$, for excitation of atomic oxygen; $\tau_3$, for inelastic collisions with molecular nitrogen ($6.5 < E < 13$ eV); $\tau_4$, for ionization of atomic oxygen and molecular nitrogen; $\tau_5$, for vibrationally exciting collisions of molecular nitrogen ($1.5 < E < 3.5$ eV).

The results in Fig. 8 show that at altitudes below approximately 300 km the energy is distributed between the loss processes so quickly that in practice all the photoelectrons are locally thermalized. At upper altitudes the only effective process is the elastic collision with ambient electrons, and all the energetic photoelectrons move along the magnetic lines according to their initial pitch angle distribution until their energy reaches the lower limit $E_o = 1.5$ eV. Below the energy $E_o$, no further drift is possible and the photoelectrons undergo local thermalization. The photoelectron production rates at altitudes below the critical level are just those computed neglecting any non-local contribution. At heights above $z_c$, due to the drift of photoelectrons, a modification of the height distribution must occur. The heating flux is also affected by the downward and the upward fluxes of photoelectrons above the critical level; finally, a flux of escaping electrons is expected at very high altitudes, let us say above 1000 km from the ground, where one can assume that no further significant collision effect is expected. The mean rate of the energy loss through elastic collisions with the ambient thermal electrons may be easily derived (see Butler and Buckingham, 1962). The electrons are guided by the magnetic field line so that if the pitch angle of the individual electron does not change, the energy loss rate in eV cm$^{-1}$ is given by:

$$\frac{dE}{dz} = -\frac{p}{\cos \alpha} \frac{N_e}{E} \tag{1}$$

where $p = 1.95 \cdot 10^{-12}/\sin I$, $I$ is the magnetic dip, $E$ is the kinetic energy in eV and $N_e$ is in cm$^{-3}$. For any given height distribution $N_e(z)$, a range $R$ can be defined as the altitude at which the energy is reduced to a minimum below which the dissipation becomes local, i.e., no further motion is possible. On the other hand, if the initial energy is sufficiently high, the upward photoelectron can escape to infinity along the field line guided toward the opposite hemisphere until thermalization in the denser atmosphere occurs.

If an electron with kinetic energy $E_k > 1.5$ eV is produced at the height $h'$ with a pitch angle $\alpha$, its residual energy $E(h, h', E_k)$ at the height $h$ is given according to eq. 1 by:

$$E(h, h', E_k) = \left[ E_k^2 - \frac{2p}{\cos \alpha} \left| \int_{h'}^{h} N_e(\zeta) \, d\zeta \right| \right]^{1/2} \tag{2}$$

Use of the absolute value of the electron density integral is required to maintain the validity of the formula in both cases $h > h'$ and $h' > h$, i.e., for both upward and downward photoelectrons. Since the energy dissipation expressed by (2) is effective only in the presence of an electron drift, there is a minimum altitude which is either at the point where $E$ becomes 1.5 eV, or at the critical level $z_c$.

Once the pitch angle distribution at the production $g(\alpha)$ is known, one can very simply compute the number of photoelectrons with initial energy $E_k$ actually stopping in the unit volume in the unit time at the height $h$, by means of the integral:

$$n_k(h, E_k) = \frac{2p}{E_k^2 - E_o^2} \int_{h_{min}}^{h_{max}} v_k(h', E_k) N_e(h') \frac{g(\alpha')}{\sin \alpha'} \, dh' \tag{3}$$

where $v_k(h', E_k)$ is the total production rate of photoelectrons of energy $E_k$ at the altitude $h'$, and $\alpha'$ is the pitch angle, so that photoelectrons with initial $\alpha = \alpha'$ are stopping at the altitude $h$. This angle $\alpha'$ is easily derived from eq. 2:

$$\alpha' = \cos^{-1}\left[\frac{p}{E_k^2 - E_o^2}\left|\int_{h'}^{h} N_e(\zeta)\,d\zeta\right|\right] \tag{4}$$

The heights $h_{max}$ and $h_{min}$ appearing in eq. 3 are either those derived from the condition $\alpha' = 0$ which means:

$$\frac{2p}{E_k^2 - E_o^2} = \int_{h}^{h_{max}} N_e(\zeta)\,d\zeta = \int_{h_{min}}^{h} N_e(\zeta)\,d\zeta \tag{5}$$

or their absolute extreme values (in our computation we took for these $h_{min} = 300$ and $h_{max} = 1000$ km).

The energy deposition per unit volume and unit time due to drifting photoelectrons with initial energy $E_k > E_o$ is similarly expressed by

$$\eta_k(h) = pN_e(h)\int_0^{\pi/2}\frac{g(\alpha)}{\cos\alpha}\,d\alpha$$

$$\left[\int_{h_{min}}^{h}\frac{v_k(h', E_k)\,dh'}{\left(E_k^2 - \frac{2p}{\cos\alpha}\int_{h'}^{h} N_e(\zeta)\,d\zeta\right)^{1/2}} + \int_{h}^{h_{max}}\frac{v_k(h', E_k)\,dh'}{\left(E_k^2 - \frac{2p}{\cos\alpha}\int_{h}^{h'} N_e(\zeta)\,d\zeta\right)^{1/2}}\right] \tag{6}$$

where the limits $h_{min}(\alpha, E_k)$ and $h_{max}(\alpha, E_k)$ are functions of both the energy $E_k$ and the pitch angle $\alpha$, according to eq. 2, satisfying the same condition considered above. The total energy deposition is then obtained as the sum of the contributions from all the energies $E_k$.

Finally, the pitch angle density flux of escaping photoelectrons at the maximum height $h^*$ is given for every initial energy $E_k$, by:

$$\frac{dF_k}{d\alpha} = g(\alpha)\int_{h_{min}}^{h^*} v_k(h', E_k)\,dh' \tag{7}$$

where $h_{min}(\alpha, E_k)$ is derived by eq. 2 in a similar way.

The total flux is obtained for each energy $E_k$ by integration of eq. 7 with respect to $\alpha$ between 0 and $\pi/2$.

Finally, the sum over all the energies $E_k$ gives the energetic integral spectrum of the escaping photoelectrons.

By means of analogous considerations, the differential and the integral energy spectrum of the escaping photoelectrons can also be computed, only taking into account the upward drifting electrons.

Although the hypotheses on which the above derivations are based can alter more or less the absolute figures, we feel that the relative height, diurnal and solar activity variations are generally reliable. As regards the pitch angle distribution of the electrons produced in the photoionization process, we have, for the isotropic case $g(\alpha) = (\sin \alpha)/2$. However, if a $\sin^2\gamma$ distribution, where $\gamma$ is the angle between the directions of the incident photon flux and the photoelectron ejection velocity vector, is taken (Mariani, 1964) then:

$$g(\alpha) = \frac{3}{4} \sin \alpha \left[ \sin^2\alpha + \sin^2\gamma \left( 1 - \frac{3}{2} \sin^2\alpha \right) \right] \tag{8}$$

where $\gamma$ is the angle between the direction of the incident photon and the geomagnetic field (Fig. 9). Some details of the computations are given in the appendix.

According to Geisler and Bowhill (1965), use of eq. 8 at noon leads to results substantially similar to those obtained when an isotropic distribution is used. However, any computation at different local times to look into the time variations or into the transfer of photoelectrons to the opposite hemisphere requires use of eq. 8.

Some unpublished results by Mariani are shown in Fig. 10 and 11, where the photoelectron production rates and the energy spectra of the escaping photoelec-

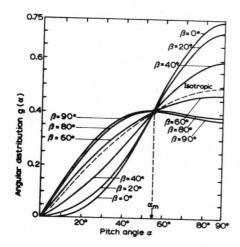

Fig. 9. The pitch angle distribution $g(\alpha)$ of the electrons produced in the photoionization process. (After Mariani, 1964.)

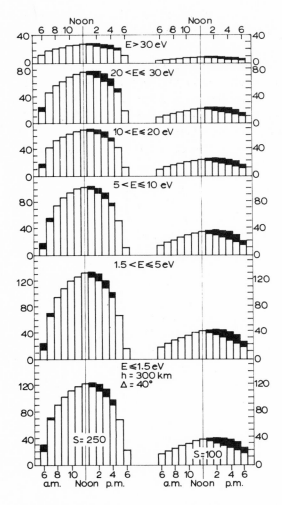

Fig. 10. Computed photoelectron production rates within different energy ranges, at a latitude of 40° and at 300 km altitude, for an equinoctial day at high (left) and medium (right) solar activity. Dashed portions represent the excess with respect to the value computed at a symmetrical time with respect to noon. The Harris and Priester (1962) time variable model of neutral atmosphere was used. (F. Mariani, unpublished data, 1963.)

trons are plotted for two typical conditions. The results obtained with scattering included by Banks and Nagy (1970) show that this effect is not exceedingly important, and therefore the results given above are substantially valid.

The heating fluxes are not too large, so that during daytime the expected effect on the electron temperature is small. However, when one end of a field line is in the

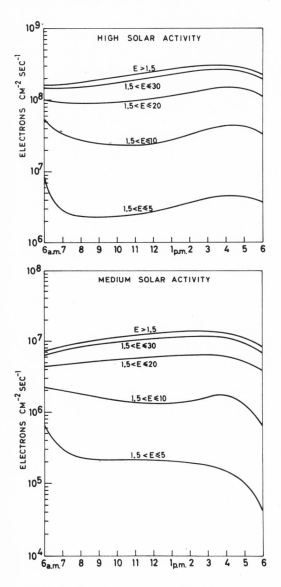

Fig. 11. Computed fluxes of escaping photoelectrons at 1000 km from the ground, at a latitude of 40°
for an equinoctial day. Energy bands are indicated on each curve. (F. Mariani, unpublished data, 1963.)

dark hemisphere and the other in the lighted hemisphere, an appreciable effect is
possible in the dark hemisphere, just as observed experimentally.

The simple theory outlined above was extended by Whitten (1968) to the case
where the neutral, the ion and the electron gases are not in thermal equilibrium.

## 5. CONCLUSIONS

Firm conclusions about the existence of corpuscular effects in the upper atmosphere cannot be made. In fact, some of the effects attributed to corpuscular radiation do not survive after a critical check. For other effects, however, like the maintenance of the nocturnal ionosphere and the diurnal correlation with the solar activity, one can only conclude that particles are contributing, since they are really present in the desired range of energy, but the relative entity of this contribution has not yet been well defined. The effects connected with the photoelectron fluxes along the magnetic field lines are the only well established ones, although the details may be worthy of further study.

The conclusion is that all possible efforts should be made to design experiments capable of separating, whenever possible, corpuscular and electromagnetic effects.

## 6. APPENDIX: COMPUTATION OF THE PITCH ANGLE DISTRIBUTION GIVEN BY EQUATION 8 OF THE TEXT

Use is made of Fig. 12, where the symbols are as follows: $\vec{H}$ = magnetic vector in $P$; $\vec{f}$ = direction of the sun; $\vartheta$ = angle of ejection of photoelectrons, around the direction $\vec{f}$; $\gamma$ = angle between $\vec{H}$ and $\vec{f}$; $\alpha$ = pitch angle. All the dotted lines are circles on the same sphere.

From the spherical triangle ABC (Fig. 12a) the angle $\varepsilon$ opposite to the side AC, is defined by:

$$\varepsilon(\vartheta, \alpha, \gamma) = \cos^{-1} \frac{\cos \alpha - \cos \vartheta \cos \gamma}{\sin \vartheta \sin \gamma} \tag{9}$$

Now, considering $\varepsilon$ as a function $\varepsilon(\vartheta)$ at fixed $\alpha$ and $\gamma$, and using the angular distribution of probability:

$$dp(\vartheta) = kf(\vartheta) \sin\vartheta d\vartheta, \text{ with } k = \int_0^\pi f(\vartheta)\sin\vartheta \, d\vartheta$$

the probability $G(\alpha)$ that one photoelectron is ejected in any direction having a pitch angle between 0 and $\alpha$ is:

$$G(\alpha) = \frac{k}{\pi} \int_{\gamma-\alpha}^{\gamma+\alpha} \varepsilon(\vartheta) f(\vartheta) \sin\vartheta \, d\vartheta \tag{10}$$

from which, using eq. 9, the differential pitch angle distribution may be obtained:

$$g(\alpha) = \frac{dG}{d\alpha} = \frac{k}{\pi} \sin\alpha \int_{\gamma-\alpha}^{\gamma+\alpha} \frac{f(\vartheta)\sin\vartheta \, d\vartheta}{[a + b \cos\vartheta - \cos^2\vartheta]^{1/2}} \tag{11}$$

where: $a = \sin^2\gamma - \cos^2\alpha$ and $b = 2 \cos\alpha \cos\gamma$.

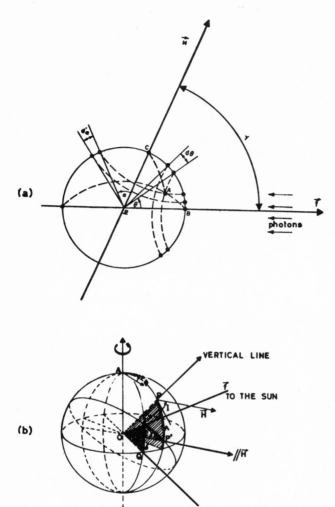

Fig. 12. The geometry of the photoionization process (for explanation see text).

In the particular case of the $\sin^2\vartheta$ angular distribution for the ejected photoelectrons, the integration of eq. 11 leads to

$$g(\alpha) = \frac{3}{4}\sin\alpha\left[\sin^2\alpha + \sin^2\gamma\left(1 - \frac{3}{2}\sin^2\alpha\right)\right] \qquad (12)$$

The relationship between the angle $\gamma$, the hourly angle $\Phi$, the geographical latitude $\Lambda$ (positive in the northern hemisphere), the solar declination $\delta$ and the magnetic inclination $I$ (positive when the field $\vec{H}$ has an upward component) is easily obtained by looking at Fig. 12b where $N$ is the northern pole and $OP'$ is a vector parallel to the magnetic field $\vec{H}$. It should be noted that the spherical angle $PP''$ is just the zenith angle of the sun.

From the spherical triangle $AP'P''$ for the case of a centered dipole magnetic field (which means that $\vec{H}$ is in the same plane as $A$, $P$ and $O$) one immediately gets:

$$\cos \gamma = -\sin \delta \cos(I + \Lambda) + \cos \delta \sin(I + \Lambda)\cos \Phi \qquad (13)$$

In this simple case the usual relationship:

$$\tan I = 2 \tan \Lambda \qquad (14)$$

is also applicable.

## REFERENCES

Antonova, L. A. and Ivanov-Kholodny, G. S., Ionisation in the night ionosphere (corpuscular hypothesis). *Space Res.*, **2**:981–992, 1961.

Banks, P. M. and Nagy, A. F., Concerning the influence of elastic scattering upon photoelectrons transport and escape. *J.Geophys.Res.*, **75**:1902–1910, 1970.

Belrose, J. S. and Thomas, L., Ionization changes in the middle latitude D region associated with geomagnetic storms. *J.Atmos.Terr.Phys.*, **30**:1397–1413, 1968.

Bowen, P. J., Boyd, R. L. F., Henderson, C. L. and Willmore, A. P., Electron temperature in the upper F-region. *Proc. R.Soc., Lond.Ser.* A, **281**:526–538. 1964.

Butler, S. T. and Buckingham, M. J., Energy loss of a fast ion in a plasma. *Phys.Rev.*, **126**:1–4, 1962.

Carlson, H. C., Ionospheric heating by magnetic conjugate-point photoelectrons. *J.Geophys.Res.*, **71**:195–199, 1966.

Cole, K. D., The predawn enhancement of 6300Å airglow. *Ann.Geophys.*, **21**:156–158, 1965.

Dalgarno, A., Corpuscular radiation in the upper atmosphere. *Ann.Geophys.*, **20**:65–74, 1964.

Duboin, M. L., Lejeune, G., Petit, M. and Weill, G., Excitation of oxygen lines and ionospheric heating by conjugate photoelectrons. *J.Atmos.Terr.Phys.*, **30**:299–304, 1968.

Fritz, T. A., High latitude outer zone boundary region for $\geqslant$ 40 keV electrons during geomagnetically quiet periods. *J.Geophys.Res.*, **72**:7245–7255, 1968.

Galperin, Yu. L., The energy sources of the upper atmosphere. *Bull.Acad.Sci.U.S.S.R.,Geophys.Ser.*, **2**:174–179, 1962.

Geisler, J. E. and Bowhill, S. A., Exchange of energy between the ionosphere and protonosphere. *J.Atmos.Terr.Phys.*, **27**:1119–1146, 1965.

Hale, L. C., Parameters of the low ionosphere at night deduced from parachute born blunt probe measurements. *Space Res.*, **7**:140–151, 1967.

Hanson, W. B., Electron temperatures in the upper atmosphere. *Space Res.*, **3**:282–302, 1963.

Hanson, W. B. and Johnson, F. S., Electron temperatures in the ionosphere. *Mem.Soc.R.Sci.Liege*, **4**:390–424, 1961.

Harris, I. and Priester, W., Time-dependent structure of the upper atmosphere. *J.Atmos.Sci.*, **19**:286–301, 1962.

Heikkila, W. J., Photoelectron escape flux observation at midlatitudes. *J.Geophys.Res.*, **75**:4877–4879, 1970. ·

Ivanov-Kholodny, G. S., Maintenance of the night ionosphere and corpuscular fluxes in the upper atmosphere. *Space Res.*, **5**:19–42, 1965.

# REFERENCES

Jacchia, L. G., Electromagnetic and corpuscular heating of the upper atmosphere. *Space Res.*, **3**:3–26, 1962.

Knudsen,W. C. and Sharp, G. W., F2 region electron concentration enhancements from inner radiation belt particles. *J.Geophys.Res.*, **73**:6275–6283, 1968.

Lauter, E. A. and Knuth, R., Precipitation of high energy particles into the upper atmosphere at medium latitudes after magnetic storms. *J.Atmos.Terr.Phys.*, **29**:411–417, 1967.

Maehlum, B. N., On the winter anomaly in the midlatitude D region. *J.Geophys.Res.*, **72**:2287–2299, 1967.

Mariani, F., Evidence for the effect of corpuscular radiation on the ionosphere. *J.Atmos.Sci.*, **20**:479–491, 1963.

Mariani, F., Pitch angle distribution of the photoelectrons and origin of the geomagnetic anomaly in the $F_2$ layer. *J.Geophys.Res.*, **69**:556–560, 1964.

McDiarmid, I. B., Burrows, J. R., Budzinski, E. E., and Wilson, M. D., Some average properties of the outer radiation zone at 1000 km. *Can.J.Phys.*, **41**:2064–2079, 1963.

Mechtly, E. A. and Smith, L. G., Growth of the D-region at sunrise. *J.Atmos.Terr.Phys.*, **30**:363–369, 1968.

Meredith, L. H., Gottlieb, M. B. and Van Allen, J. A., Direct detection of soft radiation above 50 kilometers in the auroral zone. *Phys.Rev.*, **97**:201–205, 1955.

O'Brien, B. J., Lifetimes of outer-zone electrons and their precipitation into the atmosphere. *J.Geophys.Res.*, **67**:3687–3706, 1962.

O'Brien, B. J., High-latitude geophysical studies with satellite Injun 3; 3: Precipitation of electrons into the atmosphere. *J.Geophys.Res.*, **69**:13–43, 1964.

O'Brien, B. J. and Laughlin, C. D., Electron precipitation and the outer radiation zone. *Space Res.*, **3**:399–417, 1963.

Ogawa, T. and Tohmatsu, T., Photoelectric processes in the upper atmosphere, 2: The hydrogen and helium ultraviolet glow as an origin of the night-time ionosphere. *Rep.Ionos. Space Res.Japan*, **20**:395–417, 1966.

Paetzold, H. K., Corpuscular heating of the upper atmosphere. *J.Geophys.Res.*, **67**:2741–2744, 1962.

Perkins, F. W. and Salpeter, E. E., Enhancement of plasma density fluctuations by non-thermal electrons. *Phys. Rev.*, **139**:A55–A62, 1965.

Potemra, T. A. and Zmuda, A. J., Precipitating energetic electrons as a ionization source in the midlatitude night-time D region. *J.Geophys.Res.*, **75**:7161–7167, 1970.

Rao, B. C. and Donley, J. L., Photoelectron flux in the topside ionosphere measured by retarding potential analyzers. *J.Geophys.Res.*, **74**:1715–1719, 1969.

Rao, B. C. and Maier, E. J. R., Photoelectron flux and protonospheric heating during the conjugate point sunrise. *J.Geophys.Res.*, **75**:816–822, 1970.

Rees, M. H., Auroral ionization and excitation by incident energetic electrons. *Planet. Space Sci.*, **11**:1209–1218, 1963.

Sagalyn, R. C. and Smiddy, M., Rocket investigation of the electrical structure of the lower ionosphere. *Space Res.*, **4**:371–387, 1964.

Savenko, I. A., Shavrin, P. I. and Pisarenko, N. F., Soft particle radiation at an altitude of 320 km in the latitudes near the equator. *Planet. Space Sci.*, **11**:431–436, 1963.

Thomas, L., The winter anomaly in ionospheric absorption. In *Radio Wave Absorption in the Ionosphere* (N. C. Gerson, Editor). Pergamon Press, New York, N.Y., pp. 301–317, 1962.

Whitten, R. C., Non local energy deposition in the ionosphere and the photoelectron escape flux. *J.Atmos.Terr.Phys.*, **30**:1523–1533, 1968.

Yonezawa, T. and Arima, Y., On the seasonal and non-seasonal annual variations and the semiannual variation in the noon and midnight electron densities of the F2 layer in middle latitudes. *J.Radio Res.Lab. Japan*, **6**:293–309, 1959.

# Author Index

# Subject Index